T0237617

Vibration Mechanics

Haiyan Hu

Vibration Mechanics

A Research-oriented Tutorial

 Science Press
Beijing

 Springer

Haiyan Hu
School of Aerospace Engineering
Beijing Institute of Technology
Beijing, China

ISBN 978-981-16-5459-6   ISBN 978-981-16-5457-2   (eBook)
https://doi.org/10.1007/978-981-16-5457-2

Jointly published with Science Press, Beijing, China
The print edition is not for sale in China (Mainland). Customers from China (Mainland) please order the
print book from: Science Press.

This Springer imprint is published by the registered company Springer Nature Singapore Pte Ltd.
The registered company address is: 152 Beach Road, #21-01/04 Gateway East, Singapore 189721,
Singapore

To Luna for her love and support
over a half of century

Preface

In the frame of engineering education, vibration mechanics has served as a core of several courses, such as mechanical vibrations and structural dynamics, for the undergraduate students majoring in aerospace engineering, mechanical engineering, and civil engineering. Vibration mechanics has also played an important role in a great variety of engineering fields. As such, future engineers should be able to perform the dynamic modeling, computations, and tests of industrial products.

The author has been studying the dynamics and control of aerospace structures, with an emphasis on the integrated computational and experimental methods of nonlinear dynamics, since the 1980s. As for teaching, the author taught "structural vibrations" of four credits for the undergraduate students in the program of aircraft engineering at *Nanjing University of Aeronautics and Astronautics* in the 1980s and 1990s. In recent years, the author has been teaching "fundamentals of structural dynamics" of two credits for the undergraduate students majoring in aerospace engineering at *Beijing Institute of Technology*. In 1998 and 2005, the author had two textbooks published in Chinese, which have served for teaching materials of the courses of four credits and two credits, respectively.

The above experience of teaching and researches well confirms the importance of fundamental concepts and theories in vibration mechanics. The author tried to assist readers to gain an insight into vibration mechanics in the above two textbooks. Yet, it was not possible to address some issues in detail in those textbooks because of a well-established theoretical frame of both linear vibrations and teaching materials.

The new century has witnessed a global reform of engineering education with the rapid developments of information technology. As a consequence, the above courses with a core of vibration mechanics have been subject to a great reduction of teaching periods. Many undergraduate students have paid attention to the software of computational mechanics only. Some graduates with engineering degrees have shown their weakness in fundamental concepts, theoretical foundations, and experimental skills while working in either academia or industry. They have shown a strong intention of upgrading their understanding to vibration mechanics.

The above background stimulated the author to write a new book in order to help readers improve themselves. The majority of readers of this book may be those, who have taken an elementary course, such as mechanical vibration and structural dynamics, of two credits or more. The objective of the book is to assist those readers to conduct the research-oriented study on linear vibrations, and to gain an insight into this subject via a helical upgrade of understanding level.

As a trial of the above ideas of engineering education, the book is an attempt to make a different design of contents and a new style of writing in comparison with other available textbooks. The main features of the book are as follows.

First, the book focuses on some interesting problems of linear vibrations and waves of mechanical or structural systems and does not touch upon the scope of nonlinear vibrations, random vibrations, and dynamic measurements in other textbooks of graduate levels. That is, the priority of the book is an academic depth, rather than a broad scope.

Second, the writing style of the book embodies the process of research-oriented study. That is, the book presents how to find problems in a research or engineering practice, how to clearly define the problems to be studied, how to solve the problems, and how to look forward to future studies. This style may guide readers to think independently and to conduct the research-oriented study of vibration mechanics step by step. Of course, readers need to prepare themselves adaptive to such a writing style of the book, which begins with the problems to be studied and then turns to the necessary preparations for dealing with those problems before solving them. This is a practical sequence of most researches oriented from real problems, rather than available knowledge.

The major contents of the book came from the experience of the author in education, researches, and engineering consultations over the past decades. The main body of the book looks like a collection of research papers, but serves as the teaching materials for readers who have learnt an elementary course of linear vibrations and need to upgrade their understanding to this useful subject.

The first two chapters of the book provide readers with an introduction to the problems proposed and to be solved, and then a review of elementary course of linear vibrations from a perspective of aesthetics of science. The remaining chapters are relatively independent, presenting theories and methods of linear vibrations and waves. They provide readers with some optional reading materials as an introduction to specific topics and also a guide to further studies.

The author is very grateful to Prof. Liqun Chen at *Shanghai University*, Prof. Yufeng Xing at *Beihang University*, Prof. Gàbor Stèpàn at *Budapest University of Technology and Economics*, and Prof. Earl Dowell at *Duke University*, who reviewed the book draft and made constructive suggestions.

Meanwhile, the author thanks his colleagues at both *Beijing Institute of Technology* and *Nanjing University of Aeronautics and Astronautics* for their suggestions after reading the different versions of the book draft. They are Profs. Gengkai Hu, Dongping Jin, Zaihua Wang, Yonghui Zhao, Huailei Wang, Lifeng Wang, Yanfei Jin, Li Zhang, Rui Huang, and Chunyan Zhou for the Chinese version published by *Science Press* in 2020, and Profs. Kai Luo, Shilei Han, Minghe Shan, and Jianqiao Guo for the English version to be jointly published by *Science Press* and *Springer Nature*.

The author also appreciates the great efforts of Dr. Yangfeng Xu with *Science Press* in Beijing, and Dr. Mengchu Huang and Dr. Nan Zhang with *Springer Nature* in Shanghai and Beijing to ensure the publication quality.

Beijing, China
June 2021

Haiyan Hu

Contents

Chapter 1
Start of Research-Oriented Study

Vibration refers to the oscillatory motion of a body or a mechanical/structural system about its equilibrium position. It was one of the earliest physical phenomena that ancient scholars observed and studied, and is one of the most important dynamic problems that engineers today need to solve for designing, manufacturing/constructing and operating a great variety of industrial products. As a consequence, vibration has been a kind of popular engineering problems.

The industrial products over the past decades have seen a change from the static design and dynamic validation to the dynamic design. Hence, engineers should lay a solid foundation of vibration mechanics so as to properly solve various dynamic problems of industrial products from the design stage to the operational stage, and make further innovations in future.

In engineering education, vibration mechanics has served as a core for several courses, such as mechanical vibrations and structural dynamics, for different educational programs. The textbooks of those courses look standard and conventional. In order to conduct a kind of research-oriented studies, this chapter assists readers to have a new start of learning vibration mechanics, that is, the research-oriented study starting from open problems.

1.1 Needs for Research-Oriented Study

Recent years have witnessed the rapid development of both computational mechanics and associated software. As such, the dynamic modeling, computation, analysis and design of industrial products have greatly relied on the software of computational mechanics. The above development, thus, has shortened the courses of mechanics in the engineering education of most countries. As a consequence, the courses of mechanical vibrations and structural dynamics have been simpler and simpler, with a great reduction in fundamental theory and experimental training. The above trend has produced a negative influence on the quality of engineering education.

© Science Press 2022
H. Hu, *Vibration Mechanics*,
https://doi.org/10.1007/978-981-16-5457-2_1

For example, the engineers of new generation usually do not establish the simple models of mechanics for their industrial products so as to analyze the influence of model parameters on the dynamic behaviors of the products, but rely on the software of computational mechanics and conduct a huge amount of numerical simulations. They often present many simulation results to the chief-engineer or the third party in evaluations, but are not able to well explain the simulation results from the physical mechanisms. Some recent failures in aerospace engineering, for instance, have indicated that the engineers of new generation had not gained an insight into the dynamic problems because of the weakness in fundamental concepts and theoretical foundations.

The author has studied the dynamics and control of aerospace structures over the past four decades, and participated in many consultative discussions with aerospace engineers. The above experience has enabled the author to recognize the importance of both fundamental concepts and theoretical foundations in engineering education. Many senior professors and experienced engineers feel that the current courses of mechanical vibrations and structural dynamics are not able to ensure the high quality of engineering education.

In the recent discussions about teaching reforms in China, there are two kinds of appeals from the professors teaching those courses. One is to increase the teaching periods for undergraduate students and return to the conventional model of theoretical and experimental trainings. The other is to increase the related course periods for graduate students so as to solve the problems. These two suggestions, however, may not be feasible in practice.

First, it is difficult to increase course periods for most universities in the frame of current engineering education. In addition, it is not easy for students to lay a solid foundation even though they spend more time learning more advanced textbooks. As a matter of fact, the cognitive learning law of human beings shows that the upgrade of understanding requires theoretical and experimental circulations and then gets a helical increase gradually. To gain an insight into fundamental concepts and lay a solid foundation, it may be more helpful to conduct a research-oriented study. That is, to learn a course from different views and to solve a number of comprehensive problems.

Second, the graduate students coming from a great variety of institutions have different starting levels and are subject to the pressure of researches and dissertations. Hence, it is not easy to enhance the courses associated with vibration mechanics. Furthermore, the current courses of graduate level usually cover the basic knowledge of nonlinear vibrations, random vibrations and dynamic measurements. These courses enable the graduate students, who have had basic knowledge of vibration mechanics, to broaden their scope of knowledge, rather than to lay a solid foundation and to solve dynamic problems in engineering.

The author has discussed the above issues with many world renowned professors and recognized that this is a tough problem in the current engineering education in the world. Therefore, it is necessary to explore a new way to solve the tough problem from the rules of cognitive learning and advanced studying.

The objective of this book is to assist readers to explore a new way of research-oriented study. That is, readers may arouse initial interest in some academic problems associated with engineering practice or teaching practice, then keep those problems in mind while reading the whole book, solve the problems after learning subsequent chapters and gradually upgrade their understanding to vibration mechanics.

For this purpose, the book will begin with several case studies of vibration mechanics and associated problems to be addressed, then present the vibration mechanics of discrete systems and continuous systems, respectively, with an emphasis on those problems. The case studies came from the consultative works, researches and teaching practice of the author. The remaining part of this chapter will present the background, associated problems and basic ideas of solving problems of each case study. The subsequent chapters will give detailed studies and results, as well as some further reading materials and open problems.

1.2 Case Studies and Associated Problems

This section presents six case studies of vibration mechanics, each of which gives rise to two associated problems to be solved, in order to provide a guide for reading subsequent chapters. The case studies came from various researches of the author, as well as helpful discussions with colleagues and students in teaching vibration mechanics, in the past decades. Compared with the development of vibration mechanics over the past century, the problems of concern here are "very small". They mainly serve as appetizers to arouse the interest and independent thinking of readers so that they are able to initiate the research-oriented study of vibration mechanics and solve some "big problems" in future.

1.2.1 Preliminary Study of a Tethered Satellite

1.2.1.1 Background

One of space technologies of increasing interest is a *tethered satellite system* , where a satellite is deployed and retrieved via a very thin and soft tether from a space station, a spaceship or a relatively large satellite in order to perform the space missions, such as adjusting the orbit of a spacecraft, capturing space debris and generating electric powers. It is also possible to tether several small satellites so as to construct a space-based radar system with long baselines for an observation mission of higher resolution.

Figure 1.1 presents the basic idea of a tethered satellite of mass m deployed and retrieved from a spaceship of mass M, which is flying at a constant speed in a circular orbit around the Earth. From the condition when $0 < m << M$ holds, it is reasonable to establish an orbital frame of coordinates $oxyz$ fixed on the mass

Fig. 1.1 The schematic of a spaceship-tethered satellite

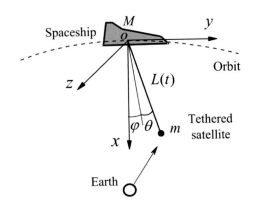

center of the spaceship. Here, direction x points towards the mass center of the Earth, direction y is along the velocity direction of the spaceship, and direction z follows the right-handed rule. As shown in Fig. 1.1, $L(t)$, $\varphi(t)$ and $\theta(t)$ describe the relative motion of the satellite in the orbital frame of coordinates $oxyz$. $L(t)$ is the tether length, $\varphi(t)$ is the angle between the tether projection and direction x, and $\theta(t)$ is the angle between the tether and the orbital plane, respectively.

The key technologies of a tethered satellite are both deployment and retrieval of the satellite. It is essential, therefore, to gain an insight into the dynamics of the satellite connected with the spaceship via a thin tether, the length of which may reach more than ten kilometers. The tether is a thin cord with distributed mass and yields the dynamics of a continuous system. The tethered satellite may exhibit nonlinear vibrations in three-dimensional space. Furthermore, the tethered satellite during deployment or retrieval is subject to Coriolis force, as well as Lorenz force when the tether made of a conductive material is moving in the magnetic field of the Earth. As such, the dynamics and control of a tethered satellite is faced with many challenges[1,2].

In the preliminary study on a tethered satellite system, engineers needed to study the system vibration when the tethered satellite was slowly deployed or retrieved. In this stage, they dealt with the vibration of a simplified model of the tethered satellite, where the above orbital frame of coordinates was assumed as an inertial frame, the satellite was a lumped mass without gyration, and the length of tether was a constant. In further stages, of course, engineers established the model of a rigid-body tethered to the spaceship so as to study the attitude dynamics of the tethered satellite.

If the simplified model exhibits a planar vibration when the tether is straight, the model is the simple pendulum studied by G. Galileo, a great Italian scientist, in 1581. Yet, the model may exhibit various planar and spatial vibrations since the tether may undergo bending vibrations in Fig. 1.2a, b, or a whirling vibration in Fig. 1.2c. Many

---

[1] Wen H, Jin D P, Hu H Y (2008). Advances in dynamics and control of tethered satellite systems. Acta Mechanica Sinica, 24(3): 229–241.

[2] Alpatov A P, Beletsky V V, Dranovskii V I, et al. (2010). Dynamics of Tethered Space Systems. Baca Raton: CRC Press.

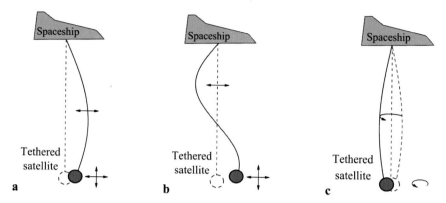

Fig. 1.2 The dynamic configurations of a tethered satellite. **a** a transverse vibration; **b** a coupled longitudinal and transverse vibration in a plane; **c** a whirling vibration out of a plane

scientists studied such a vibrating system and named it a *tether pendulum*. The tether pendulum becomes a *massive cord* with distributed mass when the tip mass vanishes. The book will deal with the tether pendulum and check the case when the tip mass is zero.

1.2.1.2 Problems of Concern

(1) Problem 1A: Dynamic modeling in a preliminary study

In the preliminary study of a tethered satellite, engineers should study the planar vibration of a tethered satellite subject to micro-gravity, or called gravity for short, shown in Fig. 1.2a, b. Hence, they are interested in the small vibration of a simplified tether pendulum, where a tether with fixed length L and linear density ρA has a lumped mass m attached at the bottom end for simplified satellite under the gravitational acceleration g.

Compared with a simple pendulum under normal gravity, the present model of a tether pendulum accounts for the effects of distributed inertia of the tether and different gravitational accelerations denoted by g. As shown in Fig. 1.3, the internal tension of the tether due to gravity varies with an increase of arc-coordinate s. That is, the tension gets the maximum at the top end and the minimum at the bottom end. The dynamic equation of the tether pendulum is a partial differential equation with variable coefficient. If engineers use a discrete model to describe the tether, they ought to select an appropriate approach among finite element method, Ritz method and the lumped parameter method.

(2) Problem 1B: Influence of system parameters on natural vibrations

In the preliminary study, engineers paid attention to the influence of some system parameters on the natural vibrations of the tethered satellite. The major parameters of concern include the gravitational acceleration g, the tether length L, the linear density ρA and the lumped mass m of the satellite. The vibration modes include both natural

Fig. 1.3 The small vibration
of a tether pendulum subject
to gravity

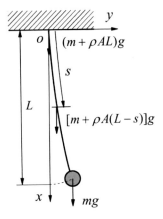

frequencies and natural mode shapes. One may ask whether the influence follows
any simple rules or not?

1.2.1.3   Basic Ideas of Study

The above two problems have a logic relation. That is, solving Problem 1A serves
for solving Problem 1B. Hence, it is necessary to account for the influence of gravitational acceleration g, tether length L, linear density ρA and lumped mass m of
satellite on the dynamics of the tethered satellite when establishing the dynamic
model of the tethered satellite.

Even though the software of finite elements enables one to model and compute
the natural vibrations of the above simplified model of the tethered satellite at ease,
it is not feasible to compute them for all combinations of the four parameters of g,
L, ρA and m when each of them comes from many given values. For instance, the
number of combination reaches 10^4 if each parameter takes 10 values.

The elementary textbooks presented several methods of discretization, such as
the lumped parameter method, Rayleigh method and Ritz method. Among them,
Rayleigh method is simplest, but only predicts the fundamental natural frequency.
Ritz method provides discrete models of lower dimensions and higher accuracy, but
may give rise to the difficulty of assuming mode shapes for a tethered satellite.

Under the above limits, it is possible to work out the following way of study. The
first step is to establish a discrete model of a tethered satellite by using the method
of lumped parameters and then to analyze the influence of the four parameters g,
L, ρA and m on the dynamic equation of the tethered satellite. The second step is
to establish a continuous model of a tethered satellite and study the computation of
natural modes. The third step is to make a comparison between the results of two
models and to discuss the influence of the four parameters on the natural vibrations.

Chapter 3 will present the first step and give some preliminary results. Chapter 5 will describe the second step and the third step, and provide detailed results. To have a quick reading, it is straightforward for readers to go to Sect. 3.1 and then to Sect. 5.1.

It is worthy to mention here that the study results of the above three steps are quite promising. The major parametric influence on the natural vibrations of a tethered satellite comes from a parameter ratio $\eta \equiv m/(\rho A L)$. That is, the mode shapes rely on the ratio η only, while the natural frequencies not only depend on the ratio η, but also proportionate to $\sqrt{g/L}$, which is the circular natural frequency of a simple pendulum. This assertion implies that when the gravitational acceleration changes, it is not necessary to compute the natural frequencies again, but only to rescale them via a factor $\sqrt{g/L}$.

1.2.2 Design of a Hydro-Elastic Vibration Isolation System

1.2.2.1 Background

Most elementary textbooks presented the basic principle of a vibration isolation system of single degree of freedom, including the influence of system damping ratio on the transmissibility of vibration isolation. In the design of a vibration isolation system of single degree of freedom, thus, it is desirable to have a large damping ratio to suppress the possible resonance in a resonant frequency band and to have a small damping ratio to reduce the transmissibility of vibration isolation in a higher frequency band. In practice, it is feasible to design and adjust the damping effect via a hydro-device.

In vehicle engineering, for example, engineers designed a kind of *hydro-elastic vibration isolation system* for vehicle engines to reduce the vibration transmitted from an engine to a vehicle body. The vibration isolation system has four hydro-elastic

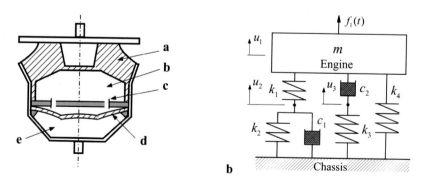

Fig. 1.4 A hydro-elastic vibration isolation system for vehicle engines. **a** the profile of a hydro-elastic mounting; **b** a model for the vibration isolation system

mountings to support an engine. Figure 1.4a shows the profile of a mounting[3], where **a** represents the primary rubber, **b** represents two oil chambers connected through a set of orifices denoted by **c**, **d** represents the secondary rubber, the deformation of which pushes oil back from the bottom oil chamber to the upper one. **e** represents an air chamber with a hole at lower steel case to provide an adequate air pressure. The intention of the design was to adjust the orifices to achieve ideal damping ratio of the vibration isolation system.

In the design stage, engineers simplified the vibration isolation system as a model in Fig. 1.4b, where m represents the mass of the engine, k_1 associated with the upper left spring represents the volumetric stiffness coefficient of the oil in two chambers, k_2 with the bottom left spring represents the stiffness coefficient of the secondary rubber, c_1 with the left dashpot represents the damping coefficient of the lamination of oil passing through the orifices, k_4 with the right spring represents the stiffness coefficient of the primary rubber. The dashpot with c_2 and the spring with k_3 in serial constitute *Maxwell's fluidic component*, which, together with the right spring with k_4, describes the viscoelastic behavior of the primary rubber.

1.2.2.2 Problems of Concern

(1) Problem 2A: Abnormal effects of damping coefficients

In the hydro-elastic mounting, the oil passes through a set of orifices and produces damping. Hence, it is possible to control damping effect by adjusting the number of orifices and the radii of orifices. Yet, both computations and experiments showed that the resonance suppression seemed almost unchanged when adjusting these parameters. Therefore, engineers felt strange why the hydro-damping did not produce expected results and contradicted to their intuition of mechanics.

(2) Problem 2B: How to determine the degree of freedom of a system

In classical dynamics, the *degree of freedom*, or the *DoF* for short, of a system is the minimal number of independent generalized coordinates of the system. Then, the degree of freedom of a lumped parameter system is the number of lumped inertial components, such as a lumped mass. The vibration system in Fig. 1.4b has a single lumped mass and seems to be a system of *single degree of freedom*, or *SDoF* for short. Yet, it is not possible to use only the vertical displacement u_1 of the lumped mass to describe all motions of the system. That is, it is necessary to use the other two vertical displacements u_2 and u_3 in Fig. 1.4b to describe the motions of two connection points of dashpots and springs. As such, the system seems to have three degrees of freedom, or 3-DoFs for short. Engineers wondered how many degrees of freedom the system possesses.

---

[3] Morello L, Rossini L R, Pia G, et al. (2011). The Automotive Body. Volume II: System Design. New York: Springer-Verlag.

Fig. 1.5 A dynamic system with a spring and a dashpot in serial

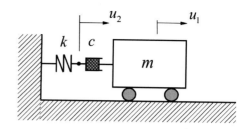

1.2.2.3 Basic Ideas of Study

In fact, the two problems come from the same origin. In elementary textbooks, most vibration systems do not have a spring and a dashpot in serial shown in Fig. 1.4b, where the two connection points do not have any lumped mass so that two degrees of freedom of the system degenerate. The contribution of the dashpot to the system damping ratio of the system, thus, changes.

To understand the above issue, consider a *dynamic serial system* shown in Fig. 1.5, where a dashpot and a spring in serial constitute Maxwell's fluidic component[4,5]. Apart from the displacement $u_1(t)$ of the mass block, one needs to use displacement $u_2(t)$ to describe the motion of the connection point between the dashpot and the spring. Hence, the system is not an SDoF system. Assume that the system has 2-DoFs, but the lumped mass at the connection point vanishes. Thus, the dynamic equations of the system satisfy

$$\begin{cases} m\dfrac{d^2 u_1(t)}{dt^2} + c\left[\dfrac{du_1(t)}{dt} - \dfrac{du_2(t)}{dt}\right] = 0 \\ c\left[\dfrac{du_2(t)}{dt} - \dfrac{du_1(t)}{dt}\right] + ku_2(t) = 0 \end{cases} \tag{1.2.1}$$

Eliminating displacement u_2 from Eq. (1.2.1) leads to an ordinary differential equation in terms of velocity $du_1(t)/dt$, namely,

$$\frac{1}{k}\frac{d^2}{dt^2}\left[\frac{du_1(t)}{dt}\right] + \frac{1}{c}\frac{d}{dt}\left[\frac{du_1(t)}{dt}\right] + \frac{1}{m}\left[\frac{du_1(t)}{dt}\right] = 0 \tag{1.2.2}$$

Equation (1.2.2) looks exactly similar to the standard dynamic equation of an SDoF system in terms of displacement $u_1(t)$. Hence, one can solve Eq. (1.2.2) for velocity $\dot{u}_1(t) \equiv du_1(t)/dt$ and then get $u_1(t)$ by integrating $\dot{u}_1(t)$ with respect to time t. This procedure implies that the dynamic serial system in Fig. 1.5 has a part of properties of an SDoF system. It is interesting that the dynamic serial system governed by

---

[4] Balachandran B, Magrab E B (2019). Vibrations. 3rd Edition. Cambridge: Cambridge University Press, 176.

[5] Hu H Y (1998). Mechanical Vibrations and Shocks. Beijing: Press of Aviation Industry, 26 (in Chinese).

Eq. (1.2.2) exhibits damped vibration when the condition $c > \sqrt{mk}/2$ holds. In this case, the system damping ratio is $\zeta \equiv \sqrt{mk}/(2c)$. As such, the larger the damping coefficient c, the smaller the system damping ratio ζ.

According to the above analysis of a simple case, the basic ideas of study are as follows. The first step is to study Problem 2B, that is, the degree of freedom of a discrete system, especially the case when a degree of freedom degenerates because of a spring and a dashpot in serial. The second step is to study the damping coefficient of a system with some degrees of freedom degenerated.

In the first step, it is necessary to notice that the second equation in Eq. (1.2.1) is a constraint relation between velocities $\dot{u}_1(t)$ and $\dot{u}_2(t)$, and can not be integrated as an algebraic equation in terms of displacements $u_1(t)$ and $u_2(t)$. This equation, thus, describes a *non-holonomic constraint* in analytical mechanics. In the hydro-elastic vibration isolation system for vehicle engines in Fig. 1.4b, the two sets of spring-dashpot in serial give rise to more complicated velocity relations, that is, more complicated non-holonomic constraints in analytical mechanics. It is necessary, therefore, to study the degeneration of degrees of freedom of a system with non-holonomic constraints.

In the second step, it is possible to extend the modal analysis of linear vibration systems since the systems of concern are linear and time invariant. Hence, it is reasonable to perform the modal analysis in both time domain and frequency domain to reveal the relation between the damping coefficients and the system dynamics.

Chapter 3 will present the first step of the study and Chap. 4 will go to the second step, solving Problem 2B and 2A, successively. To have a quick reading, one can go to Sect. 3.2 first and then to Sect. 4.1. Furthermore, the viscoelastically damped system discussed in Sect. 3.3 also exhibits the degenerated degree of freedom of a system.

1.2.3 Two Kinds of Immovable Points in a Vibration System

1.2.3.1 Background

An important feature of dynamics, different from statics, of a mechanical/structural system is the existence of some immovable points when the system vibrates. Consider, as an example, a slender beam, or called a beam for short, and let the beam be clamped at one end and subject to a static load at the other end. Then, all the points of the beam, except for the clamped cross-section, undergo static deformations. However, this cantilever beam has an immovable, but not clamped centroid in a cross-section in the natural vibration of the second order. This centroid of the cross-section is called a *node* of the mode shape of the second order. The corresponding cross-section is often named a node of the mode shape when the beam is slender enough. Furthermore, the steady-state vibration of a centroid of a cross-section of the cantilever beam to a harmonic excitation may be zero at a number of excitation frequencies. This phenomenon is called the *anti-resonance*. Similarly, the

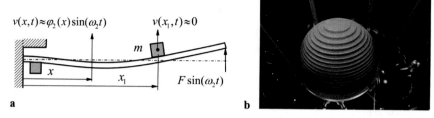

$v(x,t) \approx \varphi_2(x)\sin(\omega_2 t)$ $v(x_1,t) \approx 0$

m

x

x_1 $F\sin(\omega_2 t)$

a

b

Fig. 1.6 Utilizations of nodes and anti-resonances of vibrating structures. **a** an equipment mounted at the node of the second mode shape of a cantilever structure; **b** the dynamic vibration absorber mounted at Building 101 in Taipei, China

corresponding cross-section is said to be in anti-resonance when the beam is slender enough. In the above two dynamic cases, some un-clamped points of the cantilever beam are immovable and quite different from their behaviors in the static case.

In engineering, it is very helpful to utilize both nodes and anti-resonances of a structure to achieve the local minimum of vibration. As shown in Fig. 1.6a, an equipment, denoted by m, needs to be mounted on a cantilever slender structure under a harmonic force $F\sin(\omega_2 t)$, where ω_2 is the second natural frequency of the structure. In this case, the steady-state vibration $v(x, t)$ of the structure is dominated by the natural vibration of the second order. If no room is available near the clamped end as shown in Fig. 1.6a, the best location for mounting the equipment is at node $x = x_1$ of the second mode shape to reduce the vertical vibration of the equipment. The other example is the successful design of Building 101 in Taipei, China. As shown in Fig. 1.6b, a steel ball of weight 660 t was hung via 4 sets of stranded steal cables from the top floor of the building such that the top floor exhibits an anti-resonance of horizontal vibration when the building is subject to the distributed loads of wind. This design greatly reduces the horizontal vibrations of the top floor and becomes one of the most successful designs for tall buildings in the world.

The above concepts of two immovable points of a vibrating system have had a long history, but the recognitions of their physical mechanisms are not sufficient enough. For example, some assertions about the node number of a mode shape and the mechanism of anti-resonance in textbooks and handbooks are not accurate enough.

The author found two problems in 1980 when pursuing Bachelor of Science. The first problem came from some books published between the 1950s and the 1970s. The books dealt with the natural vibrations of rods and beams, stating that the r-th mode shape of a chain system like rods and beams always exhibit $r-1$ nodes. Some other books even claimed that this is a general rule for vibrating systems. It is easy to question the second statement since it does not apply to the mode shapes of a thin plate discussed in elementary textbooks. Nevertheless, what about the validity of the assertion for a chain system?

The second problem also came from some books, which explained the principle of a dynamic vibration absorber via the anti-resonance of a 2-DoF system. As usual, the 2-DoF system includes a primary 1-DoF system attached with a secondary 1-DoF system. In addition, only the primary system is subject to a harmonic force. When the excitation frequency of the force coincides with the natural frequency of the secondary system, the primary system exhibits anti-resonance. In this case, many books clamed that the excitation energy to the primary system transfers to the secondary system such that the primary system becomes at rest. In other words, the secondary system absorbs the excitation energy. As a consequence, the secondary system was historically referred to as the *dynamic vibration absorber*. Yet, it is easy to find that the harmonic force does not do any work since the primary system is immovable at the anti-resonance. Therefore, it is reasonable to question the viewpoints of both energy transfer and energy absorption.

1.2.3.2   Problems of Concern

(1) Problem 3A: The node number of a mode shape of a chain system
  Even though the *r*-th mode shape of some chain systems, such as a string, a rod and a beam, has *r*-1 nodes, does this conclusion apply to all chain systems?
(2) Problem 3B: The energy consumed by an un-damped vibration absorber
  The key issue of the problem is to reveal the physical mechanism of an anti-resonance of a 2-DoF system and check the energy transfer.

1.2.3.3   Basic Ideas of Study

These two problems are different from those in Sects. 1.2.1 and 1.2.2. That is, they came from curiosity, rather than engineering practice. To study the two problems, it is necessary to sort out the ideas of study even though it is almost impossible to work out a research plan in advance.
  The author, as an undergraduate student, noticed Problem 3A in 1980, but studied it when pursuing Master of Science in 1982. The consultation with the authors of associated books indicated that the major conclusions came from the Russian monograph by F. R. Gantmacher and M. G. Krein, two eminent Russian mathematicians, in 1950. However, the author could not get and read the Russian monograph at that time. In fact, the English translation and the Chinese translation of the Russian monograph were not available until 2002 and 2008, respectively. To doubt the rule of the node number of a mode shape of a chain system, thus, it was necessary for the author to construct at least one counter-example. And it was also reasonable to look for a counter-example in consideration of mechanics, instead of mathematics, since the authors of the Russian monograph were world renowned mathematicians. When reading the English translation[6] of the Russian monograph by Gantmacher and Krein 20 years later, the author found that all the counter-examples of chain

---

[6] Gantmacher F R, Krein M G (2002). Oscillation Matrices and Kernels and Small Vibrations of Mechanical Systems. 2nd Edition. Rhode Island: AMS Chelsea Publishing, 113–179.

systems constructed in 1982 do not satisfy the conditions of oscillatory matrix that Gantmacher and Krein had used to prove their theorems for the node number of a mode shape of a chain system.

The author studied Problem 3B after receiving doctorate in 1988. From the research experience at that time, the author divided the research into three steps. The first step was to study the frequency response matrix and the energy transfer of an arbitrary 2-DoF system since the vibration system with any anti-resonance must have at least 2-DoFs. The second step was to construct a 2-DoF system with adjustable parameters, examine the conditions of anti-resonance of the system, and the evolution of natural frequencies and anti-resonant frequencies with a variation of system parameters, including the anti-resonance when the system has repeated natural frequencies. The third step was to generalize the above analysis and results to the system of *multiple degrees of freedom*, or called *MDoF* for short.

Section 4.2 will deal with Problem 3A, namely, the node number of a mode shape of a chain system. Then, Sect. 4.3 will discuss Problem 3B and present the physical mechanism about the anti-resonance for an arbitrary 2-DoF system. Furthermore, Sect. 4.4 will address the anti-resonance problem of a kind of MDoF systems.

1.2.4 Identical Natural Frequencies of Different Structures

1.2.4.1 Background

During the teaching process in the 1980s, the author noticed some interesting phenomena of uniform beams. For instance, a clamped-clamped beam and a free-free beam have identical natural frequencies. So do a hinged-clamped beam and a hinged-free beam. As shown in Fig. 1.7a, a clamped-clamped beam and a free-free beam have the same fundamental natural frequency even though their corresponding mode shapes are totally different. Figure 1.7b presents the similar result for a hinged-clamped beam and a hinged-free beam.

From both vibration mechanics and engineering experience, the natural frequencies of a structure will increase if some constraints are imposed on the structure. What about the reasons for the natural frequencies of the above beams?

As discussed in elementary textbooks, the free vibrations of a clamped-clamped beam and a free-free beam satisfy the following two boundary value problems of the same partial differential equation, respectively

$$
\begin{cases}
\rho A \dfrac{\partial^2 v(x,t)}{\partial t^2} + EI \dfrac{\partial^4 v(x,t)}{\partial x^4} = 0 \\
v(0,t) = 0, \quad v_x(0,t) = 0, \quad v(L,t) = 0, \quad v_x(L,t) = 0
\end{cases}
\tag{1.2.3}
$$

$$
\begin{cases}
\rho A \dfrac{\partial^2 v(x,t)}{\partial t^2} + EI \dfrac{\partial^4 v(x,t)}{\partial x^4} = 0 \\
v_{xx}(0,t) = 0, \quad v_{xxx}(0,t) = 0, \quad v_{xx}(L,t) = 0, \quad v_{xxx}(L,t) = 0
\end{cases}
\tag{1.2.4}
$$

Fig. 1.7 Two beams under
different boundary
conditions with identical
natural frequencies. **a** a
clamped-clamped beam and
a free-free beam; **b** a
hinged-clamped beam and a
hinged-free beam

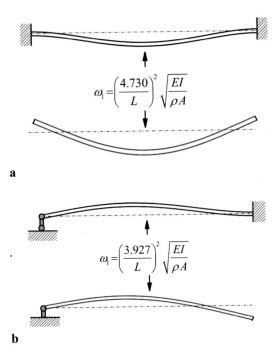

where L is the length of the beam, A is the cross-sectional area of the beam, I is the second moment of cross-sectional area of the beam, ρ is the material density, E is Young's modulus of material, $v(x, t)$ is the dynamic deflection of the beam, x is the coordinate measured from the left end of the beam, t is time. In boundary conditions, $v_x(x, t) \equiv \partial v(x, t)/\partial x$, $v_x(0, t) \equiv v_x(x, t)|_{x=0}$, and so on.

Taking the partial derivative of the dynamic equation in Eq. (1.2.4) twice with respect to x and introducing the curvature $\eta(x, t) \equiv v_{xx}(x, t)$ for the small vibration of the beam, Eq. (1.2.4) can be recast as

$$
\begin{cases}
\rho A \dfrac{\partial^2 \eta(x, t)}{\partial t^2} + EI \dfrac{\partial^4 \eta(x, t)}{\partial x^4} = 0 \\
\eta(0, t) = 0, \quad \eta_x(0, t) = 0, \quad \eta(L, t) = 0, \quad \eta_x(L, t) = 0
\end{cases}
\tag{1.2.5}
$$

The comparison of Eqs. (1.2.5) and (1.2.3) indicates that they have the same form. Hence, the two beams have identical natural frequencies if the rigid-body motion of free-free beam is not taken into account. The above treatment also applies to the relation between a hinged-clamped beam and a hinged-free beam.

The author made a comment on this issue in an elementary textbook published in 1998 and suggested readers thinking about the physical mechanism behind the phenomena[7]. Unfortunately, nobody ever discussed this comment with the author until a recent meeting, where an aerospace engineer put a question about the comment to the author. This initiated the following considerations.

1.2.4.2 Problems of Concern

(1) Problem 4A: Mechanism of the same natural frequencies for different beams
    Both deflection and curvature are geometric descriptions for the deformation of a uniform beam. Thus, the curvature dynamics of a uniform beam under free boundary conditions and the deflection dynamics of the other uniform beam under clamped boundary conditions are identical. What is the mechanism behind the phenomenon? Is it possible to observe the same phenomena for a pair of non-uniform beams and what is the corresponding mechanism?
(2) Problem 4B: Universality of these issues and their physical mechanisms
    For other one-dimensional structures, such as a non-uniform rod or a non-uniform shaft, are there any similar phenomena? If yes, what about the physical mechanisms behind the phenomena?

1.2.4.3 Basic Ideas of Study

For a uniform beam, the bending moment is proportional to the curvature, namely, $M(x, t) = EI\eta(x, t) = EIv_{xx}(x, t)$. Therefore, Eq. (1.2.5) can be recast as

$$\begin{cases} \rho\tilde{A}\dfrac{\partial^2 M(x, t)}{\partial t^2} + E\tilde{I}\dfrac{\partial^4 M(x, t)}{\partial x^4} = 0 \\ M(0, t) = 0, \quad M_x(0, t) = 0, \quad M(L, t) = 0, \quad M_x(L, t) = 0 \end{cases} \tag{1.2.6}$$

where $\rho\tilde{A} \equiv 1/EI$ and $E\tilde{I} \equiv 1/\rho A$. Thus, the bending moment dynamics of a free-free beam and the displacement dynamics of a clamped-clamped beam have the identical boundary value problems. From the viewpoint of energy analysis, either the deflection or the bending moment can be used as an unknown function to solve the problem of natural vibrations of a beam[8]. Hence, the two problems are referred to as a *dual* when the displacement dynamics of a beam and the bending moment dynamics of a beam have the identical boundary value problems. Such a dual includes both a dual of dynamic equations of two beams and a dual of their boundary conditions. That is, a clamped-clamped beam and a free-free beam form a dual in natural vibrations, and so do a hinged-clamped beam and a hinged-free beam.

---

[7] Hu H Y (1998). Mechanical Vibrations and Shocks. Beijing: Press of Aviation Industry, 159–160 (in Chinese).

[8] Hu H C (1981). Variational Principles in Elasticity and Their Applications. Beijing: Science Press, 100–101 (in Chinese).

Based on the description of bending moment, the study on a dual of uniform beams can be extended to a pair of non-uniform beams. Given a non-uniform beam and associated boundary conditions, for instance, it is feasible to construct the other non-uniform beam with different boundary conditions such that the two beams have identical natural frequencies.

Similarly, it is possible to study the dual of a pair of non-uniform rods or a pair of non-uniform shafts. For example, the description of internal force for a free-free rod gives the same description of displacement for a fixed-fixed rod, namely, a pair of dual for these two rods. It is straightforward to deal with a pair of shafts since they have the same dynamic equation of rods.

Chapter 5 will present the recent studies of the author and related results. Section 5.2 will deal with Problem 4B. That is, the dual of two rods (or two shafts) in natural vibrations, including different variations of their cross-sections, the same variations of their cross-sections and the uniform cross-sections. Section 5.3 will discuss Problem 4A, namely, the dual of two beams, including various cross-sections in Sect. 5.2. Sections 7.2 and 7.4 will also touch upon the dual problems of boundaries for rods and beams from the viewpoint of longitudinal waves and bending waves, respectively.

1.2.5   Closely Distributed Natural Modes of a Symmetric Structure

1.2.5.1   Background

There are numerous symmetric structures in engineering. For example, airplanes and ground vehicles must be symmetric about a plane parallel to their moving direction. That is, they have *mirror-symmetry*. The other example of symmetry covers the

a   **b**

Fig. 1.8 Two typical structures with cyclosymmetry. **a** the bladed disc of an aero-engine ($n = 23$); **b** the satellite Spektr-R with a radio-telescopic antenna ($n = 27$)[9]

---

[9] Chen Q F (2012). Encyclopedia of Worldwide Spacecraft. Beijing: China Space Press, 329 (in Chinese)

bladed disc of an aero-engine in Fig. 1.8a and the radio-telescopic antenna of satellite Spektru-R in Fig. 1.8b, where both structures are exactly identical after they are rotated through an angle of $2\pi/n$ about a central axis and exhibit the *cyclic symmetry* or a compound term *cyclosymmetry*. The designs of those symmetric structures not only show the beauty in appearance, but also meet the requirements of their functions.

It is well known that the system symmetry enables one to simplify the dynamic computation of the system. For a structure with mirror symmetry, it is easy to divide an arbitrary dynamic load to the structure into two parts, namely, a symmetric part and an anti-symmetric part. They excite the symmetric vibration and anti-symmetric vibration, respectively. Hence, one can take a half of the structure and compute respectively the symmetric and anti-symmetric vibrations, and then sum up the two vibrations to get the vibration due to an arbitrary dynamic load. For a structure with cyclosymmetry, the past decades since the 1970s have seen several approaches to simplify the computation of natural vibrations of the structure. The approaches enable one to compute the natural vibrations of the structure from those of one of n sectors. The bladed disc of an aero-engine or a turbo-engine usually has so many blades that $n >> 10$ holds true in practice. These approaches, therefore, can save a great amount of computational cost.

In recent years, engineers have not worried so much about the computational cost with an integration of those approaches based on structure symmetry and fast development of computational techniques. Yet, they have encountered some new problems from structure symmetry as follows.

1.2.5.2 Problems of Concern

(1) Problem 5A: Poor repeatability of modal tests of a symmetric structure

The engineering consultations that the author attended dealt with several kinds of symmetric structures. The problems often proposed by engineers were about the natural modes closely distributed in frequency domain and the poor repeatability in modal tests. For instance, the modal test of a bladed disc of a turbo-engine usually led to many sets of natural vibrations with very close frequencies but different mode shapes. One mode shape looked like the rotation of the other mode shape about the central axis by an angle. However, these mode shapes exhibited very poor repeatability in further modal test. As well known, the poor repeatability of industrial products in tests before use is a tough issue. This is particularly a big issue for any space structures in ground vibration tests before launch. Then, what is the reason of the above problem?

(2) Problem 5B: Natural vibrations of a cyclosymmetric structure with the dynamic deformation at the central axis

In the previous researches on the natural vibrations of cyclosymmetric structures, most publications dealt with annular structures shown in Fig. 1.8a, where the central axis of the structure is either clamped or hollowed, and no dynamic deformation needs to be studied. In practice, however, a cyclosymmetric structure may undergo the dynamic deformation at the central axis. A typical example is the radio-telescopic antenna of satellite Spektr-R in Fig. 1.8b. Then, it is reasonable to check the difference between the two kinds of cyclosymmetric systems, and whether the research on the natural vibrations of cyclosymmetric structures of the first kind can be extended to the second kind.

Among the early studies on the natural vibrations of cyclosymmetric structures of the first kind, D. L. Thomas in the UK analyzed the standing waves in the structures and pointed out the existence of standing waves along the circumferential direction. Many people thought that the standing waves in a cyclosymmetric structure of the second kind would be no longer those along the circumferential direction since the central DoFs of the cyclosymmetric structure of the second kind were not fixed. Therefore, it is difficult to extend the vibration analysis of cyclosymmetric structures of the first kind to those of the second kind.

1.2.5.3 Basic Ideas of Study

The nature of Problem 5A is the repeated natural frequencies due to the structural symmetry. In the circle of vibration engineering, people have briefly known this phenomenon in theory, but popularly believed that real structures are not ideally symmetric and do not have repeated natural frequencies. As a consequence, most textbooks of mechanical vibrations or structural dynamics did not pay much attention to the repeated natural frequencies. Hence, engineers know a little about the repeated natural frequencies so that they usually take them as a kind of special behaviors of the structures. The basic ideas of solving this problem include two steps. The first step is to analyze the repeated natural frequencies due to the structural symmetry and gain an insight into the non-uniqueness of normalized mode shapes. The second step is to establish a simplified model described by a pair of mode shapes of repeated natural frequencies and to study the variation of mode shapes of a symmetric system under a small disturbance. This way enables one to clearly see the reason of poor repeatability of modal tests for a symmetric system.

To solve Problem 5B, the author checked the natural vibrations of circular plates and circular annular plates first and found no essential difference between them. That is, they have the standing waves in both circumferential direction and radial direction. This fact indicated that the natural vibrations of both kinds of cyclosymmetric systems may follow similar rules. The major difference between the two kinds of cyclosymmetric structures is that any two neighboring sectors of a cyclosymmetric structure of the first kind share a single interface only, but all sectors of a cyclosymmetric structure of the second kind share the central axis. Therefore, it is necessary to establish the compatibility conditions of all sectors on the central axis, rather than the conditions of standing waves.

Chapter 6 will present the detailed research on the above two problems and their results. Section 6.1 will discuss the repeated natural frequencies of a hinged rectangular plate and demonstrate the influence of a small inertial disturbance on the mode shapes associated with Problem 5A. Sections 6.2 and 6.3 will present a systematic study on the natural vibrations of cyclosymmetric structures without or with the dynamic deformation in the central axis and solve Problem 5B.

1.2.6 Transient Response of a Slender Structure

1.2.6.1 Background

The scale of recent space structures has been larger and larger, and the stiffness of them has been lower and lower. For example, the space power station conceptually proposed in Fig. 1.9 was about 7 km in length and the area of a single solar sail reached 3 km^2. To construct a space structure of so large scale, it is necessary to use light structure configurations and light materials. As such, the natural frequencies of a large space structure are extremely low and any disturbed response of the structure decays very slowly.

In the current design of space structures, it is seldom to take the wave effects into account since the propagation speed of any elastic waves in a metal structure reaches several km/s and the elastic wave in a space structure of several meters long undergoes many reflections within a short duration. As such, the structural dynamics usually exhibits the modal responses of lower orders because the modal responses of higher orders are damped out. When the scale of a space structure reaches several kilometers, however, the elastic waves of the structure undergo a few of reflections and exhibit the property of traveling waves. Hence, it is reasonable for engineers to pay attention to the transient response of a large space structure, especially to that of a slender space structure, including the effects of traveling waves.

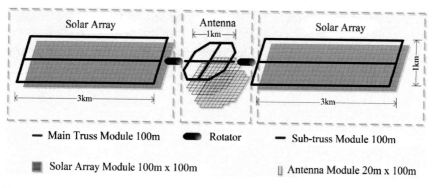

Fig. 1.9 A conceptual schematic of a space power station[10]

---

[10] Cheng Z A, Hou X B, Zhang X H, et al. (2016). In-orbit assembly mission for the space solar power station. Acta Astronautica, 129: 299–308.

1.2.6.2 Problems of Concern

(1) Problem 6A: Validity of modal superposition method for wave analysis

The propagation of an elastic wave in a space structure of limited size includes complicated reflections and refractions. The structural engineers can hardly compute and analyze the dynamic response of a slender space structure by using the method of wave analysis. It is natural, therefore, to ask whether the modal superposition method is applicable to the dynamic analysis of a slender space structure, including the dynamic computations of the structure in a narrow frequency band or in a wide frequency band.

(2) Problem 6B: Validity of modal superposition method for impact analysis

In the vibration analysis of a slender structure, the displacement disturbance and velocity disturbance seem no difference by nature. Yet the velocity disturbance due to impact load is a non-smooth dynamic process and gives rise to non-smooth or even discontinuous strain waves. In such a case, it is reasonable to question whether the modal superposition method is valid or not for the dynamic analysis of a slender space structure.

1.2.6.3 Basic Ideas of Study

The above two problems are different, but closely related. The first one focuses on the convergence of modal superposition solution, or called the modal series solution, for the dynamic response of a slender space structure, while the second one concentrates on the spatial smoothness of the dynamic response of a slender space structure. From the viewpoint of mathematics, they both depend on the property of modal series solutions since the smoothness of the dynamic response determines the convergence of modal series solution.

To deal with Problem 6A, it is better to study the dynamics of rods and beams, to compare the exact solutions and modal series solutions for checking the validity of modal superposition methods, and to discuss the influences of a variation of cross-section and transverse inertia on the disperse properties of an elastic wave in a rod so as to check the convergence of modal series solution.

To study Problem 6B related to the discontinuous velocity waves and strain waves, it is possible to check the dynamic responses of both rods and beams subject to an impact and to discuss the validity and convergence of modal superposition method both qualitatively and quantitatively.

Chapter 7 will present the detailed researches on the dynamics of one-dimensional structures, including rods and beams. Sections 7.1 and 7.2 will focus on the waves and vibrations of several kinds of rods, while Sects. 7.4 and 7.5 will deal with the waves and vibrations of two beam models, that is, the models of Euler–Bernoulli beam and Timoshenko beam, in order to address Problem 6A. Section 7.3 will present the wave analysis of a rod impacting a rigid wall, and give a comparison between the exact solution and the modal series solution so as to address Problem 6B. Section 7.5

will discuss the impact problem of a beam and provide a practical criterion for modal truncation.

1.3 Scope and Style of the Book

The design and selection of contents in the book are intended to guide readers to conduct the research-oriented study on linear vibrations. In the main body of Chap. 1, 12 problems have been proposed as the start of research-oriented study at first. Then, some necessary preparations will be made in Chap. 2 from a perspective of beauty of science. Afterwards, detailed studies will be made on 12 problems proposed in Chap. 1 from Chaps. 3 to 7. The contents of the whole book can be summarized as follows.

As seen from previous sections, this chapter has presented the start of research-oriented study, including the intention and ideas of the author to write this book. In order to assist readers to upgrade their understanding to vibration mechanics helically via a research-oriented study, the chapter has proposed 12 problems to be studied in subsequent chapters from 6 case studies of vibration mechanics. The intention of this chapter is to arouse interest of readers so as to find and propose new problems while reading the entire book.

Chapter 2 will provide readers with a new perspective of reading the book during the research-oriented study. The chapter will survey the major theoretical and methodological results of vibration mechanics covered by most elementary text-books from a perspective of beauty of science so as to assist readers to make necessary preparations for subsequent reading. The chapter looks quite different from conventional textbooks and assists readers to review the vibration mechanics in hands from a new perspective and to understand the entire book from the new perspective.

Chapter 3 will address some fundamental problems in the dynamic modeling of mechanical and structural systems, including the discrete model of a continuous system, the degree of freedom of a discrete system, and structural damping. The chapter will focus on the problems from a continuous system to a discrete system, a half degree of freedom caused by a non-holonomic constraint, and the viscoelastic damping model, which is also associated with a half degree of freedom. The chapter will give detailed researches and conclusions for Problems 1A, 1B and 2B, respectively.

Chapter 4 will be about some issues related to discrete systems of vibration, including the vibration mechanics of a discrete system with non-holonomic constraints, the node number of a mode shape of a chain system, the anti-resonance of a discrete system to a harmonic excitation, and the dynamic modification of a discrete system based on its frequency response matrix. The chapter will give detailed studies and conclusions for Problems 2A, 3A and 3B, respectively.

Chapter 5 will focus on some natural vibration problems of one-dimensional structures, including the natural vibration of a tether pendulum under gravity and the dual problems of rods and beams in their natural vibrations. That is, a pair of rods or beams has identical natural frequencies when their cross-sections and boundaries satisfy some conditions of a dual. The chapter will present detailed studies on Problems 1A, 1B, 4A and 4B, respectively.

Chapter 6 will deal with the natural vibrations of symmetric structures, such as those with mirror-symmetry and cyclosymmetry, with an emphasis on the repeated natural frequencies due to structure symmetry. The chapter will begin with the natural vibrations of a thin rectangular plate and analyze the natural vibrations of repeated frequencies, as well as the change of mode shapes with repeated natural frequencies subject to a small structural disturbance. Then, the chapter will present a systematic study on both high-efficient computation and major properties of the natural vibrations of cyclosymmetric structures. The chapter will give detailed solutions and conclusions for Problems 5A and 5B.

Chapter 7 will be about both wave and vibration problems of rods and beams, including their free responses, steady-state harmonic responses and impact responses. The chapter will focus on the wave analysis and modal analysis of the responses of both rods and beams, provide their comparisons in detail and verify the feasibility of modal superposition method for wave analysis. The chapter will present the research results and conclusions of Problems 6A and 6B.

The appendix of the book will present the wave theory of a three-dimensional elastic medium so as to provide readers with some supplementary knowledges to one-dimensional waves discussed in Chap. 7. The appendix will give the description and decomposition of three-dimensional elastic waves, the properties of planar waves and spherical waves, the detailed discussions about the reflection of three planar harmonic waves at the boundary of a medium, as well as the Rayleigh wave propagating near the boundary surface.

As a book of guiding research-oriented study, the internal structure of each chapter, from Chaps. 3 to 7, looks like a collection of research papers. That is, each section is equivalent to a paper, which starts with the research background and ends with some conclusions. Nevertheless, the internal content of each section takes the style of textbook, including detailed inferences, remarks and examples. In addition, each chapter will terminate with some materials for further reading and thinking.

The contents in the first two chapters seem simple, but need to be understood via repeated thinking. The contents in Chaps. 3 to 7 are almost independent. One may read them optionally. It is more beneficial for readers to review the first two chapters after reading all chapters or even a part of them. Such a review may enable one to initiate some new inspirations.

1.4 Further Reading and Thinking

(1) Read chapters of at least three books among the references[11,12,13,14,15,16], sort out the research history of vibrations and waves, list five milestone achievements and briefly state the key issues for the problems or methods proposed by the great investigators.

(2) Recalling the tethered satellite system in Fig. 1.2, discuss the rationality of a model of planar vibrations shown in Figs. 1.2a and 1.2b, and list the physical conditions that the model satisfies.

(3) Recalling the dynamic serial system in Fig. 1.5, discuss how to analyze the free vibration of the system under some disturbances, and how to realize those initial conditions.

(4) Following the ideas in Sect. 1.2, propose a vibration problem in engineering or daily life, and state how to verify the effectiveness of the research results.

(5) In scientific researches, both imaginary guess and logical inference are essentially important. Try to give a guess for Problems 6A or 6B, and state the reasons based on the physical intuitions.

[11] Balachandran B, Magrab E B. Vibrations (2019). 3rd Edition. Cambridge: Cambridge University Press, 1–10.

[12] Craig Jr R R (1981). Structural Dynamics. New York: John Wiley and Sons Inc., 1–12.

[13] Den Hartog J P (1984). Mechanical Vibrations. 4th Edition. New York: Dover Publications, 23–169.

[14] Graff K F (1975). Wave Motion in Elastic Solids. Columbus: Ohio State University Press, 1–8.

[15] Rao S S (2007). Vibration of Continuous Systems. Hoboken: John Wiley & Sons Inc., 1–30.

[16] Zhang C (2009). History of Machine Dynamics. Beijing: Press of Higher Education, 45–157 (in Chinese).

Chapter 2
Preparation of Research-Oriented Study

Readers may wonder how to conduct the research-oriented study in an intermediate or advanced course of vibration mechanics. The success depends not only on the previous theoretical foundation, but also on the learning methods and academic tastes. To some extent, the latter seems more important since an investigator always learns something new during his research. As such, this chapter will provide readers with necessary preparation from two aspects. One is to assist readers to review a part of important issues of vibration mechanics, covered by most elementary textbooks, so as to consolidate the theoretic foundations. The other is to help readers upgrade their academic tastes. Different from any available elementary textbooks, thus, the chapter will not follow any conventional books of vibration mechanics, but will survey the major issues of vibration mechanics from the viewpoint of aesthetics of science so as to stimulate readers to start comprehensive thinking.

2.1 Briefs of Beauty of Science

To touch upon the beauty of science, this section begins with *aesthetics*, which is a branch of philosophy concerned with the study on ideas of beauty. In general, aesthetics is an overlapped field between philosophy and arts, studying the science of sensations of humankind towards to the reality, and focusing on the aesthetic problems in practice. Specifically, aesthetics mainly deals with the existence of beauty (including the types and their relations), the sensations of humankind to beauty (including the sense of beauty, the properties and features of the sense of beauty, as well as the relations between the beauty and the sense of beauty), the creation of beauty and the aesthetic education.

The beauty of science is a part of aesthetics and covers the beauty from both phenomena of science and their rules behind. The features of the beauty of science

© Science Press 2022
H. Hu, *Vibration Mechanics*,
https://doi.org/10.1007/978-981-16-5457-2_2

include unity, harmony, simplicity, regularity, symmetry and so on[1]. For example, J. Kepler, a great German astronomer, named his celebrated monograph "The Harmony of the World" in 1619 to demonstrate the beauty in the orbital rules of planets[2]. The other example is J. H. Poincaré, a great French scientist, who emphasized unity and simplicity in his mathematical and physical studies in the beginning of the twentieth century[3]. He wrote: "it is in its aesthetic value that the justification of scientific theory is to be found, with it the justification of scientific method"[4]. In the studies on mathematics and physics, the beauty of science has received so great attention that the harmony, simplicity and symmetry of many theories and methods have been well known. Good examples may cover, but not limited to, the theory of linear operators and group theory in mathematics, Maxwell's equations of electro-magnetic fields and Einstein's formula between mass and energy in physics.

Vibration mechanics stemmed from mathematics and physics, and possesses some beautiful features of science by nature. The author first introduced the aesthetic considerations to the course of "Structural Vibrations" in the 1980s and encouraged undergraduate students to discuss and summarize the teaching materials. The practice aroused the interest of students and received their positive feedback. As a consequence, the author published a review article to address some aesthetic issues of vibration mechanics in the turn of the new century[5]. Later on, the author lectured some aesthetic considerations in engineering education in many universities and also received quite active feedback from audience.

Based on the above article and lectures, this chapter will survey the major theoretical contents covered by most elementary textbooks from the viewpoint of aesthetics of science, and discuss their relations associated with unity, simplicity, regularity, symmetry and singularity, respectively. The objective of the chapter is to help readers review their knowledge from a new perspective.

---

[1] Xu B S, Yin Q Z (2008). Aesthetic Methods in Mathematics. Dalian: Press of Dalian University of Science and Technology, (in Chinese).

[2] Aiton E J, Duncan A M, Field J V (1997). The Harmony of the World by Johannes Kepler: Translated into English with an Introduction and Notes. Philadelphia: American Philosophical Society.

[3] Ivanova M (2017). Poincaré's aesthetics of science. Syntheses, 294: 2581–2594.

[4] Curtin D W (1982). The Aesthetic Dimension of Science, 1980 Nobel Conference. New York: Philosophical Library.

[5] Hu H Y (2000). Aesthetical considerations of vibration theory and its development. Journal of Vibration Engineering, 13(2): 161–169 (in Chinese).

2.2 Beautiful Features of Vibration Mechanics

2.2.1 Unity

In the theoretical frame of vibration mechanics, unity exists between different parts of the frame, between parts and the whole frame as well, such that the frame is a unified and beautiful entirety. For instance, the dynamic equations of a discrete system are a set of ordinary differential equations of the second-order of time derivative, while the dynamic equation of a continuous system is a partial differential equation of the second-order of time derivative. If a discrete system is divided into several sub-systems, the dynamic equations of each sub-system keep as a set of ordinary differential equations similar to those of the entire system. Thus, the theory of vibration mechanics enjoys a quite unified frame. Once a breakthrough is made, it will bring the development of the whole field.

The elementary textbooks mainly dealt with the linear vibration systems in a discrete form. That is, the linear dynamic system with an input vector $f(t)$ and an output vector $u(t)$. The linear system has two fundamental properties, namely, *homogeneity* and *additivity*. Mathematically speaking, the system can be defined as a linear mapping $L: f(t) \mapsto u(t)$ from an input vector to an output vector, while these two properties can be expressed as

$$\begin{cases} L[\alpha f(t)] = \alpha L[f(t)] = \alpha u(t), & \alpha \in (-\infty, +\infty) \\ L[f_1(t) + f_2(t)] = L[f_1(t)] + L[f_2(t)] = u_1(t) + u_2(t) \end{cases} \tag{2.2.1}$$

In the community of mechanics, the two properties are named the *principle of superposition*. That is, when the multiple inputs, including initial disturbance and multiple external loads, are applied to a linear system, the entire output, namely, the dynamic response, is the superposition of all outputs, each of which is caused by an individual input. The theory of linear vibrations based on the principle of superposition exhibits the beautiful unity.

To be more specific, consider a linear n-DoF system, which yields the following initial value problem governed by a set of linear ordinary differential equations

$$\begin{cases} M\ddot{u}(t) + C\dot{u}(t) + Ku(t) = f(t) \\ u(0) = u_0, \quad \dot{u}(0) = \dot{u}_0 \end{cases} \tag{2.2.2}$$

where $f: t \mapsto \mathbb{R}^n$ is the external force vector, $u: t \mapsto \mathbb{R}^n$ is the system displacement vector, $u_0 \in \mathbb{R}^n$ is the initial displacement vector, $\dot{u}_0 \in \mathbb{R}^n$ is the initial velocity vector, $\mathbb{R}^n$ is a real vector space of n dimensions and named the *configuration space* of the system. In addition, $M \in \mathbb{R}^{n \times n}$ is the positive definite mass matrix, $K \in \mathbb{R}^{n \times n}$ is the positive or semi positive definite stiffness matrix, $C \in \mathbb{R}^{n \times n}$ is the symmetric damping matrix, and $\mathbb{R}^{n \times n}$ is the space of real square matrix of order n.

According to the principle of superposition, the dynamic response of the linear system governed by Eq. (2.2.2) yields[6]

$$u(t) = U(t)u_0 + V(t)\dot{u}_0 + \int_0^t h(t - \tau)f(\tau)d\tau, \quad t \geq 0 \qquad (2.2.3)$$

where $U{:}t \mapsto \mathbb{R}^{n \times n}$ is the system response matrix under a vector of initial unit displacements, $V{:}t \mapsto \mathbb{R}^{n \times n}$ is the system response matrix under a vector of initial unit velocities, $h{:}t \mapsto \mathbb{R}^{n \times n}$ is the system response matrix under a vector of ideal unit impulses. That is, the system response includes two parts. The first part is the response of the system without any external excitation but subject to the initial disturbance, the second part is the response of the initially rest system subject to the external excitation. The second part is also named *Duhamel integral*.

Example 2.2.1 For an SDoF system, Eq. (2.2.3) can be simplified to a scalar form as follows

$$u(t) = U(t)u_0 + V(t)\dot{u}_0 + \int_0^t h(t - s)f(s)ds, \quad t \geq 0 \qquad (a)$$

with

$$\begin{cases} h(t) = \dfrac{1}{m\omega_n\sqrt{1 - \zeta^2}} \exp(-\zeta\omega_n t) \sin(\omega_n\sqrt{1 - \zeta^2}t) \\[2mm] U(t) = \exp(-\zeta\omega_n t)\left[\cos(\omega_n\sqrt{1 - \zeta^2}t) + \dfrac{\zeta}{\sqrt{1 - \zeta^2}} \sin(\omega_n\sqrt{1 - \zeta^2}t)\right] \\[2mm] V(t) = \exp(-\zeta\omega_n t)\dfrac{\sin(\omega_n\sqrt{1 - \zeta^2}t)}{\omega_n\sqrt{1 - \zeta^2}}, \quad t \geq 0 \end{cases} \qquad (b)$$

where m is the system mass, ω_n is the circular natural frequency of the system, and ζ is the damping ratio of the system. When damping disappears, Eq. (b) becomes

$$h(t) = \frac{1}{m\omega_n} \sin(\omega_n t), \quad U(t) = \cos(\omega_n t), \quad V(t) = \frac{\sin(\omega_n t)}{\omega_n}, \quad t \geq 0 \qquad (c)$$

It may be enjoyable for readers to examine the unity of vibration mechanics further while reading subsequent chapters, in particular, checking Sects. 3.2, 4.1, 5.2 and 5.3, respectively.

---

[6] Hu H Y (1998). Mechanical Vibrations and Shocks. Beijing: Press of Aviation Industry, 115–127 (in Chinese).

2.2.2 Simplicity

The theoretical frame of vibration mechanics has a clear outlines, the minimal concepts of component elements, and a very simple structure. For instance, the theoretical frame of linear vibrations is based on the principle of superposition and gives rise to a few simple, but significant theorems and formulae. Among them, Eq. (2.2.3) is a good example.

In addition, the vibration problems often have the possibility of decomposition. As stated in Sect. 2.2.1, for instance, the response of a linear system includes two parts. One is the system response subject to the initial disturbance, and the other is the response subject to the external excitation.

The other example is the complicated time history of quasi-periodic vibration $w(t)$ shown in Fig. 2.1a, which can be decomposed, via Fourier transform, into three harmonic vibrations in frequency domain. Figure 2.1b presents the simple picture of the amplitude-frequency relation $|W(f)|$.

The main thread in the study of vibration problems is to decompose a complicated problem into a number of simpler ones. The main thread has led to a number of well known methods, such as the modal analysis, the Duhamel integral in time domain, the spectral analysis in frequency domain, and the dynamic sub-structuring in space domain. Now, we turn to the *modal analysis* of a linear MDoF system to explain the main thread as follows.

Let $\boldsymbol{u}_r(t) = \boldsymbol{\varphi}_r \sin(\omega_r t + \theta_r)$, $r = 1, 2, \ldots, n$ be the natural vibrations of the linear system governed by Eq. (2.2.2). They satisfy the following solutions of an eigenvalue problem

$$(\boldsymbol{K} - \omega_r^2 \boldsymbol{M})\boldsymbol{\varphi}_r = \boldsymbol{0}, \quad r = 1, 2, \ldots, n \qquad (2.2.4)$$

If all the natural frequencies are distinct, namely, $0 \le \omega_1 < \omega_2 < \cdots < \omega_n$, the mode shape vectors $\boldsymbol{\varphi}_r \in \mathbb{R}^n$, $r = 1, 2, \ldots, n$ satisfy the *orthogonal relations*, or called the *orthogonality*, as follows

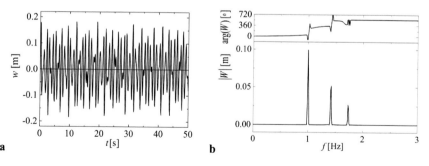

Fig. 2.1 The quasi-periodic vibration $w(t) = 0.1\,[\sin(2\pi t) + 0.6\sin(2\pi\sqrt{2}\,t) + 0.3\sin(2\pi\sqrt{3}\,t)]$.
a the time history; **b** the Fourier spectrum

$$\varphi_r^{\mathrm{T}} M \varphi_s = \begin{cases} M_r, & r = s \\ 0, & r \neq s; \end{cases} \quad \varphi_r^{\mathrm{T}} K \varphi_s = \begin{cases} K_r, & r = s \\ 0, & r \neq s; \end{cases} \quad r, s = 1, 2, \ldots, n \quad (2.2.5)$$

where $M_r > 0$ is the r-th modal mass coefficient, and $K_r = M_r \omega_r^2 \geq 0$ is the r-th modal stiffness coefficient.

The above orthogonal relations show that the mode shape vectors φ_r, $r = 1, 2, \ldots, n$ are linearly independent and can serve as a set of base vectors for the configuration space $\mathbb{R}^n$. Hence, it is reasonable to express the system displacement vector $u(t)$ as a linear combination of the mode shape vectors as follows

$$u(t) = \sum_{r=1}^{n} \varphi_r q_r(t) \tag{2.2.6}$$

This equation has received several names, such as *modal expansion* and *modal superposition*.

When the damping matrix is proportional to the linear combination of mass matrix and stiffness matrix, the mode shape vectors satisfy the orthogonal relations with respect to the damping matrix[7], namely,

$$\varphi_r^{\mathrm{T}} C \varphi_s = \begin{cases} C_r, & r = s \\ 0, & r \neq s; \end{cases} \quad r, s = 1, 2, \ldots, n \tag{2.2.7}$$

where $C_r \geq 0$ is the r-th modal damping coefficient. Substituting Eq. (2.2.6) into Eq. (2.2.2), replacing subscript r with s, and pre-multiplying both sides of the result by φ_r^{T}, $r = 1, 2, \ldots, n$, we derive a set of ordinary differential equations by using the orthogonal relations in Eqs. (2.2.5) and (2.2.7), that is,

$$\begin{cases} M_r \ddot{q}_r(t) + C_r \dot{q}_r(t) + K_r q_r(t) = f_r(t) \equiv \varphi_r^{\mathrm{T}} f(t) \\ q_r(0) = \dfrac{\varphi_r^{\mathrm{T}} M u_0}{M_r}, \quad \dot{q}_r(0) = \dfrac{\varphi_r^{\mathrm{T}} M \dot{u}_0}{M_r}, \quad r = 1, 2, \ldots, n \end{cases} \tag{2.2.8}$$

This is a set of decoupled initial value problems of scalar differential equations and describes the dynamic responses of n SDoF systems subject to both initial disturbance and external excitation. From Eqs. (a) and (b) in Example 2.2.1, it is easy to get the dynamic response $q_r(t)$, $r = 1, 2, \ldots, n$ of Eq. (2.2.8). Then, substitution of them into Eq. (2.2.6) leads to the dynamic response $u(t)$ of the linear MDoF system.

For the dynamic problem of an SDoF system subject to an arbitrary excitation force $f_r(t)$, two typical solution methods have been available as follows.

One method is to compute Duhamel integral in time domain, or called the *convolution* in time domain. It enables one to decompose the excitation force as an infinite number of ideal impulses in time domain, and get the free vibration due to each ideal impulse and sum up all those free vibrations.

---

[7] Craig Jr R R (1981). Structural Dynamics. New York: John Wiley and Sons Inc., 447–451.

The other method is to compute the product of the frequency response function of the system and the spectrum of excitation force in frequency domain and apply the inverse Fourier transform to the product. That is, to decompose the excitation force into an infinite number of harmonic forces, to compute the steady-state vibration of the system under a harmonic force, and to sum up all those steady-state vibrations.

In summary, both methods based on the principle of superposition follow the main thread: to decompose an arbitrary excitation into a number of simpler ones first, compute the system response under a simple excitation, and then sum up the results to get the system response.

Readers are suggested discussing and summarizing the decomposition and superposition procedures of vibration analysis in time domain, frequency domain and space domain when reading Sects. 4.1, 7.3 and 7.5, respectively.

2.2.3 Regularity

The term "regularity" usually means the repeatability of identical or similar shapes in geometry. For instance, the mast of satellite NuSTAR for detecting black-holes in universe was composed of many identical and deployable cells as shown in Fig. 2.2. These identical cells enabled the mast to be stowed into a very small volume before the launch of space mission and then deployed with a telescope at one end in orbit, exhibiting the beautiful regularity.

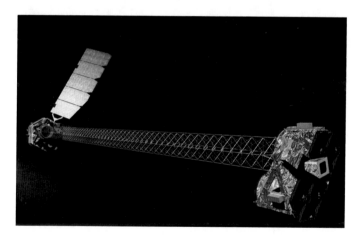

Fig. 2.2 Satellite NuSTAR with a deployed mast[8]

---

[8] Chen Q F (2012). Encyclopedia of Spacecrafts in the World. Beijing: China Astronautics Press, 359 (in Chinese).

During the elementary course of vibrations, one has touched upon many regular patterns. For example, the periodic vibration in Fig. 2.3 appears almost everywhere, exhibiting the regular time history in Fig. 2.3a and the beautiful phase portrait in Fig. 2.3b. The other example is the quasi-periodic vibration shown in Fig. 2.4a. This vibration consists of two harmonic vibrations with very close frequencies and exhibits a regular beat phenomenon in the time history.

In the theory of linear vibrations, many formulae show regularity. For example, if the mode shape vectors in Eq. (2.2.4) are normalized with respect to the modal mass coefficients first and then used to define a mode shape matrix as follows

$$\overline{\boldsymbol{\Phi}} \equiv \left[\overline{\boldsymbol{\varphi}}_1 \; \overline{\boldsymbol{\varphi}}_2 \; \cdots \; \overline{\boldsymbol{\varphi}}_n\right] \in \mathbb{R}^{n \times n}, \quad \overline{\boldsymbol{\varphi}}_r \equiv \frac{1}{\sqrt{M_r}} \boldsymbol{\varphi}_r, \quad r = 1, 2, \ldots, n \qquad (2.2.9)$$

it is easy to express Eq. (2.2.5) as a regular form as follows

$$\overline{\boldsymbol{\Phi}}^\mathrm{T} \boldsymbol{M} \overline{\boldsymbol{\Phi}} = \boldsymbol{I}_n \in \mathbb{R}^{n \times n}, \quad \overline{\boldsymbol{\Phi}}^\mathrm{T} \boldsymbol{K} \overline{\boldsymbol{\Phi}} = \boldsymbol{\Omega}^2 \equiv \underset{1 \le r \le n}{\mathrm{diag}} \left[\omega_r^2\right] \in \mathbb{R}^{n \times n} \qquad (2.2.10)$$

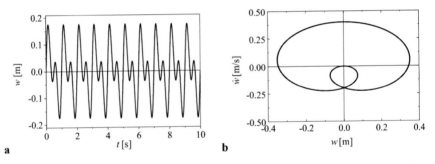

Fig. 2.3 The periodic vibration $w(t) = 0.1 \left[\sin(2\pi t) + \sin(4\pi t)\right]$. **a** the time history; **b** the phase portrait

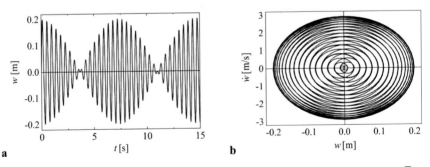

Fig. 2.4 The beat phenomenon of a quasi-periodic vibration $w(t) = 0.1 \left[\sin(15 t) + \sin(10\sqrt{2} t)\right]$. **a** the time history; **b** the phase portrait

Here, I_n is the identity matrix of order n, diag represents a diagonal matrix, where the diagonal entries are listed according to subscript r. Equation (2.2.10) implies that the kinetic energies of the natural vibrations of different orders are decoupled, so are the potential energies of those natural vibrations. The regular relations like Eq. (2.2.10) are so impressive and unforgettable.

It is interesting that the regularity of the formulae may remain unchanged in the vibration analysis. To see the assertion, analyze the free vibration of an un-damped MDoF system. Let the modal expansion of Eq. (2.2.6) be in a matrix form, namely,

$$u(t) = \overline{\Phi}\, q(t) \tag{2.2.11}$$

Substituting Eq. (2.2.11) into the initial conditions in Eq. (2.2.2), we have the initial conditions of the system in the space of modal displacements and the space of modal velocities respectively

$$q(0) = \overline{\Phi}^{-1} u(0) = \overline{\Phi}^{-1} u_0, \quad \dot{q}(0) = \overline{\Phi}^{-1} \dot{u}(0) = \overline{\Phi}^{-1} \dot{u}_0 \tag{2.2.12}$$

In view of Eq. (c) in Example 2.2.1, it is similar to derive the displacement under an initial unit displacement and the velocity under an initial unit velocity as follows

$$U_r(t) = \cos(\omega_r t), \quad V_r(t) = \frac{\sin(\omega_r t)}{\omega_r}, \quad t \geq 0, \quad r = 1, 2, \ldots, n \tag{2.2.13}$$

Substitution of Eqs. (2.2.12) and (2.2.13) into Eq. (2.2.11) leads to the free vibration of the system in a matrix form, namely,

$$
\begin{aligned}
u(t) &= \overline{\Phi}[q_r(t)] = \overline{\Phi}[U_r(t)q_r(0) + V_r(t)\dot{q}_r(0)] \\
&= \overline{\Phi} \operatorname*{diag}_{1 \leq r \leq n} [\cos(\omega_r t)] \overline{\Phi}^{-1} u_0 + \overline{\Phi} \operatorname*{diag}_{1 \leq r \leq n} [\sin(\omega_r t)/\omega_r] \overline{\Phi}^{-1} \dot{u}_0 \\
&= U(t)u_0 + V(t)\dot{u}_0, \quad t \geq 0
\end{aligned}
\tag{2.2.14}
$$

where

$$U(t) \equiv \overline{\Phi} \operatorname*{diag}_{1 \leq r \leq n} [\cos(\omega_r t)] \overline{\Phi}^{-1}, \quad V(t) \equiv \overline{\Phi} \operatorname*{diag}_{1 \leq r \leq n} [\sin(\omega_r t)/\omega_r] \overline{\Phi}^{-1}, \quad t \geq 0 \tag{2.2.15}$$

They are the system displacement matrix under a vector of initial unit displacements and the system velocity matrix under a vector of initial unit velocities, exhibiting a regular form. It is similar to derive the matrix form of free vibrations of a proportionally damped MDoF system from Eq. (b) in Example 2.2.1.

To demonstrate the regularity of natural vibrations of one-dimensional structures, Table 2.1 presents the natural modes of one string, two rods and three beams under

Table 2.1 Wave numbers, natural frequencies and mode shapes of six one-dimensional structures

Type	κ_1	κ_2	κ_r	ω_r	$\varphi_r(x)$
Fixed-fixed string	$\dfrac{\pi}{L}$	$\dfrac{2\pi}{L}$	$\dfrac{r\pi}{L}$	$\kappa_r\sqrt{\dfrac{T}{\rho A}}$	$\sin(\kappa_r x)$
Fixed-fixed rod	$\dfrac{\pi}{L}$	$\dfrac{2\pi}{L}$	$\dfrac{r\pi}{L}$	$\kappa_r\sqrt{\dfrac{E}{\rho}}$	$\sin(\kappa_r x)$
Fixed-free rod	$\dfrac{\pi}{2L}$	$\dfrac{3\pi}{2L}$	$\dfrac{(2r-1)\pi}{2L}$	$\kappa_r\sqrt{\dfrac{E}{\rho}}$	$\sin(\kappa_r x)$
Hinged-hinged beam	$\dfrac{\pi}{L}$	$\dfrac{2\pi}{L}$	$\dfrac{r\pi}{L}$	$\kappa_r^2\sqrt{\dfrac{EI}{\rho A}}$	$\sin(\kappa_r x)$
Clamped-clamped beam	$\dfrac{4.730}{L}$	$\dfrac{7.853}{L}$	$\dfrac{(2r+1)\pi}{2L}$	$\kappa_r^2\sqrt{\dfrac{EI}{\rho A}}$	$\cosh(\kappa_r x)-\cos(\kappa_r x)$ $-c_r[\sinh(\kappa_r x)-\sin(\kappa_r x)],$
					$c_r = \dfrac{\cosh(\kappa_r L)-\cos(\kappa_r L)}{\sinh(\kappa_r L)-\sin(\kappa_r L)}$
Clamped-free beam	$\dfrac{1.875}{L}$	$\dfrac{4.694}{L}$	$\dfrac{(2r-1)\pi}{2L}$	$\kappa_r^2\sqrt{\dfrac{EI}{\rho A}}$	$\cosh(\kappa_r x)-\cos(\kappa_r x)$ $-c_r[\sinh(\kappa_r x)-\sin(\kappa_r x)],$
					$c_r = \dfrac{\cosh(\kappa_r L)+\cos(\kappa_r L)}{\sinh(\kappa_r L)+\sin(\kappa_r L)}$

typical boundaries, which have been covered by most elementary textbooks[9,10]. Here, κ_r, ω_r and $\varphi_r(x)$ are the r-th wave number, natural frequency and mode shape, respectively, while L is the length of the string/rod/beam, $\overline{T}$ is the tensional force of the string, A is the cross-sectional area of the string/rod/beam, I is the second moment of cross-sectional area of the beam, ρ and E are the material density and Young's modulus, respectively.

Readers will enjoy the regularity while understanding more advanced vibration mechanics. Many complicated vibration phenomena exhibit regularity in a new level of knowledge. For instance, the time history of a quasi-periodic vibration seems not regular, but the corresponding Fourier spectrum in frequency domain exhibits regularity. In recent years, advanced spectral analysis and wavelet analysis have provided people with more and more powerful tools to reveal the inherent regularity of more complicated vibration signals.

Readers are suggested checking the modal theory of a linear system with non-holonomic constraints, and the dual problems of one-dimensional structures from the perspective of regularity while reading Sects. 4.1, 5.2 and 5.3, respectively.

---

[9] Timoshenko S, Young D H, Weaver Jr W (1974). Vibration Problems in Engineering. 4th Edition. New York: John Wiley and Sons Inc., 369–371, 409–431.

[10] Zhou S D, Heylen W, Liu L (2016). Structural Dynamics. Beijing: Press of Beijing Institute of Technology, 20–29.

2.2.4 Symmetry

Symmetry is a kind of special regularity. Many structures exhibit symmetry from the viewpoint of geometry or mechanics. For example, airplanes and cars are symmetric about one plane parallel to their moving direction, while the Eiffel Tower in France and Pyramids in Egypt are symmetric about two planes. The bladed disc of an aero-engine in Fig. 2.5 is cyclosymmetric. That is, the bladed disc is identical when it rotates through an angle of $2\pi/n$ about the central axis. In recent years, the symmetric lattice structures shown in Fig. 2.6 have been popular since they have very good performance in mechanics. The symmetries of the above structures do not only provide special functions, but also show their beauty.

Most elementary textbooks dealt with some symmetric structures, such as a fixed-fixed uniform rod, a hinged-hinged uniform beam and a thin rectangular plate with four edges hinged. In the studies on vibration problems, the symmetry of a structure not only exhibits the beautiful view, but also brings some special problems in their dynamic modeling, computation, analysis and design.

Example 2.2.2 Consider a hinged-hinged uniform beam under distributed harmonic forces in Fig. 2.7a and discuss the vibration properties of the beam about the beam mirror-symmetry.

Fig. 2.5 A bladed disc of aero-engine ($n = 24$)

Fig. 2.6 Two lattice structures of light weight

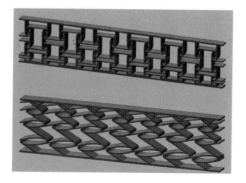

Fig. 2.7 The forced
vibration of a symmetric
beam. **a** a hinged-hinged
uniform beam under
distributed harmonic forces;
b the first four mode shapes
(thick solid line: 1st order,
thin solid line: 2nd order,
dashed line: 3rd order, dot
line: 4th order)

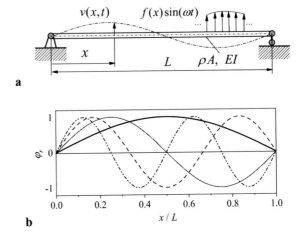

The dynamics of the beam in Fig. 2.7a yields the following dynamic boundary
value problem

$$\begin{cases} \rho A \dfrac{\partial^2 v(x,t)}{\partial t^2} + \eta E I \dfrac{\partial^5 v(x,t)}{\partial t \partial x^4} + E I \dfrac{\partial^4 v(x,t)}{\partial x^4} = f(x)\sin(\omega t) \\ v(0,t) = 0, \quad v_{xx}(0,t) = 0, \quad v(L,t) = 0, \quad v_{xx}(L,t) = 0 \end{cases} \tag{a}$$

where L is the length of the beam, A is the cross-sectional area of the beam, I is the
second moment of cross-sectional area of the beam, ρ is the material density, E is
Young's modulus of material, $\eta > 0$ is the damping factor of the beam system, ω is
the excitation frequency, $f(x)$ describes the distribution of excitation amplitude. In
addition, $v_{xx}(x,t) \equiv \partial^2 v(x,t)/\partial x^2$, $v_{xx}(0,t) \equiv v_{xx}(x,t)|_{x=0}$, and so forth.

The method of separation of variables in elementary textbooks enables one to get
the natural vibrations of the hinged-hinged beam as follows

$$\begin{cases} v(x,t) = \varphi_r(x)\sin(\omega_r t), \\ \omega_r = \kappa_r^2 \sqrt{\dfrac{EI}{\rho A}}, \quad \kappa_r = \dfrac{r\pi}{L}, \quad \varphi_r(x) = \sin(\kappa_r x), \quad r = 1, 2, \dots \end{cases} \tag{b}$$

where ω_r, κ_r and $\varphi_r(x)$ are the r-th natural frequency, wave number and mode shape,
corresponding to those in Table 2.1. Figure 2.7b shows the first four mode shapes,
where the mode shapes of the odd order are symmetric about the plane $x = L/2$,
while those of the even order are anti-symmetric about $x = L/2$.

To determine the dynamic response of the beam to harmonic forces, let the mode
shapes be the base functions and define the modal transform similar to Eq. (2.2.6) as

$$v(x,t) = \sum_{s=1}^{+\infty} \varphi_s(x) q_s(t) \tag{c}$$

From the orthogonal relations of mode shapes, it is easy to transform the partial differential equation in Eq. (a) into a set of ordinary differential equations as follows

$$M_r \ddot{q}_r(t) + \eta K_r \dot{q}_r(t) + K_r q_r(t) = f_r \sin(\omega_0 t), \quad r = 1, 2, \ldots \quad (d)$$

where the modal mass coefficients and modal stiffness coefficients in Eq. (d) respectively satisfy

$$M_r = \rho A \int_0^L \varphi_r^2(x) dx, \quad K_r = EI \int_0^L \left[\frac{d^2 \varphi_r(x)}{dx^2} \right]^2 dx, \quad r = 1, 2, \ldots \quad (e)$$

Because $\eta K_r > 0$ holds, the transient response of Eq. (d) approaches zero and the steady-state response plays an essential role in practice. In this case, the amplitude of modal response $q_r(t)$ in Eq. (d) depends on the amplitude of the following *modal excitation*, namely,

$$f_r \equiv \int_0^L \varphi_r(x) f(x) dx \quad (f)$$

Equations. (b) and (f) lead to the following assertions.

(1) When r is an odd number, then $\varphi_r(x)$ is symmetric about $x = L/2$. In this case, if $f(x)$ is anti-symmetric about $x = L/2$, then $f_r = 0$ and $q_r(t) = 0$ hold true.

(2) When r is an even number, then $\varphi_r(x)$ is anti-symmetric about $x = L/2$. Now, if $f(x)$ is symmetric about $x = L/2$, then $f_r = 0$ and $q_r(t) = 0$ hold true.

From the viewpoint of mechanics, Eq. (f) describes the work done by the distributed force on the mode shape of order r. Hence, even though the above two assertions come from a hinged-hinged uniform beam subject to a harmonic excitation, they are universal. That is, for a symmetric system of concern, the symmetric loads can not drive any anti-symmetric response, and the anti-symmetric load can not excite any symmetric response.

If a system is symmetric, it can be divided at least two identical sub-systems. As the identical sub-systems carry abundant information, it is possible to remove the abundant information and simplify the dynamic analysis of the entire system. To deal with the dynamics of a symmetric system, the group theory in mathematics enables one to greatly reduce the cost in dynamic modeling, computations and tests.

Example 2.2.3 Figure 2.8 presents the measured natural modes of a bladed disc model with 6 short blades. The model is identical after it rotates through an angle $2\pi/6$ about the central axis. Using the language of group theory, the bladed disc model is symmetric in cyclic group C_6. From the linear representative theory of group C_6, one needs to take only one of 6 sectors to make the dynamic analysis and

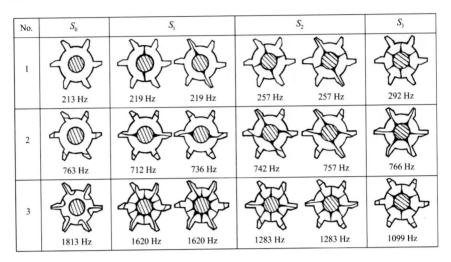

No.	S_0	S_1		S_2		S_3
1	213 Hz	219 Hz	219 Hz	257 Hz	257 Hz	292 Hz
2	763 Hz	712 Hz	736 Hz	742 Hz	757 Hz	766 Hz
3	1813 Hz	1620 Hz	1620 Hz	1283 Hz	1283 Hz	1099 Hz

Fig. 2.8 The classification of measured natural modes of a bladed disc model with symmtry in cyclic group C_6

then can construct the natural modes of the entire bladed disc model. This procedure can dramatically reduce the computational cost.

In addition, the linear representative theory of group C_6 enables one to classify the experimental results of the bladed disc model in modal tests and construct a table of natural modes in Fig. 2.8[11]. It is very helpful to use the table to identify the natural vibrations of different types and greatly reduce the blindness in modal tests. Section 6.3 will present the corresponding theoretical and experimental studies in detail.

Furthermore, the symmetry also appears in many important formulae of vibration mechanics, making them more impressive.

To describe the dynamic behaviors of a linear discrete system, for example, it is helpful to define two mappings from $\omega \in \mathbb{R}^1$ to the space $\mathbb{C}^{n \times n}$ of complex matrix. They are the *dynamic stiffness matrix* and the *dynamic flexibility matrix* as follows

$$\mathbf{Z}(\omega) \equiv \mathbf{K} + i\omega\mathbf{C} - \omega^2\mathbf{M}, \quad \mathbf{H}(\omega) \equiv \mathbf{Z}^{-1}(\omega) \tag{2.2.16}$$

where $i \equiv \sqrt{-1}$. As the mass matrix, damping matrix and stiffness matrix are symmetric, the entries of these two matrices satisfy

$$Z_{ij}(\omega) = Z_{ji}(\omega), \quad H_{ij}(\omega) = H_{ji}(\omega), \quad i, j = 1, 2, \ldots, n \tag{2.2.17}$$

where entry $H_{ij}(\omega)$ is the frequency response computed or measured at the i-th DoF of the system subject to an excitation at the j-th DoF only, while $H_{ji}(\omega)$ is the

---

[11] Hu H Y, Cheng D L (1988). Investigation on modal characteristics of cyclosymmetric structures. Chinese Journal of Applied Mechanics, 5(3): 1–8 (in Chinese).

frequency response computed or measured at the j-th DoF of the system subject to an excitation at the i-th DoF only. Equation (2.2.17) implies that the reciprocal theorem of displacements in structure mechanics holds true for structural dynamics. In a vibration test for a column of frequency response functions of a linear system, hence, it is possible to use either a single excitation force and the multiple response sensors or the multiple excitation forces and a single response sensor.

Readers are suggested carefully reading Chap. 6, where the natural vibrations of both mirror-symmetric systems and cyclosymmetric systems will be systematically studied, with an emphasis on the mode shapes with repeated natural frequencies. The applications of group theory in Sects. 6.2 and 6.3 will show a powerful tool for solving practical problems, like Example 2.2.3, both theoretically and experimentally.

2.2.5   Singularity

Singularity here refers to an extreme case of strange phenomenon. In the history of mechanics, the earliest singularity, which received much attention, might be either the static buckling of a rod under compression or the resonance of a suspended bridge. These singular phenomena did not exhibit any evidence in advance, but occurred abruptly. Later on, engineers encountered many similar phenomena, such as the flutter of an aircraft wing and the gallop of a steel bridge of long span.

Most singular phenomena brought about disasters to humankind. However, once the physical mechanism behind a singular phenomenon is clear, the utilizations of the singular phenomenon enable people to enjoy the beauty of singularity.

Example 2.2.4   Consider an SDoF system subject to a harmonic excitation of basement and define the displacement transmissibility T_d for the system as the ratio of displacement amplitude of the system to the displacement amplitude of the basement. Figure 2.9 shows the transmissibility T_d with an increase of dimensionless excitation frequency $\lambda \equiv \omega/\omega_n$ for several damping ratio ζ of the system.

In history, people observed the resonance of such a system when the excitation frequency ω approached the natural frequency ω_n of the system, namely, a peak at

Fig. 2.9   The displacement transmissibility of an SDoF system under a harmonic excitation of basement

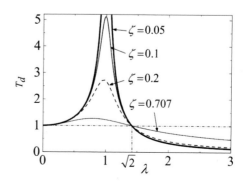

$\lambda \approx 1$ in Fig. 2.9. As most structures had a tiny damping ratio, say, $\zeta \in (0.001, \ 0.05)$, the resonant peak was very high and gave rise to disasters. When people got to know the rule of displacement transmissibility of the above system, they could not only avoid or reduce the resonance, but also utilize the lower displacement transmissibility in a frequency range of $\lambda > \sqrt{2}$ in order to design various vibration isolation systems.

In the studies of vibration problems, some singularities come from the assumptions for modeling a real system. For instance, the un-damped model of a real system exhibits infinitely large resonance, and an idealized system with exact symmetry has repeated natural frequencies. The following example demonstrates such a singularity from a simplified system with symmetry.

Example 2.2.5 As shown in Fig. 2.10a, the system consists of a thin circular plate fixed at its center and three lumped masses $m_1 = m_2 = m_3 = m$, which are installed on the plate and trisected along the circumference. Under the assumption of negligible inertia of the plate, analyze the natural vibrations, normal to the plate, of the system.

In the frame of coordinates in Fig. 2.10a, the dynamic equations of the system in a matrix form yield

$$m\ddot{w}(t) + Kw(t) = 0, \quad w \equiv \begin{bmatrix} w_1 & w_2 & w_3 \end{bmatrix}^{\mathrm{T}} \tag{a}$$

In view of the system symmetry, the stiffness matrix of the system looks like

$$K = k \begin{bmatrix} 1 & \beta & \beta \\ \beta & 1 & \beta \\ \beta & \beta & 1 \end{bmatrix}, \quad \beta < 0 \tag{b}$$

where k and βk represent the direct stiffness coefficient and cross stiffness coefficient, respectively.

Assume that the natural vibration of the system yields

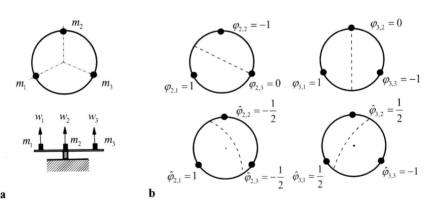

Fig. 2.10 A symmetric system of 3-DoFs and the mode shapes of repeated natural frequencies. **a** the schematic of the system; **b** the nodal lines of two pairs of mode shapes $[\varphi_2, \ \varphi_3]$ and $[\hat{\varphi}_2, \ \hat{\varphi}_3]$

$$w(t) = \varphi \sin(\omega t), \quad \varphi \in \mathbb{R}^3 \tag{c}$$

Substitution of Eqs. (b) and (c) into Eq. (a) leads to the following generalized eigenvalue problem

$$\begin{bmatrix} k - m\omega^2 & \beta k & \beta k \\ \beta k & k - m\omega^2 & \beta k \\ \beta k & \beta k & k - m\omega^2 \end{bmatrix} \varphi = 0 \tag{d}$$

Solving Eq. (d) for eigenvalues, we have the natural frequencies of the system as follows

$$\omega_1 = \sqrt{\frac{(1 + 2\beta)k}{m}}, \quad \omega_2 = \omega_3 = \sqrt{\frac{(1 - \beta)k}{m}} \tag{e}$$

Substitution of $\omega_1 = \sqrt{(1 + 2\beta)k/m}$ into Eq. (d) gives a set of linear homogeneous algebraic equations of rank 2, which leads to the mode shape $\varphi_1 = [\,1\ 1\ 1\,]^T$. In this case, the three lumped masses exhibit identical vibrations.

Upon substituting repeated natural frequencies $\omega_2 = \omega_3 = \sqrt{(1 - \beta)k/m}$ into Eq. (d), we have a set of linear homogeneous algebraic equations of rank 1 as follows

$$k \begin{bmatrix} \beta & \beta & \beta \\ \beta & \beta & \beta \\ \beta & \beta & \beta \end{bmatrix} \varphi = 0 \tag{f}$$

whereby we have a pair of linearly independent mode shapes φ_2 and φ_3. Yet, there are an infinite number of linearly independent pairs of mode shapes for the above repeated natural frequencies. Two pairs of them, as an example, are as follows

$$\varphi_2 = \begin{bmatrix} 1 & -1 & 0 \end{bmatrix}^T, \quad \varphi_3 = \begin{bmatrix} 1 & 0 & -1 \end{bmatrix}^T \tag{g}$$

$$\hat{\varphi}_2 = \begin{bmatrix} 1 & -1/2 & -1/2 \end{bmatrix}^T, \quad \hat{\varphi}_3 = \begin{bmatrix} 1/2 & 1/2 & -1 \end{bmatrix}^T \tag{h}$$

Figure 2.10b presents the nodal lines of two pairs of mode shapes. In this case, given a pair of linearly independent mode shapes φ_2 and φ_3, their arbitrary linear combination

$$\varphi = c_2 \varphi_2 + c_3 \varphi_3 \tag{i}$$

yields Eq. (d) and is also a mode shape. From the viewpoint of linear algebra, the eigenvectors of a pair of repeated eigenvalues $\omega_2^2 = \omega_3^2$ form a linear subspace of

Fig. 2.11 The dimensionless
natural frequencies of a
3-DoF system with a
variation of mass ratio

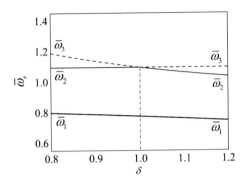

two dimensions. Hence, it is possible to select any two linearly independent eigen-vectors φ_2 and φ_3 as a pair of base vectors for this subspace. As a result, their linear combination in Eq. (i) falls into the subspace.

The above phenomenon of repeated natural frequencies is a kind of singularity cased by system symmetry. Once the system symmetry is broken, $\omega_2 \neq \omega_3$ holds so that the system returns to a normal case. If the system symmetry is slightly broken, the phenomenon of close natural frequencies $\omega_2 \approx \omega_3$ appears. This is the so-called *symmetry breaking* in physics.

To intuitively understand the phenomenon of symmetry breaking of the above system, adjust the second lumped mass in Fig. 2.10a as $m_2 = \delta m$ and keep other system parameters unchanged. Now, we compute the variation of three natural frequencies of the system with respect to the mass ratio δ so as to discuss the singularity. Figure 2.11 presents the variations of dimensionless natural frequencies $\overline{\omega}_r \equiv \omega_r / \sqrt{k/m}$, $r = 1, 2, 3$ with respect to mass ratio δ for $\beta = -0.2$.

In Fig. 2.11, the first frequency branch $\overline{\omega}_1$ monotonically decreases for $\delta \in [0.8, 1.2]$. The second frequency branch $\overline{\omega}_2$ remains unchanged and the third frequency branch $\overline{\omega}_3$ goes down for $\delta \in [0.8, 1.0)$. When $\delta = 1$, both frequency branches $\overline{\omega}_2$ and $\overline{\omega}_3$ intersect each other. That is, the two natural frequencies get repeated. For $\delta \in (1.0, 1.2]$, the frequency branch $\overline{\omega}_2$ decreases monotonically, but the frequency branch $\overline{\omega}_3$ keeps constant. The above phenomenon is called a *bifurcation* when two solution branches intersect each other.

It is interesting that either frequency branch $\overline{\omega}_2$ or frequency branch $\overline{\omega}_3$ remains unchanged with an increase of mass ratio δ. As a matter of fact, the mode shape associated with such a frequency branch is either $\varphi_2 = [1\ 0\ -1]^{\mathrm{T}}$ for $\delta < 1$ or $\varphi_3 = [1\ 0\ -1]^{\mathrm{T}}$ for $\delta > 1$. They have a node at the position of m_2. As shown in the upper right mode shape in Fig. 2.10b, the mode shape has nothing to do with the change of mass ratio δ for m_2.

In order to describe and analyze singularities, many mathematical tools and engineering theories have been established and developed. Among them, the theory of catastrophe, the theory of singularities, and the theory of bifurcations are a few good examples to name. These theories not only enable one to solve singular problems, but also embody the fancy in research ideas. The research results of singularities often show aesthetic structures. For instance, the singularity study on the equilibrium of a smooth nonlinear system shows that it is possible to use very simple polynomials to describe all possible local bifurcations[12].

Readers are suggested understanding the singularities of vibration problems associated with the symmetry breaking of a thin square plate and the discontinuous waves of an impacting rod in Sects. 6.1 and 7.3, respectively.

2.3 Enlightenments of Beauty of Vibration Mechanics

The worldwide industrialization has greatly promoted the studies on vibrations over the past century. Most breakthroughs in the studies came from the new academic ideas and new research methods. Both educations and researches of vibration mechanics, hence, need to emphasize the science ideology and research methodology.

2.3.1 Methods of Thinking

From the existence of beauty in the theoretical frame of vibration mechanics and the above aesthetic features, readers may understand the importance of finding and enjoying beauty while studying vibration mechanics. In what follows, several issues are discussed for the research-oriented study of the vibration mechanics.

2.3.1.1 Enhancing Creative Thinking

The core of creation in scientific researches is the creative thinking, which is different from logic inference and necessary to jump from the scope of available knowledge. Hence, the creative thinking mainly comes from imaginations and inspirations. In the scientific researches, the breakthrough depends on whether one can grasp and reveal the rules behind complicated and irregular phenomena. These rules are certainly regular and beautiful. The new fields of vibration researches since the twentieth century have shown the jumps in creative thinking of investigators and supported the above statement.

For example, the vibration of a tall building subject to a wind flow exhibits irregularity due to the uncertainty of the wind flow at different moments and positions, and so does the vibration of an aircraft panel under the turbulent air pressure. These two vibrations are the so-called random vibrations, which look quite irregular and do not have any rules behind them. In 1905, however, A. Einstein, a great physicist in Swiss then, found the nature of a random phenomenon of the Brownian motions

---

[12] Troger H, Steindl A (1991). Nonlinear Stability and Bifurcation Theory. Wienna: Springer-Verlag, 143–286.

of pollen grains floating on the fluid surface[13]. This seminal study in physics initiated the research of random vibrations of a dynamic system and has seen numerous subsequent studies on the random vibrations in aerospace engineering, mechanical engineering and civil engineering.

The other example associated with uncertainties is chaos. The first half of the twentieth century witnessed many advances in nonlinear vibrations, but the progress in this field became more and more slowly. In 1963, E. N. Lorenz, an American meteorologist, observed the random evidence in several numerical computations of a Benard convection problem simplified from the study of weather forecast. He found the influence of small initial disturbance on the long-term computational results of a set of nonlinear ordinary differential equations with three unknowns only. This is the so-called chaotic behavior, which exhibits an intrinsic random mechanism embedded in a deterministic problem. Meanwhile almost, S. Smale, an American mathematician, constructed a horseshoe mapping, which also exhibits chaotic behavior[14], and revealed the beautiful mathematical structure of the mapping. Their revolutionary ideas led the studies on nonlinear vibrations into a new era, which has witnessed not only numerous achievements in the theory of nonlinear vibrations, but also many successful applications of the theory to the design of mechanical and structural systems, such as the energy harvesting devices in recent years.

The above research processes demonstrate the unconventional thinking from simplicity to complexity or the other way around, whereas the final results of the thinking are some beautiful rules behind complicated phenomena. Therefore, the elegant taste plays an important role in creative researches since the aesthetic education enables one to improve the imagination and inspiration and to receive balanced development of logic thinking and image thinking.

2.3.1.2 Cultivating Elegant Tastes

In academia, there has been a popular saying: "the scientists of the first grade made a complicated problem simple, but those of the third grade made a simple problem complicated". Their difference comes from different levels of academic taste, while the taste greatly depends on the interest of researches.

A good example to support the above assertion is Rayleigh quotient. Lord Rayleigh won the Nobel Prize of Physics because of his findings in inert gases, but his early fame came from his seminal studies on vibration and acoustics, such as Rayleigh quotient for estimating the fundamental natural frequency of a complicated vibration system. Before his pioneering work, little study had dealt with the energy analysis of vibration systems. Rayleigh quotient is an extremely simple formula, but universal

---

[13] Einstein A (1956). Investigation on the theory of Brownian movement. English Translation of Einstein's Papers. New York: Dover Publication.

[14] Wiggins S (1990). Introduction to Applied Nonlinear Dynamical Systems and Chaos. New York: Spring-Verlag, 420–437.

to all vibration systems. This contribution fully reflects the elegant taste of Lord Rayleigh in scientific researches.

From the examples in Sects. 2.2.4 and 2.2.5, it is possible to see the symmetry, symmetry breaking and their relations. The similar properties embody the universal, simple and beautiful features of science, and have stimulated many important studies. In the history of vibration analysis, a well known case study is the mode shapes of a thin plate. In 1787, E. F. F. Chladni, a German scientist, observed that a thin rectangular plate under excitation exhibited a great variety of symmetric nodal lines demonstrated by different sand patterns. This observation announced in Paris stimulated the great interest of S. Germain, a French scientist. She made constant efforts over years and established the dynamic equation of a thin rectangular plate. Afterwards, several European scientists made contributions to this topic and finally established the beautiful vibration theory of thin plates. In the history of other branches of sciences, there have been also many successful examples of pursuing beauty of science[15].

In general, the representative works of masters in academia always revealed some fundamental rules or proposed some universal methods, featuring the beauty of science, such as unity, simplicity and regularity. These masterpieces are the permanent references in science. To make any achievements like them or similar, one should reach a higher grade in both academic taste and aesthetic interest. As such, one may have great passion in study and enjoy a long-term research. Otherwise, some important findings and conclusions may slip away.

2.3.1.3   Having Mastery via a Comprehensive Study

The unity of vibration mechanics discussed in Sect. 2.2.1 may help readers check the unity of mechanics, the unity of engineering science and so on in order to master the theoretical frame of engineering science. More specifically, it is possible to understand the vibration mechanics first, and then carry the feeling to enjoy the beauty of science and achieve mastery via a comprehensive study of other branches of engineering science on one hand. On the other hand, it is also possible to understand the beauty of other branches of engineering science first, and then focus on some vibration problems, and achieve mastery of this topic.

For instance, the dynamic analysis of a linear vibration system in time domain or in frequency domain, as well as the modal analysis of a linear vibration system, stemmed from the superposition principle of a linear system and is applicable to the linear electrical systems and linear control systems on one hand. On the other hand, the methods developed in the frame of automatic control have provided many successful solutions to vibration problems. The intersection and integration of different branches of engineering science have shown even brighter future.

In addition, many scientists engaged in the researches of vibration problems have made contributions to other fields, and vice versa. In the last century, Lord Rayleigh in the UK and A. A. Andronov in the Soviet Union made great achievements in both

---

[15] Stewart I, Golubitsky M (1992). Fearful Symmetry. Oxford: Blackwell Publisher, 26–242.

vibrations and other fields. In China, H. C. Hu made a celebrated contribution to the generalized variational principle in his early career, and then important contributions to the structural vibrations. W. X. Zhong found the similarity between structure mechanics and automatic control after his achievements in computational mechanics. Their successes have shown the importance of achieving mastery via a comprehensive study of several disciplines. It is possible to predict that the vibration research needs more interdisciplinary studies and requires an increasing number of scholars to be engaged in those studies.

2.3.1.4 Selecting Research Fields

It is significantly important for an individual or a research group to select a problem to be studied or a field to be engaged in.

If one examines vibration mechanics from the viewpoint of aesthetics, even the well-established branches, such as nonlinear vibrations and random vibrations, are far from perfect status, and exhibit many open and tough problems. In engineering practice, the complicated vibration problems associated with acoustic-vibration systems and fluid–solid interaction systems are challenging to engineers everyday. For instance, the explanations to vibration mechanism of some failures in aerospace projects have faced with arguments time to time since they have not received fully supports from available theory. These open problems are the candidates for future studies.

One possible way of breakthrough is to guess some theoretical explanations and mechanisms of those problems in view of unity, simplicity, regularity, singularity and symmetry in vibration mechanics, and to encourage young talents and students to explore possible solutions.

The other way of breakthrough may be out of the scope of current studies on vibration mechanics. For example, the life science has been one of the most charming branches since the new century. The rhythms of life are a kind of oscillations. Compared with the oscillatory phenomena in mechanical systems and electrical systems, the causes and rules of the rhythms of life are much more complicated. For instance, the study on electrophysiology has provided big data and led to very complicated mathematical models. It is interesting that the rhythms of life often look like a kind of relaxation vibrations of a simple mechanical system partially because of the self-organization of a life system[16]. From the simplicity in the beauty of science, then, is it possible to establish some simple models and propose some reduction methods to analyze the rhythms of a life system?

---

[16] Glass L, Mackey M C (1988). From Clock to Chaos - the Rhythms of Life. Princeton: Princeton University Press, 1–248.

2.3.1.5 Being Confident in Researches

Harmony, simplicity and economy are the universal rules in the evolution of nature. Both scientists and philosophers got agreement that nature does not do anything unnecessary. For instance, the motion of a physical system always follows the principle of the minimal action. It is reasonable to have a strong intention, thus, to explore the beauty of science, such as harmony and simplicity if one believes that they are the fundamental features of nature.

For example, most elementary textbooks presented the applications of Fourier transform to the frequency analysis of a linear system. This transform has been playing an essential role in many other fields, such as signal processing and automatic control. To be more specific, the dynamic signal $w(t)$ in time domain and $W(\omega)$ in frequency domain share a beautiful pair of *Fourier transforms*, namely,

$$
\begin{cases}
W(\omega) \equiv \mathscr{F}[w(t)] \equiv \dfrac{1}{\sqrt{2\pi}} \displaystyle\int_{-\infty}^{+\infty} w(t)\exp(-\mathrm{i}\omega t)\mathrm{d}t, & \omega \in (-\infty, +\infty) \\[2mm]
w(t) = \mathscr{F}^{-1}[W(t)] = \dfrac{1}{\sqrt{2\pi}} \displaystyle\int_{-\infty}^{+\infty} W(\omega)\exp(\mathrm{i}\omega t)\mathrm{d}\omega, & t \in (-\infty, +\infty), \quad \mathrm{i} \equiv \sqrt{-1}
\end{cases}
$$

$$(2.3.1)$$

Equation (2.3.1) has the discrete form as follows

$$
\begin{cases}
W(\omega_k) = \dfrac{\Delta t}{\sqrt{2\pi}} \displaystyle\sum_{j=1}^{N} w(t_j)\exp(-\mathrm{i}\omega_k t_j), & k = 1, 2, \ldots, N \\[2mm]
w(t_j) = \dfrac{\Delta\omega}{\sqrt{2\pi}} \displaystyle\sum_{k=1}^{N} W(\omega_k)\exp(\mathrm{i}\omega_k t_j), & j = 1, 2, \ldots, N
\end{cases}
$$

$$(2.3.2)$$

Equation (2.3.2) has been more useful than Eq. (2.3.1) since the era of computer technology. Because both indices j and k run from 1 to N in computing Eq. (2.3.2), the computational cost is proportional to N^2 and very time consuming. The fast computation of Eq. (2.3.2) had drawn much attention, but did not achieve any success until 1966.

In the 1960s, J. W. Cooley and T. W. Tukey, two young engineers in the USA, believed that the periodicity and symmetry of the discrete Fourier transform implied simplicity so that a kind of fast algorithm should exist. After constant efforts, they found the right way to the *fast Fourier transform*, called *FFT* for short, via the periodicity and symmetry. Their algorithm has a beautiful structure like a butterfly and can reduce the computational cost from level N^2 to level $N \log_2 N$. In the vibration analysis, the time series usually has the shortest length of $N = 2^{10} = 1024$. Obviously, the ratio $N^2/(N \log_2 N) = 102.4$ implies that the computation accelerates more than one hundred times!

2.3.2 Aesthetic Literacy

The engineering education over the past decades has emphasized the literacy, which plays a general and long-term role for the growth of talents, in order to eliminate conventional one-sidedness. In what follows, some issues associated with the improvement of aesthetic literacy are addressed further.

2.3.2.1 Establishing Aesthetic Consciousness

It is a reasonable logic that one ought to have elementary knowledge of aesthetics in order to establish aesthetic consciousness and beauty appreciation. The author introduced some aesthetic features of vibration mechanics when teaching "Structural Vibrations" in the 1980s and 1990s so as to initiate the interest of students in aesthetics. Meanwhile, the author introduced some popularized books of aesthetics to students and arranged them to discuss the aesthetic features in the vibration theory of linear MDoF systems. During the course, a number of students began to love the beauty of science.

As such, this book is intended to take the beauty of science as an important methodology of thinking in the research-oriented study and to help readers review the major issues in vibration mechanics covered by most elementary textbooks and read the subsequent chapters from the viewpoint of aesthetics of science.

2.3.2.2 Studying via Explorations

The central role of aesthetic education is students themselves. Hence, no matter whether the teaching process in class or the writing style of a textbook are concerned with, they should enable students or readers to feel enjoyable. The practice of repeated presentations of aesthetics may not have good results. The intention of heuristic teaching and research-oriented study is to let learners be in a central position so that they are able to enjoy finding something new. This is a process of self-discovery and self-transcendence. In the traditional ideas of education in East Asia, the highest realm is the unity and perfect. As a consequence, many previous teaching processes and textbooks have pursued the completeness. A practice like them greatly deviates from the novelty of learning process and makes students or readers reduce their interest in those courses or textbooks.

In order to change the above mentioned status, the author is trying to design the book as a guide of research-oriented study for vibration mechanics. The book does not cover many topics and issues, but features novelty instead. Hence, Chap. 1 has presented some case studies of vibration mechanics and associated problems to be studied. These problems are not difficult, but worthy to be studied. The remaining part of the book after Chap. 2 will present not only the theories and methods associated with the above problems, but also the research processes and results. The purpose of the book is to assist readers to understand how to solve the problems step by step, and to feel the beauty of vibration mechanics.

2.3.2.3 Taking Part in Practice

The final, but most important issue of aesthetic education is practice. Readers are suggested having at least two practices as follows. The first one is to summarize the theoretical results of each section, prove and discuss some details, which have not been completed in the book, so as to enhance their feelings to the beauty of science. The second one is to pay enough attention to the sections of further reading and thinking by the end of each chapter, read the references and solve at least two or three problems given there, so as to increase their aesthetic appreciation.

2.4 Concluding Remarks

This section surveyed the major theoretical and methodological issues of vibration mechanics in an elementary level from the aesthetical considerations to help readers prepare themselves for a research-oriented study of vibration mechanics. As a trial, the section discussed about the enlightenments of beauty of vibration mechanics and made relevant suggestions to readers. Accordingly, it is quite optional for each individual to enjoy beauty of science. To conclude this interesting but cutting-edge topic, some remarks are made as follows.

(1) The existence of beauty of science is objective and possesses absoluteness. In each historical stage, however, the cognition degree to beauty of science has relativity. Hence, the cognition to beauty of science is an endless and progressive process, and has never stopped. In the history of science, some scientists, even great figures, had an obsession with their perfect theories and delayed the discovery of new theories. For instance, there had been a popular conclusion that the output of a deterministic system to a deterministic input must be deterministic since the time of P. S. Laplace, a great French mathematician, in the eighteenth century. All scientists worldwide favored this conclusion until the discovery of chaos in 1963. Because of the conventional thinking set, this imperfect finding received acceptance of academia even several years later. Yet, the successive studies revealed that the discoveries of chaos and fractal show a new side of the beauty of science.

(2) Today, we have enjoyed more and more advanced computational and experimental facilities. The deep understanding, explorations and discussions of computational and experimental results mainly depend on the humankind even though the novel techniques of artificial intelligence may help us. Therefore, it is the scientists with elegant taste of science and literacy who can feel the beauty of science from complicated phenomena and make new breakthrough.

That is, the younger generation of scientists and engineers can reach a new peak of vibration mechanics through a helical road, on which they appreciate the beauty of vibration mechanics, find some imperfect issues, take new challenges and explore more advanced beauty of vibration mechanics.

2.5 Further Reading and Thinking

(1) Make a comparison between any two of the following courses: "Mechanical Vibrations", "Automatic Control" and "Theory of Linear Systems", and give more than three examples to demonstrate the unity of these courses.

(2) Read chapters of the references[17,18] and give two examples to support the simplicity of vibration mechanics.

(3) Read one of the references[19,20] and discuss the aesthetic features of vibration mechanics presented in Sect. 2.2.

(4) Read the reference[21], make some improvements for Sect. 2.2.4 in view of symmetry in other branches of natural science.

(5) Read the reference[22] and discuss a possible relation between the vibration mechanics and the rhythms of life.

[17] Balachandran B, Magrab E B. (2019). Vibrations. 3rd Edition. Cambridge: Cambridge University Press, 344–544.

[18] Craig Jr R R (1981). Structural Dynamics. New York: John Wiley and Sons Inc., 237–379.

[19] Radman Z (2004). Towards aesthetics of science. Journal of the Faculty Letters, University of Tokyo, Aesthetics, 29: 1–16.

[20] Zhang S Y (2012). Beauty and Reasoning in Mathematics. 2nd Edition. Beijing: Press of Peking University, 37–87 (in Chinese).

[21] Stewart I, Golubitsky M (1992). Fearful Symmetry. Oxford: Blackwell Publisher, 26–242.

[22] Glass L, Mackey M C (1988). From Clock to Chaos - the Rhythms of Life. Princeton: Princeton University Press, 1–248.

Chapter 3
Models of Vibration Systems

Vibration mechanics is the first branch of science to establish the model for a real problem and conduct the study based on the model. G. Galileo, a great Italian scientist, is the first who started the research based on a model in 1581. Galileo observed the same swing period of candle lamps hung from the top ceiling in a church, and then used a tethered stone to model the candle lamps and checked the influence of different stones on the swing period. He found that the swing period had nothing to do with the stones, revealing the mechanism behind the natural vibration of a simple pendulum. This seminal exploration created a new normal form of scientific researches, that is, the model-based research.

In the dynamic analysis and design of an engineering system today, it has been a normal procedure to establish the dynamic model for the system first, and then conduct the dynamic computation, analysis and design on the basis of the model. As such, the dynamic modeling of a vibration system plays an essential role in the entire process of designing and manufacturing industrial products. Yet, it is not an easy task to establish a proper and simple dynamic model since the success of the task relies not only on engineering experience, but also on the right concepts, theories and methods of vibration mechanics.

As shown in Sects. 1.2.1 and 1.2.2, no matter whether the tethered satellite or the hydro-elastic vibration isolation system for vehicle engines was concerned with, the dynamic modeling procedure had some open problems. The major problem for studying the tethered satellite is how to establish a proper discrete model for the continuous tether. The key issue for studying the hydro-elastic vibration isolation system is how to determine the degree of freedom, i.e. the DoF, and to understand the role of system parameters. In addition, it is a tough problem to describe the system damping, which has not been discussed in depth in most elementary textbooks.

This chapter, hence, begins with how to simplify a continuous system with distributed inertia to a discrete model, including the modeling process of a non-uniform rod and associated theoretical issues and the discretization of a tethered satellite. Then, the chapter turns to the study of degeneration of some DoFs of a discrete system due to non-holonomic constraints. Finally, the chapter presents

© Science Press 2022
H. Hu, *Vibration Mechanics*,
https://doi.org/10.1007/978-981-16-5457-2_3

how to study the model of structural damping, which is also associated with the degeneration of DoFs of a discrete system.

3.1 Continuous Systems and Their Discrete Models

All engineering systems in reality are continuous systems. That is, their inertia, elasticity and damping are continuously distributed in a range of space. The description of motion and deformation of a continuous system, thus, needs to use continuous coordinates in space and time. As such, the dynamic equations of a continuous system are a set of partial differential equations with respect to both time and space coordinates.

Most elementary textbooks of vibration mechanics only deal with uniform rods, uniform beams and thin rectangular plates. Thus, the dynamic modeling is based on some principles of mechanics in differential form. For example, the dynamic modeling of a uniform beam begins with analyzing the forces to an infinitesimal segment of the beam and then using Newton's law or d'Alembert's principle to establish the dynamic equation of the beam. For more complicated continuous systems, such a modeling procedure may be a tough work.

Hence, this section presents the principle of mechanics in integral form, which enables one to establish both dynamic equations and boundary conditions of a continuous system. In addition, solving the above partial differential equations is not easy and only leads to exact solutions for uniform rods, uniform beams and a few of thin plates. It is reasonable to simplify a continuous system into a discrete system via the lumped parameter method, Ritz method and the method of finite elements presented in elementary textbooks. The section will discuss some insightful issues about the preconditions of discretization and discrete models for non-uniform rods and beams, as well as a kind of variable stiffness systems.

3.1.1 Dynamic Models of Continuous Systems

In the frame of continuous mechanics, the principles of mechanics have several integral forms. The most popular one among them is the extended variational principle named after W. R. Hamilton, a great Irish scientist. Theoretically speaking, the extended variational principle is equivalent to Newton's law or d'Alembert's principle. From the viewpoint of computations, however, the solution of the dynamic equation of a continuous system derived from Newton's law or d'Alembert's principle is a *strong solution*, which should satisfy the smoothness required by the dynamic equation, while the solution directly determined from the extended variational principle is a *weak solution*, which may not be smooth enough, and only approximately yields the dynamic equation in an average sense of integration. This subsection presents how to use the extended variational principle to establish the dynamic equations and their boundary conditions of a continuous system.

3.1.1.1 Extended Hamilton's Principle

To well address the extended variational principle, we begin with three types of configurations of a system. Type A is the *real configuration* denoted by Ψ_R, which yields the dynamic equations, boundary conditions and initial conditions of the system, and spends time for realization. Type B is the *admissible configuration* denoted by Ψ_A, which satisfies the geometric boundary conditions of both displacements and rotation angles of the system but may not yield the dynamic equations of the system, and is realized via infinitesimal displacements. Type C is the *virtual configuration* denoted by Ψ_V, which is among the admissible configurations and does not spend time for realization.

From the concepts of these configurations, the real and virtual configurations yield the same constraints if the system is only subject to *scleronomic constraints* that are independent of time. Otherwise, the system is subject to the variation of *rhenomic constraints* which are dependent of time. The admissible configurations include both real and virtual configurations, but the realization of a real configuration spends time. The difference of any two virtual configurations is a *virtual displacement*, which is also named a *contemporaneous variation* with the corresponding operation denoted by δ. If a system moves from one virtual configuration to the other virtual configuration, the work done by both internal and external forces of the system is named the *virtual work* denoted by δW.

In 1834, Hamilton proposed how to identify the real configuration of a conservative system from all admissible configurations of the system via a variational idea. His seminal idea was named Hamilton's principle and developed later to the extended Hamilton's principle for an arbitrary system as follows.

The extended Hamilton's principle: For a given time interval $t \in [t_1, t_2]$, let δT be the variation of kinetic energy of the system, and δW be the virtual work done by both internal and external forces of the system. Then, the real configuration Ψ_R of the system among all admissible configurations Ψ_A of the system yields

$$\int_{t_1}^{t_2} (\delta T + \delta W)\mathrm{d}t = 0 \qquad (3.1.1)$$

If a part of internal forces comes from a potential function denoted by V, the corresponding virtual work $-\delta V$ can be separated from the entire virtual work δW such that Eq. (3.1.1) satisfies

$$\int_{t_1}^{t_2} (\delta T - \delta V + \delta W_n)\mathrm{d}t = 0, \quad \delta W_n \equiv \delta W + \delta V \qquad (3.1.2)$$

Here, δW_n is the summation of virtual works done by the external forces and the internal non-potential forces.

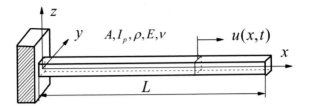

Fig. 3.1 A three-dimensional model for a fixed-free uniform rod

If the system of concern is conservative, $\delta W_n = 0$ holds true for Eq. (3.1.2). In particular, if the conservative system is subject to holonomic constraints only, the variation and integration in Eq. (3.1.2) is exchangeable so that a stationary condition holds as follows

$$\delta \int_{t_1}^{t_2} (T - V)\mathrm{d}t = 0 \qquad (3.1.3)$$

When the conservative system has a non-holonomic constraint, Eq. (3.1.3) may not hold[1]. If this is the case, one simply returns to Eq. (3.1.2) with $\delta W_n = 0$.

Remark 3.1.1 Compared with the variational principle for a static problem in elasticity or structure mechanics, the extended Hamilton's principle here is a functional of integration over a time interval in Eq. (3.1.1). From the viewpoint of mechanics, Eq. (3.1.1) requires that admissible configuration Ψ_A be identical to real configuration Ψ_R at any two moments t_1 and t_2 to determine the real configuration via the variation. For instance, if the admissible configuration of the system is $u(x, t) : \mathfrak{R}^3 \otimes \mathfrak{R}^+ \mapsto \mathfrak{R}^3$, the following relation holds true for any two moments t_1 and t_2

$$\delta u(x, t_1) = \delta u(x, t_2) = 0 \qquad (3.1.4)$$

The following example shows how to use the extended Hamilton's principle to establish the dynamic equation and the boundary conditions of a continuous system.

Example 3.1.1 In 1944, A. E. H. Love, a British scientist, established a simplified model of a three-dimensional rod to account for the *effect of transverse inertia* on the natural vibrations. It is relatively simple to use the extended Hamilton's principle, compared with the force analysis of an infinitesimal segment of the rod, to establish the dynamic equation of the problem.

Figure 3.1 presents a three dimensional-model for a fixed-free uniform rod. Here, L is the rod length, A is the cross-sectional area of the rod, I_p is the polar moment of cross-sectional area of the rod, ρ is the material density, E is Young's modulus of material and ν is Poisson's ratio of material.

---

[1] Greenwood D T (2003). Advanced dynamics. Cambridge: Cambridge University Press, 292–296.

Assume that the dynamic deformation of any cross-section of the rod remains plane and $u(x, t)$ is the dynamic longitudinal displacement of cross-section x of the rod in the frame of coordinates in Fig. 3.1 and is independent of y and z. In addition, the rod is subject to an axial stress component only, namely, $\sigma_x(x, t) \neq 0$ and $\sigma_y(x, t) = \sigma_z(x, t) = 0$. From the above assumptions and the generalized Hook's law, the longitudinal and transverse strains satisfy

$$
\begin{cases}
\varepsilon_x(x, t) = u_x(x, t) = \dfrac{\sigma_x(x, t)}{E}, & u_x(x, t) \equiv \dfrac{\partial u(x, t)}{\partial x} \\[2mm]
\varepsilon_y(x, t) = \varepsilon_z(x, t) = -\nu \dfrac{\sigma_x(x, t)}{E} = -\nu u_x(x, t)
\end{cases}
\tag{a}
$$

Equation (a) gives the transverse displacement components of the rod as follows

$$
v(x, t) = \varepsilon_y(x, t)y = -\nu y u_x(x, t), \quad w(x, t) = \varepsilon_y z = -\nu z u_x(x, t) \tag{b}
$$

For brevity, the independent variables in $u(x, t)$ do not show up and the subscripts of u stand for the corresponding partial derivatives of $u(x, t)$ hereinafter.

From Eq. (b), it is easy to derive the kinetic energy of the rod as

$$
\begin{aligned}
T &= \frac{\rho}{2}\int_0^L \left[\int_A (u_t^2 + v_t^2 + w_t^2)\,dA\right]dx = \frac{\rho}{2}\int_0^L \left\{\int_A [u_t^2 + v^2(y^2 + z^2)u_{xt}^2]\,dA\right\}dx \\
&= \frac{\rho A}{2}\int_0^L (u_t^2 + v^2 r_p^2 u_{xt}^2)\,dx
\end{aligned}
\tag{c}
$$

where r_p is the radius of gyration of the polar moment of cross-sectional area of the rod, or called the *polar radius of gyration* for short, defined as

$$
r_p^2 \equiv \frac{I_p}{A}, \quad I_p \equiv \int_A (y^2 + z^2)\,dA \tag{d}
$$

If the shear stress components on the cross-section of the rod are negligible, the elastic strain energy of the rod reads

$$
V = \frac{1}{2}\int_0^L \left(\int_A \sigma_x \varepsilon_x\,dA\right)dx = \frac{E}{2}\int_0^L \left(\int_A u_x^2\,dA\right)dx = \frac{EA}{2}\int_0^L u_x^2\,dx \tag{e}
$$

From the extended Hamilton's principle given in Eq. (3.1.2) with $\delta W_n = 0$, the free vibration of the rod in an arbitrary time interval $[t_1, t_2]$ yields

$$
\int_{t_1}^{t_2} (\delta T - \delta V)\,dt = 0 \tag{f}
$$

Calculation of the two terms in Eq. (f) via integration by parts leads to

$$\int_{t_1}^{t_2} \delta V \, dt = \frac{EA}{2} \int_{t_1}^{t_2} \left(\delta \int_0^L u_x^2 dx \right) dt = EA \int_{t_1}^{t_2} \left(\int_0^L u_x \delta u_x dx \right) dt$$

$$= EA \int_{t_1}^{t_2} \left(u_x \delta u \Big|_0^L - \int_0^L u_{xx} \delta u \, dx \right) dt \tag{g}$$

$$\int_{t_1}^{t_2} \delta T \, dt = \frac{\rho A}{2} \int_{t_1}^{t_2} \left[\delta \int_0^L (u_t^2 + v^2 r_p^2 u_{xt}^2) \, dx \right] dt$$

$$= \rho A \int_{t_1}^{t_2} \left[\int_0^L (u_t \delta u_t + v^2 r_p^2 u_{xt} \delta u_{xt}) \, dx \right] dt \tag{h}$$

In calculating two variation terms in Eq. (h), it is beneficial to exchange the sequence of integrals and utilize $\delta u(x, t_1) = \delta u(x, t_2) = 0$ from Eq. (3.1.4). As such, we have

$$\int_{t_1}^{t_2} \int_0^L u_t \delta u_t dx \, dt = \int_0^L \left(u_t \delta u \Big|_{t_1}^{t_2} - \int_{t_1}^{t_2} u_{tt} \delta u dt \right) dx = - \int_{t_1}^{t_2} \left(\int_0^L u_{tt} \delta u dx \right) dt \tag{i}$$

$$\int_{t_1}^{t_2} \left(\int_0^L u_{xt} \delta u_{xt} \, dx \right) dt = \int_0^L \left(u_{xt} \delta u_x \Big|_{t_1}^{t_2} - \int_{t_1}^{t_2} u_{xtt} \delta u_x \, dt \right) dx$$

$$= - \int_{t_1}^{t_2} \left(\int_0^L u_{xtt} \delta u_x \, dx \right) dt$$

$$= \int_{t_2}^{t_1} \left(-u_{xtt} \delta u \Big|_0^L + \int_0^L u_{xxtt} \delta u \, dx \right) dt \tag{j}$$

Substituting Eqs. (g) and (h), with the help of Eqs. (i) and (j), into Eq. (f), we have

$$\int_{t_1}^{t_2} \left[\int_0^L (\rho A v^2 r_p^2 u_{xxtt} - \rho A u_{tt} + EA u_{xx}) \delta u dx \right] dt$$

$$- \int_{t_1}^{t_2} \left[(\rho A v^2 r_p^2 u_{xtt} + EA u_x) \delta u \Big|_L^0 \right] dt = 0 \tag{k}$$

As $\delta u(x, t)$ is [3] dynamic equation and the corresponding boundary conditions of the rod must satisfy

$$\begin{cases} \rho A v^2 r_p^2 u_{xxtt} - \rho A u_{tt} + EA u_{xx} = 0, & x \in [0, L], \quad t \in [t_1, t_2] \\ (\rho A v^2 r_p^2 u_{xtt} + EA u_x) \Big|_L^0 = 0 \text{ or } u \Big|_0^L = 0 \end{cases} \tag{l}$$

For a fixed-free rod shown in Fig. 3.1, Eq. (l) becomes

$$\begin{cases} \dfrac{\partial^2 u(x, t)}{\partial t^2} = c_0^2 \dfrac{\partial^2 u(x, t)}{\partial x^2} + v^2 r_p^2 \dfrac{\partial^4 u(x, t)}{\partial x^2 \partial t^2}, & c_0 \equiv \sqrt{\dfrac{E}{\rho}} \\ u(0, t) = 0, \quad c_0^2 u_x(L, t) + v^2 r_p^2 u_{xtt}(L, t) = 0 \end{cases} \tag{m}$$

This is the dynamic equation and boundary conditions of a fixed-free uniform rod with transverse inertia taken into account. This rod model is called *Love's rod model* or *Love's rod* for short. In Eq. (m), the forth-order partial derivative and the third-order partial derivative give the influence of transverse inertia on the longitudinal dynamics of the rod. Yet, it is not easy to analyze the above influence via the force analysis of an infinitesimal segment of the rod.

3.1.1.2 Mixed Initial-Boundary Value Problem of a Continuous System

To discuss the dynamic problem of a continuous system more specifically, consider a non-uniform rod shown in Fig. 3.2. Let L be the rod length, ρ be the material density, E be Young's modulus of material, η be the material damping coefficient, and they all are positive. In the frame of coordinates in Fig. 3.2, let $A(x) > 0$ be the cross-sectional area at coordinate x of the rod, and $u(x, t)$ be the longitudinal dynamic displacement of cross-section x of the rod, and $f(x, t)$ be the distributed dynamic forces to the cross-section x of the rod, respectively.

It is feasible to use either the extended Hamilton's principle or the force analysis to establish the dynamic equation of the rod as following

$$\rho A(x)\frac{\partial^2 u(x, t)}{\partial t^2} - \frac{\partial}{\partial x}\left[EA(x)\frac{\partial u(x, t)}{\partial x} + \eta EA(x)\frac{\partial^2 u(x, t)}{\partial t \partial x}\right] = f(x, t) \quad (3.1.5)$$

The fixed boundary and free boundary of the rod at its two ends read

$$u(0, t) = 0, \quad EA(L)u_x(L, t) = 0 \quad (3.1.6)$$

The initial displacement and velocity of the rod at moment $t = 0$ are

$$u(x, 0) = u_0(x), \quad u_t(x, 0) = \dot{u}_0(x) \quad (3.1.7)$$

Equations (3.1.5), (3.1.6) and (3.1.7) compose the *initial-boundary value problem* for a non-uniform rod under dynamic forces, and enable one to determine the dynamic response of the rod for $t > 0$.

The recent advances in additive manufacturing and materials make it possible to produce a rod having the axially variable cross-section and material in coordinate x. For such a rod, let the material density and Young's modulus be $\rho(x) > 0$ and $E(x) > 0$. Then, we have the dynamic equation of the rod as follows

Fig. 3.2 A non-uniform rod subject to distributed dynamic forces

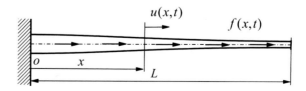

$$\rho(x)A(x)\frac{\partial^2 u(x,t)}{\partial t^2} - \frac{\partial}{\partial x}\left[E(x)A(x)\frac{\partial u(x,t)}{\partial x} + \eta E(x)A(x)\frac{\partial^2 u(x,t)}{\partial t \partial x}\right] = f(x,t)$$

$$(3.1.8)$$

Now, define a transform of coordinate x as

$$y \equiv \int_0^x \sqrt{\frac{\rho(s)}{E(s)}}\, ds \qquad\qquad (3.1.9)$$

Equation (3.1.9) implies that $dy/dx = \sqrt{\rho(x)/E(x)} > 0$ holds. That is, y monotonically increases with an increase of x so that an inverse transform $x \equiv x(y)$ exists. Substitution of Eq. (3.1.9) into Eq. (3.1.8) leads to

$$\sqrt{\frac{\rho(x)}{E(x)}}\left\{A(x)\sqrt{\rho(x)E(x)}\frac{\partial^2 u(y,t)}{\partial t^2} - \frac{\partial}{\partial y}\left[A(x)\sqrt{\rho(x)E(x)}\frac{\partial u(y,t)}{\partial y}\right.\right.$$
$$\left.\left. + \eta A(x)\sqrt{\rho(x)E(x)}\frac{\partial^2 u(y,t)}{\partial t \partial y}\right]\right\} = f(y,t) \qquad (3.1.10)$$

From the inverse transform $x \equiv x(y)$, it is easy to recast Eq. (3.1.10) as

$$\begin{cases} \overline{A}(y)\dfrac{\partial^2 u(y,t)}{\partial t^2} - \dfrac{\partial}{\partial y}\left[\overline{A}(y)\dfrac{\partial u(y,t)}{\partial y} + \eta\overline{A}(y)\dfrac{\partial^2 u(y,t)}{\partial t \partial y}\right] = \overline{f}(y,t) \\[2mm] \overline{A}(y) \equiv A[x(y)]\sqrt{\rho[x(y)]E[x(y)]}, \quad \overline{f}(y,t) \equiv \sqrt{\dfrac{E[x(y)]}{\rho[x(y)]}}\, f[x(y),t] \end{cases} \qquad (3.1.11)$$

The partial differential equations in Eqs. (3.1.11) and (3.1.5) have the same form. Substituting Eq. (3.1.9) into Eqs. (3.1.6) and (3.1.7) leads to the same boundary conditions and initial conditions. Hence, one only needs to study the mixed initial-boundary value problem given in Eqs. (3.1.5), (3.1.6) and (3.1.7) for a non-uniform rod when the material properties vary along the rod axis.

It is worthy to point out that Eq. (3.1.5) is a linear partial differential equation with variable coefficient $A(x)$ and only solvable for exact solutions in a few of special cases. This is also the case for non-uniform beams and non-uniform plates.

In addition, Eq. (3.1.6) describes the *homogeneous boundary conditions*. If a lumped mass m or a spring of axial stiffness coefficient k is installed at the right end of the rod in Fig. 3.2, the right boundary conditions of the rod become the *inhomogeneous boundary conditions* as follows

$$m u_{tt}(L,t) + EA(L)u_x(L,t) = 0 \qquad\qquad (3.1.12a)$$

$$EA(L)u_x(L, t) - ku(L, t) = 0 \qquad (3.1.12b)$$

The inhomogeneous boundary conditions definitely increase the complexity of the dynamic problem of the rod. In the case of non-uniform beams and plates, the inhomogeneous boundaries give rise to more difficult problems. As a matter of fact, even the natural vibration problems of a uniform plate with non-hinged boundaries have not been fully solved[2].

The above difficulties of solving the dynamic problems of continuous systems have made people turn to the numerical studies on the dynamic problems of discrete systems over the past decades.

3.1.2 Preconditions of Discretization

Many elementary textbooks presented how to discretize a continuous system to a discrete system by means of lumped parameter method, Ritz method and finite element methods so that one can study the dynamics of a continuous system according to the vibration theory of discrete systems. In other words, it is beneficial to discretize a partial differential equation to a set of ordinary differential equations so as to simplify the dynamic computation and analysis. Yet, readers may wonder whether the discretization is always feasible or not?

From the viewpoint of mechanics, the natural frequencies of a structure with limited size appear discretely in frequency domain, and the dynamic problem of the structure falls into the category of discrete spectrum of partial differential equations. In this case, the structure has an infinite, but countable mode shapes satisfying their orthogonal relations so that a proper linear combination of mode shapes can well approximate the structural dynamics. Hence, the discretization of a structure is feasible when the Ritz functions and the linear combinations of interpolation functions of finite elements can well approach the mode shapes with the natural frequencies discretely distributed in frequency domain.

3.1.2.1 Continuous and Discrete Spectrum Problems

Now, we discuss the continuous spectrum and discrete spectrum of a continuous system in frequency domain. For simplicity, consider, as examples, an infinitely long uniform rod and a finitely long uniform rod and discuss their spectral problems.

Example 3.1.2 Study the problem of free vibrations of a semi-infinitely long uniform rod fixed at the left end.

---

[2] Xing Y F, Sun Q Z, Li B, et al. (2018). The overall assessment of closed-form solution methods for free vibrations of rectangular thin plates. International Journal of Mechanical Sciences, 140: 455–470.

According to Eqs. (3.1.5) and (3.1.6), the dynamic boundary value problem of the rod is as follows

$$
\begin{cases}
\dfrac{\partial^2 u(x,t)}{\partial t^2} - c_0^2 \dfrac{\partial^2 u(x,t)}{\partial x^2} = 0, \quad c_0 \equiv \sqrt{\dfrac{E}{\rho}} > 0 \\[2ex]
u(0,t) = 0
\end{cases}
\tag{a}
$$

where c_0 is the *longitudinal wave speed* of the uniform rod. Let the solution of Eq. (a) be in a form of separated variables

$$
u(x,t) = \tilde{u}(x)q(t)
\tag{b}
$$

Substitution of Eq. (b) into the partial different equation in Eq. (a) leads to

$$
\frac{c_0^2}{\tilde{u}(x)} \frac{d^2\tilde{u}(x)}{dx^2} = \frac{1}{q(t)} \frac{d^2 q(t)}{dt^2}
\tag{c}
$$

Equation (c) holds if and only if both sides of Eq. (c) are identically equal to a constant γ. This results in two differential equations as following

$$
\begin{cases}
\dfrac{d^2\tilde{u}(x)}{dx^2} - \dfrac{\gamma}{c_0^2}\tilde{u}(x) = 0, \quad \tilde{u}(0) = 0 \\[2ex]
\dfrac{d^2 q(t)}{dt^2} - \gamma q(t) = 0
\end{cases}
\tag{d}
$$

The value of γ has three possibilities to be discussed as follows.
  Case (1): When $\gamma = 0$, the solutions of Eq. (d) satisfy

$$
\begin{cases}
\tilde{u}(x) = ax \\
q(t) = b_1 t + b_2
\end{cases}
\tag{e}
$$

where a, b_1 and b_2 are integration constants. Substitution of Eq. (e) into Eq. (b) leads to the following sub-cases. If $a \neq 0$ and $x \to +\infty$, either $b_1^2 + b_2^2 > 0$ makes $|u(x,t)| \to +\infty$ or $b_1^2 + b_2^2 = 0$ makes $u(x,t) = 0$. Otherwise, $a = 0$ gives $u(x,t) = 0$. None of these sub-cases has a meaningful or expected solution.
  Case (2): When $\gamma > 0$, the solutions of Eq. (d) are

$$
\begin{cases}
\tilde{u}(x) = a \sinh\left(\dfrac{\sqrt{\gamma}x}{c_0}\right) \\[2ex]
q(t) = b_1 \cosh(\sqrt{\gamma}t) + b_2 \sinh(\sqrt{\gamma}t)
\end{cases}
\tag{f}
$$

Substituting Eq. (f) into Eq. (b) leads to the following sub-cases. That is, solution $u(x,t)$ is either divergent or a null when $x \to +\infty$ and $t \to +\infty$. Hence, Eq. (f) does not result in any expected solution.

Case (3): When $\gamma < 0$, let $\gamma = -\omega^2$ and get the bounded non-zero solution of Eq. (d) as follows

$$\begin{cases} \tilde{u}(x) = a\sin(\kappa x) \\ q(t) = b_1\cos(\omega t) + b_2\sin(\omega t), \quad \kappa \equiv \omega/c_0 \end{cases} \tag{g}$$

where a, b_1 and b_2 are integration constants, and $\kappa > 0$ is the *wave number*. Substitution of Eq. (g) into Eq. (b) leads to

$$\begin{aligned} u(x, t) &= ab_1\sin(\kappa x)\cos(\omega t) + ab_2\sin(\kappa x)\sin(\omega t) \\ &= \frac{ab_1}{2}\sin[\kappa(c_0 t + x)] - \frac{ab_1}{2}\sin[\kappa(c_0 t - x)] \\ &\quad + \frac{ab_2}{2}\cos[\kappa(c_0 t - x)] - \frac{ab_2}{2}\cos[\kappa(c_0 t + x)] \\ &= u_L(c_0 t + x) + u_R(c_0 t - x) \end{aligned} \tag{h}$$

where

$$\begin{cases} u_L(c_0 t + x) \equiv \dfrac{ab_1}{2}\sin[\kappa(c_0 t + x)] - \dfrac{ab_2}{2}\cos[\kappa(c_0 t + x)] \\[2mm] u_R(c_0 t - x) \equiv \dfrac{ab_2}{2}\cos[\kappa(c_0 t - x)] - \dfrac{ab_1}{2}\sin[\kappa(c_0 t - x)] \end{cases} \tag{i}$$

It is easy to see that $u_L(c_0 t + x)$ and $u_R(c_0 t - x)$ are traveling waves with wave speed c_0 in the left and right directions respectively. At the left end of the rod, the left traveling wave and the right traveling wave counteract each other so that the left boundary condition holds, namely,

$$u(0, t) = u_L(c_0 t) + u_R(c_0 t) = 0 \tag{j}$$

As the right end of the rod is infinitely far, the right traveling wave is not subject to any reflection and can not form a standing wave. Thus, $\kappa > 0$ does not need to satisfy the condition of a standing wave and can take an arbitrary value. As such, the frequency $\omega = \kappa c_0$ determined by the wave number $\kappa > 0$ takes continuous values in frequency domain $[0, +\infty)$. That is, the semi-definitely long uniform rod fixed at one end has a continuous frequency spectrum.

Example 3.1.3 Study the free vibrations of a fixed-free uniform rod of length L, which yields the following dynamic boundary value problem

$$\begin{cases} \dfrac{\partial^2 u(x, t)}{\partial t^2} - c_0^2\dfrac{\partial^2 u(x, t)}{\partial x^2} = 0 \\[2mm] u(0, t) = 0, \quad u_x(L, t) = 0 \end{cases} \tag{a}$$

where the notations are the same as those in Example 3.1.2.

From the natural modes of a fixed-free rod in Table 2.1, it is easy to derive the free vibration of the fixed-free rod via the modal superposition method as follows

$$
\begin{cases}
u(x, t) = \sum_{r=1}^{+\infty} \sin(\kappa_r x)[b_{1r} \cos(\omega_r t) + b_{2r} \sin(\omega_r t)] \\
\kappa_r = \dfrac{(2r - 1)\pi}{2L}, \quad \omega_r = \kappa_r c_0, \quad r = 1, 2, \ldots
\end{cases}
\tag{b}
$$

Here, the coefficients $b_{1r}, b_{2r}, r = 1, 2, \ldots$ are to be determined from the initial conditions of the rod. In Eq. (b), both wave number κ_r and natural frequency ω_r take discrete values, forming two infinite sequences.

A comparison between two examples shows that the dynamics of a fixed-free uniform rod of a finite length has a *discrete spectrum* including wave numbers κ_r, $r = 1, 2, \ldots$ and discrete natural frequencies $\omega_r = \kappa_r c_0, r = 1, 2, \ldots$, whereas the dynamic problem of a fixed-free rod with semi-infinite length has a *continuous spectrum*, where wave number κ takes an arbitrary real number and so does the free vibration frequency $\omega = \kappa c_0$.

Remark 3.1.2 For the dynamic problem of a continuous system with a discrete spectrum, it is feasible to establish a model of lower order by approaching the spectrum at a lower frequency band. Yet, no similar model applies to the dynamic problem of any continuous system with a continuous spectrum. In this case, the approximate model established in a discretized space domain usually has much higher dimensions.

3.1.2.2   Singular Point of a Structure

As shown in above examples, the method of separation of variables is a useful tool to simplify the dynamic equation of a non-uniform rod to two decoupled linear differential equations, which have solutions $q(t)$ and $\varphi(x)$, respectively. As studied in mathematics[3], the solution $\varphi(x)$ of a linear differential equation may have the continuous wave number κ in a real number interval if the variable coefficient of the linear differential equation has a *singular point* $x = x_s$. In this case, the vibration frequency $\omega = \kappa c_0$ of solution $q(t)$ takes the continuous values in a frequency band and the dynamic problem of the rod exhibits continuous spectrum.

From the viewpoint of vibration mechanics, the natural vibration of a structure is a standing wave, which yields a number of specific wave numbers of traveling waves reflected at the structure boundary. Thus, these wave numbers are discrete in a real number interval and so are the natural frequencies. As a result, no observations have been available for any continuous spectrum.

Now, consider, as an example, a non-uniform rod in Fig. 3.3 and verify that the natural frequencies of the rod only have a discrete spectrum even though the linear differential equation governing the mode shapes has a singular point at the free end of the rod. Furthermore, we give the asymptotic behavior of the natural vibration at the singular point.

---

[3] Courant R, Hilbert D (1989). Methods of Mathematical Physics. Volume I. New York: John Wiley & Sons Inc., 339–343.

Fig. 3.3 A non-uniform rod with a singular point at the free end

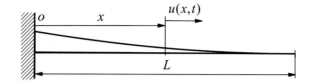

Example 3.1.4 Study the natural vibrations of a fixed-free non-uniform rod as shown in Fig. 3.3, in which the cross-sectional area of the rod yields

$$A(x) = A_0(x - L)^2 > 0, \quad A_0 > 0, \quad x \in [0, L) \tag{a}$$

where the cross-sectional area approaches zero at the free end of the rod.

Substituting Eq. (a) into Eqs. (3.1.5) and (3.1.6), we have the dynamic boundary value problem of the free vibrations of the rod

$$
\begin{cases}
\rho A_0(x - L)^2 \dfrac{\partial^2 u(x, t)}{\partial t^2} - E A_0 \dfrac{\partial}{\partial x}\left[(x - L)^2 \dfrac{\partial u(x, t)}{\partial x}\right] = 0 \\
u(0, t) = 0, \quad \lim_{x \to L^-} E A_0 (x - L)^2 u_x(x, t) = 0
\end{cases} \tag{b}
$$

Let the solution of Eq. (b) be in the following form of separated variables

$$u(x, t) = \varphi(x) q(t) \tag{c}$$

Substitution of Eq. (c) into Eq. (b) yields

$$
\begin{cases}
\dfrac{d^2 \varphi(x)}{dx^2} + \dfrac{2}{x - L} \dfrac{d\varphi(x)}{dx} + \left(\dfrac{\omega}{c_0}\right)^2 \varphi(x) = 0, \quad c_0 \equiv \sqrt{\dfrac{E}{\rho}} \\
\varphi(0) = 0, \quad \lim_{x \to L^-} E A_0 (x - L)^2 u_x(x, t) = 0
\end{cases} \tag{d}
$$

The coefficient of the linear differential equation in Eq. (d) has a singular point at $x = L$, but one can verify that Eq. (d) has a discrete spectrum as follows
By using the following transform

$$\varphi(x) = \frac{\psi(x)}{x - L}, \quad x \in [0, \ L) \tag{e}$$

it is easy to simplify the linear differential equation in Eq. (d) as

$$\frac{d^2 \psi(x)}{dx^2} + \kappa^2 \psi(x) = 0, \quad \kappa \equiv \frac{\omega}{c_0} > 0 \tag{f}$$

The corresponding boundary conditions in Eq. (d) satisfy

$$\begin{cases} \psi(0) = 0 \\ \lim_{x \to L^-} E A_0 (x - L)^2 \varphi_x(x) = E A_0 \lim_{x \to L^-} [(x - L)\psi_x(x) - \psi(x)] = 0 \end{cases} \quad \text{(g)}$$

The solution of Eq. (f) reads

$$\psi(x) = c_1 \cos(\kappa x) + c_2 \sin(\kappa x) \tag{h}$$

Substitution of Eq. (h) into Eq. (g) leads to

$$\begin{cases} c_1 = 0 \\ \lim_{x \to L^-} E A_0 [(x - L)\psi_x(x) - \psi(x)] = -c_2 E A_0 \sin(\kappa L) = 0 \end{cases} \quad \text{(i)}$$

whereby we have the discrete eigenvalues and the corresponding eigenvectors

$$\kappa_r = \frac{r\pi}{L}, \quad \psi_r(x) = \sin\left(\frac{r\pi x}{L}\right), \quad r = 1, 2, \ldots \tag{j}$$

Therefore, the non-uniform rod has an infinite, but countable, number of natural frequencies and mode shapes

$$\omega_r = \frac{r\pi}{L} c_0, \quad \varphi_r(x) = \frac{\psi_r(x)}{x - L} = \frac{1}{x - L} \sin\left(\frac{r\pi x}{L}\right), \quad r = 1, 2, \ldots \tag{k}$$

As shown in Eq. (k), the natural vibrations of the rod have a discrete spectrum even though the coefficient of Eq. (d) has a singular point at $x = L$.

To understand the natural vibrations of the rod at its free end, calculating both limits of the r-th mode shape $\varphi_r(x)$ and the internal force $N_r(x) \equiv E A(x)\varphi_{rx}(x)$ at the singular point $x = L$, we obtain

$$\begin{cases} \varphi_r(L) = \lim_{x \to L^-} \left[\frac{1}{x - L} \sin\left(\frac{r\pi x}{L}\right) \right] \\ \qquad = \frac{r\pi \cos(r\pi)}{L} \lim_{x \to L^-} \left\{ \frac{L}{r\pi(x - L)} \sin\left[\frac{r\pi(x - L)}{L}\right] \right\} = \frac{r\pi \cos(r\pi)}{L} \\ N_r(L) = \lim_{x \to L^-} E A(x)\varphi_{rx}(x) \\ \qquad = E A_0 \lim_{x \to L^-} \left\{ \frac{r\pi(x - L)}{L} \cos\left[\frac{r\pi(x - L)}{L}\right] - \sin\left[\frac{r\pi(x - L)}{L}\right] \right\} = 0, r = 1, 2, \ldots \end{cases}$$

$$\tag{l}$$

Equation (l) indicates that the singular point of the non-uniform rod is a removable discontinuity point of the mode shapes and the continuous point of the internal force associated with the mode shape. Hence, the singular point does not give rise to any problem to the vibrations of the rod although the computation of mode shapes seems to touch upon the singularity.

Remark 3.1.3 As studied for ordinary differential equations, the singular point $x = L$ in this example is a regular singularity of the differential equations and does not affect the solution. If the cross-sectional area of the rod is a linear function

$A(x) = A_0(x - L)$, the singular point $x = L$ is still regular even though the mode shapes of the rod becomes a linear combination of Bessel functions, rather than any harmonic functions[4].

3.1.3 Case Studies of Discretization

When the continuous system has a discrete spectrum, many approaches are applicable to the discretization of the continuous system and each of them has their own features. For instance, the lumped parameter method is simple but not accurate, and greatly depends on some artificial factors. Ritz method has a clear picture of physics and higher accuracy, but the selection of assumed mode functions of higher order is hard. The methods of finite elements are suitable for a great variety of complicated problems, but not convenient for their parametric studies. The following two examples demonstrate the first two methods.

3.1.3.1 Ritz Method

Example 3.1.5 Given the cross-sectional area of the rod in Fig. 3.3 as below

$$A(x) = A_0(x - L)^2 > 0, \quad A_0 > 0, \quad x \in [0, L) \tag{a}$$

solve the first two natural vibrations of the rod by using Ritz method.

By taking the Ritz functions, which satisfy the boundary conditions of the rod in Fig. 3.3, and defining the linear transform matrix as follows

$$P(x) \equiv \begin{bmatrix} p_1(x) & p_2(x) \end{bmatrix} \equiv \begin{bmatrix} x^2 - 2Lx & 2x^3 - 3Lx^2 \end{bmatrix} \tag{b}$$

we write the dynamic displacement of the rod approximated as

$$u(x, t) = P(x)q(t) = \begin{bmatrix} x^2 - 2Lx & 2x^3 - 3Lx^2 \end{bmatrix} \begin{bmatrix} q_1(t) \\ q_2(t) \end{bmatrix} \tag{c}$$

where vector $q(t)$ includes the projection coefficients of $u(x, t)$ onto the Ritz functions. According to the cross-sectional area of the rod in Eq. (a), we derive the kinetic and potential energies of the rod as follows

---

[4] Courant R, Hilbert D (1989). Methods of Mathematical Physics. Volume I. New York: John Wiley & Sons Inc., 466–501.

$$T = \frac{1}{2} \int_0^L \rho A(x) \left[\frac{\partial u(x,t)}{\partial t} \right]^2 dx$$

$$= \frac{\rho A_0}{2} \int_0^L (x-L)^2 \left[(x^2 - 2Lx)\dot{q}_1(t) + (2x^3 - 3Lx^2)\dot{q}_2(t) \right]^2 dx$$

$$= \frac{\rho A_0 L^7}{2} \left[\frac{8}{105} \dot{q}_1^2(t) + \frac{19L}{210} \dot{q}_1(t)\dot{q}_2(t) + \frac{19L^2}{630} \dot{q}_2^2(t) \right] \tag{d}$$

$$V = \frac{1}{2} \int_0^L EA(x) \left[\frac{\partial u(x,t)}{\partial x} \right]^2 dx$$

$$= \frac{EA_0}{2} \int_0^L (x-L)^2 \left[2(x-L)q_1(t) + 6(x^2 - Lx)q_2(t) \right]^2 dx$$

$$= \frac{EA_0 L^5}{2} \left[\frac{4}{5} q_1^2(t) + \frac{4L}{5} q_1(t)q_2(t) + \frac{12L^2}{35} q_2^2(t) \right] \tag{e}$$

Substitution of Eqs. (d) and (e) into Lagrange's equation of the second kind leads to the dynamic equations of a discrete model of the rod, namely,

$$\begin{bmatrix} 8L^2/105 & 19L^3/420 \\ 19L^3/420 & 19L^4/630 \end{bmatrix} \begin{bmatrix} \ddot{q}_1(t) \\ \ddot{q}_2(t) \end{bmatrix} + c_0^2 \begin{bmatrix} 4/5 & 4L/10 \\ 4L/10 & 12L^2/35 \end{bmatrix} \begin{bmatrix} q_1(t) \\ q_2(t) \end{bmatrix} = 0 \tag{f}$$

where $c_0 \equiv \sqrt{E/\rho}$. Let the solution of Eq. (f) be expressed as

$$q(t) \equiv \begin{bmatrix} q_1(t) \\ q_2(t) \end{bmatrix} = \begin{bmatrix} \tilde{q}_1 \sin(\omega t) \\ \tilde{q}_2 \sin(\omega t) \end{bmatrix} \equiv \tilde{q} \sin(\omega t) \tag{g}$$

Substituting Eq. (g) into Eq. (f) gives a generalized eigenvalue problem as follows

$$\left\{ c_0^2 \begin{bmatrix} 4/5 & 4L/10 \\ 4L/10 & 12L^2/35 \end{bmatrix} - \omega^2 \begin{bmatrix} 8L^2/105 & 19L^3/420 \\ 19L^3/420 & 19L^4/630 \end{bmatrix} \right\} \tilde{q} = 0 \tag{h}$$

Solving Eq. (h) leads to the first two approximate natural frequencies of the rod while the eigenvectors of Eq. (h) are the projections of the first two mode shapes onto the Ritz functions, namely,

$$\begin{cases} \omega_1 = \dfrac{3.1417}{L} c_0, \quad \tilde{q}_1 = \begin{bmatrix} 0.9706 & 1.0000 \end{bmatrix}^T \\[4mm] \omega_2 = \dfrac{6.7874}{L} c_0, \quad \tilde{q}_2 = \begin{bmatrix} -0.6214 & 1.0000 \end{bmatrix}^T \end{cases} \tag{i}$$

From Eqs. (c) and (i), we obtain the first two mode shapes approximated as

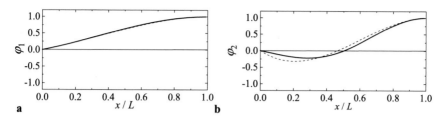

Fig. 3.4 The mode shapes of the non-uniform rod with a singular point at free end (solid line: exact solution, dashed line: approximate solution). **a** the first mode shape; **b** the second mode shape

$$\begin{cases} \varphi_1(x) = \boldsymbol{P}(x)\tilde{\boldsymbol{q}}_1 = 0.9706(x^2 - 2Lx) + (2x^3 - 3Lx^2) \\ \varphi_2(x) = \boldsymbol{P}(x)\tilde{\boldsymbol{q}}_2 = 0.6214(x^2 - 2Lx) - (2x^3 - 3Lx^2) \end{cases} \tag{j}$$

A comparison between Eq. (i) and the exact solution in Example 3.1.4 indicates that the first approximate natural frequency is very accurate and the relative error is only 0.0035%, yet the second approximate natural frequency has a relative error of 8.0246% which is still acceptable in engineering. Figure 3.4 shows the comparison between the exact mode shapes and the approximate mode shapes. The error of the first approximate mode shape is almost invisible, while the error of the second one looks obvious but not significant.

Remark 3.1.4 Compared with the numerical results of finite element methods, the approximate results of Ritz method here include the rod length L and wave speed c_0. The results, hence, are more suitable for parametric studies.

3.1.3.2 Lumped Parameter Method

Section 1.2.1 presented the engineering background of the planar vibration of a tethered satellite and attributed it to the planar vibration of a tether pendulum under gravity. Following the basic ideas of solving Problems 1A and 1B proposed in Sect. 1.2.1, the first step here is to establish the dynamic equation of the tether pendulum and discuss the influence of system parameters on the natural vibrations of the system briefly.

Example 3.1.6 Study how to establish the dynamic model of a tether pendulum via the lumped parameter method and check the influence of system parameters on the natural vibrations of the tether pendulum.

As shown in Fig. 3.5, the tether is divided as N identical segments of length $a \equiv L/N$. The distributed inertia of the tether is lumped via two ways so as to get model A and model B, respectively.

(1)   Model A: In view of the lumped parameter method for finite elements, let the mass of each segment be lumped to the two ends of segment and get an N-DoF system shown in Fig. 3.5a, where $\overline{m} \equiv \rho AL/N$, $\hat{m} \equiv m + \overline{m}/2 = \rho AL(\eta + 1/2N)$, and $\eta \equiv m/\rho AL$ is the ratio of the tip mass to the tether mass. In this model, the lumped mass $\overline{m}/2$ at the top segment is immovable and has nothing to do with the system dynamics.

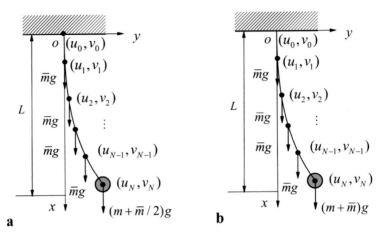

Fig. 3.5 Two discrete models of a tether pendulum under gravity. **a** Modal A; **b** model B

(2) Model B: Let the lumped mass $\overline{m} = \rho AL/N$ be assigned to the bottom end of each segment. In this model shown in Fig. 3.5b, the lumped mass $\overline{m}/2$ at the top segment in Model A moves to the bottom end of the tether and enhances the influence of the tip mass.

In what follows, we mainly study Model A and list the results of Model B from similarity. According to Fig. 3.5a, the vertical and horizontal displacements of a lumped mass respectively satisfy

$$\begin{cases} u_i = u_{i-1} + a\cos\theta_i \\ v_i = v_{i-1} + a\sin\theta_i, \quad i = 1,\ldots,N \end{cases} \tag{a}$$

where $u_0 = 0$, $v_0 = 0$, $\theta_0 = 0$, and θ_i presents a counter-clockwise angle between the top end of segment i and the vertical line. Differentiating Eq. (a) with respect to time gives the corresponding velocities

$$\begin{cases} \dot{u}_i = \dot{u}_{i-1} - a\dot{\theta}_i\sin\theta_i \\ \dot{v}_i = \dot{v}_{i-1} + a\dot{\theta}_i\cos\theta_i, \quad i = 1,\ldots,N \end{cases} \tag{b}$$

As the subsequent analysis needs accelerations, let Eq. (b) be differentiated with respect to time and then be linearized, under the assumption of small angles for θ_i, $i = 1, \ldots, N$, as

$$\begin{cases} \ddot{u}_i = \ddot{u}_{i-1} - a\ddot{\theta}_i\sin\theta_i - a\dot{\theta}_i^2\cos\theta_i \approx \ddot{u}_{i-1} \\ \ddot{v}_i = \ddot{v}_{i-1} + a\ddot{\theta}_i\cos\theta_i - a\dot{\theta}_i^2\sin\theta_i \approx \ddot{v}_{i-1} + a\ddot{\theta}_i, \quad i = 1,\ldots,N \end{cases} \tag{c}$$

From Eqs. (a) and (b), we derive the kinetic energy of model A and the potential energy with zero position assumed at the equilibrium, namely,

$$T = \frac{\overline{m}}{2} \sum_{i=1}^{N-1} \left[(\dot{u}_{i-1} - a\dot{\theta}_i \sin \theta_i)^2 + (\dot{v}_{i-1} + a\dot{\theta}_i \cos \theta_i)^2 \right]$$

$$+ \frac{\hat{m}}{2} \left[(\dot{u}_{N-1} - a\dot{\theta}_N \sin \theta_N)^2 + (\dot{v}_{N-1} + a\dot{\theta}_N \cos \theta_N)^2 \right]$$

$$= \frac{\overline{m}}{2} \sum_{i=1}^{N-1} (\dot{u}_{i-1}^2 + \dot{v}_{i-1}^2 + a^2\dot{\theta}_i^2 - 2a\dot{u}_{i-1}\dot{\theta}_i \sin \theta_i + 2a\dot{v}_{i-1}\dot{\theta}_i \cos \theta_i)$$

$$+ \frac{\hat{m}}{2} (\dot{u}_{N-1}^2 + \dot{v}_{N-1}^2 + a^2\dot{\theta}_N^2 - 2a\dot{u}_{N-1}\dot{\theta}_N \sin \theta_N + 2a\dot{v}_{N-1}\dot{\theta}_N \cos \theta_N) \qquad (d)$$

$$V = -\overline{m}g \sum_{i=1}^{N-1} (u_{i-1} + a \cos \theta_i) - \hat{m}g(u_{N-1} + a \cos \theta_N) \qquad (e)$$

It is worthy to point out that the direct substitution of Eqs. (d) and (e) into Lagrange's equation of the second kind gives very complicated nonlinear differential equations. To simplify the subsequent analysis, it is better to linearize Eqs. (d) and (e) under the assumption of small angles for $\theta_i, i = 1, \ldots, N$ during the substitution and to make use of the following expressions

$$\frac{\partial \dot{u}_i}{\partial \dot{\theta}_j} \approx \frac{\partial \dot{u}_{i-1}}{\partial \dot{\theta}_j} = 0, \quad \frac{\partial \dot{v}_i}{\partial \dot{\theta}_j} \approx \frac{\partial (\dot{v}_{i-1} + a\dot{\theta}_i)}{\partial \dot{\theta}_j} = \begin{cases} a, & j \le i \\ 0, & j > i, \end{cases} \quad i, j = 1, 2, \ldots, N$$

$$(f)$$

Consider, as an example, the deduction of the first dynamic equation, we have

$$\left\{ \begin{aligned} \frac{d}{dt} \frac{\partial T}{\partial \dot{\theta}_1} - \frac{\partial T}{\partial \theta_1} &\approx \overline{m} \sum_{i=1}^{N-1} (a\ddot{v}_{i-1} + a^2\ddot{\theta}_i) + \hat{m}(a\ddot{v}_N + a^2\ddot{\theta}_N) \\ &= \overline{m}a^2 \left[\ddot{\theta}_1 + (\ddot{\theta}_1 + \ddot{\theta}_2) + \cdots + (\ddot{\theta}_1 + \ddot{\theta}_2 + \cdots + \ddot{\theta}_{N-1}) \right] \\ &\quad + \hat{m}a^2(\ddot{\theta}_1 + \ddot{\theta}_2 + \cdots + \ddot{\theta}_N) \\ &= a^2 \{ \left[(N-1)\overline{m} + \hat{m} \right]\ddot{\theta}_1 + \left[(N-2)\overline{m} + \hat{m} \right]\ddot{\theta}_2 \\ &\quad + \cdots + (\overline{m} + \hat{m})\ddot{\theta}_{N-1} + \hat{m}\ddot{\theta}_N \} \\ \frac{\partial V}{\partial \theta_1} &= -\overline{m}g \sum_{i=1}^{N-1} \frac{\partial u_i}{\partial \theta_1} - \hat{m}g \frac{\partial u_{N-1}}{\partial \theta_1} = ga \left[(N-1)\overline{m} + \hat{m} \right]\theta_1 \end{aligned} \right.$$

$$(g)$$

In the case of $N = 4$, the dynamic equations of model A yield

$$\begin{cases} a\left[(3\overline{m}+\hat{m})\ddot{\theta}_1 + (2\overline{m}+\hat{m})\ddot{\theta}_2 + (\overline{m}+\hat{m})\ddot{\theta}_3 + \hat{m}\ddot{\theta}_4\right] + (3\overline{m}+\hat{m})g\theta_1 = 0 \\ a\left[(2\overline{m}+\hat{m})\ddot{\theta}_1 + (2\overline{m}+\hat{m})\ddot{\theta}_2 + (\overline{m}+\hat{m})\ddot{\theta}_3 + \hat{m}\ddot{\theta}_4\right] + (2\overline{m}+\hat{m})g\theta_2 = 0 \\ a\left[(\overline{m}+\hat{m})\ddot{\theta}_1 + (\overline{m}+\hat{m})\ddot{\theta}_2 + (\overline{m}+\hat{m})\ddot{\theta}_3 + \hat{m}\ddot{\theta}_4\right] + (\overline{m}+\hat{m})g\theta_3 = 0 \\ a\left[\hat{m}\ddot{\theta}_1 + \hat{m}\ddot{\theta}_2 + \hat{m}\ddot{\theta}_3 + \hat{m}\ddot{\theta}_4\right] + \hat{m}g\,\theta_4 = 0 \end{cases} \qquad \text{(h)}$$

Noting that the coefficients in Eq. (h) satisfy

$$i\overline{m} + \hat{m} = \frac{i\rho AL}{N} + \left(\eta + \frac{1}{2N}\right)\rho AL = (2i + 1 + 2N\eta)\frac{\rho AL}{2N}, \quad i = 1, \ldots, N-1 \tag{i}$$

we recast Eq. (h), with the help of $\overline{\eta} \equiv 2N\eta$ and $a = L/N$, as a set of dynamic equations in a more regular form as follows

$$\begin{cases} L\left[(7+\overline{\eta})\ddot{\theta}_1 + (5+\overline{\eta})\ddot{\theta}_2 + (3+\overline{\eta})\ddot{\theta}_3 + (1+\overline{\eta})\ddot{\theta}_4\right] + (7+\overline{\eta})Ng\theta_1 = 0 \\ L\left[(5+\overline{\eta})\ddot{\theta}_1 + (5+\overline{\eta})\ddot{\theta}_2 + (3+\overline{\eta})\ddot{\theta}_3 + (1+\overline{\eta})\ddot{\theta}_4\right] + (5+\overline{\eta})Ng\theta_2 = 0 \\ L\left[(3+\overline{\eta})\ddot{\theta}_1 + (3+\overline{\eta})\ddot{\theta}_2 + (3+\overline{\eta})\ddot{\theta}_3 + (1+\overline{\eta})\ddot{\theta}_4\right] + (3+\overline{\eta})Ng\theta_3 = 0 \\ L\left[(1+\overline{\eta})\ddot{\theta}_1 + (1+\overline{\eta})\ddot{\theta}_2 + (1+\overline{\eta})\ddot{\theta}_3 + (1+\overline{\eta})\ddot{\theta}_4\right] + (1+\overline{\eta})Ng\theta_4 = 0 \end{cases} \qquad \text{(j)}$$

Assume that the natural vibrations of the system satisfy

$$\theta_i(t) = \tilde{\theta}_i \sin(\omega t), \quad i = 1, 2, \ldots, N \tag{k}$$

Substitution of Eq. (k) into Eq. (j) leads to an eigenvalue problem of natural vibrations of model A when $N = 4$ as follows

$$\left\{ \begin{bmatrix} 7+\overline{\eta} & 0 & 0 & 0 \\ 0 & 5+\overline{\eta} & 0 & 0 \\ 0 & 0 & 3+\overline{\eta} & 0 \\ 0 & 0 & 0 & 1+\overline{\eta} \end{bmatrix} - \frac{\omega^2 L}{gN} \begin{bmatrix} 7+\overline{\eta} & 5+\overline{\eta} & 3+\overline{\eta} & 1+\overline{\eta} \\ 5+\overline{\eta} & 5+\overline{\eta} & 3+\overline{\eta} & 1+\overline{\eta} \\ 3+\overline{\eta} & 3+\overline{\eta} & 3+\overline{\eta} & 1+\overline{\eta} \\ 1+\overline{\eta} & 1+\overline{\eta} & 1+\overline{\eta} & 1+\overline{\eta} \end{bmatrix} \right\} \begin{bmatrix} \tilde{\theta}_1 \\ \tilde{\theta}_2 \\ \tilde{\theta}_3 \\ \tilde{\theta}_4 \end{bmatrix} = \begin{bmatrix} 0 \\ 0 \\ 0 \\ 0 \end{bmatrix} \tag{l}$$

To study model B, one simply modifies $\hat{m}$ as $\hat{m} = m + \rho AL/N$, and replaces Eq. (i) with the following relation

$$i\overline{m} + \hat{m} = \frac{i\rho AL}{N} + \frac{\rho AL}{N} + \eta\rho AL$$

$$= (i + 1 + N\eta)\frac{\rho AL}{N}, \quad i = 1, \ldots, N-1 \tag{m}$$

By means of $\hat{\eta} \equiv N\eta$, one arrives at an eigenvalue problem of model B when $N = 4$ as follows

$$\left\{ \begin{bmatrix} 4+\hat{\eta} & 0 & 0 & 0 \\ 0 & 3+\hat{\eta} & 0 & 0 \\ 0 & 0 & 2+\hat{\eta} & 0 \\ 0 & 0 & 0 & 1+\hat{\eta} \end{bmatrix} - \frac{\omega^2 L}{gN} \begin{bmatrix} 4+\hat{\eta} & 3+\hat{\eta} & 2+\hat{\eta} & 1+\hat{\eta} \\ 3+\hat{\eta} & 3+\hat{\eta} & 2+\hat{\eta} & 1+\hat{\eta} \\ 2+\hat{\eta} & 2+\hat{\eta} & 2+\hat{\eta} & 1+\hat{\eta} \\ 1+\hat{\eta} & 1+\hat{\eta} & 1+\hat{\eta} & 1+\hat{\eta} \end{bmatrix} \right\} \begin{bmatrix} \tilde{\theta}_1 \\ \tilde{\theta}_2 \\ \tilde{\theta}_3 \\ \tilde{\theta}_4 \end{bmatrix} = \begin{bmatrix} 0 \\ 0 \\ 0 \\ 0 \end{bmatrix}$$

(n)

Following the regularity in Eqs. (l) and (n), it is easy to write out the mass matrix and stiffness matrix of the system with more DoFs taken into account.

Remark 3.1.5 As shown in Eqs. (l) and (n), the mode shapes of model A depend on parameter $\bar{\eta} = 2N\eta$ and the corresponding natural frequencies rely on parameter $2N\eta$ and are proportional to $\sqrt{gN/L}$, whereas the natural mode shapes of model B depend on parameter $\hat{\eta} \equiv N\eta$ and the natural frequencies rely on parameter $N\eta$ and are proportional to $\sqrt{gN/L}$. As the selection of number N is optional, the mode shapes only rely on the mass ratio η and the natural frequencies rely on the mass ratio η and are proportional to $\sqrt{g/L}$. These assertions provide important information for the preliminary study of a tethered satellite.

Remark 3.1.6 Although the eigenvectors in Eqs. (l) and (n) multiplied by any real constant keeps similar, the eigenvectors in computation needs to be scaled so that all angles $\tilde{\theta}_i$, $i = 1, \ldots, N$ are small enough and the mode shape determined by Eq. (a) satisfy the precondition of a small vibration.

In Sect. 5.1, we shall establish the dynamic equation of the continuous model of the tether pendulum, derive the exact solutions of natural vibrations and validate models A and B in order to complete the study on Problems 1A and 1B proposed in Chap. 1.

3.1.4 Concluding Remarks

This section briefly presented how to establish the dynamic equation of a continuous system according to the extended Hamilton's principle and how to discretize a continuous system. To help readers understand the nature of associated mechanics in a simple way and solve Problems 1A and 1B, the section demonstrated the above processes through the dynamic models of several non-uniform rods and a tether pendulum, and led to the major conclusions of the section as follows.

(1) For a rod with the axially variable cross-section and material, the dynamic equation of the rod can be attributed to that of a uniform rod with homogeneous material. The dynamic equation of a non-uniform rod has discrete spectrum and can be approximated by a simplified model of lower order. The dynamic equation of a non-uniform rod with an infinite or semi-infinite length has continuous spectrum and can hardly be approximated by a simplified model of lower order.

(2)   The mode shape of some vibration systems, such as a non-uniform rod and a tether pendulum, yields an ordinary differential equation having a variable coefficient with a singular point. Yet, the singular point has regular singularity such that the ordinary differential equation describing the mode shape still has the discrete spectrum. As a consequence, the mode shape has non-singular property at the singular point.

(3)   Each discretization method has its own features. For example, Ritz method provided a simple model of lower orders and higher accuracy for studying the vibrations of a non-uniform rod, whereas the lumped parameter method to model the tether pendulum showed the advantages over others.

(4)   Compared with finite element methods, the above methods enabled one to establish the dynamic equations of a system with major system parameters and to reveal the nature of vibration mechanics. In the dynamic modeling of a tether pendulum under gravity, for instance, the model established via lumped parameter method showed the simple relation between system parameters and natural modes. The natural mode shapes only rely on the tip mass ratio and the natural frequencies are proportional to the natural frequency of a simple pendulum. This has briefly demonstrated the effect of gravitational acceleration on the natural vibrations of a tether pendulum.

3.2   A Half Degree of Freedom of Discrete Systems

The description of the dynamic configuration of a discrete system requires a finite number of independent coordinates. The number is an integer, named the *degree of freedom* or the *DoF*, for many discrete systems. Yet, the study in this section will show that an integer DoF may not be suitable to describe some discrete systems.

The section begins with an example of vibration system having an interval mass parameter. When the mass parameter approaches zero in the interval, one DoF in the vibration system degenerates and results in a non-holonomic constraint. On the basis of geometric mechanics, the section reveals that the current definition of DoF has a shortcoming with the effect of a non-holonomic constraint over-estimated. The section presents the influence of a non-holonomic constraint on the accessibility of the system motion in the displacement configuration space and in the state configuration space, respectively. Afterwards, the section introduces a new definition of DoF for discrete systems. The section gives several case studies to show the rationality of the new concept. These studies based on a recent paper of the author[5] well address Problem 2B proposed in Chap. 1.

---

[5] Hu H Y (2018). On the degree of freedom of a mechanical system. Chinese Journal of Theoretical and Applied Mechanics, 50(5): 1135–1144 (in Chinese).

3.2.1 Degeneration of a Degree of Freedom

Example 3.2.1 The vibration system in Fig. 3.6a consists of two lumped masses m_1 and m_2, a dashpot with viscous damping coeffiient c, two springs with stiffness coefficients k_1 and k_2, respectively. The system vibrates vertically under the excitation force $f_1(t)$ to lumped mass m_1. Study the direct frequency response function of the system at u_1 when lumped mass m_2 approaches zero.

The dynamic equations of the vibration system in Fig. 3.6a satisfy

$$\begin{cases} m_1\ddot{u}_1(t) + c[\dot{u}_1(t) - \dot{u}_2(t)] + k_1u_1(t) = f_1(t) \\ m_2\ddot{u}_2(t) + c[\dot{u}_2(t) - \dot{u}_1(t)] + k_2u_2(t) = 0 \end{cases} \tag{a}$$

Applying Fourier transform to both sides of Eq. (a) leads to

$$\begin{cases} (k_1 - m_1\omega^2 + ic\omega)U_1(\omega) - ic\,\omega U_2(\omega) = F_1(\omega) \\ -ic\omega U_1(\omega) + (k_2 - m_2\omega^2 + ic\omega)U_2(\omega) = 0 \end{cases} \tag{b}$$

There follows the direct frequency response function of m_1 to the excitation $f_1(t)$ as follows

$$H_{11}(\omega) \equiv \frac{U_1(\omega)}{F_1(\omega)} = \frac{k_2 - m_2\omega^2 + ic\omega}{(k_1 - m_1\omega^2 + ic\omega)(k_2 - m_2\omega^2 + ic\omega) + c^2\omega^2} \tag{c}$$

Let the system parameters be $m_1 = 10$ kg, $k_1 = k_2 = 40$ kN/m and $c = 50$ N·s/m. Figure 3.7 shows the evolution of $|H_{11}(f)|$, where $\omega = 2\pi f$, when $m_2 \in [0$ kg, 6 kg$]$. As shown in Fig. 3.7, $|H_{11}(f)|$ exhibits two resonant peaks for $m_2 = 6$ kg. The second resonant peak of $|H_{11}(f)|$ decreases and the second resonant frequency goes up rapidly when $m_2 \to 0^+$kg. Finally, $|H_{11}(f)|$ has one resonant peak only.

Once $m_2 = 0$ kg is set, the 2-DoF vibration system in Fig. 3.6a degenerates to the vibration system in Fig. 3.6b and the corresponding dynamic equations are

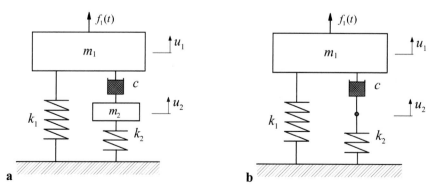

Fig. 3.6 The degeneration of a DoF. **a** a system of 2-DoFs; **b** a system with a degenerated DoF

Fig. 3.7 The evolution of $|H_{11}(f)|$, the amplitude of the direct frequency response function of m_1, with a variation of mass $m_2 \in [0\ \text{kg},\ 6\ \text{kg}]$

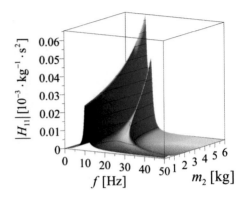

$$\begin{cases} m_1\ddot{u}_1(t) + c[\dot{u}_1(t) - \dot{u}_2(t)] + k_1u_1(t) = f_1(t) \\ c[\dot{u}_2(t) - \dot{u}_1(t)] + k_2u_2(t) = 0 \end{cases} \tag{d}$$

Now, the second differential equation of the second order in Eq. (a) degenerates to a differential equation of the first order, which is a velocity constraint in Eq. (d). As $|H_{11}(f)|$ exhibits a single resonant peak for $m_2 = 0$ kg, the vibration system in Fig. 3.6b behaves like a 1-DoF system, but one can not describe the motion of the connection point between the dashpot and the spring in serial by using u_1 only.

This example shows the degeneration of one DoF in a vibration system. That is, the vibration system in Fig. 3.6a is subject to the degeneration of one DoF and becomes the system in Fig. 3.6b when $m_2 \rightarrow 0^+$kg. To well understand the above degeneration, it is necessary to examine the current concept of DoF.

3.2.2 Conventional Concept of Degree of Freedom

In classical mechanics, the *degree of freedom* or abbreviated as DoF, of a discrete system is defined as the number of independent coordinates to describe the displacement configuration of the system, or as the dimension of displacement configuration space of the system. This definition has a clear picture of mechanics and has seen wide applications. For instance, IFToMM adopted this definition for the DoF of a mechanism or a kinematical chain.

From the nature of mechanics, the above definition of the DoF of a system focuses on the displacement configuration of the system only, falls into the category of the system kinematics, rather than the system dynamics. In analytical mechanics, which mainly deals with dynamic problems, the concept of the DoF of a system was extended to the number of independent contemporaneous variations (or independent virtual displacements)[6]. If a system has no constraints or has holonomic constraints only, the number of independent coordinates and the number of independent contemporaneous variations are the same. Thus, the definitions of the DoF of a system in analytical mechanics and classical mechanics are identical. When the system has any non-holonomic constraints, nevertheless, the velocity constraints impose some relations on the contemporaneous variations so that the number of independent contemporaneous variations decreases. In this case, the DoF of a system defined in analytical mechanics is less than the number of independent coordinates. This subsection analyzes the concept of DoF in analytical mechanics from the viewpoint of geometric mechanics.

We first establish a fixed frame of coordinates in three-dimensional Euclidian space $\mathfrak{R}^3$ and study the kinematics of a dynamic system composed of N free rigid bodies. According to geometry mechanics[7], the motion of the i-th rigid body falls into a six-dimensional Lee group $SE(3) \equiv SO(3) \otimes \mathfrak{R}^3$, which describes a translation described by vector $\boldsymbol{u}_i \in \mathfrak{R}^3$ and a rotation governed by an attitude matrix $\boldsymbol{R}(\boldsymbol{\theta}_i) \in SO(3)$, where the entries in $\boldsymbol{R}(\boldsymbol{\theta}_i)$ rely on the vector of Eulerian angles defined as $\boldsymbol{\theta}_i$ and satisfy the definition of group $SO(3)$, that is, $\boldsymbol{R}^{\mathrm{T}}(\boldsymbol{\theta}_i)\boldsymbol{R}(\boldsymbol{\theta}_i) = \boldsymbol{I}_3$ and $\det \boldsymbol{R}(\boldsymbol{\theta}_i) = 1$.

The above vector $\boldsymbol{u}_i$ and entries in $\boldsymbol{R}(\boldsymbol{\theta}_i)$ can be assembled into a vector of generalized displacements, i.e., $\boldsymbol{v} \in \mathbb{R}^n \equiv [SE(3)]^N$, where $n = 6N$ holds. If all rigid bodies become lumped masses without any rotations taken into account, there follows $\boldsymbol{v} \in \mathbb{R}^n = \mathfrak{R}^{3N} = (\mathfrak{R}^3)^N$. This section uses vector $\boldsymbol{v}$ to describe the motions of the unconstrained rigid bodies and lumped masses, and defines $\mathbb{R}^n$ as the configuration space of the system. As vector $\boldsymbol{v}$ has n independent entries, the contemporaneous variation $\delta\boldsymbol{v}$ also has n independent entries. Thus, the DoF of the system is n and the same as the dimensions of $\mathbb{R}^n$.

Remark 3.2.1 From the property of Lee group $SE(3)$, $\mathbb{R}^n$ is not a Euclidian space if the system undergoes any rotations. The analysis of this section focuses on the accessible manifold governed by the system constraints in $\mathbb{R}^n$ and only checks the local behaviors of any constraint. This is the standard way of studying a dynamic system with constraints[8], especially when the constraints are locally defined in the configuration space $\mathbb{R}^n$.

Now, let r holonomic constraints and s non-holonomic constraints be imposed on the above system and denote the system as $S_{r,s}^n$. In particular, according to this notation, $S_{0,0}^n$ is an unconstrained system, $S_{r,0}^n$ is a system with r holonomic constraints, and $S_{0,s}^n$ is a system with s non-holonomic constraints. In practice, the system $S_{r,0}^n$ is called a *holonomic system* while both systems $S_{r,s}^n$ and $S_{0,s}^n$ are named *non-holonomic systems*.

---

[6] Mei F X (1991). Advanced Analytical Mechanics. Beijing: Press of Beijing Institute of Technology, 25–28 (in Chinese).

[7] Lee T, Leok M, McClamroch N H (2018). Global Formulations of Lagrangian and Hamiltonian Dynamics on Manifolds. New York: Springer-Verlag, 1–388.

[8] Chen B (2012). Analytical Dynamics. 2nd Edition. Beijing: Peking University Press, 37–45 (in Chinese).

For simplicity, assume that the r *holonomic constraints* are bilateral, scleronomic and linearly independent, and satisfy

$$\boldsymbol{\varphi}(\boldsymbol{v}) = \boldsymbol{0}, \quad \boldsymbol{\varphi} : \mathbb{R}^n \mapsto \mathbb{R}^r \tag{3.2.1}$$

Now, the number of independent entries in displacement vector $\boldsymbol{v}$ becomes $n - r$. So is the number of independent entries in $\delta\boldsymbol{v}$. As such, the DoF of system $S_{r,0}^n$ is defined as $D(S_{r,0}^n) = n - r$ in analytical mechanics.

Similarly, let the s *non-holonomic constraints* be bilateral, scleronomic and linear independent so that they yield

$$\begin{cases} \boldsymbol{A}(\boldsymbol{v})\dot{\boldsymbol{v}} + \boldsymbol{a}_0(\boldsymbol{v}) = \boldsymbol{0} \\ \boldsymbol{A} : \mathbb{R}^n \mapsto \mathbb{R}^{s \times n}, \quad \boldsymbol{a}_0 : \mathbb{R}^n \mapsto \mathbb{R}^s \end{cases} \tag{3.2.2}$$

Furthermore, assume that any linear combination of s non-holonomic constraints results in no integrable constraints and so does any combination of the above non-holonomic and holonomic constraints. In view of Eq. (3.2.2), let the contemporaneous variation $\delta\boldsymbol{v}$ satisfy the following *Chetaev condition*[9]

$$\boldsymbol{A}(\boldsymbol{v})\delta\boldsymbol{v} = \boldsymbol{0} \tag{3.2.3}$$

Now, the number of independent entries in the contemporaneous variation $\delta\boldsymbol{v}$ becomes $n - r - s$. From analytical mechanics, the DoF of system $S_{r,s}^n$ reduces to $D(S_{r,s}^n) = n - r - s$. Because the non-holonomic constraints in Eq. (3.2.2) are not integrable, it is necessary to use $n - r$ independent coordinates to describe the configuration of system $S_{r,s}^n$.

It is worthy to point out that the non-holonomic constraints in Eq. (3.2.3) make a part of entries in the contemporaneous variation $\delta\boldsymbol{v}$ linearly dependent, but they do not change the relation of the entries in the displacement vector $\boldsymbol{v}$. That is, holonomic constraints and non-holonomic constraints have different influences on the DoF of a system.

3.2.3 Degree of Freedom Based on Accessible Manifolds

3.2.3.1 Accessible Displacement Manifold

When the motion of system $S_{r,0}^n$ satisfies r holonomic constraints in Eq. (3.2.1), it can access to a super-surface Φ^{n-r} or a sheet of Φ^{n-r} in the configuration space $\mathbb{R}^n$, but rather than the whole $\mathbb{R}^n$. As the holonomic constraints are linearly independent, it is

---

[9] Guo Y X, Mei F X (1998). Integrability for Pfaffian constrained systems: a connection theory. Acta Mechanica Sinica, 14(1): 85–91.

easy to prove that the dimension of super-surface Φ^{n-r} is $n - r$[10]. From differential geometry[11], the smooth super-surface Φ^{n-r} is a differential manifold in $\mathbb{R}^n$. Thus, let $Q \equiv \Phi^{n-r}$ be defined as the *accessible displacement manifold* with dimension $\dim(Q) = n - r$.

As all linear combinations of non-holonomic constraints in Eq. (3.2.2) are not integrable, they can not generate any super-surface in $\mathbb{R}^n$ to limit the motion of system $S_{r,s}^n$. Thus, the motion of system $S_{r,s}^n$ can access to any point on the accessible displacement manifold Q, but the corresponding velocity must yield Eq. (3.2.2). That is, the contemporaneous variation δv yields the s non-holonomic constraints in Eq. (3.2.3). As a result, when system $S_{r,0}^n$ with r holonomic constraints are subject to s non-holonomic constraints, the accessible displacement manifold Q of new system $S_{r,s}^n$ remains unchanged.

3.2.3.2 Accessible State Manifold

According to Newton's principle of determination, the dynamic evolution of a system depends not only on the initial displacements of the system, but also the initial velocities of the system. That is, the system dynamics is the dynamic evolution of the system state by nature. The non-holonomic constraints of a system limit the system velocities and affect the dynamic evolution of the system. Hence, it is necessary to study the role of non-holonomic constraints in the *state space* $\mathbb{R}^{2n}$.

The theory of differential geometry enables one to establish a local frame of coordinates on the differential manifold Q with dimension $n - r$, and denote an arbitrary point q in Q as an $(n-r)$-dimensional vector of generalized displacements $q \in Q$, and generate the tangential space TQ_q with dimension $n - r$ for $q \in Q$. From differential geometry, define the union of all tangential spaces TQ_q for $q \in Q$ as a *tangential bundle* and denote it as

$$TQ \equiv \bigcup_{q \in Q} TQ_q \qquad (3.2.4)$$

The tangential bundle is a differential manifold in state space $\mathbb{R}^{2n}$ and its dimension is $\dim(TQ) = 2(n - r)$. Here, the differential bundle is defined as the *accessible state manifold* of holonomic system $S_{r,0}^n$.

To clearly understand the geometric picture of the above terms, consider a super-surface Φ^{n-r} determined by the holonomic constraints in the configuration space $\mathbb{R}^n$ of system $S_{r,0}^n$. When we turn to the state space $\mathbb{R}^{2n}$ of the system, the super-surface Φ^{n-r} automatically becomes a super-surface Φ^{2n-r} in $\mathbb{R}^{2n}$. As Eq. (3.2.1) has nothing to do with velocity vector $\dot{v}$, Φ^{2n-r} is a super cylindrical surface along velocity vector $\dot{v}$. Taking the partial derivative of Eq. (3.2.1) with respect to time leads to

---

[10] Chen B (2012). Analytical Dynamics. 2nd Edition. Beijing: Peking University Press, 37–44 (in Chinese).

[11] Marsden J E, Ratiu T S (1999). Introduction to Mechanics and Symmetry. 2nd Edition. New York: Springer-Verlag, 121–180.

$$\boldsymbol{\varphi}_{v^\mathrm{T}}(\boldsymbol{v})\dot{\boldsymbol{v}} = 0 \qquad\qquad (3.2.5)$$

where $\boldsymbol{\varphi}_{v^\mathrm{T}}:\mathbb{R}^n \mapsto \mathbb{R}^{r\times n}$ is the Jacobian of column vector $\boldsymbol{\varphi}(\boldsymbol{v})$ with respect to row vector $\boldsymbol{v}^\mathrm{T}$. Equation (3.2.5) linearly depends on velocity vector $\dot{\boldsymbol{v}}$, and can be referred to as a super ruled surface Ψ^{2n-r} in the state space $\mathbb{R}^{2n}$. Thus, the two super surfaces Φ^{2n-r} and Ψ^{2n-r} must intersect to each other. As such, the motion of holonomic system $S_{r,0}^n$ is constrained at the intersection set of two super surfaces Φ^{2n-r} and Ψ^{2n-r} in the state space $\mathbb{R}^{2n}$. This set is the tangential bundle TQ of dimension $2n - 2r$.

To be more specific, check the role of a single constrain $\Phi^{2-1} \equiv v_1^2 + v_2^2 - 1 = 0$ in the configuration space $\mathbb{R}^2$ of a system. In this case, Φ^{2-1} describes one-dimensional accessible displacement manifold Q. In the state space $\mathbb{R}^4$ of the system, let the super cylindrical surface Φ^{4-1} be projected to an arbitrary plane $\dot{v}_2 = \mathrm{const}$ and get the circular cylindrical surface shown in Fig. 3.8, where Eq. (3.2.5) leads to $\Psi^{4-1} \equiv 2(v_1\dot{v}_1 + v_2\dot{v}_2) = 0$. If $\dot{v}_2 \neq 0$ holds, Ψ^{4-1} can be recast as $v_2 = -v_1\dot{v}_1/\dot{v}_2$, which is a hyperbolic paraboloid in the frame of coordinates $(v_1, v_2, \dot{v}_1)$. Figure 3.8 presents a pair of hyperbolic paraboloids for $\dot{v}_2 = \pm 1$. For different values of $\dot{v}_2$, there follows a cluster of hyperbolic paraboloids, the intersections of which with the circular cylindrical surface are the tangential bundle TQ in Fig. 3.8.

From differential geometry, when Q is a unit circle in $\mathbb{R}^2$, the trivial case of its tangential bundle TQ is the circular cylindrical surface in $\mathbb{R}^3$ as shown in Fig. 3.8, while the non-trivial case is a twisting Möbius belt in $\mathbb{R}^3$. The second case does not have a fixed direction and is not taken into consideration here.

To study the accessible state configuration of non-holonomic system $S_{r,s}^n$, we establish a local frame of coordinates in the tangential bundle TQ of holonomic system $S_{r,0}^n$ and define a vector of $[\,\boldsymbol{q}^\mathrm{T}\ \dot{\boldsymbol{q}}^\mathrm{T}\,]^\mathrm{T} \in TQ$ for the generalized state of system $S_{r,0}^n$. Let $\boldsymbol{v} = \tilde{\boldsymbol{v}}(\boldsymbol{q})$ be the displacement vector of system $S_{0,0}^n$ where $\tilde{\boldsymbol{v}} : Q \mapsto \mathbb{R}^n$. Substitution of $\boldsymbol{v} = \tilde{\boldsymbol{v}}(\boldsymbol{q})$ into Eqs. (3.2.1) and (3.2.2) leads to

Fig. 3.8 The accessible displacement manifold in the state space and its tangential bundle

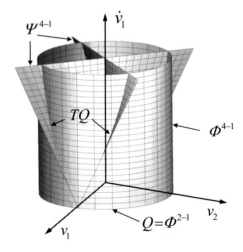

$$\varphi(v) = \varphi\big[\tilde{v}(q)\big] \equiv \tilde{\varphi}(q) = 0 \qquad (3.2.6)$$

$$\begin{cases} A\big[\tilde{v}(q)\big]\tilde{v}_{q^\top}(q)\dot{q} + a_0\big[\tilde{v}(q)\big] \equiv \tilde{A}(q)\dot{q} + \tilde{a}_0(q) = 0 \\ \tilde{A} : Q \mapsto \mathbb{R}^{s\times(n-r)}, \quad \tilde{a}_0 : Q \mapsto \mathbb{R}^s \end{cases} \qquad (3.2.7)$$

Equation (3.2.6) holds true for any $q \in Q$ such that the state of non-holonomic system $S_{r,s}^n$ in the tangential state bundle TQ is only constrained by the non-holonomic constraints in Eq. (3.2.7) and the number of independent entries in vector $\dot{q}$ is reduced to $n - r - s$. Hence, the accessible state manifold of non-holonomic system $S_{r,s}^n$ can be denoted as a sub-manifold Ω with dimension $\dim(\Omega) = 2n - 2r - s$ in the tangential state bundle TQ.

Remark 3.2.2 In view of Gibbs–Appell equations in analytical mechanics[12], it is feasible to use $n - r - s$ quasi-velocities and $n - r$ generalized displacements to describe the dynamic states of non-holonomic system $S_{r,s}^n$. This is equivalent to a proof that the dimensions of the accessible state manifold of non-holonomic system $S_{r,s}^n$ is $2n - 2r - s$.

Remark 3.2.3 For a dynamic system with rhenomic constraints, the accessible state manifold of the system has similar property.

3.2.3.3 New Definition of Degree of Freedom

The concept of the DoF of a discrete system stemmed from the number of independent coordinates for system kinematics. As a consequence, the original definition of DoF depends on the system configuration and agrees with the dimension of the accessible displacement configuration space of the system. It is natural, thus, to believe that the dimension of the accessible state space of a system is the doubled DoFs of the system in dynamics.

For a holonomic system $S_{r,0}^n$, the dimension of the accessible state manifold TQ is always double to the dimension of the accessible displacement manifold Q. For a non-holonomic system $S_{r,s}^n$, however, the dimension of the accessible state manifold Ω is $2n - 2r - s$, rather than the doubled DoFs, i.e., $2(n - r - s)$, for the system in analytical mechanics. The subsequent analysis shows that the DoF defined in analytical mechanics will impose over-constraints on the DoF.

Now, we propose a *new definition* for the degree of freedom of a system. That is, the DoF of a system is a half dimension of the accessible state manifold of the system. Consider, as an example, the non-holonomic system $S_{r,s}^n$ with r holonomic constraints and s non-holonomic constraints, the DoF of the system is

$$\tilde{D}(S_{r,s}^n) \equiv \frac{\dim(\Omega)}{2} = n - r - \frac{s}{2} \qquad (3.2.8)$$

---

[12] Ardema M D (2005). Analytical Dynamics: Theory and Applications. New York: Kluwer Academic Publisher, 245–259.

It is obvious to see that the new definition coincides with the conventional definition of the DoF for an unconstrained system $S_{0,0}^n$ and a holonomic system $S_{r,0}^n$, but gives a larger value to the DoF of a non-holonomic system $S_{r,s}^n$ than $D(S_{r,s}^n) = n - r - s$ defined in analytical mechanics.

3.2.4 Demonstrative Systems with a Non-Holonomic Constraint

In view of the new definition of the DoF of a system, a half DoF appears if the system has an odd number of non-holonomic constraints. The following two examples will demonstrate how to understand the mechanics behind such a phenomenon.

3.2.4.1 Degeneration of a DoF of a Vibration System

Example 3.2.2 Following the viewpoint in Sect. 3.2.3, discuss the degeneration of a DoF in the vibration system from Fig. 3.6.

In Fig. 3.6a, the two lumped masses of the system only move vertically and require two independent coordinates u_1 and u_2 to describe the system. Hence, the dimension of configuration space of the system is $\mathbb{R}^2 = (\mathfrak{R}^1)^2$. Let the system be denoted as $S_{0,0}^2$ and the DoF of the system be $D(S_{0,0}^2) = 2$. From Eq. (a) in Example 3.2.1, the dynamic equations of the system are

$$\begin{cases} m_1 \ddot{u}_1(t) + c[\dot{u}_1(t) - \dot{u}_2(t)] + k_1 u_1(t) = f_1(t) \\ m_2 \ddot{u}_2(t) + c[\dot{u}_2(t) - \dot{u}_1(t)] + k_2 u_2(t) = 0 \end{cases} \tag{a}$$

Let the lumped mass $m_2 \to 0$, we get the system in Fig. 3.6b and the dynamic equations of the degenerated system satisfy

$$\begin{cases} m_1 \ddot{u}_1(t) + c[\dot{u}_1(t) - \dot{u}_2(t)] + k_1 u_1(t) = f_1(t) \\ c[\dot{u}_2(t) - \dot{u}_1(t)] + k_2 u_2(t) = 0 \end{cases} \tag{b}$$

where the second equation in Eq. (b) is a linear scleronomic velocity constraint and is not integrable. That is, it is a non-holonomic constraint. As such, the degenerated system is a non-holonomic system denote by $S_{0,1}^2$.

The accessible displacement configuration of the degenerated system $S_{0,1}^2$ has an accessible configuration manifold $Q = \mathbb{R}^2$ and a tangential bundle $TQ = \mathbb{R}^4$. Recast Eq. (b) as a set of differential equations with a constraint in tangential bundle $TQ = \mathbb{R}^4$, namely,

$$\begin{cases} \dot{q}_1(t) = q_3(t) \\ \dot{q}_2(t) = q_4(t) \\ \dot{q}_3(t) = [-k_1q_1(t) - cq_3(t) + cq_4(t) + f_1(t)]/m_1 \end{cases} \tag{c}$$

$$c[q_4(t) - q_3(t)] + k_2q_2(t) = 0 \tag{d}$$

Now, the non-holonomic constraint in Eq. (d) gives the accessible state manifold $\Omega \subset TQ$ for system $S_{0,1}^2$ and leads to $\dim(\Omega) = 3$.

Solving Eq. (d) gives $q_4(t) = [cq_3(t) - k_2q_2(t)]/c$. Substituting this result into Eq. (c), we have the dynamic equations of system $S_{0,1}^2$ in Ω as follows

$$\begin{cases} \dot{q}_1(t) = q_3(t) \\ \dot{q}_2(t) = [cq_3(t) - k_2q_2(t)]/c \\ \dot{q}_3(t) = [-k_1q_1(t) - k_2q_2(t) + f_1(t)]/m_1 \end{cases} \tag{e}$$

The free vibration of system $S_{0,1}^2$ yields the following characteristic equation of the third order

$$\det \begin{bmatrix} -\lambda & 0 & 1 \\ 0 & -k_2/c - \lambda & 1 \\ -k_1/m_1 & -k_2/m_1 & -\lambda \end{bmatrix}$$

$$= \frac{1}{m_1c}[m_1c\lambda^3 + m_1k_2\lambda^2 + c(k_1 + k_2)\lambda + k_1k_2] = 0 \tag{f}$$

According to the Routh-Hurwitz criterion, the real parts of three roots of Eq. (f) are negative since all coefficients in Eq. (f) are positive. Usually, two roots are a pair of conjugate complex roots $\lambda_{1,2} = -\gamma \pm i\omega$, and the third root is a real one $\lambda_3 < 0$, where $\gamma > 0$ and $\omega > 0$. Thus, the free vibration of system $S_{0,1}^2$ subject to an initial disturbance is the superposition of two kinds of motions. One is a decaying vibration with exponential function $\exp(-\gamma t)$ as the envelop and the other is a decaying motion described by $\exp(-|\lambda_3|t)$.

Now, we turn to the other viewpoint to check Eq. (b). Eliminating $\dot{u}_2(t)$ and $\ddot{u}_2(t)$ from both the time derivative of Eq. (b) and the first equation in Eq. (b), we get a third-order differential equation in terms of displacement $u_1(t)$, namely,

$$m_1c\frac{d^3u_1(t)}{dt^3} + m_1k_2\frac{d^2u_1(t)}{dt^2} + c(k_1 + k_2)\frac{du_1(t)}{dt} + k_1k_2u_1(t)$$

$$= c\frac{df_1(t)}{dt} + k_2f_1(t) \tag{g}$$

Equation (g) is equivalent to the dynamic equations in Eq. (e) for system $S_{0,1}^2$ in Ω. From Eq. (g), it is easier to derive the characteristic equation of the free vibration of system $S_{0,1}^2$ in Eq. (f).

In what follows, more detailed discussions are made about the DoF of system $S_{0,1}^2$.

First, the non-holonomic constraint in Eq. (b) gives the relation of two contemporaneous variations of the system, that is, $c[\delta u_2(t) - \delta u_1(t)] = 0$. As the DoF of the system defined in analytical mechanics is $D(S_{0,1}^2) = 2 - 0 - 1 = 1$, system $S_{0,1}^2$ has only one DoF. Yet, the motion of system $S_{0,1}^2$ is different from that of a 1-DoF system and requires two independent coordinates for description. For example, one requires three initial conditions $q_1(0) = q_{10}$, $q_2(0) = q_{20}$ and $q_3(0) = q_{30}$ to determine the free vibration of system $S_{0,1}^2$ in Ω from Eq. (e). Similarly, one needs three initial conditions $u_1(0) = u_{10}$, $\dot{u}_1(0) = \dot{u}_{10}$ and $\ddot{u}_1(0) = \ddot{u}_{10}$ so as to complete the same task from Eq. (g). Furthermore, the free vibration of system $S_{0,1}^2$ to an initial disturbance includes not only decaying vibration of a damped 1-DoF system, but also a decaying motion. As such, the degenerated system $S_{0,1}^2$ is really different from any 1-DoF systems.

Second, the degenerated system $S_{0,1}^2$ differs from a normal 2-DoF system. For the original 2-DoF system $S_{0,0}^2$, it has two pairs of complex conjugate eigenvalues when lumped masses m_2 and m_1 are in the same quantity levels, and any free vibration of the system to a disturbance consists of two decaying vibrations. When $m_2 \to 0$, however, the system has an eigenvalue $\lambda_4 < 0$ with very large norm while the other three eigenvalues are $\lambda_{1,2} = -\gamma \pm i\omega$ and $\lambda_3 < 0$. In this case, the free vibration of the system to a disturbance is the superposition of one decaying vibration and two non-oscillatory motions with different decaying rates. Thus, the free vibrations of systems $S_{0,0}^2$ and $S_{0,1}^2$ are quite different. In addition, the determinations of the free vibrations of systems $S_{0,0}^2$ and $S_{0,1}^2$ require four and three initial conditions, respectively. That is, the free vibration of system $S_{0,1}^2$ has nothing to do with initial condition $\dot{u}_2(0)$. Once $\dot{u}_1(0)$ and $u_2(0)$ are available, the non-holonomic constraint in Eq. (b) leads to $\dot{u}_2(0) = \dot{u}_1(0) - (k_2/c)u_2(0)$.

As discussed above, the dynamics of system $S_{0,1}^2$ is between those of a 1-DoF system and a 2-DoF system. Hence, it is appropriate to define the DoF of system $S_{0,1}^2$ as $\tilde{D}(S_{0,1}^2) = 2 - 0 - 1/2 = 1.5$ from Eq. (3.2.8). The half DoF comes from the contribution of a non-holonomic constraint since it requires that the damping force of dashpot c and the elastic force of spring k_2 are always identical. From the viewpoint of dynamics, a 1.5-DoF system exhibits more complicated behavior than a 1-DoF system, but simpler behavior than a 2-DoF system.

Remark 3.2.4 The dynamic analysis of a 1.5-DoF system began with the integrability of a Hamiltonian system studied by S. L. Ziglin, a mathematician in Soviet Union, in 1981[13] and has been going on sine then. The study here, however, comes from a 1.5-DoF vibration system and shows a clear picture of mechanics. As shown in Fig. 3.6b, spring k_1, spring k_2 and dashpot c form a standard linear viscoelastic model[14], or called *Kelvin's viscoelastic model*, in the theory of viscoelasticity.

---

[13] Ziglin S L (1981). Self-intersection of the complex separatrices and the nonexistence of the integrals in the Hamiltonian-systems with one-and-half degrees of freedom. PMM Journal of Applied Mathematics and Mechanics, 45(3): 411–413.

[14] Fung Y C (1965). Foundations of Solid Mechanics. Englewood Cliffs: Prentice-Hall, 20–29.

Section 3.3 will present how to use Kelvin's viscoelastic model to describe the system with structural damping and remove the non-causality of time response of the system due to frequency-invariant damping. In recent years, C. Q. Min and his co-authors in Germany used a mechanical system model of 1.5-DoFs to study the vibration control[15].

Remark 3.2.5 It is also possible to construct the vibration system in Fig. 3.6b from a simple system with a lumped mass m_1 and a spring k_1, and add Maxwell's fluidic component with a dashpot c and a spring k_2 in serial. Maxwell's fluidic component yields the second equation in Eq. (b), that is, the non-holonomic constraint equation. This ordinary equation of the first order describes the dynamic equation of a half DoF. The theory of viscoelasticity usually deals with the half DoF as an *intrinsic variable*. In the theory of viscoelasticity, the system $S_{0,1}^2$ with Maxwell's fluidic component is said to have 1-DoF, but an extra note should be added. That is, the free vibration of the system includes not only a damped vibration, but also a decaying motion, or named a *creep mode*.

Remark 3.2.6 In solid mechanics, there are numerous problems involving intrinsic variables, but only a part of them fall into the category of non-holonomic constraints for system velocities. For instance, the system in Fig. 3.6b would describe a nonlinear vibration isolation system with memory or a more arbitrary elasto-plastic 1-DoF system if Maxwell's fluidic component was replaced with an ideal plastic compo-nent[16] or Bouc-Wen's plastic component[17]. Yet, the intrinsic variables in those two systems were plastic forces and did not have a meaning of velocity.

Remark 3.2.7 When the condition of $0 < m_2 << m_1$ holds, it is also possible to study the 2-DoF system $S_{0,0}^2$ directly, but the initial value problem of a set of differential equations for system $S_{0,0}^2$ is usually ill-conditioned in computation. For example, the Rung-Kutta integrator in software Maple or MATLAB fails to compute the free vibration of system $S_{0,0}^2$ when $m_2/m_1 < 10^{-5}$. As such, it is necessary to simplify the 2-DoF system $S_{0,0}^2$ with almost singular mass matrix to a 1.5-DoF system $S_{0,1}^2$ so as to avoid computational problems.

3.2.4.2 A Sleigh Moving on an Inclined Plane

As the concept of DoF is essential in mechanics, it is reasonable to examine whether the new definition of DoF in Eq. (3.2.8) is universal even though this book focuses

[15] Min C Q, Dahlmann M, Sattel T (2017). A concept for semi-active vibration control with a serial-stiffness-switch system. Journal of Sound and Vibration, 405: 234–250.

[16] Hu H Y, Li Y F (1989). Parametric identification of nonlinear vibration isolators with memory. Journal of Vibration Engineering, 2(2): 17–27 (in Chinese).

[17] Wen Y K (1976). Method for random vibration of hysteretic systems. ASCE Journal of Engineering Mechanics, 12(1): 249–263.

on linear vibrations. For this purpose, we discuss a nonlinear dynamic system, which is the most popular example in analytical mechanics.

Example 3.2.3 A sleigh is moving on an inclined ice plane $\mathfrak{R}^2$ due to the force component $mg \sin \alpha$ of gravity as shown in Fig. 3.9, where the sleigh is simplified as a rigid rod with length L and mass m. Study the rationality of the DoF of the sleigh.

In the frame of coordinates in Fig. 3.9, we use displacement vector $[u_1(t) \ u_2(t)]^T$ to describe the motion of the mass center of the rod and use orientation angle $\theta(t)$ to describe the counter-clockwise rotation of the rod. Thus, the configuration space of the system is $\mathbb{R}^3 = SO(1) \otimes \mathfrak{R}^2$.

Because of the constraint of the ice plane, the velocity vector of the rod denoted by $\dot{u}$ is always parallel to the rod such that the components of vector $\dot{u}$ satisfy the following non-holonomic constraint equation

$$\dot{u}_1(t) \sin[\theta(t)] - \dot{u}_2(t) \cos[\theta(t)] = 0 \tag{a}$$

Hence, the sleigh system is a non-holonomic system $S_{0,1}^3$.

For brevity, we use Gibbs–Appell equations in analytical mechanics to establish the dynamic equations of the sleigh with the non-holonomic constraint eliminated and skip the detailed analysis of tangential bundle and accessible state manifold. For this purpose, define two quasi-velocities as follows

$$\sigma_1(t) = \dot{u}_1(t)/\cos[\theta(t)], \quad \sigma_2(t) = \dot{\theta}(t) \tag{b}$$

From Eqs. (a) and (b), all velocity components of the sleigh are expressed in terms of two quasi-velocities as follows

$$\dot{u}_1(t) = \sigma_1(t) \cos[\theta(t)], \quad \dot{u}_2(t) = \sigma_1(t) \sin[\theta(t)], \quad \dot{\theta}(t) = \sigma_2(t) \tag{c}$$

Thus, the virtual power done by the gravity in terms of quasi-velocities reads

$$\delta P = (mg \sin \alpha)\delta \dot{u}_2(t) = mg \sin \alpha \sin[\theta(t)]\delta \sigma_1(t) \tag{d}$$

There follow the generalized forces associated with quasi-velocities

Fig. 3.9 A sleigh moving on an inclined plane under gravity

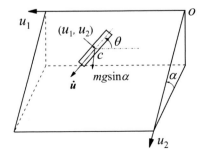

$$g_{\sigma 1}(t) = mg \sin \alpha \sin[\theta(t)], \quad g_{\sigma 2}(t) = 0 \tag{e}$$

Differentiating Eq. (c) with respect to time leads to the vector of accelerations of the mass center and the vector of angular accelerations as follows

$$\boldsymbol{a}_c(t) = \begin{bmatrix} \ddot{u}_1(t) \\ \ddot{u}_2(t) \end{bmatrix} = \begin{bmatrix} \dot{\sigma}_1(t)\cos[\theta(t)] - \sigma_1(t)\sigma_2(t)\sin[\theta(t)] \\ \dot{\sigma}_1(t)\sin[\theta(t)] + \sigma_1(t)\sigma_2(t)\cos[\theta(t)] \end{bmatrix},$$

$$\dot{\boldsymbol{\omega}}(t) = \begin{bmatrix} 0 \\ 0 \\ \ddot{\theta}(t) \end{bmatrix} = \begin{bmatrix} 0 \\ 0 \\ \dot{\sigma}_2(t) \end{bmatrix} \tag{f}$$

Substitution Eq. (f) into the *Gibbs function*[18] gives

$$G \equiv \frac{m}{2}\boldsymbol{a}_c^{\mathrm{T}}(t)\boldsymbol{a}_c(t) + \frac{J}{2}\dot{\boldsymbol{\omega}}^{\mathrm{T}}\dot{\boldsymbol{\omega}} = \frac{m}{2}\dot{\sigma}_1^2(t) + \frac{mL^2}{24}\dot{\sigma}_2^2(t) \tag{g}$$

Recalling analytical mechanics leads to Gibbs–Appell equations as follows

$$\frac{\partial G}{\partial \dot{\sigma}_i} = g_{\sigma_i}, \quad i = 1, 2 \tag{h}$$

Substituting Eqs. (e) and (g) into Eq. (h) and collecting Eq. (c) together, we derive the initial value problem of the dynamic equations of the sleigh as follows

$$\begin{cases} \dot{\sigma}_1(t) = g \sin \alpha \sin[\theta(t)], \quad \dot{\sigma}_2(t) = 0 \\ \dot{u}_1(t) = \sigma_1(t)\cos[\theta(t)], \quad \dot{u}_2(t) = \sigma_1(t)\sin[\theta(t)], \quad \dot{\theta}(t) = \sigma_2(t) \\ \sigma_1(0) = \dot{u}_{10}/\cos\theta_0, \quad \sigma_2(0) = \omega_0 \\ u_1(0) = u_{10}, \quad u_2(0) = u_{20}, \quad \theta(0) = \theta_0 \end{cases} \tag{i}$$

The differential equations in Eq. (i) have five unknown functions, but are decoupled. It is easy to integrate these differential equations in the following order: $\sigma_2(t) \to \theta(t) \to \sigma_1(t) \to [u_1(t), \ u_2(t)]^{\mathrm{T}}$. From the first initial condition in Eq. (i), the initial orientation angle should be $\theta_0 \neq \pm\pi/2$. As the initial angular velocity ω_0 plays an important role in the sleigh dynamics, the solution procedure of Eq. (i) is classified as two cases.

If the initial angular velocity $\omega_0 \neq 0$ holds, integration of Eq. (i) leads to the motion of the sleigh as follows

---

[18] Ardema M D (2005). *Analytical Dynamics: Theory and Applications*. New York: Kluwer Academic Publisher, 245–259.

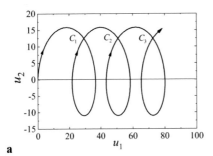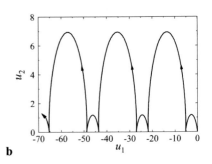

a   b

Fig. 3.10 Two trajectories of the mass center of the sleigh under different initial conditions. **a** $\alpha = \pi/4$, $u_{10} = 0$, $u_{20} = 0$, $\dot{u}_{10} = 1$, $\theta_0 = \pi/2.1$, $\omega_0 = -1$; **b** $\alpha = \pi/4$, $u_{10} = 0$, $u_{20} = 0$, $\dot{u}_{10} = 0$, $\theta_0 = 2$, $\omega_0 = 1$

$$
\begin{cases}
\begin{aligned}
u_1(t) &= \frac{g\sin\alpha}{2\omega_0^2}[\sin\theta_0\cos\theta_0 - \sin(\omega_0 t + \theta_0)\cos(\omega_0 t + \theta_0) - \omega_0 t] \\
&\quad + \left(\frac{\dot{u}_{10}}{\omega_0\cos\theta_0} + \frac{g\sin\alpha}{\omega_0^2}\cos\theta_0\right)[\sin(\omega_0 t + \theta_0) - \sin\theta_0] + u_{10} \\
u_2(t) &= \frac{g\sin\alpha}{2\omega_0^2}[\cos^2(\omega_0 t + \theta_0) - \cos^2\theta_0] \\
&\quad + \left(\frac{\dot{u}_{10}}{\omega_0\cos\theta_0} + \frac{g\sin\alpha}{\omega_0^2}\cos\theta_0\right)[\cos\theta_0 - \cos(\omega_0 t + \theta_0)] + u_{20} \\
\theta(t) &= \omega_0 t + \theta_0
\end{aligned}
\end{cases}
\tag{j}
$$

In this case, the trajectory of the mass center of the sleigh is the superposition of a rectilinear motion in u_1 direction and a periodic motion. Figure 3.10 presents two typical trajectories of the mass center of the sleigh, showing the energy conservation. The sleigh is moving in direction $u_1 \geq 0$ or $u_1 \leq 0$ when $\omega_0 < 0$ or $\omega_0 > 0$ holds, respectively.

When the initial angular velocity is $\omega_0 = 0$, the orientation angle of the sleigh keeps θ_0 and the sleigh is moving in a straight-line. Solving Eq. (i) leads to

$$
\begin{cases}
u_1(t) = \dfrac{g\sin\alpha}{2}(\sin\theta_0\cos\theta_0)t^2 + \dot{u}_{10}t + u_{10} \\[2mm]
u_2(t) = \dfrac{g\sin\alpha}{2}(\sin\theta_0\sin\theta_0)t^2 + \dfrac{\dot{u}_{10}}{\cos\theta_0}t + u_{20} \\[2mm]
\theta(t) = \theta_0
\end{cases}
\tag{k}
$$

In view of the DoF defined in analytical mechanics, the DoF of the sleigh is $D(S_{0,1}^3) = 3 - 1 = 2$. Following the concept of 2-DoFs, one can understand the rectilinear motion of the sleigh for $\omega_0 = 0$, but can hardly explain the curvilinear motion in Fig. 3.10, especially the case when the trajectories have self intersections in

Fig. 3.10a, where the mass center of the sleigh passes through points $C_1, C_2, C_3, \ldots$, but the orientation angles of the trajectory are different.

According to the DoF defined in Eq. (3.2.8), the DoF of the sleigh is $\tilde{D}(S_{0,1}^3) = 3 - 1/2 = 2.5$. In this example, the non-holonomic constraint equation in Eq. (a) does not change the accessible state configuration and keeps the motion of mass center independent of the orientation angle, but imposes a relation on the velocity direction of the mass center and the orientation angle. That is, the DoF of system $S_{0,1}^3$ is between a 2-DoF system $S_{1,0}^3$ and a 3-DoF of system $S_{0,0}^3$. Hence, it is appropriate to define 2.5 DoFs for the sleigh system $S_{0,1}^3$.

3.2.5 Demonstrative Systems with Two Non-holonomic Constraints

Having the idea of a half DoF accepted, readers may question whether two non-holonomic constraints make the reduction of one DoF for a system. This subsection focuses on this issue.

As analyzed for the accessible state manifold of system $S_{r,s}^n$ in Sect. 3.2.3, when $s = 2$, the dimension of accessible state manifold Ω is $2n - 2r - 2$ and the DoF of system $S_{r,2}^n$ is $\tilde{D}(S_{r,s}^n) = n - r - 1$. The issue looks reasonable in form, yet the further analysis shows some complexities.

As a matter of fact, the accessible velocity manifold does not exist independently, but depends on the accessible displacement manifold. Consider, as an example, a holonomic system $S_{r+1,0}^n$. It has the accessible displacement manifold Q of dimension $n - r - 1$ and the tangential bundle TQ of dimension $2n - 2r - 2$, the accessible velocity manifold is the complementary set of Q in TQ and has dimension $n - r - 1$. In addition, we can construct cotangential bundle T^*Q of dimension $2(n - r - 1)$, and study the Hamiltonian dynamics of the system. For a system $S_{r,2}^n$ with 2 non-holonomic constraints, the dimensions of accessible displacement manifold and accessible velocity manifold are $n - r$ and $n - r - 2$, respectively although $\tilde{D}(S_{r,2}^n) = n - r - 1$ is an integer. Therefore, system $S_{r,2}^n$ and system $S_{r+1,0}^n$ may not have equivalent dynamics. The following two examples will demonstrate this assertion.

3.2.5.1 A Hydro-Elastic Vibration Isolation System

Example 3.2.4 Section 1.2.2 presented the dynamic model of a hydro-elastic vibration isolation system for vehicle engines as shown in Fig. 3.11[19] and pointed out both problems of DoFs and damping effect that engineers questioned. Study the rationality of DoFs of the system in Fig. 3.11.

---

[19] Simionatto V G S, Miyasato H H, de Melo F M, et al. (2011). Singular mass matrices and half degrees of freedom: a general method for system reduction. Natal: The 21st International Congress of Mechanical Engineering.

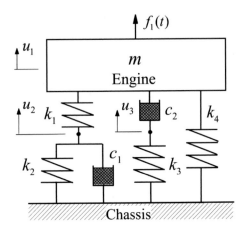

As the hydro-elastic vibration isolation system in Fig. 3.11 only moves vertically, it
is straightforward to establish the dynamic equations, in terms of three displacements
in Fig. 3.11, in the configuration space $\mathbb{R}^3 = (\mathfrak{R}^1)^3$ as follows

$$
\begin{cases}
m\ddot{u}_1(t) + c_2[\dot{u}_1(t) - \dot{u}_3(t)] + (k_1+k_4)u_1(t) - k_1u_2(t) = f_1(t) \\
c_1\dot{u}_2(t) - k_1u_1(t) + (k_1 + k_2)u_2(t) = 0 \\
c_2[\dot{u}_3(t) - \dot{u}_1(t)] + k_3u_3(t) = 0
\end{cases}
\tag{a}
$$

where

$$
\begin{cases}
m = 20 \text{ kg}, \quad c_1 = 300 \text{ N} \cdot \text{s/m}, \quad c_2 = 100 \text{ N} \cdot \text{s/m} \\
k_1 = 500 \text{ N/m}, \quad k_2 = 500 \text{ N/m}, \quad k_3 = 1000 \text{ N/m}, \quad k_4 = 5000 \text{ N/m}
\end{cases}
\tag{b}
$$

As shown in Eq. (a), the system requires three independent coordinates and has
two linear non-holonomic constraints. Thus, the system can be denoted as $S_{0,2}^3$. The
DoF of system $S_{0,2}^3$ defined in analytical mechanics is $D(S_{0,2}^3) = 3 - 0 - 2 = 1$,
yet the new definition in Eq. (3.2.8) leads to the DoF of system $S_{0,2}^3$ satisfying
$\tilde{D}(S_{0,2}^3) = 3 - 0 - 2/2 = 2$.

To study the free vibration of system $S_{0,2}^3$, substituting Eq. (b) into Eq. (a) and
solving the corresponding eigenvalue problem, we get three eigenvalues

$$
\lambda_{1,2} = -0.68972 \mp 17.706\text{i}, \quad \lambda_3 = -3.1742, \quad \lambda_4 = -8.7796.
\tag{c}
$$

As implied by Eq. (c), the disturbed motion of system $S_{0,2}^3$ includes a decaying
vibration and two fast decaying motions. That is, the system looks similar to a 2-
DoF system composed of an under-damped oscillator and an over-damped oscillator.
The dynamics of the system well supports the new definition of DoF in Eq. (3.2.8),
instead of the DoF defined in analytical mechanics.

From the two non-holonomic constraint equations in Eq. (a), let displacement $u_1(t)$ and velocity $\dot{u}_1(t)$ of the lumped mass m be expressed as

$$u_1(t) = [c_1\dot{u}_2(t) + (k_1 + k_2)u_2(t)]/k_1, \quad \dot{u}_1(t) = [c_2\dot{u}_3(t) + k_3u_3(t)]/c_2 \quad \text{(d)}$$

Taking the time derivative of Eq. (d) yields

$$\dot{u}_1(t) = [c_1\ddot{u}_2(t) + (k_1 + k_2)\dot{u}_2(t)]/k_1, \quad \ddot{u}_1(t) = [c_2\ddot{u}_3(t) + k_3\dot{u}_3(t)]/c_2 \quad \text{(e)}$$

Substitution of the second equation in Eq. (d) into the first one in Eq. (e) leads to

$$c_1c_2\ddot{u}_2(t) + c_2(k_1 + k_2)\dot{u}_2(t) - c_2k_1\dot{u}_3(t) - k_1k_3u_3(t) = 0. \quad \text{(f)}$$

Let the second equation in Eqs. (e) and (d) be substituted into the first one in Eq. (a), we get

$$mc_2k_1\ddot{u}_3(t) + c_1c_2(k_1{+}k_4)\dot{u}_2(t) + mk_1k_3\dot{u}_3(t)$$
$$+c_2\big[(k_1{+}k_4)(k_1 + k_2) - k_1^2\big]u_2(t) + c_2k_1k_3u_3(t) = c_2k_1f_1(t) \quad \text{(g)}$$

Equations (f) and (g) are the dynamic equations of system $S_{0,2}^3$ in terms of $u_2(t)$ and $u_3(t)$, and have the following matrix form

$$\begin{bmatrix} c_1c_2 & 0 \\ 0 & mc_2k_1 \end{bmatrix}\begin{bmatrix} \ddot{u}_2(t) \\ \ddot{u}_3(t) \end{bmatrix} + \begin{bmatrix} c_2(k_1 + k_2) & -c_2k_1 \\ c_1c_2(k_1 + k_4) & mk_1k_3 \end{bmatrix}\begin{bmatrix} \dot{u}_2(t) \\ \dot{u}_3(t) \end{bmatrix}$$
$$+ \begin{bmatrix} 0 & -k_1k_3 \\ c_2[(k_1 + k_2)(k_1 + k_4) - k_1^2] & c_2k_1k_3 \end{bmatrix}\begin{bmatrix} u_2(t) \\ u_3(t) \end{bmatrix} = \begin{bmatrix} 0 \\ c_2k_1f_1(t) \end{bmatrix} \quad \text{(h)}$$

As shown in Eq. (h), system $S_{0,2}^3$ looks like a 2-DoF system, but the corresponding damping matrix and stiffness matrix are not symmetric. To properly understand the behaviors of system $S_{0,2}^3$, we denote it as $S_{1,0}^3$, rather than $S_{0,0}^3$, since we should determine the system state from Eq. (d). Thus, the two non-holonomic constraint equations in Eq. (a) serve as a set of output equations of a dynamic control system. Furthermore, we get the same eigenvalues as those in Eq. (c) by substituting Eq. (b) into Eq. (h). That is, system $S_{0,2}^3$ and system $S_{1,0}^3$ exhibit the same dynamics.

Based on the above analysis, we briefly discuss the damping problem of the hydro-elastic vibration isolation system for vehicle engines. In the dynamic equations of system $S_{0,2}^3$ in terms of $u_2(t)$ and $u_3(t)$ in Eq. (h), the damping coefficient c_1 associated with oil via the orifices appears not only in the coefficient matrix of velocity vector, but also in the coefficient matrix of acceleration vector. Hence, an increase of damping coefficient c_1 enlarges not only the damping force $-c_1c_2(k_1 + k_4)\dot{u}_2(t)$, but also the inertial force $-c_1c_2\ddot{u}_2(t)$. As such, a larger damping coefficient c_1 may not increase the damping ratio of the system and reduce the resonant peaks of the system. This issue will be further addressed in Sect. 4.1.

Fig. 3.12 A knife-edged circular disc subject to a pure rolling on a horizontal rough floor

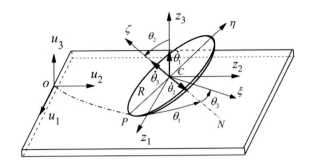

Remark 3.2.8 As analyzed above, the two linear non-holonomic constraints in the hydro-elastic vibration isolation system $S_{0,2}^3$ described by Eq. (a) really reduces one DoF of the system so that the system dynamics is equivalent to that of a holonomic system $S_{1,0}^3$.

Remark 3.2.9 The dimension of the accessible state manifold of system $S_{0,2}^3$ is four, but the dimension of the accessible displacement manifold is three. Thus, it is impossible to establish the tangential bundle and cotangential bundle[20] for system $S_{0,2}^3$. For the equivalent holonomic system $S_{1,0}^3$ described by Eq. (h), yet it is possible to use $[u_2(t)\ u_3(t)]^T$ as a displacement vector $\boldsymbol{q} \in Q = \mathbb{R}^2$ and establish the four-dimensional tangential bundle TQ and cotangential bundle T^*Q in order to study the Hamiltonian dynamics of the equivalent system $S_{1,0}^3$.

3.2.5.2   Pure Rolling of a Knife-Edged Circular Disc on a Horizontal Floor

To demonstrate whether the above assertions apply not only to linear vibration systems such as Example 3.2.4, but also to nonlinear dynamic systems, we check a well known problem in analytical mechanics.

Example 3.2.5 A knife-edged circular disc of radius R is rolling, without any slip, on a horizontal rough floor as shown in Fig. 3.12[21]. Study the dynamic equations and the DoFs of the rolling disc.

As shown in Fig. 3.12, we first establish a fixed frame of coordinates $ou_1u_2u_3$ and let vector $\boldsymbol{u}_C(t) \equiv [u_1(t)\ u_2(t)\ u_3(t)]^T \in \mathfrak{R}^3$ denote the displacements of the mass center of the rolling disc. Then, we set a translational frame of coordinates $Cz_1z_2z_3$ and a rotational frame of coordinates $C\xi\eta\zeta$, both attached to the mass center C of the rolling disc in Fig. 3.12, where the intersection line CN of a horizontal plane

[20] Marsden J E, Ratiu T S (1999). Introduction to Mechanics and Symmetry. 2nd Edition. New York: Springer-Verlag, 121–180.

[21] Greenwood D T (2003). Advanced Dynamics. Cambridge: Cambridge University Press, 222–225.

passing through the mass center C is called the *line of nodes*. Thus, we can define a vector $[\theta_1(t) \;\; \theta_2(t) \;\; \theta_3(t)]^\mathsf{T} \in SO(3)$ for Eulerian angles in Fig. 3.12 to describe the orientation of the rolling disc. Here, $\theta_1(t)$ is the procession angle, $\theta_2(t)$ is the nutation angle, and $\theta_3(t)$ is the spinning angle. As such, the configuration space of the rolling disc is $\mathbb{R}^6 = SO(3) \otimes \mathfrak{R}^3$.

As the rolling disc is not slipping, the velocity vector $\dot{u}_P(t)$ of the disc at the contact point P yields

$$\dot{u}_P(t) = \dot{u}_C(t) + \dot{u}_{CP}(t) = 0, \quad \dot{u}_{CP}(t) \equiv \omega(t) \times r_{CP} \tag{a}$$

where $\dot{u}_C(t)$ is the velocity vector of mass center C, $\dot{u}_{CP}(t)$ is the relative velocity vector, $\omega(t)$ is the angular velocity vector and r_{CP} is the vector from mass center C to contact point P. To specify Eq. (a) in the fixed frame of $ou_1u_2u_3$, it is helpful to use the Résal frame of coordinates, which is the frame $C\xi\eta\zeta$ when the spinning angle is zero. In the Résal frame, angular velocity vector $\omega(t)$, skew-symmetric matrix $\tilde{\omega}(t)$ corresponding to vector $\omega(t)$, vector r_{CP} and relative velocity vector $\dot{u}_{CP}(t)$ satisfy

$$\begin{cases} \omega(t) = \begin{bmatrix} \dot{\theta}_2 \\ \dot{\theta}_1 \sin\theta_2 \\ \dot{\theta}_1 \cos\theta_2 + \dot{\theta}_3 \end{bmatrix}_C, & \tilde{\omega}(t) = \begin{bmatrix} 0 & -(\dot{\theta}_1 \cos\theta_2 + \dot{\theta}_3) & \dot{\theta}_1 \sin\theta_2 \\ \dot{\theta}_1 \cos\theta_2 + \dot{\theta}_3 & 0 & -\dot{\theta}_2 \\ -\dot{\theta}_1 \sin\theta_2 & \dot{\theta}_2 & 0 \end{bmatrix}_C \\[2em] r_{CP} = \begin{bmatrix} 0 \\ -R \\ 0 \end{bmatrix}_C & \dot{u}_{CP}(t) = \tilde{\omega}(t)r_{CP} = \begin{bmatrix} R(\dot{\theta}_1 \cos\theta_2 + \dot{\theta}_3) \\ 0 \\ -R\dot{\theta}_2 \end{bmatrix}_C \end{cases} \tag{b}$$

where time t does not show up for brevity. Applying the rotation transform between the Résal frame and the fixed frame to relative velocity vector $\dot{u}_{CP}(t)$ and substituting the result into the first equation in Eq. (a), we get three constraint equations under the condition of no slipping as follows

$$\dot{u}_P(t) = \begin{bmatrix} \dot{u}_1 \\ \dot{u}_2 \\ \dot{u}_3 \end{bmatrix}_o + \begin{bmatrix} \cos\theta_1 & -\sin\theta_1 & 0 \\ \sin\theta_1 & \cos\theta_1 & 0 \\ 0 & 0 & 1 \end{bmatrix} \begin{bmatrix} 1 & 0 & 0 \\ 0 & \cos\theta_2 & -\sin\theta_2 \\ 0 & \sin\theta_2 & \cos\theta_2 \end{bmatrix} \begin{bmatrix} R(\dot{\theta}_1 \cos\theta_2 + \dot{\theta}_3) \\ 0 \\ -r\dot{\theta}_2 \end{bmatrix}_C$$

$$= \begin{bmatrix} \dot{u}_1 + R\cos\theta_1(\dot{\theta}_1 \cos\theta_2 + \dot{\theta}_3) - R\dot{\theta}_2 \sin\theta_1 \sin\theta_2 \\ \dot{u}_2 + R\sin\theta_1(\dot{\theta}_1 \cos\theta_2 + \dot{\theta}_3) + R\dot{\theta}_2 \cos\theta_1 \sin\theta_2 \\ \dot{u}_3 - R\dot{\theta}_2 \cos\theta_2 \end{bmatrix}_o = 0 \tag{c}$$

In Eq. (c), the integration of the third equation leads to $u_3(t) - R\sin[\theta_2(t)] = 0$. Hence, the rolling disc in $\mathbb{R}^6$ is subject to one holonomic constraint and two non-holonomic constraints such that the rolling disc is a system $S_{1,2}^6$. The DoF of the rolling disc defined in analytical mechanics is $D(S_{1,2}^6) = 6 - 1 - 2 = 3$, while the new definition of DoF gives $\tilde{D}(S^6_{1,2}) = 6 - 1 - 2/2 = 4$. Now, we check what definition is more reasonable.

In a similar way of using Gibbs–Appell equations in Example 3.2.3, we define three quasi-velocities $\sigma_1(t)$, $\sigma_2(t)$ and $\sigma_3(t)$, and derive the dynamic equations of the rolling disc as follows

$$\begin{cases} \dot{\theta}_1(t) = \sigma_2(t)/\sin[\theta_2(t)] \\ \dot{\theta}_2(t) = \sigma_1(t), \\ \dot{\theta}_3(t) = \sigma_3(t) - \sigma_2(t)\cot[\theta_2(t)] \\ \dot{\sigma}_1(t) = \{\sigma_2^2(t)\cot[\theta_2(t)] - 6\sigma_1(t)\sigma_2(t) - 4g\sin[\theta_2(t)]/R\}/5 \\ \dot{\sigma}_2(t) = 2\sigma_2(t)\sigma_3(t) - \sigma_1(t)\sigma_2(t)\cot[\theta_2(t)] \\ \dot{\sigma}_3(t) = 2\sigma_1(t)\sigma_2(t)/3 \end{cases} \qquad (d)$$

Here, Eq. (d) is a set of closed differential equations of the first order, which have six unknown functions. Hence, the attitude dynamic equations of the rolling disc are independent from the motion of the mass center of the disc, look like the dynamic equations of a 3-DoF system in state space.

However, the configuration of the rolling disc includes the motion of the mass center. One needs to solve Eq. (d) for all Eulerian angles and corresponding angular velocities, and substitute the results into Eq. (c) and get the displacement vector $u_C(t) \equiv [u_1(t) \ u_2(t) \ u_3(t)]^T$ via integration. That is, the non-holonomic constraints in Eq. (c) impose a unidirectional coupling on the attitude dynamics to the dynamics of mass center. This coupling described by two differential equations of the first order in Eq. (c) is different from the algebraic equations for the dynamic outputs in Example 3.2.4.

As indicated by the two non-holonomic constraint equations in Eq. (c), the horizontal velocity components $\dot{u}_1(t)$ and $\dot{u}_2(t)$ of the rolling disc are not free, but are the linear combinations of angular velocities of the disc, and also the functions of Eulerian angles of the disc. From Eqs. (c) and (d), the description of the pure rolling of disc requires five independent generalized coordinates and three independent quasi-velocities. The conventional definition of DoF does not account for the motions of the mass center of the disc. The new definition, therefore, looks more reasonable.

It is worthy to notice the difference between Examples 3.2.4 and 3.2.5. That is, it is extremely hard to select a four-dimensional displacement manifold to construct the tangential bundle and cotangential bundle for the rolling disc due to the complicated structure of two non-holonomic constraints even though the DoF of the rolling disc is $\tilde{D}(S^6_{1,2}) = 4$. Hence, it seems impossible to replace the notation $S^6_{1,2}$ for the rolling disc by $S^6_{2,0}$. According to Eq. (d), however, we have a six-dimensional accessible state manifold $\Omega \subset TQ$ as a part of tangential buddle for the rolling disc.

As a matter of fact, the pure rolling of a knife-edged circular disc on a horizontal floor falls into the category of Chaplygin systems in analytical mechanics and possesses the above feature[22].

Remark 3.2.10 It is interesting for readers to check the DoFs of a rigid wheelset subject to a constant velocity along two parallel rails and pure rolling on the rails in $\mathbb{R}^6$. Under the assumption that each wheel has a single point contact with a rail, Antali et al.[23] pointed out that the wheelset has three holonomic constraints and three non-holonomic constraints. As such, the DoFs of the pure rolling wheelset on rails is $\tilde{D}(S_{3,3}^6) = 6 - 3 - 3/2 = 1.5$.

3.2.5.3 Extension of Two Examples

To extend the result of Example 3.2.4, we study the role of two non-holonomic constraint equations of system $S_{r,2}^n$ in the tangential bundle TQ with dimension $2(n - r)$. According to Eq. (3.2.7), let the two non-holonomic constraint equations be expressed as

$$\begin{cases} \sum_{j=1}^{n-r-1} \tilde{a}_{1j}\big[q_1(t), \ldots, q_{n-r-1}(t)\big]\dot{q}_j(t) + a_{1n-r}q_{n-r}(t) = 0 \\ \sum_{j=1}^{n-r-1} \tilde{a}_{2j}\big[q_1(t), \ldots, q_{n-r-1}(t)\big]\dot{q}_j(t) + a_{2n-r}\dot{q}_{n-r}(t) = 0 \end{cases} \tag{3.2.9}$$

If $a_{1n-r}a_{2n-r} \neq 0$ holds in Eq. (3.2.9), it is easy to prove, similar to the analysis in Example 3.2.4, that the DoF of system $S_{r,2}^n$ is $\tilde{D}(S_{r,2}^n) = n - r - 1$ and system $S_{r,2}^n$ is equivalent to system $S_{r+1,0}^n$. Thus, it is possible to construct a $2(n - r - 1)$-th dimensional tangential bundle and a cotangential bundle in order to study the Hamiltonian dynamics of system $S_{r+1,0}^n$. The generalized displacement $q_{n-r}(t)$ and generalized velocity $\dot{q}_{n-r}(t)$ behave like the dynamic output of system $S_{r+1,0}^n$. If system $S_{r,s}^n$ has more even number of non-holonomic constraints like Eq. (3.2.9), it is similar to combine each pair of them as a holonomic constraint.

To extend the result of Example 3.2.5, let the vector of locally generalized displacements of system $S_{r,s}^n$ in $(n-r)$-dimensional accessible state manifold Q be expressed as

$$q(t) = \begin{bmatrix} q_\alpha^T(t) & q_\beta^T(t) \end{bmatrix}^T \tag{3.2.10}$$

---

[22] Liu C, Liu S X, Guo Y X (2011). Inverse problem for Chaplygin's nonholonomic systems. Science in China E, 54(8): 2100–2106.

[23] Antali M, Stepan G, Hogan S J (2015). Kinematic oscillations of railway wheelsets. Multibody System Dynamics, 34(1): 259–274.

where $q_\alpha(t)$ and $q_\beta(t)$ are an $(n-r-s)$-th dimensional vector and an s dimensional vector, respectively. In addition, let the $(n-r-s)$-th dimensional vector of quasi-velocities $\sigma(t)$ satisfy

$$\sigma(t) = C[q_\alpha(t)]\dot{q}_\alpha(t) \tag{3.2.11}$$

Here, matrix $C[q_\alpha(t)]$ is invertible. If the dynamic equations can be decoupled as two sets of differential equations of the first order, the first set of differential equations are $n-r-s$ Gibbs–Appell equations and $n-r-s$ equations of quasi-velocities derived from Eq. (3.2.11). They have the following vector form

$$\begin{cases} B[q_\alpha(t)]\dot{\sigma}(t) = f(t) \\ \dot{q}_\alpha(t) = C^{-1}[q_\alpha(t)]\sigma(t) \end{cases} \tag{3.2.12}$$

where $B[q_\alpha(t)]\dot{\sigma}(t) \equiv \partial G/\partial\dot{\sigma}$, G is the Gibbs function of the system, $f(t)$ is the vector of generalized forces associated with $\sigma(t)$. The second set of differential equations is s non-holonomic constraint equations recast from Eq. (3.2.7), the vector form of which reads

$$\tilde{A}[q_\alpha(t), q_\beta(t)]\left[\dot{q}_\alpha^\mathrm{T}(t) \quad \dot{q}_\beta^\mathrm{T}(t)\right]^\mathrm{T} + \tilde{a}_0[q_\alpha(t), q_\beta(t)] = 0 \tag{3.2.13}$$

Hence, we can select an $(n-r-s)$-th dimensional accessible displacement manifold Q for non-holonomic system $S_{r,s}^n$, and then construct a $2(n-r-s)$-th dimensional tangential bundle TQ and a cotangential bundle T^*Q. The solution of Eq. (3.2.12) describes the dynamics of system $S_{r,s}^n$ in TQ and T^*Q, and looks similar to the dynamics of a holonomic system $S_{r+s,0}^n$. Let the solution be substituted into Eq. (3.2.13) and be integrated, we get the dynamics of non-holonomic system $S_{r,s}^n$ on the $(2n-2r-s)$-th dimensional accessible state manifold Ω.

It is worthy to mention that the above extension is similar to the study for the Chaplygin system in analytical mechanics. That is, if an appropriate part of state variables in both dynamic equations and non-holonomic constraint equations are decoupled, it is possible to get the dynamic equations of a holonomic system via state variable reductions. Furthermore, it is feasible to establish the tangential bundle and the cotangential bundle first and then to study the Hamiltonian dynamics of the reduced holonomic system.

3.2.6 Concluding Remarks

The section presented a detailed study on the degree of freedom, i.e., the DoF, of a discrete system from the viewpoint of geometric mechanics. The section proposed a new definition of DoF, gave four examples to validate the new definition and solved Problem 2B proposed in Chap. 1. The major conclusions of the section are as follows.

(1)   The concept of the DoF of a discrete system stemmed from the number of independent coordinates to describe the system configuration, and belongs to the kinematics in classical mechanics. In analytical mechanics, when an n-DoF system is subject to r holonomic constraints, the DoF of the system is defined as $n - r$. If such a system is subject to s non-holonomic constraints further, the DoF of the system is defined as $n - r - s$ since the number of independent contemporaneous variations reduces by s. The above definition, however, does not clearly distinguish the influences of holonomic constraints and non-holonomic constraints on the system state and over-estimates the influence of non-holonomic constraints.

(2)   The study in this section dealt with the DoF of a discrete system from the viewpoint of dynamics, checked the different influences of two kinds of constraints on the accessible state manifold of the system and discussed the dimension of the accessible state manifold with both kinds of constraints. As a consequence, the section suggested that the DoF of a discrete system be $n - r - s/2$, that is, a half of dimension of the accessible state manifold, so as to eliminate the above over-estimation of conventional definition.

(3)   Two examples in the section gave a clear picture of physics for a half DoF due to a non-holonomic constraint, including the relation between a half DoF and two adjacent integer DoFs. The other two examples in the section demonstrated the possible reduction of one DoF due to two non-holonomic constraints, including the discussions about the tangential bundle and cotangential bundle for the system after reduction.

(4)   It is straightforward to extend the above studies and conclusions to the dynamic systems subject to rhenomic bilateral constraints even though the study in this section is confined to the dynamic systems with scleronomic bilateral constraints.

3.3   Structural Damping

In the early decades of the twentieth century, a number of steady-state vibration tests of metal structures under a harmonic excitation showed that the macro damping force caused by the internal friction in metallic materials had a weak dependence of the response frequency. This experimental evidence led to the model of structural damping in macro sense[24]. The structural damping model also refers to a *hysteretic damping model* or a *complex stiffness model*[25]. The most important feature of the structural damping model is the independence of response frequency or called the *frequency-invariance* in a wide frequency band. The model of structural damping

---

[24] Robertson J M, Yorgiadis A J (1946). Internal friction in engineering materials. ASME Journal of Applied Mechanics, 13(1): 173–180.

[25] Myklestad N O (1952). The concept of complex damping. ASME Journal of Applied Mechanics, 19(3): 284–286.

exhibits a simple form and has found numerous applications to the harmonic vibration analysis in structural dynamics and aeroelastic mechanics.

Many people tried to apply the structural damping model in the harmonic vibration analysis to the study of other vibration problems from the 1950s to 1980s. Yet, their attempts led to counter-examples against the causality of physics even in studying the free vibration of an SDoF system[26]. The discussions about this issue aroused some arguments in previous publications. In 1993, the author sorted out the arguments, revealed the nature of frequency-invariant structural damping of a vibration system and proposed two approximate methods for analyzing the vibration system with structural damping[27]. This section presents the major contents of the study of the author and makes some supplementary results based on Sect. 3.2.

3.3.1   Frequency-Invariant Damping and Its Limitations

The classical model of structural damping came from the experimental observations in the harmonic vibration tests of simple metal structures. The model fit the experimental data where the damping force of the structure had a small variation of amplitude and an opposite phase of velocity in a wide frequency band. That is, the model of structural damping is frequency-invariant.

To begin with, consider the forced vibration of a simple structure described by an SDoF system as follows

$$m\ddot{u}(t) + ku(t) = f(t) + f_d[u(t), \dot{u}(t), t] \tag{3.3.1}$$

where m is the mass, k is the stiffness coefficient, $f(t)$ is the external excitation force, and $f_d[u(t), \dot{u}(t), t]$ is the damping force, respectively.

For brevity, we use complex functions to describe the harmonic excitation force and the steady-state harmonic vibration of the system. After the system under a harmonic excitation force $f(t) = F(\omega)\exp(\mathrm{i}\omega t)$ arrives at the steady-state vibration $u(t) = U(\omega)\exp(\mathrm{i}\omega t)$, the damping force satisfies, in a relatively wide frequency band, the following relation

$$f_d[u(t), \dot{u}(t), t] = -\mathrm{i}k\eta_0 u(t), \quad \omega \in (\omega_a, \ \omega_b) \tag{3.3.2}$$

where $\mathrm{i} \equiv \sqrt{-1}$, and $\eta_0 \in (0, \ 1)$ is named the *loss factor*, which keeps constant in a frequency band $(\omega_a, \ \omega_b)$. Therefore, Eq. (3.3.1) becomes

---

[26] Reid T J (1956). Free vibration and hysteretic damping. Journal of Royal Aeronautical Society, 60(544): 283.

[27] Hu H Y (1993). Structural damping model and system response in time domain. Chinese Journal of Applied Mechanics, 10(1): 34–43 (in Chinese).

$$m\ddot{u}(t) + k(1 + i\eta_0)u(t) = F(\omega)\exp(i\omega t) \tag{3.3.3a}$$

$$u(t) = U(\omega)\exp(i\omega t) \tag{3.3.3b}$$

Now, we review previous studies and make discussions about Eq. (3.3.3) as follows.

Although Eq. (3.3.3a) looks like the dynamic equation of the system in time domain, it is indeed an equilibrium equation of all forces in complex forms in frequency domain according to the steady-state vibration in Eq. (3.3.3b), namely,

$$\left[k(1 + i\eta_0) - m\omega^2\right]U(\omega) = F(\omega), \quad \omega \in (\omega_a, \ \omega_b) \tag{3.3.4}$$

Equation (3.3.4) implies that the frequency response function of the system in a frequency band $(\omega_a, \ \omega_b)$ reads

$$H(\omega) \equiv \frac{U(\omega)}{F(\omega)} = \frac{1}{k(1 + i\eta_0) - m\omega^2}, \quad \omega \in (\omega_a, \ \omega_b) \tag{3.3.5}$$

In 1970, R. H. Scanlan in the USA extended Eq. (3.3.3a) into the dynamic equation of a system as follows[28]

$$m\ddot{u}(t) + k(1 + i\eta_0)u(t) = 0 \tag{3.3.6}$$

This extension dropped off the condition of steady-state response in Eq. (3.3.3b) and intended to use Eq. (3.3.6) to describe the free vibration of a system with structural damping. However, the free vibration calculated from Eq. (3.3.6) always exhibits the divergent response. Some subsequent studies tried to explain the response divergence and improve Eq. (3.3.6) in other forms. As a matter of fact, the damped free vibration is no longer a harmonic vibration and the corresponding damping force does not yield Eq. (3.3.2) at all. As a consequence, Eq. (3.3.6) is not able to describe the free vibration of a system with structural damping. This section will show that the free vibration of a system with proper structural damping yields a differential-convolution equation.

Some other people tried to give up the condition of frequency band in Eq. (3.3.5) and extended the frequency response of a system with structural damping force from Eq. (3.3.5) to the following expression

$$H(\omega) = \frac{1}{k\left[1 + i\eta_0\mathrm{sgn}(\omega)\right] - m\omega^2}, \quad \omega \in (-\infty, \ +\infty) \tag{3.3.7}$$

---

[28] Scanlan R H (1970). Linear damping models and causality in vibrations. Journal of Sound and Vibration, 13(4): 499–509.

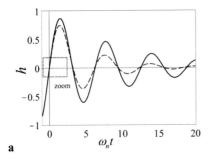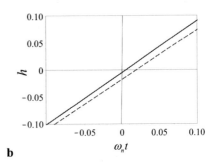

Fig. 3.13 The dimensionless displacement histories of a structurally damped system to a unit impulse (solid line: $\eta_0 = 0.4$, dashed line: $\eta_0 = 0.2$). **a** the global view; **b** the zoomed view

Applying the inverse Fourier transform to both sides of Eq. (3.3.7), we obtain the *unit impulse response*[29]

$$h(t) = \mathscr{F}^{-1}[H(\omega)] = \frac{1}{2\pi} \int_{-\infty}^{+\infty} H(\omega) \exp(i\omega t) \, d\omega \tag{3.3.8}$$

In Eq. (3.3.7), function $\text{sgn}(\omega)$ was introduced in order to make $\text{Re}[H(\omega)]$ and $\text{Im}[H(\omega)]$ be even function and odd function respectively such that $h(t)$ is a real function in view of the inverse Fourier transform. Now, let Eq. (3.3.8) be recast as the following generalized integral, which can be completed by using software Maple, namely,

$$\begin{aligned}
h(t) &= \frac{1}{2\pi} \int_{-\infty}^{+\infty} \{\text{Re}[H(\omega)] + i\text{Im}[H(\omega)]\}[\cos(\omega t) + i\sin(\omega t)] \, d\omega \\
&= \frac{1}{\pi} \int_{0}^{+\infty} \left[\frac{(k - m\omega^2)\cos(\omega t)}{(k - m\omega^2)^2 + (\eta_0 k)^2} + \frac{\eta_0 k \sin(\omega t)}{(k - m\omega^2)^2 + (\eta_0 k)^2} \right] d\omega
\end{aligned} \tag{3.3.9}$$

Figure 3.13 presents the dimensionless responses of the system with $\eta_0 = 0.2$ and $\eta_0 = 0.4$ respectively to a unit impulse calculated from Eq. (3.3.9). It is a pity to observe from the figure that $h(t) \neq 0$, $t \in (-\infty, 0]$ holds. That is, the above system response violates the causality of a physical system. In fact, it is feasible to use the contour integral in complex analysis to complete the following integral[30]

$$h(0) = \frac{1}{\pi} \int_{0}^{+\infty} \frac{k - m\omega^2}{(k - m\omega^2)^2 + (\eta_0 k)^2} \, d\omega = -\frac{1}{2\sqrt{mk}} \sqrt{\frac{\sqrt{1 + \eta_0^2} - 1}{2(1 + \eta_0^2)}} < 0 \tag{3.3.10}$$

[29] Milne H K (1985). The impulse response function of a single degree of freedom system with hysteretic damping. Journal of Sound and Vibration, 100(4): 590–593.

[30] Hu H Y (1993). Structural damping model and system response in time domain. Chinese Journal of Applied Mechanics, 10(1): 34–43 (in Chinese).

As any real system must satisfy *causality*, that is, $h(t) = 0$, $t \in (-\infty, \ 0]$, the system response under a unit impulse in Eq. (3.3.9) fails to describe a real system. The problem comes from the extension from Eqs. (3.3.5) to (3.3.7) in mathematics since the extension contradicts to the experimental data of frequency-invariant damping force in a finite frequency band $(\omega_a, \ \omega_b)$.

Hence, no matter whether the structural damping force of complex form in a finite frequency band is extended to time domain or to entire frequency domain, the extension contradicts to physics. Theoretically speaking, the frequency-invariant damping force only holds in a finite frequency band.

3.3.2 Frequency-Variant Damping Model and System Response

Even though the amplitude of structural damping force of a metal structure looks quite flat in a wide frequency band, the model of damping force should be a frequency-variant function in frequency domain as a whole. That is, the loss factor of structural damping is an odd continuous function, namely,

$$\eta \equiv \eta(\omega) = -\eta(-\omega), \quad \omega \in (-\infty, +\infty) \tag{3.3.11}$$

As for the structures made of polymers and composite materials, their loss factors usually exhibit obvious frequency-variance, and so do their elastic modulus.

In this case, we introduce the following *complex stiffness coefficient* $K^*(\omega)$, or called the *complex modulus*, for both elastic force and damping force of the system to a harmonic excitation, namely,

$$K^*(\omega) \equiv K(\omega)[1 + i\eta(\omega)], \quad \omega \in (-\infty, +\infty) \tag{3.3.12}$$

where $K(\omega)$ and $\eta(\omega)$ are the real part and the tangent function of phase of the complex stiffness $K^*(\omega)$, respectively. They are an even function and an odd function of frequency ω, respectively. The model described by Eq. (3.3.12) covers the model in Eq. (3.3.11).

For the damping model in Eq. (3.3.12), the equilibrium equation of all forces of an SDoF system in frequency domain reads

$$\left[K^*(\omega) - m\omega^2\right]U(\omega) = F(\omega), \quad \omega \in (-\infty, +\infty) \tag{3.3.13}$$

Equation (3.3.13) is equivalent to

$$U(\omega) = H(\omega)F(\omega), \quad H(\omega) \equiv \frac{1}{K^*(\omega) - m\omega^2}, \quad \omega \in (-\infty, +\infty) \tag{3.3.14}$$

where $H(\omega)$ is the response function of the SDoF system.

Both experimental and theoretical studies have shown that the frequency-variant complex stiffness $K^*(\omega)$ represents the viscoelastic property of materials or structures. In practice, it is feasible to obtain $K^*(\omega)$ in a form of rational fraction by fitting experimental data[31]. In this case, the frequency response function $H(\omega)$ is also a rational fraction. Applying the inverse Fourier transform to both $K^*(\omega)$ and $H(\omega)$ leads to two real functions in time domain as follows

$$k^*(t) = \mathscr{F}^{-1}[K^*(\omega)] = \frac{1}{2\pi} \int_{-\infty}^{+\infty} K^*(\omega)\exp(i\omega t)d\omega, \quad t \geq 0 \qquad (3.3.15)$$

$$h(t) = \mathscr{F}^{-1}[H(\omega)] = \frac{1}{2\pi} \int_{-\infty}^{+\infty} H(\omega)\exp(i\omega t)d\omega, \quad t \geq 0 \qquad (3.3.16)$$

where $k^*(t)$ is the *convolution kernel function* of the system and $h(t)$ is the unit impulse response function of the system, which looks similar to Eq. (3.3.8). Yet, both $k^*(t)$ and $h(t)$ are causal functions in time domain according to Paley-Wiener theorem of Fourier transform.

From the causality of Eqs. (3.3.15) and (3.3.16), as well as the convolution theorem of the inverse Fourier transform, applying Fourier transform to both sides of Eq. (3.3.13) leads to the differential convolution equation for the system dynamics, namely,

$$m\ddot{u}(t) + \int_0^t k^*(t - \tau)u(\tau)\,d\tau = f(t), \quad t \geq 0 \qquad (3.3.17)$$

That is, the complex stiffness $K^*(\omega)$ corresponds to the convolution kernel function $k^*(t)$ in Eq. (3.3.17) in time domain.

The above Fourier transform is double sides and does not account for the initial conditions of the system. Given the initial conditions of the system as follows

$$u(0) = u_0, \quad \dot{u}(0) = \dot{u}_0 \qquad (3.3.18)$$

the single-side Fourier spectrum for Eq. (3.3.17) is

$$m[-\omega^2 U(\omega) - i\omega u_0 - \dot{u}_0] + K^*(\omega)U(\omega) = F(\omega), \quad \omega \geq 0 \qquad (3.3.19)$$

Solving Eq. (3.3.19) for the single-side Fourier spectrum of the system response gives

---

[31] Braun S, Ewins D, Rao S S (2002). Encyclopedia of Vibration. San Diego: Academic Press, 327–342.

$$U(\omega) = \frac{1}{K^*(\omega) - m\omega^2}[F(\omega) + im\omega u_0 + m\dot{u}_0]$$

$$= H(\omega)[F(\omega) + i\omega u_0 + \dot{u}_0], \quad \omega \geq 0$$

(3.3.20)

Applying the single-side Fourier transform to Eq. (3.3.20) and using Eq. (3.3.16), we get the dynamic response of the system to initial disturbance and external excitation, namely,

$$u(t) = mu_0 \dot{h}(t) + m\dot{u}_0 h(t) + \int_0^t h(t - \tau)f(\tau)\,d\tau, \quad t \geq 0$$

(3.3.21)

A comparison between Eq. (3.3.21) and Eq. (a) in Example 2.2.1 indicates the same physical sense behind.

Theoretically speaking, the frequency-variant damping model avoids the physical contradiction due to the frequency-invariant structural damping, but it is not convenient in applications as follows.

The complex stiffness $K^*(\omega)$ is usually a rational fraction of higher order so that the convolution kernel function $k^*(t)$ in Eq. (3.3.15) is complicated and it is hard to solve the differential convolution equation with $k^*(t)$ in Eq. (3.3.17). For instance, a standard way of solving Eq. (3.3.17) is to transform it to a set of differential equations, the number of which depends on the order of the rational fraction of complex stiffness $K^*(\omega)$. In addition, the solving procedure requires the initial conditions associated with not only the initial displacement and velocity, but also initial jerk and so forth.

An alternative way is to compute Eq. (3.3.21) involving the unit impulse response function $h(t)$. Thus, one needs to substitute $K^*(\omega)$, through Eq. (3.3.14), into Eq. (3.3.16), perform the inverse Fourier transform in Eq. (3.3.16) and then compute the convolution in Eq. (3.3.21). In general, this procedure relies on numerical computations.

To simplify the dynamic analysis of an SDoF system with structural damping under the precondition of system causality, the following subsections present two approaches.

3.3.3 An Approximate Viscous Damping Model

In practice, the major frequency components of the dynamic response of a linear system fall into either a few resonant frequency bands or the excitation frequency bands. These frequency bands are referred to as dominant frequency bands for short. When a linear system with structural damping has one or several dominant frequency bands, it is reasonable to take a viscous damping model to approximate the structural damping in each dominant frequency band. In this approximation, the viscous damping is frequency-variant, but the major contribution falls into the dominant frequency band and dose not produce large errors. Hence, the equivalent viscous damping for the r-th dominant frequency band is as follows

$$c_r = \frac{\eta_0 k}{\omega_r}, \quad r = 1, 2, \ldots \tag{3.3.22}$$

where ω_r is the central frequency of the band.

Consider the free vibration of an SDoF system with structural damping. It has a single dominant frequency band centered at the natural frequency $\omega_n = \sqrt{k/m}$. Thus, we take

$$c_1 = \frac{k\eta_0}{\omega_n} = \frac{k\eta_0}{\sqrt{k/m}} = \eta_0 \sqrt{mk} \tag{3.3.23}$$

The free vibration of the system with structural damping approximately yields

$$m\ddot{u}(t) + \eta_0 \sqrt{mk}\,\dot{u}(t) + ku(t) = 0 \tag{3.3.24}$$

Solving Eq. (3.3.24) for $u(t)$ leads to

$$u(t) = a \exp\left(-\frac{\eta_0 \omega_n t}{2}\right) \cos\left(\sqrt{1 - \frac{\eta_0^2}{4}}\,\omega_n t + \theta\right) \tag{3.3.25}$$

where constants a and θ are to be determined from the initial state of the system.

Remark 3.3.1 From Eq. (3.3.24), an SDoF system with structural damping of loss factor η_0 is equivalent to a viscously damped SDoF system with damping ratio $\zeta = \eta_0/2$. Although Eqs. (3.3.24) and (3.3.25) look similar to those in previous studies, the approach here avoids logic confusions, and clearly states that the viscous damping is an approximate equivalence for a dominant frequency band. In fact, this approximation has been hidden in the current free vibration analysis and modal parameter identifications. The following example well validates the approximation.

Example 3.3.1 Make a comparison between the model of frequency-invariant structural damping and that of viscous damping for an SDoF system.

We first compute the response of an SDoF system to a unit impulse for $\eta_0 < 0.4$, i.e., $\zeta < 0.2$, from Eqs. (3.3.9) and (3.3.25). Figure 3.14 presents a comparison between the dimensionless displacement histories of the SDoF system with two damping models for $\eta_0 = 0.4$. Here, the displacement of the system with frequency-invariant structural damping becomes negative when $t < 0$ and does not satisfy the causality in physics. The system with approximate viscous damping yields causality, but the period of free vibration is a little bit longer.

Then, we compute the frequency response functions of the system with two damping models when $\eta_0 = 0.4$ as shown in Fig. 3.15, where the difference of the displacement frequency responses appears at a lower frequency band, while that of the acceleration frequency responses comes at both middle and higher frequency bands. In other words, the viscous damping only approximates the frequency-invariant structural damping in the dominant frequency band selected.

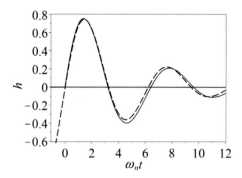

Fig. 3.14 The dimensionless displacement histories of an SDoF system with two damping models under a unit impulse (solid line: the approximate viscous damping model, dashed line: the structural damping model)

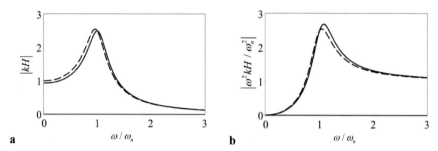

Fig. 3.15 The dimensionless frequency response functions of an SDoF system for two damping models (solid line: structural damping, dashed line: approximate viscous damping). **a** the displacement responses; **b** the acceleration responses

3.3.4 An Approximate Viscoelastic Damping Model

As the viscous damping is frequency-variant, the approximation in Sect. 3.3.3 works well only in dominant frequency bands. This subsection improves the damping approximation by using *Kelvin's viscoelastic model*. The vibration system with such a damping model is called a system with *viscoelastic damping*.

3.3.4.1 A Vibration System with Viscoelastic Damping

Figure 3.16 presents a vibration system with viscoelastic damping, where the spring with stiffness coefficient k and Maxwell's fluidic component composed of a spring with stiffness coefficient μk and a dashpot with damping coefficient c in serial constitute Kelvin's viscoelastic model of three parameters. That is, Figs. 3.16 and 3.6b are the same. In view of the study in Sect. 3.2.4, this system has one lumped mass, but 1.5-DoFs since Maxwell's fluidic component has a half DoF by nature.

Fig. 3.16 A vibration
system with viscoelastic
damping

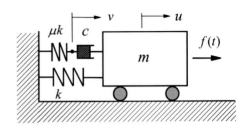

From Eq. (b) in Example 3.2.2, the dynamic equations of the vibration system
with viscoelastic damping in Fig. 3.16 satisfy

$$\begin{cases} m\ddot{u}(t) + c[\dot{u}(t) - \dot{v}(t)] + ku(t) = f(t) \\ c[\dot{v}(t) - \dot{u}(t)] + \mu k v(t) = 0, \quad t \in [0, +\infty) \end{cases} \tag{3.3.26}$$

where $v(t)$ describes an intrinsic DoF. Applying Fourier transform to both sides of
Eq. (3.3.26) leads to

$$\begin{cases} -m\omega^2 U(\omega) + ic\omega[U(\omega) - V(\omega)] + kU(\omega) = F(\omega) \\ ic\omega[V(\omega) - U(\omega)] + \mu k V(\omega) = 0, \quad \omega \in (-\infty, +\infty) \end{cases} \tag{3.3.27}$$

Solving the second equation in Eq. (3.3.27) for $V(\omega)$ and substituting the solution
of $V(\omega)$ into the first equation in Eq. (3.3.27), we get the equilibrium equation of all
forces of the system in frequency domain, namely,

$$\left[\frac{\mu k^2 + ick(1 + \mu)\omega}{(\mu k + ic\omega)} - m\omega^2 \right] U(\omega) = F(\omega), \quad \omega \in (-\infty, +\infty) \tag{3.3.28}$$

A comparison between Eqs. (3.3.28) and (3.3.13) gives the complex stiffness of
viscoelastic damping as follows

$$K^*(\omega) = \frac{\mu k^2 + ick(1 + \mu)\omega}{\mu k + ic\omega} = k\frac{1 + ia\omega}{1 + ib\omega} = K_R(\omega) + iK_I(\omega) \tag{3.3.29}$$

where

$$K_R(\omega) \equiv \frac{k(1 + ab\omega^2)}{1 + b^2\omega^2}, \quad K_I(\omega) \equiv \frac{k(a - b)\omega}{1 + b^2\omega^2}, \quad a \equiv \frac{c(1 + \mu)}{\mu k}, \quad b \equiv \frac{c}{\mu k} \tag{3.3.30}$$

In engineering, $K_R(\omega)$ and $K_I(\omega)$ are named the *storage modulus* and the *loss
modulus* of complex stiffness $K^*(\omega)$, respectively. From Eqs. (3.3.29) and (3.3.30),
it is easy to prove their properties as follows.

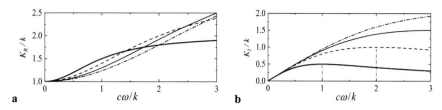

Fig. 3.17 The dimensionless complex stiffness vs. dimensionless frequency (thick solid line: $\mu = 1$, dashed line: $\mu = 2$, thin solid line: $\mu = 3$, dot line: $\mu = 4$). **a** the amplitude of storage modulus; **b** the amplitude of loss modulus

(1) $K_R(\omega)$ is an even function of $\omega \in (-\infty, +\infty)$. It increases monotonically for $\omega \in [0, +\infty)$ and yields $K_R(0) = k$ and $K_R(+\infty) = k(1 + \mu)$.

(2) $K_I(\omega)$ is an odd function of $\omega \in (-\infty, +\infty)$. It has a unique peak $K_I(1/b) = \mu k/2$ for $\omega \in [0, +\infty)$ and yields $K_I(0) = K_I(+\infty) = 0$.

Figure 3.17 depicts the above two moduli for stiffness ratios $\mu = 1, 2, 3, 4$. In practice, it is feasible to select stiffness ratio μ and damping coefficient c to determine the required damping model for a given stiffness parameter k.

Example 3.3.2 In Example 3.3.1, the free vibration period of an SDoF system with approximate viscous damping under a unit impulse is a little bit longer and the deviation of frequency response function of such a vibration system appears out of the resonant frequency band. Use the viscoelastic damping model to improve the damping approximation for the vibration system.

We first qualitatively analyze the vibration system with viscoelastic damping in Fig. 3.16. It is feasible to change the contribution of the dashpot to the equivalent damping ratio of the system by adjusting the spring stiffness coefficient ηk. In particular, the viscoelastic damping model becomes a viscous damping model when $\mu \to +\infty$ holds. Hence, it is reasonable to begin with the damping coefficient $c = \eta_0 \sqrt{mk}$ for the system and then adjust the stiffness ratio μ to achieve a desirable damping approximation for a frequency-invariant structural damping model.

Now, we check the dimensionless frequency response functions of acceleration for the system with different damping models. Figure 3.18 presents a comparison among the frequency-invariant structural damping model for $\eta_0 = 0.4$, the viscous damping model for $c = \eta_0 \sqrt{mk}$, and the viscoelastic damping model when $c = \eta_0 \sqrt{mk}$ and $\mu = 3$. The zoomed view in Fig. 3.18b shows that the viscoelastic damping model is better than the viscous damping model with regard to the approximation for the frequency-invariant structural damping model.

Figure 3.19a shows a comparison among the displacement histories of the system with above three damping models computed by applying the inverse Fourier transform to the frequency response functions of the system. The zoomed view in Fig. 3.19b shows that the viscoelastic damping model is also superior to the viscous damping model.

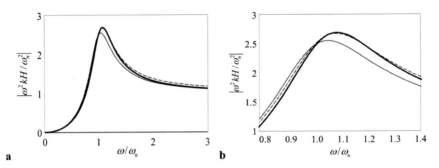

Fig. 3.18 The dimensionless frequency response functions of acceleration of the system for three damping models (thick solid line: structural damping, thin solid line: viscous damping, dashed line: viscoelastic damping). **a** the global view; **b** the zoomed view around the peak

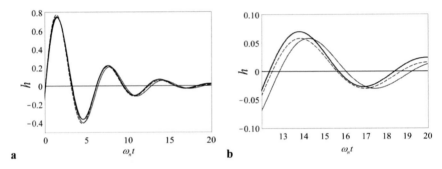

Fig. 3.19 The displacement histories of the vibration system to a unit impulse for three damping models (thick solid line: structural damping, thin solid line: viscous damping, dashed line: viscoelastic damping). **a** the global view; **b** the zoomed view around the second positive peak

Example 3.3.3 Determine the damping coefficient in the viscoelastic damping model with a given stiffness ratio μ so as to approximate the frequency-invariant structural damping model with a given η_0 for the vibration system in the resonant frequency band.

For a given stiffness ratio μ, select the corresponding curve K_I/k in Fig. 3.17b, plot a horizontal line $K_I/k = \eta_0$ to intersect the curve and a vertical line passing through this intersection point, respectively. Then, we get $c\omega/k$ from the intersection point of the vertical line and the abscissa.

Consider, as an example, $\mu = 5$ and $\eta_0 = 0.4$. The above procedure leads to $c\omega/k = 0.4026$. Taking the centered frequency of the resonant frequency band as $\omega_1 = \omega_n = \sqrt{k/m}$, we get $c = 0.4026k/\omega_n = 0.4026\sqrt{mk}$. This result is very close to $c = \eta_0\sqrt{mk} = 0.4\sqrt{mk}$ selected in Example 3.3.2.

From Fig. 3.17, the viscoelastic damping model is able to approximate a great variety of damping, including experimental damping data, in a wide frequency band. Hence, it is useful to gain an insight into the free vibration of the system with viscoelastic damping in Fig. 3.16, and get the explicit relations between the three parameters of the damping model and the vibration properties in order to well understand the viscoelastic damping model.

3.3.4.2 Free Vibrations

According to Eq. (f) in Example 3.2.2, we have the corresponding characteristic equation as follows

$$mc\lambda^3 + m\mu k\lambda^2 + k(1+\mu)c\lambda + \mu k^2 = 0 \qquad (3.3.31)$$

From Routh-Hurwitz theorem, the three eigenvalues of Eq. (3.3.31) have negative real parts since all coefficients in Eq. (3.3.31) are positive. They are usually a pair of conjugate complex eigenvalues and a real eigenvalue, but their expressions are somewhat complicated.

To get simple expressions for those eigenvalues, we recast Eq. (3.3.31) as

$$\varepsilon\lambda^3 + \lambda^2 + \varepsilon(1+\mu)\omega_n^2\lambda + \omega_n^2 = 0, \quad \varepsilon = \frac{c}{\mu k}, \quad \omega_n^2 = \frac{k}{m} \qquad (3.3.32)$$

Based on the discussion in Example 3.3.2, we solve Eq. (3.3.32) for approximate eigenvalues under the condition of $0 < \varepsilon << 1$. When $\varepsilon = 0$, Eq. (3.3.32) has a pair of pure imaginary roots denoted as $\lambda_{10} = i\omega_n$ and $\lambda_{20} = -i\omega_n$. When $0 < \varepsilon << 1$ holds, let the eigenvalues of Eq. (3.3.32) be expressed as follows

$$\lambda_r \approx \lambda_{r0} + \varepsilon\lambda_{r1} + \varepsilon^2\lambda_{r2}, \quad r = 1, 2 \qquad (3.3.33)$$

Substituting Eq. (3.3.33) into Eq. (3.3.32) and equating the same order of ε lead to

$$\lambda_{11} = \lambda_{21} = -\frac{\mu}{2}\omega_n^2, \quad \lambda_{12} = -\lambda_{22} = -\frac{i\mu(\mu-4)}{8}\omega_n^3 \qquad (3.3.34)$$

Hence, we derive a pair of approximate complex eigenvalues as follows

$$\lambda_{1,2} \approx -\frac{\mu\varepsilon}{2}\omega_n^2 \pm i\omega_n\left[1+\frac{\mu(4-\mu)\varepsilon^2}{8}\omega_n^2\right] = -\frac{c}{2k}\omega_n^2 \pm i\omega_n\left[1+\frac{(4-\mu)c^2}{8\mu k^2}\omega_n^2\right] \qquad (3.3.35)$$

As the coefficient of λ^3 in Eq. (3.3.32) is a small parameter ε, it is impossible to get an approximate real eigenvalue from the degenerated equation for $\varepsilon = 0$. To solve the problem, we recast Eq. (3.3.32) as

$$\omega_n^2 s^3 + s^2 + \varepsilon(1+\mu)\omega_n^2 s + \varepsilon = 0, \quad s \equiv 1/\lambda \qquad (3.3.36)$$

Equation (3.3.36) has a real root $s_0 = 0$ when $\varepsilon = 0$ holds. Let the real eigenvalue of Eq. (3.3.36) be expressed as the second-order expansion, namely,

$$s \approx s_0 + \varepsilon s_1 + \varepsilon^2 s_2 \tag{3.3.37}$$

Substituting Eq. (3.3.37) into Eq. (3.3.36) and equating the same order of ε give

$$s_1 = -1, \quad s_2 = 0 \tag{3.3.38}$$

There follows the approximate real eigenvalue of Eq. (3.3.32), namely,

$$\lambda_3 = \frac{1}{s} \approx -\frac{1}{\varepsilon} = -\frac{\mu k}{c} \tag{3.3.39}$$

From the above three eigenvalues, we have the free vibration of the system as follows

$$u(t) = \sum_{r=1}^{3} a_r \exp(\lambda_r t) \tag{3.3.40}$$

where constants a_r, $r = 1, 2, 3$ are to be determined from initial conditions. According to the approximate eigenvalues in Eqs. (3.3.35) and (3.3.39), the first two terms in Eq. (3.3.40) exhibit a decaying vibration, similar to the behavior of a viscously damped SDoF system, but the vibration frequency can be adjusted around ω_0 via different stiffness ratio μ. The third term in Eq. (3.3.40) exhibits no oscillation, but decays rapidly, implying a creep mode or a half DoF caused by viscoelasticity.

Remark 3.3.2 Let $c = \eta_0 \sqrt{mk}$, then we can write the three approximate eigenvalues as

$$\lambda_{1,2} = -\frac{\eta_0 \omega_n}{2} \pm i\omega_n \left[1 + \frac{(4-\mu)\eta_0^2}{8\mu}\right], \quad \lambda_3 = -\frac{\mu \omega_0}{\eta_0} \tag{3.3.41}$$

When $\mu \to +\infty$, Eq. (3.3.41) becomes

$$\lambda_{1,2} = -\frac{\eta_0 \omega_n}{2} \pm i\omega_n \left(1 - \frac{\eta_0^2}{8}\right), \quad \lambda_3 = -\infty \tag{3.3.42}$$

The real part and imaginary part of eigenvalues λ_1 and λ_2 are respectively close to the decaying rate and vibration frequency in Eq. (3.3.25), which describes the free vibration of an SDoF system with approximate viscous damping.

3.3.4.3 System Response Under a Unit Impulse

According to the theorem of momentum, the system response under a unit impulse is the free vibration of the system under the following initial conditions

$$u(0) = 0, \quad \dot{u}(0) = \frac{1}{m}, \quad \ddot{u}(0) = 0 \tag{3.3.43}$$

Substituting Eq. (3.3.40) into Eq. (3.3.43) leads to

$$\sum_{r=1}^{3} a_r = 0, \quad \sum_{r=1}^{3} a_r \lambda_r = \frac{1}{m}, \quad \sum_{r=1}^{3} a_r \lambda_r^2 = 0 \tag{3.3.44}$$

Solving Eq. (3.3.44) for a_r, $r = 1, 2, 3$, we have

$$\begin{cases} a_1 = \dfrac{\lambda_2 + \lambda_3}{m(\lambda_3 - \lambda_1)(\lambda_1 - \lambda_2)} \\[3mm] a_2 = \dfrac{\lambda_1 + \lambda_3}{m(\lambda_1 - \lambda_2)(\lambda_2 - \lambda_3)} \\[3mm] a_3 = \dfrac{\lambda_1 + \lambda_2}{m(\lambda_2 - \lambda_3)(\lambda_3 - \lambda_1)} \end{cases} \tag{3.3.45}$$

It is easy to verify that a_1 and a_2 in Eq. (3.3.45) are conjugate and the system response under a unite impulse reads

$$h(t) = 2\mathrm{Re}\big[a_1 \exp(\lambda_1 t)\big] + a_3 \exp(\lambda_3 t) \tag{3.3.46}$$

Taking $c = \eta_0 \sqrt{mk}$ and using approximate eigenvalues in Eq. (3.3.41), we have a more specific expression, namely,

$$h(t) = 2|a_1| \exp\left(-\frac{\eta_0 \omega_n t}{2}\right) \cos\left\{\left[1 + \frac{(4 - \mu)\eta_0^2}{8\mu}\right]\omega_n t + \theta\right\} + a_3 \exp\left(-\frac{\mu \omega_0 t}{\eta_0}\right) \tag{3.3.47}$$

where $\theta = \tan^{-1}[\mathrm{Im}(a_1)/\mathrm{Re}(a_1)]$ is the initial phase.

To compute the system response to an arbitrary excitation, one can substitute Eq. (3.3.46) or Eq. (3.4.47) into Eq. (3.3.21) and compute the convolution. The computation of free vibration under an initial velocity disturbance shows that the viscoelastic damping model is more accurate than the viscous damping model.

3.3.5 Concluding Remarks

This section analyzed the preconditions for the frequency-invariant structural damping model, and showed how to use two frequency-variant damping models to approximate such a structural damping model. The studies in the section led to the major conclusions as follows.

(1) The structural damping model came from the experimental observations of the steady-state harmonic vibration of an SDoF system to a harmonic excitation in order to describe the frequency-invariance of the structural damping in a relatively wide frequency band. Nevertheless, the extension of the frequency-invariance of the structural damping to the whole frequency domain contradicts to the physical observations and gives rise to some troubles. For instance, the inverse Fourier transform of the frequency response of the system leads to the system response in time domain, but violates the causality in physics. Furthermore, it is wrong to get the free vibration of the system by removing the harmonic excitation.

(2) The amplitude of structural damping force slightly varies in a wide frequency band and can be described by using a complex stiffness model. The inverse Fourier transform for the frequency response of an SDoF system with complex stiffness leads to a differential convolution equation, which describes the system dynamics in time domain.

(3) To avoid solving a tough differential convolution equation, it is feasible to use viscous damping model to approximate the structural damping in a number of dominant frequency bands. This approach is simple, but may give inaccurate results in a wide frequency band.

(4) To eliminate the shortcomings of the viscous damping model, the section presented how to use a viscoelastic damping model of three parameters to approximate the structural damping in a wide frequency band. The viscoelastic damping model involves a dashpot and a spring in serial such that the vibration system with viscoelastic damping has an extra half DoF as newly defined in Sect. 3.2. Hence, the dynamic response of a vibration system with viscoelastic damping exhibits a creep mode decaying rapidly.

3.4 Further Reading and Thinking

(1) For a clamped-free Timoshenko beam, establish the dynamic equation and boundary conditions according to extended Hamilton's principle.

(2) The deployable mast of a satellite in Fig. 2.2 consists of numerous identical cells. Figure 3.20 presents a simplified planar mast model for bending vibrations. The rods in an arbitrary cell in Fig. 3.20 are hinged to each and have the same linear density ρA and the same tensional stiffness EA. Take the simplified mast model as an equivalent beam from the viewpoint of energy,

Fig. 3.20 A simplified planar mast model composed of an infinite number of identical cells

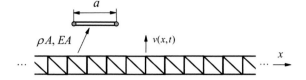

and establish the dynamic equation of the bending vibration of the equivalent beam.

(3) From the singularity in the beauty of science, discuss the degeneration of DoFs in Sects. 3.2 and 3.3. Read the references[32,33] and discuss the influence of non-holonomic constraints on the accessibility of a linear MDoF system in the configuration space and state space, respectively. Read the references[34,35] and discuss when one requires the concept of differential manifolds to study the above accessibility of a dynamic system.

(4) Given a thin rectangular plate with two free edges of length a and two hinged edges of length b, discuss the difference of natural vibrations between the plate and a hinged-hinged slender beam when $0 < b << a$ holds.

(5) Section 3.3 presented how to use a viscoelastic damping model to approximate the structural damping and adjust the damping model in a wide frequency band. The viscoelastic damping model gives rise to memory problem. Read the reference[36] and discuss the memory property of viscoelastic damping model based on fractional derivatives. Read the reference[37] and discuss the computation of the response of a viscoelastic MDoF system to a unit impulse.

[32] Greenwood D T (2003). Advanced Dynamics. Cambridge: Cambridge University Press, 34–39, 110–117.

[33] Ardema M D (2005). Analytical Dynamics: Theory and Applications. New York: Kluwer Academic Publisher, 54–62.

[34] Arnold V I (1978). Mathematical Methods of Classical Mechanics. New York: Springer-Verlag, 75–135.

[35] Marsden J E, Ratiu T S (1999). Introduction to Mechanics and Symmetry. 2nd Edition. New York: Springer-Verlag, 121–180.

[36] Du M L, Wang Z H, Hu H Y (2013). Measuring memory with the order of fractional derivative. Scientific Reports, 3: 3431.

[37] Chen Q, Zhu D M (1990). Vibrational analysis theory and applications to elastic–viscoelastic composite structure. Computers and Structures, 37(4): 585–595.

Chapter 4
Vibration Analysis of Discrete Systems

Since the era of computer technology, it has become a routine practice for engineers to model a continuous system in vibration studies via a discrete system and study the vibration of the discrete system numerically and experimentally. As a consequence, the vibration theory of discrete systems has been playing an essential and central role in the vibration mechanics. On one hand, the theory fully presents the scientific beauty of vibration mechanics. On the other hand, the theory, integrated with finite element methods and dynamic measurement techniques, provides many powerful tools for engineers to solve the vibration problems coming from a great variety of fields. Hence, the understanding level of engineers to the vibration theory of discrete systems determines the design and manufacturing levels of industrial products.

Most elementary textbooks of vibration mechanics presented how to analyze the linear un-damped or viscously damped discrete systems, while the monographs of vibration mechanics covered advanced topics of damped discrete systems selected for different fields, such as aerospace engineering[1,2]. This chapter looks like an intermediate between them.

The objective of the chapter is to help readers gain an insight into the vibration theory of discrete systems and discuss some problems in detail. Thus, the content selection of the chapter comes from two considerations. One is to assist readers to master the vibration analysis of discrete systems in time domain and frequency domain, while the other is to help readers perform the dynamic designs and modifications.

The chapter begins with the vibration analysis of a discrete system with non-holonomic constraints, and reveals the dynamic behavior of such a system in the accessible state manifold. This is not only a reasonable extension of Sect. 3.2, but also the content beyond the available textbooks. Afterwards, the chapter addresses several interesting vibration problems of discrete systems, including the node number of a

---

[1] Wijker J (2004). Mechanical Vibrations in Spacecraft Design, Berlin: Springer-Verlag.

[2] Zheng G T (2016). Sequel of Structural Dynamics: Applications to Flight Vehicles. Beijing: Science Press, (in Chinese).

© Science Press 2022
H. Hu, *Vibration Mechanics*,
https://doi.org/10.1007/978-981-16-5457-2_4

natural mode shape, the anti-resonance of a system undergoing harmonically forced vibration, and the dynamic modification of a vibration system so as to demonstrate how to understand, adjust and design the dynamic characteristics of a vibration system.

4.1   Vibration Systems with Non-holonomic Constraints

Based on the study of the degree of freedom, or the DoF for short, of a system with non-holonomic constraints in Sect. 3.2, this section presents the dynamic analysis of such a system in time domain and frequency domain, respectively. The section focuses on the degeneration influence of a part of DoFs on the system dynamics so as to solve problem 2A proposed in Chap. 1.

Let the vibration system of concern be in an n-dimensional configuration space $\mathbb{R}^n$ and be subject to p non-holonomic constraints. In this case, the system has $2n$-dimensional state space $\mathbb{R}^{2n}$, while the p non-holonomic constraints keep the vibration of the system in an accessible state manifold Ω of dimensions $2n - p$. That is, the DoF of the system is $n - p/2$. When the vibration system is linear, the accessible state manifold is $\Omega = \mathbb{R}^{2n-p}$, which is named the *accessible state space*. This section presents the dynamic analysis of such a linear vibration system in time domain and frequency domain successively, as well as the dynamic relation between the two domains.

4.1.1   Dynamic Analysis in Time Domain

4.1.1.1   Dynamic Equations in an Accessible State Space

In the configuration space $\mathbb{R}^n$, the dynamic equations and non-holonomic constraint equations in a matrix form are

$$\begin{cases} M_{aa}\ddot{u}_a(t) + C_{aa}\dot{u}_a(t) + C_{ab}\dot{u}_b(t) + K_{aa}u_a(t) + K_{ab}u_b(t) = f_a(t) \\ C_{ba}\dot{u}_a(t) + C_{bb}\dot{u}_b(t) + K_{ba}u_a(t) + K_{bb}u_b(t) = 0 \end{cases}$$

$$(4.1.1a, b)$$

In Eq. (4.1.1), the system displacement vector is partitioned into two vectors $u_a(t)$ and $u_b(t)$. Vector $u_a : t \mapsto \mathbb{R}^{n-p}$ produces the inertial force vector, i.e., $-M_{aa}\ddot{u}_a(t)$, so that the corresponding degrees of freedom are named the *inertial degrees of freedom*, or the *inertial DoFs* for short, while $f_a : t \mapsto \mathbb{R}^{n-p}$ is the external force vector applied to the inertial DoFs. Vector $u_b : t \mapsto \mathbb{R}^p$ does not produce inertial forces directly and the corresponding degrees of freedom are named the *non-inertial degrees of freedom*, or the *non-inertial DoFs* for short. As studied in Sect. 3.2, each non-inertial DoF plays a role of a half DoF. In addition, the dimensions of matrices in Eq. (4.1.1) are consistent with the vectors multiplied, and both matrices M_{aa} and C_{bb} are invertible.

For the dynamic equations in Eq. (4.1.1) in $\mathbb{R}^n$, the system acceleration vectors $\ddot{u}_a(t)$ and $\ddot{u}_b(t)$ absent from in Eq. (4.1.1b) result in a singular mass matrix in a global sense and lead to the problem of both eigenvalue computation and modal analysis as stated in Sect. 3.2. Following the idea of dealing with the degeneration of DoFs in Sect. 3.2, it is feasible to eliminate the degenerated DoFs due to the non-holonomic constraint equations in Eq. (4.1.1b) and establish the dynamic equations of the system in the accessible state space Ω as follows.

In the state space $\mathbb{R}^{2n}$, Eq. (4.1.1) becomes

$$\begin{cases} \dot{u}_a(t) = v_a(t) \\ \dot{u}_b(t) = v_b(t) \\ \dot{v}_a(t) = M_{aa}^{-1}\left[-C_{aa}v_a(t) - C_{ab}v_b(t) - K_{aa}u_a(t) - K_{ab}u_b(t) + f_a(t)\right] \end{cases}$$

$$(4.1.2a)$$

$$C_{ba}v_a(t) + C_{bb}v_b(t) + K_{ba}u_a(t) + K_{bb}u_b(t) = 0 \qquad (4.1.2b)$$

Equation (4.1.2b) enables one to eliminate $v_b(t)$ and study the system dynamics in the accessible state space Ω as follows. Solving Eq. (4.1.2b) leads to

$$v_b(t) = C_{bb}^{-1}[-C_{ba}v_a(t) - K_{ba}u_a(t) - K_{bb}u_b(t)] \qquad (4.1.3)$$

Substitution of Eq. (4.1.3) into Eq. (4.1.2a) gives

$$\begin{cases} \dot{u}_a(t) = v_a(t), \\ \dot{u}_b(t) = C_{bb}^{-1}[-K_{ba}u_a(t) - K_{bb}u_b(t) - C_{ba}v_a(t)] \\ \dot{v}_a(t) = M_{aa}^{-1}\left[(C_{ab}C_{bb}^{-1}K_{ba} - K_{aa})u_a(t) + (C_{ab}C_{bb}^{-1}K_{bb} - K_{ab})u_b(t) \right. \\ \left. +(C_{ab}C_{bb}^{-1}C_{ba} - C_{aa})v_a(t) + f_a(t)\right] \end{cases}$$

$$(4.1.4)$$

It is easy, hence, to recast Eq. (4.1.4) as the following dynamic equations in terms of the system state vector, namely,

$$\dot{w}(t) = Aw(t) + g(t) \qquad (4.1.5a)$$

where $A \in \mathbb{R}^{(2n-p)\times(2n-p)}$ is the system matrix, $w : t \mapsto \Omega$ is the system state vector and $g : t \mapsto \Omega$ is the external force vector. They can be partitioned as

$$w(t) \equiv \begin{bmatrix} u_a(t) \\ u_b(t) \\ v_a(t) \end{bmatrix}, \quad g(t) \equiv \begin{bmatrix} 0 \\ 0 \\ M_{aa}^{-1} f_a(t) \end{bmatrix} \tag{4.1.6}$$

The system matrix A is a real un-symmetric matrix with the following sub-matrices

$$\begin{cases} A_{aa} = 0, & A_{ba} \equiv -C_{bb}^{-1} K_{ba}, \ A_{ca} \equiv M_{aa}^{-1}(C_{ab}C_{bb}^{-1} K_{ba} - K_{aa}) \\ A_{ab} = 0, & A_{bb} \equiv -C_{bb}^{-1} K_{bb}, \ A_{cb} \equiv M_{aa}^{-1}(C_{ab}C_{bb}^{-1} K_{bb} - K_{ab}) \\ A_{ac} = I_{n-p}, & A_{bc} \equiv -C_{bb}^{-1} C_{ba}, \ A_{cc} \equiv M_{aa}^{-1}(C_{ab}C_{bb}^{-1} C_{ba} - C_{aa}) \end{cases} \tag{4.1.7}$$

From Eq. (4.1.6), the initial system state vector is written as

$$w_0 \equiv w(0) = \begin{bmatrix} u_a(0) \\ u_b(0) \\ v_a(0) \end{bmatrix} \equiv \begin{bmatrix} u_{a0} \\ u_{b0} \\ v_{a0} \end{bmatrix} \equiv \begin{bmatrix} u_0 \\ v_{a0} \end{bmatrix} \in \Omega \tag{4.1.5b}$$

To this end, Eqs. (4.1.5a) and (4.1.5b) give an initial value problem of the system dynamics in the accessible state space Ω.

Example 4.1.1 Given the hydro-elastic vibration isolation system for vehicle engines as shown in Fig. 4.1, establish the dynamic equations of the system in the accessible state space Ω.

According to Sect. 3.2.5, the dynamic equation of the above system in the configuration space $\mathbb{R}^3$ reads

$$m\ddot{u}_1(t) + c_2[\dot{u}_1(t) - \dot{u}_3(t)] + (k_1 + k_4)u_1(t) - k_1 u_2(t) = f_1(t) \tag{a}$$

Fig. 4.1 A hydro-elastic
vibration isolation system for
vehicle engines

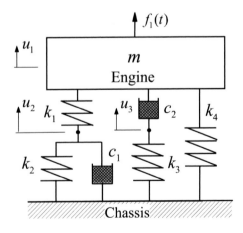

In addition, the system has two non-holonomic constraint equations to describe the relations of springs and dash-pots in Fig. 4.1 as follows

$$
\begin{cases}
c_1\dot{u}_2(t) - k_1 u_1(t) + (k_1 + k_2)u_2(t) = 0 \\
c_2[\dot{u}_3(t) - \dot{u}_1(t)] + k_3 u_3(t) = 0
\end{cases}
\tag{b}
$$

In the state space $\mathbb{R}^6$, the two non-holonomic constraints determine the accessible state space $\Omega = \mathbb{R}^4$. Then, the DoF of the system is $\dim(\Omega)/2 = 2$. To establish the dynamic equations of the system in Ω, define the system state vector in $\mathbb{R}^6$, and write Eqs. (a) and (b) as a matrix form in Eq. (4.1.2), where

$$
\begin{cases}
\boldsymbol{M}_{aa} = [m], \quad \boldsymbol{C}_{aa} = [c_2], \quad \boldsymbol{C}_{ba}^{\mathrm{T}} = \boldsymbol{C}_{ab} = [0 \ -c_2], \quad \boldsymbol{C}_{bb} = \begin{bmatrix} c_1 & 0 \\ 0 & c_2 \end{bmatrix} \\
\boldsymbol{K}_{aa} = [k_1 + k_4], \quad \boldsymbol{K}_{ba}^{\mathrm{T}} = \boldsymbol{K}_{ab} = \begin{bmatrix} -k_1 & 0 \end{bmatrix}, \quad \boldsymbol{K}_{bb} = \begin{bmatrix} k_1 + k_2 & 0 \\ 0 & k_3 \end{bmatrix} \\
\boldsymbol{u}_a(t) = [u_1(t)], \boldsymbol{u}_b(t) = \begin{bmatrix} u_2(t) \\ u_3(t) \end{bmatrix}, \boldsymbol{v}_a = [v_1(t)], \boldsymbol{v}_b = \begin{bmatrix} v_2(t) \\ v_3(t) \end{bmatrix}, \boldsymbol{f}_a = [f_1(t)]
\end{cases}
\tag{c}
$$

Substituting Eq. (c) into Eq. (4.1.3) leads to the system velocity vector to be eliminated as follows

$$
\begin{aligned}
\boldsymbol{v}_b(t) &= \begin{bmatrix} v_2(t) \\ v_3(t) \end{bmatrix} = \begin{bmatrix} c_1 & 0 \\ 0 & c_2 \end{bmatrix}^{-1} \left[\begin{bmatrix} 0 \\ c_2 \end{bmatrix} v_1(t) + \begin{bmatrix} k_1 \\ 0 \end{bmatrix} u_1(t) - \begin{bmatrix} k_1 + k_2 & 0 \\ 0 & k_3 \end{bmatrix} \begin{bmatrix} u_2(t) \\ u_3(t) \end{bmatrix} \right] \\
&= \begin{bmatrix} k_1 u_1(t)/c_1 - (k_1 + k_2)u_2(t)/c_1 \\ v_1(t) - k_3 u_3(t)/c_2 \end{bmatrix}
\end{aligned}
\tag{d}
$$

Substitution of Eq. (c) into Eq. (4.1.7) gives the sub-matrices of system matrix $\boldsymbol{A}$ in Eq. (4.1.5a), namely,

$$
\begin{cases}
\boldsymbol{A}_{aa} = [0], \quad \boldsymbol{A}_{ab} = \begin{bmatrix} 0 & 0 \end{bmatrix}, \quad \boldsymbol{A}_{ac} = [1] \\
\boldsymbol{A}_{ba} = \begin{bmatrix} k_1/c_1 \\ 0 \end{bmatrix}, \quad \boldsymbol{A}_{bb} = -\begin{bmatrix} (k_1 + k_2)/c_1 & 0 \\ 0 & k_3/c_2 \end{bmatrix}, \quad \boldsymbol{A}_{bc} = \begin{bmatrix} 0 \\ 1 \end{bmatrix} \\
\boldsymbol{A}_{ca} = [-(k_1 + k_4)/m], \quad \boldsymbol{A}_{cb} = \begin{bmatrix} k_1/m & -k_3/m \end{bmatrix}, \quad \boldsymbol{A}_{cc} = [0]
\end{cases}
\tag{e}
$$

Hence, the detailed expression of Eq. (4.1.5a) reads

$$
\begin{bmatrix} \dot{u}_1(t) \\ \dot{u}_2(t) \\ \dot{u}_3(t) \\ \dot{v}_1(t) \end{bmatrix} = \begin{bmatrix} 0 & 0 & 0 & 1 \\ k_1/c_1 & -(k_1+k_2)/c_1 & 0 & 0 \\ 0 & 0 & -k_3/c_2 & 1 \\ -(k_1+k_4)/m & k_1/m & -k_3/m & 0 \end{bmatrix} \begin{bmatrix} u_1(t) \\ u_2(t) \\ u_3(t) \\ v_1(t) \end{bmatrix} + \begin{bmatrix} 0 \\ 0 \\ 0 \\ f_1(t)/m \end{bmatrix}
$$

$$(f)$$

If one does not care about the above matrix manipulations, it is straightforward to derive velocities $v_2(t) = \dot{u}_2(t)$ and $v_3(t) = \dot{u}_3(t)$ from Eq. (b). Substitution of them into Eqs. (a) and (b) leads to the same dynamic equations in Eq. (c), namely,

$$
\begin{cases}
\dot{u}_1(t) = v_1(t) \\[2mm]
\dot{u}_2(t) = \dfrac{k_1}{c_1} u_1(t) - \dfrac{k_1+k_2}{c_1} u_2(t) \\[3mm]
\dot{u}_3(t) = -\dfrac{k_3}{c_2} u_3(t) + v_1(t) \\[3mm]
\dot{v}_1(t) = -\dfrac{k_1+k_4}{m} u_1(t) + \dfrac{k_1}{m} u_2(t) - \dfrac{k_3}{m} u_3(t) + \dfrac{f_1(t)}{m}
\end{cases}
$$

$$(g)$$

4.1.1.2   Modal Vibrations

According to the theory of linear differential equations, let $w(t) = \varphi \exp(\lambda t)$ be the solution of Eq. (4.1.5a) and $\varphi \in \mathbb{R}^{2n-p}$ be a vector to be determined. Substituting $w(t) = \varphi \exp(\lambda t)$ into Eq. (4.1.5a) leads to an eigenvalue problem

$$(A - \lambda I_{2n-p})\varphi = 0 \tag{4.1.8}$$

The corresponding characteristic equation reads

$$\det(A - \lambda I_{2n-p}) = 0 \tag{4.1.9}$$

As system matrix A is a real matrix, Eq. (4.1.9) is an algebraic equation with real coefficients. Thus, the solutions of Eq. (4.1.9) are either real eigenvalues or conjugate complex eigenvalues. For an asymptotically stable system, all the eigenvalues have negative real parts.

Without loss of generality, assume that Eq. (4.1.9) has $n - p$ pairs of conjugate complex eigenvalues, as well as p real eigenvalues coming from the DoF degeneration due to non-holonomic constraints. Let the above eigenvalues be expressed as

$$\begin{cases} \lambda_r = \overline{\lambda}_{r+n-p} = -\gamma_r + i\omega_{dr}, \quad \gamma_r > 0, \quad \omega_{dr} > 0, \quad r = 1, 2, \ldots, n - p \\ \lambda_s = -\gamma_s < 0, \quad s = 2(n - p) + 1, \ 2(n - p) + 2, \ldots 2n - p \end{cases}$$

$$(4.1.10)$$

Thus, the solution $\boldsymbol{w}(t) = \boldsymbol{\varphi} \exp(\lambda t)$ of Eq. (4.1.5a) becomes

$$\boldsymbol{w}_r(t) = \boldsymbol{\varphi}_r \exp(\lambda_r t), \quad r = 1, 2, \ldots, 2n - p \tag{4.1.11}$$

Hereinafter, Eq. (4.1.11) is named the *modal vibration* of the system described by Eq. (4.1.5).

In view of the category of the above eigenvalues, the modal vibration corresponding to a pair of conjugate complex eigenvalues yields

$$\boldsymbol{w}_r(t) = \boldsymbol{\varphi}_r \exp(\lambda_r t) + \overline{\boldsymbol{\varphi}}_r \exp(\overline{\lambda}_r t)$$
$$= 2 \exp(-\gamma_r t) \big[\mathrm{Re}(\boldsymbol{\varphi}_r) \cos(\omega_{dr} t) - \mathrm{Im}(\boldsymbol{\varphi}_r) \sin(\omega_{dr} t) \big] \tag{4.1.12}$$

This is a decaying vibration with damped natural frequency ω_{dr} and is named the *complex modal vibration* of order r, and λ_r is called the *complex frequency*.

To check the relation of different DoFs in the complex vibration of order r, recast Eq. (4.1.12) as

$$\boldsymbol{w}_r(t) = \exp(-\gamma_r t) \begin{bmatrix} a_{1,r} \cos(\omega_{dr} t + \theta_{1r}) \\ \vdots \\ a_{2n-p,r} \cos(\omega_{dr} t + \theta_{2n-p,r}) \end{bmatrix} \tag{4.1.13}$$

where the amplitudes and phase angles satisfy

$$\begin{cases} a_{i,r} = 2\sqrt{\mathrm{Re}^2(\varphi_{ir}) + \mathrm{Im}^2(\varphi_{ir})} \\ \theta_{i,r} = \tan^{-1}[\mathrm{Im}(\varphi_{ir})/\mathrm{Re}(\varphi_{ir})], \quad i = 1, 2, \ldots, 2n - p \end{cases} \tag{4.1.14}$$

As shown in Eq. (4.1.13), the phase angles $\theta_{i,r}$ are distinct in general. Hence, the DoFs in a complex vibration simultaneously reach neither the equilibrium position of the system nor the maximal vibration. As such, the vibration shapes of the system at different moments are not similar. The above behaviors look similar to the complex mode shapes of a non-proportionally damped system.

In addition, corresponding to p negative real eigenvalues, the system exhibits the over-damped non-oscillatory motions as follows

$$\boldsymbol{w}_s(t) = \boldsymbol{\varphi}_s \exp(-\gamma_s t), \quad s = 2(n - p) + 1, 2(n - p) + 2, \ldots 2n - p \tag{4.1.15}$$

This is a kind of modal motions caused by non-holonomic constraints. They are the non-oscillatory motions due to non-inertial DoFs and called the *response of creep modes*.

As matrix A is an un-symmetric real matrix, define the corresponding *left eigenvalue problem*, namely, the *adjoint eigenvalue problem* in linear algebra, as

$$\boldsymbol{\psi}^{\mathrm{T}}(A - \lambda I_{2n-p}) = 0 \quad \Leftrightarrow \quad (A^{\mathrm{T}} - \lambda I_{2n-p})\boldsymbol{\psi} = 0 \tag{4.1.16}$$

Accordingly, Eq. (4.1.8) is the *right eigenvalue problem*. As well known, the determinants of a matrix and its transpose are the same. Thus, Eqs. (4.1.16) and (4.1.8) have the same characteristic equation. Hence, they have the same set of eigenvalues, but the *left eigenvector* $\boldsymbol{\psi}$ and the *right eigenvector* $\boldsymbol{\varphi}$ are different.

Now, we prove that if any two eigenvalues of real matrix A are distinct, the above left eigenvector and right eigenvector satisfy the following two sets of *orthogonal relations*

$$\boldsymbol{\psi}_s^{\mathrm{T}}\boldsymbol{\varphi}_r = 0, \quad \boldsymbol{\psi}_s^{\mathrm{T}}A\boldsymbol{\varphi}_r = 0, \quad \lambda_r \neq \lambda_s, \quad r, s = 1, 2, \ldots, N \tag{4.1.17}$$

For this purpose, take the r-th eigen-relation in Eq. (4.1.8) and the s-th relation in Eq. (4.1.16), namely,

$$A\boldsymbol{\varphi}_r = \lambda_r \boldsymbol{\varphi}_r, \quad A^{\mathrm{T}}\boldsymbol{\psi}_s = \lambda_s \boldsymbol{\psi}_s \tag{4.1.18}$$

Pre-multiplying the first equation in Eq. (4.1.18) by $\boldsymbol{\psi}_s^{\mathrm{T}}$ and the second equation in Eq. (4.1.18) by $\boldsymbol{\varphi}_r^{\mathrm{T}}$ leads to

$$\boldsymbol{\psi}_s^{\mathrm{T}}A\boldsymbol{\varphi}_r = \lambda_r \boldsymbol{\psi}_s^{\mathrm{T}}\boldsymbol{\varphi}_r, \quad \boldsymbol{\varphi}_r^{\mathrm{T}}A^{\mathrm{T}}\boldsymbol{\psi}_s = \lambda_s \boldsymbol{\varphi}_r^{\mathrm{T}}\boldsymbol{\psi}_s \tag{4.1.19}$$

Subtracting the first equation and the transpose of the second equation in Eq. (4.1.19) gives

$$(\lambda_r - \lambda_s)\boldsymbol{\psi}_s^{\mathrm{T}}\boldsymbol{\varphi}_r = 0 \tag{4.1.20}$$

As any two eigenvalues are distinct, Eq. (4.1.20) leads to the first set of orthogonal relations in Eq. (4.1.17). Substituting this result into Eq. (4.1.19) gives the second set of orthogonal relations. This completes the proof.

When $r = s$ holds, let Eq. (4.1.8) be pre-multiplied by $\boldsymbol{\psi}_r^{\mathrm{T}}$ and substituted with eigenvalue λ_r and right eigenvector $\boldsymbol{\varphi}_r$, we have

$$\boldsymbol{\psi}_r^{\mathrm{T}}A\boldsymbol{\varphi}_r - \lambda_r \boldsymbol{\psi}_r^{\mathrm{T}}\boldsymbol{\varphi}_r = 0, \quad r, s = 1, 2, \ldots, 2n - p \tag{4.1.21}$$

It is easy to recast Eq. (4.1.21) as

$$\beta_r - \alpha_r \lambda_r = 0, \quad \beta_r \equiv \boldsymbol{\psi}_r^{\mathrm{T}} \boldsymbol{A} \boldsymbol{\varphi}_r, \quad \alpha_r \equiv \boldsymbol{\psi}_r^{\mathrm{T}} \boldsymbol{\varphi}_r \neq 0, \quad r, s = 1, 2, \ldots, 2n - p \tag{4.1.22}$$

As a result, the combination of Eqs. (4.1.17) and (4.1.22) gives the orthogonal relations of left and right eigenvectors as follows

$$\boldsymbol{\psi}_s^{\mathrm{T}} \boldsymbol{\varphi}_r = \alpha_r \delta_{rs}, \quad \boldsymbol{\psi}_s^{\mathrm{T}} \boldsymbol{A} \boldsymbol{\varphi}_r = \alpha_r \lambda_r \delta_{rs}, \quad r, s = 1, 2, \ldots, 2n - p \tag{4.1.23}$$

where δ_{rs} is the *Kronecker symbol* defined as

$$\delta_{rs} \equiv \begin{cases} 1, \ r = s \\ 0, \ r \neq s \end{cases} \tag{4.1.24}$$

The orthogonal relations in Eq. (4.1.23) are also named the *biorthogonality of complex mode shapes* because the orthogonality holds for a pair of left and right eigenvectors with distinct eigenvalues. The biorthogonality will serve as the basis for subsequent modal analysis.

4.1.1.3 Free Vibrations

According to the biorthogonality of the eigenvectors, the $2n - p$ eigenvectors are linearly independent and can serve as the base vectors for the accessible state space Ω such that the system state vector can be expressed as

$$\boldsymbol{w}(t) = \sum_{s=1}^{2n-p} \boldsymbol{\varphi}_s q_s(t) \tag{4.1.25}$$

To study the free vibration of the system, substituting both $\boldsymbol{g}(t) = \boldsymbol{0}$ and Eq. (4.1.25) into Eq. (4.1.5a) leads to

$$\sum_{s=1}^{2n-p} \boldsymbol{\varphi}_s \dot{q}_s(t) = \boldsymbol{A} \sum_{s=1}^{2n-p} \boldsymbol{\varphi}_s q_s(t) \tag{4.1.26}$$

Pre-multiplying Eq. (4.1.26) by $\boldsymbol{\psi}_r^{\mathrm{T}}$ and utilizing the biorthogonality in Eq. (4.1.23), we have a set of decoupled ordinary differential equations as follows

$$\dot{q}_r(t) = \lambda_r q_r(t), \quad r = 1, 2, \ldots, 2n - p \tag{4.1.27a}$$

Let Eq. (4.1.5b) be pre-multiplied by $\boldsymbol{\psi}_r^{\mathrm{T}}$ and substituted by $t = 0$, we derive, using the biorthogonality in Eq. (4.1.23), the initial conditions of Eq. (4.1.27a) as following

$$q_r(0) = \frac{1}{\alpha_r} \boldsymbol{\psi}_r^{\mathrm{T}} \boldsymbol{w}_0, \quad r = 1, 2, \ldots, 2n - p \tag{4.1.27b}$$

Solving the initial value problem in Eq. (4.1.27) leads to

$$q_r(t) = q(0) \exp(\lambda_r t) = \frac{\boldsymbol{\psi}_r^{\mathrm{T}} \boldsymbol{w}_0}{\alpha_r} \exp(\lambda_r t), \quad t \geq 0, \quad r = 1, 2, \ldots, 2n - p \tag{4.1.28}$$

Let Eq. (4.1.28) be substituted into Eq. (4.1.25) and subscript s be replaced by r, we get the free vibration of the system in Ω, namely,

$$\boldsymbol{w}(t) = \sum_{r=1}^{2n-p} \frac{\boldsymbol{\varphi}_r \boldsymbol{\psi}_r^{\mathrm{T}} \boldsymbol{w}_0}{\alpha_r} \exp(\lambda_r t), \quad t \geq 0 \tag{4.1.29}$$

Here, parameter α_r presents the contribution of the r-th mode to the free vibration, and is named the r-th *modal participation factor*. It is a complex number for the conjugate complex eigenvalue.

In order to derive the system response under the initial unit displacement and the initial unit velocity, the square matrix $\boldsymbol{\varphi}_r \boldsymbol{\psi}_r^{\mathrm{T}}$ of order $2n - p$ in Eq. (4.1.29) is partitioned as follows

$$\boldsymbol{\varphi}_r \boldsymbol{\psi}_r^{\mathrm{T}} \equiv \begin{bmatrix} [\boldsymbol{\varphi}_r \boldsymbol{\psi}_r^{\mathrm{T}}]_{nn} & [\boldsymbol{\varphi}_r \boldsymbol{\psi}_r^{\mathrm{T}}]_{na} \\ [\boldsymbol{\varphi}_r \boldsymbol{\psi}_r^{\mathrm{T}}]_{an} & [\boldsymbol{\varphi}_r \boldsymbol{\psi}_r^{\mathrm{T}}]_{aa} \end{bmatrix}$$

$$\equiv \begin{bmatrix} [\boldsymbol{\varphi}_r \boldsymbol{\psi}_r^{\mathrm{T}}]_n & [\boldsymbol{\varphi}_r \boldsymbol{\psi}_r^{\mathrm{T}}]_a \end{bmatrix}, \quad r = 1, 2, \ldots, 2n - p \tag{4.1.30}$$

Then, Eq. (4.1.29) can be recast as

$$\boldsymbol{u}(t) = \sum_{r=1}^{2n-p} \frac{[\boldsymbol{\varphi}_r \boldsymbol{\psi}_r^{\mathrm{T}}]_{nn} \boldsymbol{u}_0}{\alpha_r} \exp(\lambda_r t) + \sum_{r=1}^{2n-p} \frac{[\boldsymbol{\varphi}_r \boldsymbol{\psi}_r^{\mathrm{T}}]_{na} \boldsymbol{v}_{a0}}{\alpha_r} \exp(\lambda_r t)$$

$$= \left\{ \sum_{r=1}^{2n-p} \frac{[\boldsymbol{\varphi}_r \boldsymbol{\psi}_r^{\mathrm{T}}]_{nn}}{\alpha_r} \exp(\lambda_r t) \right\} \boldsymbol{u}_0 + \left\{ \sum_{r=1}^{2n-p} \frac{[\boldsymbol{\varphi}_r \boldsymbol{\psi}_r^{\mathrm{T}}]_{na}}{\alpha_r} \exp(\lambda_r t) \right\} \boldsymbol{v}_{a0}$$

$$= \boldsymbol{U}(t) \boldsymbol{u}_0 + \boldsymbol{V}(t) \boldsymbol{v}_{a0}, \quad t \geq 0 \tag{4.1.31}$$

It is easy to derive the system response matrix $\boldsymbol{U}(t)$ under the vector of initial unit displacements and the system response matrix $\boldsymbol{V}(t)$ under the vector of initial unit velocities as follows

$$U(t) \equiv \sum_{r=1}^{2n-p} \frac{[\boldsymbol{\varphi}_r \boldsymbol{\psi}_r^{\mathrm{T}}]_{nn}}{\alpha_r} \exp(\lambda_r t), \quad V(t) \equiv \sum_{r=1}^{2n-p} \frac{[\boldsymbol{\varphi}_r \boldsymbol{\psi}_r^{\mathrm{T}}]_{na}}{\alpha_r} \exp(\lambda_r t), \quad t \geq 0$$

(4.1.32)

Example 4.1.2 The hydro-elastic vibration isolation system for vehicle engines discussed in Example 4.1.1 has the following system parameters

$$\begin{cases} m = 20 \text{ kg}, \quad c_1 = 300 \text{ N} \cdot \text{s/m}, \quad c_2 = 100 \text{ N} \cdot \text{s/m} \\ k_1 = 500 \text{ N/m}, \quad k_2 = 500 \text{ N/m}, \quad k_3 = 1000 \text{ N/m}, \quad k_4 = 5000 \text{ N/m} \end{cases}$$ (a)

Study the free vibration of the system under the following initial disturbance

$$u_1(0) = u_2(0) = u_3(0) = 0.01 \text{ m}, \quad v_1(0) = 0.0 \text{ m/s}$$ (b)

Substituting Eq. (a) into system matrix A in Example 4.1.1 and solving the eigenvalue problem via software Maple, we get the following eigenvalues

$$\lambda_1 = \bar{\lambda}_2 = -0.690 + 17.706 \text{ i}, \quad \lambda_3 = -3.174, \quad \lambda_4 = -8.780$$ (c)

For sake of brevity, the right eigenvectors and the modal participation factors are not listed here. Substitution of these results into Eq. (4.1.29) leads to the displacement response of the system under the above initial disturbance, namely,

$$\begin{aligned} u_1(t) = \{&(4.743 - 0.165 \text{ i}) \exp[(-0.690 + 17.706 \text{ i})t] \\ &+(4.743 + 0.165 \text{ i}) \exp[(-0.690 - 17.706 \text{ i})t] \\ &+0.933 \exp(-3.174t) - 0.418 \exp(-8.780t)\} \times 10^{-3} \text{ m}, \quad t \geq 0 \end{aligned}$$ (d)

The corresponding real expression reads

$$\begin{aligned} u_1(t) = \{&\exp(-0.690t)[9.485 \cos(17.706t) + 0.329 \sin(17.706t)] \\ &+ 0.933 \exp(-3.174t) - 0.418 \exp(-8.780t)\} \times 10^{-3} \text{ m}, \quad t \geq 0 \end{aligned}$$ (e)

Figure 4.2 shows a comparison between the solution in Eq. (e) and the numerical solution via Ruge-Kutta integrator in software Maple. The two solutions get very good agreement. The dashed line in Fig. 4.2 presents the displacement corresponding to the two negative real eigenvalues, namely, the response of creep modes, which occupies only a very small percentage in the entire response of the system.

Remark 4.1.1 As each dashpot is in serial with a spring in the system, the increase of damping coefficient does not promote the decaying rate of the complex modal

Fig. 4.2 The free vibration of the hydro-elastic vibration isolation system for vehicle engines under an initial disturbance (solid line: the solution in Eq. (e), solid circle: numerical solution, dashed line: the response of creep modes)

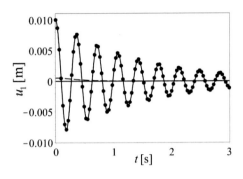

Fig. 4.3 The variations of real parts of eigenvalues of the hydro-elastic vibration isolation system for vehicle engines with an increase of damping coefficient c_1 (square: $\mathrm{Re}(\lambda_{1,2})$, rhombus: λ_3, circle: λ_4)

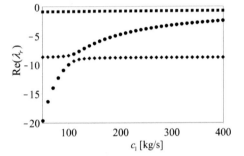

vibration, but may even reduce the decaying rate of the response of creep modes. Consider, as an example, the design of the left dashpot, the damping coefficient c_1 is adjustable via the orifice. Figure 4.3 presents the variation of the real parts of all eigenvalues with increasing damping coefficient c_1. It is interesting to notice that with an increase of damping coefficient c_1, only the negative real eigenvalue λ_4 greatly increases, but the real parts $\mathrm{Re}(\lambda_{1,2})$ of a pair of conjugate eigenvalues remain almost unchanged. That is, increasing damping coefficient c_1 is not able to suppress the system vibration. This gives a reasonable solution to Problem 2A proposed in Chap. 1.

4.1.1.4 Forced Vibrations

To begin with, we analyze the system response to an ideal unit impulse. Let the system be initially at rest, apply an ideal unit impulse $\delta(t)$, i.e., the Dirac delta function, to the j-th inertial DoF of the system at moment $t = 0$ and denote the external force vector as

$$\boldsymbol{f}_a(t) = \delta(t)\boldsymbol{\delta}_j, \quad \boldsymbol{\delta}_j \equiv [0 \cdots 0 \underbrace{1}_{j} \ 0 \cdots 0]^{\mathrm{T}} \in \mathbb{R}^{n-p} \qquad (4.1.33)$$

In view of the theorem of momentum, the system velocity vector $v_a(0^+)$ of the inertial DoFs at moment $t = 0^+$ yields

$$M_{aa} v_a(0^+) = \delta_j \tag{4.1.34}$$

The initial conditions of the system in Ω, thus, satisfy

$$w_0 = [u_a^T(0^+) \quad u_b^T(0^+) \quad v_a^T(0^+)]^T = [0^T \quad 0^T \quad M_{aa}^{-1}\delta_j]^T \tag{4.1.35}$$

From Eq. (4.1.29), the free vibration of the system to excitation $\delta(t)\delta_j$ in Ω is

$$w(t) = \sum_{r=1}^{2n-p} \frac{\varphi_r \psi_r^T w_0}{\alpha_r} \exp(\lambda_r t) = \sum_{r=1}^{2n-p} \frac{[\varphi_r \psi_r^T]_a M_{aa}^{-1}\delta_j}{\alpha_r} \exp(\lambda_r t), \quad t \geq 0 \tag{4.1.36}$$

where $[\varphi_r \psi_r^T]_a$ consists of the last $n - p$ columns in the square matrix $\varphi_r \psi_r^T$ of order $2n - p$, namely, the second partition for $\varphi_r \psi_r^T$ in Eq. (4.1.30).

To check the system displacement vector in the configuration space, one simply takes the first n rows in Eq. (4.1.36) and gets

$$u(t) = \sum_{r=1}^{2n-p} \frac{[\varphi_r \psi_r^T]_{na} M_{aa}^{-1}\delta_j}{\alpha_r} \exp(\lambda_r t), \quad t \geq 0 \tag{4.1.37}$$

where $[\varphi_r \psi_r^T]_{na}$ comes from the first partition in Eq. (4.1.30).

To emphasize the j-th inertial DoF to the excitation in Eq. (4.1.33), denote $u(t)$ in Eq. (4.1.37) as $u_j(t)$. Let subscript j run from 1 to $n - p$ and assemble all column vectors $u_j(t)$, $j = 1, 2, \ldots, n - p$ as a matrix $h : t \mapsto \mathbb{R}^{n \times (n-p)}$. This is the system displacement matrix subject to an initial unit impulse at each inertial DoF, or called the *unit impulse response matrix* of the system for short. As $n - p < n$ holds for the system with non-holonomic constraints, $h(t)$ is not a square matrix, but a "high matrix" instead.

Noting that $u_j(t)$ is the j-th column of matrix $h(t)$, we get the modal expression of matrix $h(t)$, from Eq. (4.1.37), as follows

$$h(t) = \sum_{r=1}^{2n-p} \frac{[\varphi_r \psi_r^T]_{na} M_{aa}^{-1}}{\alpha_r} \exp(\lambda_r t), \quad t \geq 0 \tag{4.1.38}$$

A comparison between the second equation in Eqs. (4.1.32) and (4.1.38) leads to the following relation

$$\boldsymbol{h}(t) = \boldsymbol{V}(t)\boldsymbol{M}_{aa}^{-1}, \quad t \geq 0 \tag{4.1.39}$$

According to the superposition principle of linear systems, the system displacement vector of an initially rest system subject to external force vector $\boldsymbol{f}_a(t)$ yields the following Duhamel integral

$$\boldsymbol{u}(t) = \int_0^t \boldsymbol{h}(t - \tau)\boldsymbol{f}_a(\tau)\mathrm{d}\tau, \quad t \geq 0 \tag{4.1.40}$$

In view of Eqs. (4.1.32) and (4.1.40), the superposition principle of linear systems leads to the system displacement vector subject to both initial disturbance vectors $\boldsymbol{u}_0$ and $\boldsymbol{v}_{a0}$, and external force vector $\boldsymbol{f}_a(t)$ as follows

$$\boldsymbol{u}(t) = \boldsymbol{U}(t)\boldsymbol{u}_0 + \boldsymbol{V}(t)\boldsymbol{v}_{a0} + \int_0^t \boldsymbol{h}(t - \tau)\boldsymbol{f}_a(\tau)\mathrm{d}\tau, \quad t \geq 0 \tag{4.1.41}$$

4.1.2 Dynamic Analysis in Frequency Domain

4.1.2.1 Frequency Response Matrix

Applying Fourier transform to both sides of Eq. (4.1.1) leads to

$$\begin{cases} (\boldsymbol{K}_{aa} - \omega^2 \boldsymbol{M}_{aa} + \mathrm{i}\omega \boldsymbol{C}_{aa})\boldsymbol{U}_a(\omega) + (\boldsymbol{K}_{ab} + \mathrm{i}\omega \boldsymbol{C}_{ab})\boldsymbol{U}_b(\omega) = \boldsymbol{F}_a(\omega) \\ (\boldsymbol{K}_{ba} + \mathrm{i}\omega \boldsymbol{C}_{ba})\boldsymbol{U}_a(\omega) + (\boldsymbol{K}_{bb} + \mathrm{i}\omega \boldsymbol{C}_{bb})\boldsymbol{U}_b(\omega) = \boldsymbol{0} \end{cases} \tag{4.1.42}$$

Solving the second equation in Eq. (4.1.42) for $\boldsymbol{U}_b(\omega)$ gives

$$\boldsymbol{U}_b(\omega) = -(\boldsymbol{K}_{bb} + \mathrm{i}\omega \boldsymbol{C}_{bb})^{-1}(\boldsymbol{K}_{ba} + \mathrm{i}\omega \boldsymbol{C}_{ba})\boldsymbol{U}_a(\omega) \tag{4.1.43}$$

Substitution of Eq. (4.1.43) into the first equation in Eq. (4.1.42) leads to

$$\boldsymbol{Z}_{aa}(\omega)\boldsymbol{U}_a(\omega) = \boldsymbol{F}_a(\omega) \tag{4.1.44}$$

where $\boldsymbol{Z}_{aa} : \omega \mapsto \mathbb{R}^{(n-p)\times(n-p)}$ is the *displacement impedance matrix of inertial DoFs* of the system and yields

$$\boldsymbol{Z}_{aa}(\omega) \equiv (\boldsymbol{K}_{aa} - \omega^2 \boldsymbol{M}_{aa} + \mathrm{i}\omega \boldsymbol{C}_{aa})$$
$$- (\boldsymbol{K}_{ab} + \mathrm{i}\omega \boldsymbol{C}_{ab})(\boldsymbol{K}_{bb} + \mathrm{i}\omega \boldsymbol{C}_{bb})^{-1}(\boldsymbol{K}_{ba} + \mathrm{i}\omega \boldsymbol{C}_{ba}) \tag{4.1.45}$$

Define the *frequency response matrix of inertial DoFs* of the system as

$$H_{aa}(\omega) \equiv Z_{aa}^{-1}(\omega) \tag{4.1.46}$$

Then, we have the frequency response vector of the inertial DoFs of the system, namely,

$$U_a(\omega) = H_{aa}(\omega)F_a(\omega) \tag{4.1.47}$$

Substitution of Eq. (4.1.47) into Eq. (4.1.43) leads to a new expression of vector $U_b(\omega)$ in terms of vector $F_a(\omega)$. Hence, the frequency response vector of the entire system yields

$$U(\omega) \equiv \begin{bmatrix} U_a(\omega) \\ U_b(\omega) \end{bmatrix} = H(\omega)F_a(\omega) \tag{4.1.48}$$

where $H(\omega)$ is the *frequency response matrix of the system* and reads

$$H(\omega) \equiv \begin{bmatrix} H_{aa}(\omega) \\ -(K_{bb} + i\omega C_{bb})^{-1}(K_{ba} + i\omega C_{ba})H_{aa}(\omega) \end{bmatrix} \tag{4.1.49}$$

Remark 4.1.2 The non-holonomic constraints lead to a singular mass matrix of linear system (4.1.1) in a global sense, but do not produce any influence on the dynamic analysis in frequency domain. In fact, the solution of Eq. (4.1.42) has nothing to do with the inertial sub-matrices corresponding to the system acceleration vectors $\ddot{u}_a(t)$ and $\ddot{u}_b(t)$ in the second equation. Thus, the singularity of the mass matrix does not affect the dynamic analysis in frequency domain.

Remark 4.1.3 Theoretically speaking, one can also obtain the above result in frequency domain via applying Fourier transform to both sides of Eq. (4.1.4) in the accessible state space. This process, however, deals with the inverse Fourier transform of $v_a(t)$ and becomes complicated. In the theory of linear control, hence, it is seldom to make the frequency response analysis of a system from the dynamic equations in state space.

Example 4.1.3 Given the hydro-elastic vibration isolation system for vehicle engines studied in Example 4.1.1 and Example 4.1.2, compute and analyze the frequency response function of the inertial DoF of the system.

According to Example 4.1.1, the dynamic equations of the system are as follows

$$\begin{cases} m\ddot{u}_1(t) + c_2[\dot{u}_1(t) - \dot{u}_3(t)] + (k_1 + k_4)u_1(t) - k_1u_2(t) = f_1(t) \\ c_1\dot{u}_2(t) - k_1u_1(t) + (k_1 + k_2)u_2(t) = 0 \\ c_2[\dot{u}_3(t) - \dot{u}_1(t)] + k_3u_3(t) = 0 \end{cases} \tag{a}$$

From Eq. (c) in Example 4.1.1, the sub-matrices to be used are

$$
\begin{cases}
\boldsymbol{M}_{aa} = [m], \quad \boldsymbol{C}_{aa} = [c_2], \quad \boldsymbol{C}_{ba}^{\mathrm{T}} = \boldsymbol{C}_{ab} = [0 \; -c_2], \quad \boldsymbol{C}_{bb} = \begin{bmatrix} c_1 & 0 \\ 0 & c_2 \end{bmatrix} \\[12pt]
\boldsymbol{K}_{aa} = [k_1 + k_4], \quad \boldsymbol{K}_{ba}^{\mathrm{T}} = \boldsymbol{K}_{ab} = [-k_1 \; 0], \quad \boldsymbol{K}_{bb} = \begin{bmatrix} k_1 + k_2 & 0 \\ 0 & k_3 \end{bmatrix}
\end{cases}
\tag{b}
$$

Substituting Eq. (b) into Eq. (4.1.45), we derive the displacement impedance function of the inertial DoF of the system as follows

$$
\begin{aligned}
Z_{aa}(\omega) &= k_1 + k_4 - m\omega^2 + ic_2\omega \\
&\quad - [-k_1 - ic_2\omega] \begin{bmatrix} \dfrac{1}{k_1 + k_2 + ic_1\omega} & 0 \\ 0 & \dfrac{1}{k_3 + ic_2\omega} \end{bmatrix} [-k_1 - ic_2\omega] \\
&= k_1 + k_4 - m\omega^2 + ic_2\omega - \dfrac{k_1^2}{k_1 + k_2 + ic_1\omega} + \dfrac{c_2^2\omega^2}{k_3 + ic_2\omega}
\end{aligned}
\tag{c}
$$

The corresponding frequency response function is

$$
\begin{cases}
H_{aa}(\omega) = \dfrac{1}{Z_{aa}(\omega)} = \dfrac{(k_1 + k_2 + ic_1\omega)(k_3 + ic_2\omega)}{\Delta(i\omega)} \\[10pt]
\Delta(i\omega) \equiv (k_1 + k_2 + ic_1\omega)(k_3 + ic_2\omega)\left[k_1 + k_4 + m(i\omega)^2 + ic_2\omega\right] \\[6pt]
\quad - k_1^2(k_3 + ic_2\omega) + c_2^2\omega^2(k_1 + k_2 + ic_1\omega).
\end{cases}
\tag{d}
$$

Substitution of the system parameters of Eq. (a) in Example 4.1.2 into Eq. (d) gives the amplitude of frequency response function of inertial DoF of the system shown in Fig. 4.4a. The subfigure exhibits a single peak and looks like the amplitude of frequency response function of an SDoF system. The analysis in Sect. 3.2.5,

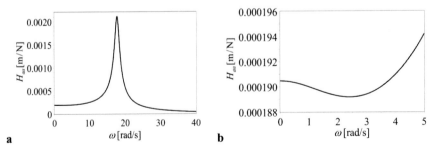

a   b

Fig. 4.4 The amplitude of frequency response function of the hydro-elastic vibration isolation system for vehicle engines. **a** the global view; **b** the zoomed view for a lower frequency band

however, pointed out that the system behaves like a 2-DoF system. How should one understand this contradiction?

As a matter of fact, the frequency response function $H_{aa}(\omega)$ in Eq. (d) becomes a transfer function $H_{aa}(\sigma)$ if the independent variable $i\omega$ in Eq. (d) is replaced with a complex independent variable σ. To this end, the numerator of $H_{aa}(\sigma)$ is a second-order polynomial of σ, while the denominator $\Delta(\sigma)$ of $H_{aa}(\sigma)$ is a fourth-order polynomial of σ. Hence, the transfer function $H_{aa}(\sigma)$ has two zeros and four poles in the complex plane so that the system has 2-DoFs as discussed in Sect. 3.2.5. For the system parameters used here, however, the system has only one pair of conjugate complex eigenvalues. That is, the transfer function $H_{aa}(\sigma)$ has only one pair of conjugate poles, which is $\pm\text{Im}(\lambda_{1,2}) = \pm 17.706$ i computed in Example 4.1.2, on the imaginary axis $s = i\omega$. Hence, the frequency response function in Fig. 4.4a has a single peak and the system looks like an SDoF system. That is, the number of resonant peaks of a linear system with non-holonomic constraints may be reasonably less than the DoFs of the system.

The careful check of the amplitude of frequency response function, however, finds a small pit in the lower frequency band in Fig. 4.4b. This behavior differs from that of an SDoF system. That is, it is possible to make the frequency response exhibit a valley, like the anti-resonance of a 2-DoF system, in a lower frequency band by adjusting the system parameters. The above discussion verifies the study in Sect. 3.2.5 on Problem 2B proposed in Chap. 1.

4.1.2.2 Modal Expansion of a Frequency Response Matrix

Applying Fourier transform to both sides of Eq. (4.1.38), we derive the modal expansion of the frequency response matrix as follows

$$H(\omega) = \sum_{r=1}^{2n-p} \frac{[\boldsymbol{\varphi}_r \boldsymbol{\psi}_r^{\text{T}}]_{na} M_{aa}^{-1}}{\alpha_r} \mathscr{F}[\exp(\lambda_r t)] = \sum_{r=1}^{2n-p} \frac{[\boldsymbol{\varphi}_r \boldsymbol{\psi}_r^{\text{T}}]_{na} M_{aa}^{-1}}{\alpha_r} \frac{1}{i\omega - \lambda_r}$$

(4.1.50)

Equation (4.1.50) is equivalent to Eq. (4.1.49), but clearly shows the contribution of each complex mode to the frequency response matrix.

Example 4.1.4 Given the hydro-elastic vibration isolation system for vehicle engines in Example 4.1.3, compute the impulse response function of the system from the frequency response function of the system.

Applying the inverse Fourier transform to both sides of Eq. (d) in Example 4.1.3, we get the impulse response function of the inertial DoF of the system, i.e.,

$$\begin{aligned}
h_{aa}(t) = \{&(-0.934 - 14.498 \text{ i}) \exp[(-0.0690 + 17.706 \text{ i})t] \\
&+ (-0.934 + 14.498 \text{ i}) \exp[(-0.0690 - 17.706 \text{ i})t] \\
&+ 1.564 \exp(-8.780t) + 0.303 \exp(-3.174t)\} \times 10^{-4} \text{ m}, \quad t \geq 0 \quad \text{(a)}
\end{aligned}$$

Fig. 4.5 The impulse response functions of the inertial DoF of the hydro-elastic vibration isolation system for vehicle engines (solid line: result from Eq. (4.1.38), solid circle: result in Eq. (b))

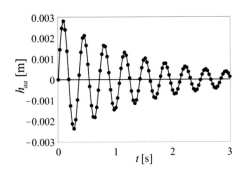

The corresponding real expression reads

$$h_{aa}(t) = \{\exp(-0.0690t)[28.996\sin(17.706t) - 1.867\cos(17.706t)]$$
$$+ 1.564\exp(-8.790t) + 0.303\exp(-3.174t)\} \times 10^{-4}\ \text{m},\ \ t \geq 0 \quad \text{(b)}$$

This is equivalent to the deduction from Eq. (4.1.49) to Eq. (4.1.38).

Figure 4.5 gives a comparison of the computational results from Eqs. (4.1.38) and (b) here, and shows very good coincidence. This verifies again that both Eqs. (4.1.49) and (4.1.50) are the Fourier transform of Eq. (4.1.38).

4.1.3 Concluding Remarks

Based on the study of the degree of freedom of a discrete system in Sect. 3.2, this section presented the dynamic analysis of a vibration system with non-holonomic constraints in the accessible state space. The analysis has solved Problem 2A proposed in Chap. 1 and led to the major conclusions of the section as follows.

(1) Compared with a linear viscously damped MDoF system, the linear MDoF system with non-holonomic constraints exhibits similar modal properties. That is, the biorthogonality holds true for any left complex eigenvector and any right complex eigenvector provided that their eigenvalues are distinct. The biorthogonality of left and right eigenvectors enables one to perform the complex modal analysis of the system and derive the complex modal expansions for both impulse response matrix and frequency response matrix of the system subject to excitations at inertial DoFs.

(2) As shown in the study, the dynamic analysis of a system with non-holonomic constraints in frequency domain is simpler than that in time domain. Furthermore, the number of resonant peaks of such a system in frequency domain may

be less than the DoFs of the system since some of complex modes, under the influence of the non-holonomic constraints, represent the response of creep modes, rather than any complex modal vibrations.

4.2 Node Number of a Natural Mode Shape

The concept of nodes of a mode shape came from some immovable points of a fixed-fixed string in natural vibrations. These immovable points of the mode shape are not fully fixed as two fixed boundaries of the string, but remain rest in a natural vibration of higher order. In general, the immovable, but not fully fixed, points of an elastic body, such as a string, a slender beam and a thin plate, in a natural vibration are called *nodes* or *nodal lines*. For discrete systems, the immovable, but not fully fixed, points are named nodes though they may exhibit one or even two-dimensional properties. The concept of nodes plays an important role in engineering and serves as criteria for identifying mode orders, arranging sensors and shakers, and reducing undesirable vibrations in industrial design.

The early observation of resonance of some elastic systems indicated that the order of a mode shape and the node number had a relation, which called great attention from academia since the nineteenth century. For instance, E. J. Routh, a British mathematician, pointed out in 1892 that the r-th mode shape of a string has $r - 1$ nodes[3]. Later on, J. C. F. Sturm in Switzerland and J. Liouville in France studied the node number of a longitudinal mode shape for a straight rod and found a similar rule of the node number[4]. In the 1940s, F. R. Gantmacher and M. G. Krein, two eminent mathematicians in the Soviet Union, systematically studied the node number problem. They pointed out, in their Russian monograph in 1950, that the r-th mode shape of a chain system, such as a rod and a beam, has $r - 1$ nodes. The first English translation of their work appeared in 1961, but was not common until 2002 when the revised version became available[5]. A number of books cited the above conclusion and claimed its validity for other vibration systems.

In fact, Gantmacher and Krein mainly studied the discrete chain systems by using flexibility method and the continuous chain systems via the influence coefficient method. They established the theory of oscillatory matrix for discrete chain systems and the theory of oscillatory kernel for continuous chain systems in order to study the change of signs of an eigenvector or an eigen-function, and called them the oscillatory property of a vibration.

Many linear vibration systems exhibit the oscillatory property and follow the rule of the node number of a mode shape, but there are also some exceptions. Consider, as an example, the natural vibrations of a circular ring. The frequency of an extensional mode shape without any nodes is much higher than those of the flexural mode shapes

---

[3] Routh E J (1892). Dynamics of a System of Rigid Bodies. London: Macmillan Publisher.

[4] Rayleigh L (1945). The Theory of Sound. 2nd Edition. New York: Dover Publications.

[5] Gantmacher F R, Krein M G (2002). Oscillation Matrices and Kernels and Small Vibrations of Mechanical Systems. 2nd Edition. Rhode Island: AMS Chelsea Publishing, 113–179.

with nodes[6]. Unfortunately, many books only cited the results of Gantmacher and Krein, but did not present their conditions of oscillatory properties. Thanks to the recent studies by D. J. Wang and his coworkers in China[7], they sorted out the results of Gantmacher and Krein, and confirmed the oscillatory properties for the natural vibrations of rods and beams subject to under-constrained boundaries.

The author doubted the assertion generalized from the results of Gantmacher and Krein, proposed several counter-examples of chain systems in 1984 and documented them as a university report in 1988. After reading the English translation of the monograph by Gantmacher and Krein, and the monograph by Wang et al. in recent years, the author improved and published the early study[8]. This section, based on the recent paper of the author, deals with the node number of a mode shape of discrete systems and corrects some popular but incorrect assertions. The section reveals the rules of the node number of an arbitrary 2-DoF system and also presents a discussion on the mode shape of a kind of MDoF systems in order to demonstrate the design feasibility of the nodes in a mode shape.

4.2.1   Reexamination of Current Results

4.2.1.1   Briefing the Rule of Node Numbers

The study of Gantmacher and Krein began with the so-called Sturm systems, such as a thin rod and a circular shaft. The dynamic equation of a Sturm system is a partial differential equation of the second order, and the corresponding discrete model has a diagonal mass matrix and a stiffness matrix in the form of *Jacobi matrix*, where all diagonal entries are positive, all off-diagonal entries are negative and other entries are zeros. They proved that a Sturm system, no matter whether it undergoes rigid-body motions or not, does not have repeated natural frequencies and the r-th mode shape has $r - 1$ nodes. In fact, the Sturm system has, at most, one rigid-body mode, which corresponds to the first natural frequency being zero and does not have any nodes.

For a slender beam or a slender beam with lumped masses mounted, the dynamic equation is a partial differential equation of the fourth order, and the stiffness matrix of the corresponding discrete model is no longer a Jacobi matrix. This is also the case for more general vibration systems. Under the assumption of no rigid-body motions, Gantmacher and Krein converted the continuous system to a discrete system by using the influence function method and established the eigenvalue problem in terms of flexibility matrix. They proved that if the flexibility matrix is an *oscillatory matrix*

[6] Timoshenko S, Young D H, Weaver Jr W (1974). Vibration Problems in Engineering. 4th Edition. New York: John Wiley & Sons Inc., 476–481.

[7] Wang D J, Wang Q S, He B C (2019). Qualitative Theory in Structural Mechanics. Singapore: Springer-Nature, 33–269.

[8] Hu H Y (2018). On the node number of a natural mode-shape. Journal of Dynamics and Control, 16(3): 193–200 (in Chinese).

and the system does not have repeated natural frequencies, the r-th mode shape has $r - 1$ nodes. Their essential contribution is to find the *criterion of oscillatory matrix* for any square matrix[9]. That is, the square matrix yields the following conditions: (1) it is not singular, (2) the sub-determinant of any order is not negative, and (3) the off-diagonal entries are positive. The above conditions of non-singularity exclude the case when the flexibility matrix is singular, or equivalently, when the beam undergoes any rigid-body motion.

For the beams undergoing rigid-body motions, Wang et al. introduced the concept of conjugate beam and studied the relation between the order of mode shape and the node number[10]. The hinged-free beam has one rigid-body mode shape and the similar property to that of a Sturm system. It is worthy to notice that the free-free beam has two rigid-body mode shapes corresponding to repeated natural frequencies being zero. Thus, the mode shapes are two linearly independent rigid-body motions. It is quite natural to select the translation of the beam in bending direction as the first mode shape which has no node, and the rotation of the beam about the section of mid-span of the beam as the second mode shape which has one node. Then, ranking the elastic mode shapes of the beam from $r = 3$ also gives the rule that the r-th mode shape has $r - 1$ nodes.

4.2.1.2 Problem of Generalized Coordinates

In the vibration analysis of an MDoF system, the mode shapes are described by the eigenvectors determined from a pencil of mass matrix and stiffness matrix. It is reasonable to judge the existence of a node when two adjacent entries of an eigenvector change signs. In particular, when an entry of an eigenvector is zero, the origin of corresponding generalized coordinate of the system is a node of the corresponding mode shape.

Nevertheless, it is worthy to notice that both mass matrix and stiffness matrix rely on the frame of coordinates and the selection of frame of coordinates is so flexible and greatly affects the expression of a mode shape.

Example 4.2.1 Figure 4.6a shows a 2-DoF vibration system, where mass m of rigid bar AB is uniformly distributed, and two massless springs have the same stiffness coefficient k. Take the vertical displacements of any two points on the bar as generalized coordinates and discuss the relation between the node and the selection of coordinate.

We first select the vertical displacements u_A and u_B at two ends A and B in Fig. 4.6a to describe the vibration system. It is easy to select a symmetric mode shape and an anti-symmetric mode shape from the system symmetry, and then get natural frequencies, namely,

[9] Gantmacher F R, Krein M G (2002). Oscillation Matrices and Kernels and Small Vibrations of Mechanical Systems. 2nd Edition. Rhode Island: AMS Chelsea Publishing, 113–179.

[10] Wang D J, Wang Q S, He B C (2019). Qualitative Theory in Structural Mechanics. Singapore: Springer-Nature, 237–243.

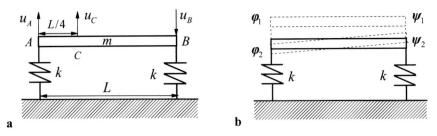

Fig. 4.6 A vibration problem of 2-DoF system. **a** the schematic of system; **b** two mode shapes

$$
\begin{cases}
\omega_1 = \sqrt{2k/m}, \ \varphi_1 = \begin{bmatrix} 1 & -1 \end{bmatrix}^{\mathrm{T}} \\
\omega_2 = \sqrt{6k/m}, \ \varphi_2 = \begin{bmatrix} 1 & 1 \end{bmatrix}^{\mathrm{T}}
\end{cases}
\tag{a}
$$

As displacements u_A and u_B are opposite, the two entries in vector φ_1 have different signs but the mode shape does not have any node, while the two entries in vector φ_2 have the same sign but the mode shape has a node at the mass center of the bar.

Next, we take displacement u_A at the left end A of the bar and displacement u_C at point C measured as $L/4$ from point A, and derive the two natural modes as follows

$$
\begin{cases}
\omega_1 = \sqrt{2k/m}, \ \boldsymbol{\psi}_1 = \begin{bmatrix} 1 & 1 \end{bmatrix}^{\mathrm{T}} \\
\omega_2 = \sqrt{6k/m}, \ \boldsymbol{\psi}_2 = \begin{bmatrix} 1 & 1/2 \end{bmatrix}^{\mathrm{T}}
\end{cases}
\tag{b}
$$

Although the two entries in vector $\boldsymbol{\psi}_2$ have the same sign, the mode shape has a node at the mass center of the bar. In fact, vectors φ_r and $\boldsymbol{\psi}_r$, $r = 1, 2$ here correspond to the same mode shapes in Fig. 4.6b.

Mode shape is a geometric shape and does not depend on the frame of coordinates by nature. To study a mode shape conveniently, however, one needs a frame of coordinates. As shown in the above coordinate selections, it is dangerous to determine whether a mode shape has any nodes or not only according to the change of signs of two adjacent entries in the eigenvector. Hence, it is necessary to check the geometry of the mode shape in the frame of coordinates when making a conclusion about the node number of a mode shape.

Remark 4.2.1 The selection of frame of coordinates is flexible and gives difficulty in studying the node number of a mode shape. Hence, Gantmacher and Krein only studied one-dimensional vibration systems in a frame of coordinates in the same direction. In this case, the change of signs of two adjacent entries in an eigenvector implies the existence of a node, but only serves as a sufficient condition. As demonstrated in the second case in Example 4.2.1, the mode shape may have a node somewhere when all entries of an eigenvector have the same sign.

4.2.1.3 Problem of Oscillatory Matrix

Example 4.2.2 Figure 4.7a presents a 3-DoF vibration system, where three lumped masses have the same value m, the bar AC is a massless rigid-body, the left and middle massless springs have the same stiffness coefficient k, and the stiffness coefficient of the right massless spring is γk with $\gamma > 0$. Study the node number of each mode shape of the system for $\gamma = 1$ and $\gamma = 100$, respectively.

In the frame of coordinates in Fig. 4.7a, it is easy to derive the kinetic energy and potential energy of the system as follows

$$
\begin{cases}
T = \dfrac{m}{2}\dot{u}_A^2(t) + \dfrac{m}{2}\dot{u}_B^2(t) + \dfrac{m}{2}\dot{u}_D^2(t) \\[2mm]
V = \dfrac{k}{2}u_A^2(t) + \dfrac{k}{2}u_B^2(t) + \dfrac{\gamma k}{2}\left\{u_D(t) - \left[u_B(t) + \dfrac{u_B(t) - u_A(t)}{L}\cdot L\right]\right\}^2 \\[2mm]
\quad = \dfrac{k}{2}\left\{u_A^2(t) + u_B^2(t) + \gamma[u_A(t) - 2u_B(t) + u_D(t)]^2\right\}
\end{cases}
\tag{a}
$$

Equation (a) enables one to derive the mass matrix, stiffness matrix and flexibility matrix of the system, namely,

$$
\begin{cases}
M = m\begin{bmatrix} 1 & 0 & 0 \\ 0 & 1 & 0 \\ 0 & 0 & 1 \end{bmatrix}, \quad
K = k\begin{bmatrix} 1+\gamma & -2\gamma & \gamma \\ -2\gamma & 1+4\gamma & -2\gamma \\ \gamma & -2\gamma & \gamma \end{bmatrix} \\[6mm]
F = K^{-1} = k^{-1}\begin{bmatrix} 1 & 0 & -1 \\ 0 & 1 & 2 \\ -1 & 2 & 5+1/\gamma \end{bmatrix}
\end{cases}
\tag{b}
$$

Solving the eigenvalue problem of these mass and stiffness matrices for $\gamma = 1$ and $\gamma = 100$ respectively leads to the natural modes of the system as follows

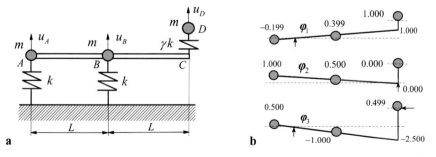

Fig. 4.7 The vibration problem of a 3-DoF system. **a** the schematic of the system; **b** the three mode shapes for $\gamma = 100$ (dot line: equilibrium position, arrow: node position)

$$\gamma = 1: \quad \begin{cases} \omega_1 = 0.382\sqrt{k/m}, & \varphi_1 = [-0.171 \quad 0.342 \quad 1.000]^T \\ \omega_2 = 1.000\sqrt{k/m}, & \varphi_2 = [1.000 \quad 0.500 \quad 0.000]^T \\ \omega_3 = 2.618\sqrt{k/m}, & \varphi_3 = [0.500 \quad -1.000 \quad 0.427]^T \end{cases} \quad \text{(c)}$$

$$\gamma = 100: \quad \begin{cases} \omega_1 = 0.408\sqrt{k/m}, & \varphi_1 = [-0.199 \quad 0.399 \quad 1.000]^T \\ \omega_2 = 1.000\sqrt{k/m}, & \varphi_2 = [1.000 \quad 0.500 \quad 0.000]^T \\ \omega_3 = 24.51\sqrt{k/m}, & \varphi_3 = [0.500 \quad -1.000 \quad 0.499]^T \end{cases} \quad \text{(d)}$$

A comparison between Eqs. (c) and (d) shows that the stiffness of the right spring slightly affects the mode shapes of the system. For $\gamma = 100$, Fig. 4.7b presents the mode shapes, where the first mode shape has a node on the bar, the second mode shape has nodes at the right end and the above oscillator, and the third mode shape has two nodes on the bar and the right spring, respectively.

Remark 4.2.2 The vibration system in Example 4.2.2 looks like a chain and the mass matrix is diagonal. Yet, the stiffness matrix here is not a Jacobi matrix such that the theory of Sturm systems fails to predict the number of nodes. Furthermore, the off-diagonal entries in the flexibility matrix F include zeros and do not satisfy the condition of oscillatory matrix. Hence, the criterion of Gantmacher and Krein fails to work.

4.2.1.4 Problem of Inertial Coupling

Example 4.2.3 Study the vibration system in Fig. 4.7a when $\gamma \to +\infty$.

In this limit case, the right lumped mass is directly mounted at the right end of the rigid bar so that the system has only 2-DoFs. In the frame of coordinates in Fig. 4.7a, one derives the kinetic and potential energies of the system, namely,

$$\begin{cases} T = \dfrac{m}{2}\dot{u}_A^2(t) + \dfrac{m}{2}\dot{u}_B^2(t) + \dfrac{m}{2}\left[\dot{u}_B(t) + \dfrac{\dot{u}_B(t) - \dot{u}_A(t)}{L} \cdot L\right]^2 \\ \quad = \dfrac{m}{2}\left\{\dot{u}_A^2(t) + \dot{u}_B^2(t) + [2\dot{u}_B(t) - \dot{u}_A(t)]^2\right\} \\ V = \dfrac{k}{2}u_A^2(t) + \dfrac{k}{2}u_B^2(t) \end{cases} \quad \text{(a)}$$

Equation (a) leads to both mass matrix and stiffness matrix as follows

$$M = m\begin{bmatrix} 2 & -2 \\ -2 & 5 \end{bmatrix}, \quad K = k\begin{bmatrix} 1 & 0 \\ 0 & 1 \end{bmatrix} \quad \text{(b)}$$

Solving the eigenvalue problem of above two matrices gives the natural modes

$$\begin{cases} \omega_1 = 0.408\sqrt{k/m}, & \boldsymbol{\varphi}_1 = [-0.500 \quad 1.000]^{\mathrm{T}} \\ \omega_2 = \sqrt{k/m}, & \boldsymbol{\varphi}_2 = [1.000 \quad 0.500]^{\mathrm{T}} \end{cases} \tag{c}$$

A comparison between Eq. (c) here and Eq. (c) in Example 4.2.2 shows that the two mode shapes here are very close to the first two mode shapes there. It is interesting that each mode shape here has a node.

Remark 4.2.3 In the study of Gantmacher and Krein, the mass matrix of concern was a diagonal one. Their study, hence, is not valid for the system in Example 4.2.3, where the mass matrix in Eq. (b) is not diagonal. In general, the inertial coupling in mass matrix makes the rule of the node number of a mode shape very complicated. For instance, each mode shape of a 2-DoF system in this example has one node. The next subsection will discuss this issue in detail.

4.2.2 Rules of Node Numbers of 2-DoF Systems

4.2.2.1 Theoretical Analysis

Given an arbitrary 2-DoF system, define two generalized coordinates with the same direction to describe the system dynamics, and let $\boldsymbol{M} \in \mathbb{R}^{2\times2}$ and $\boldsymbol{K} \in \mathbb{R}^{2\times2}$ be the mass matrix and stiffness matrix of the system. In addition, define the matrix of natural frequencies squared and the matrix of normalized mode shapes as follows

$$\boldsymbol{\Omega}^2 \equiv \begin{bmatrix} \omega_1^2 & 0 \\ 0 & \omega_2^2 \end{bmatrix} \in \mathbb{R}^{2\times2}, \quad \boldsymbol{\Phi} \equiv [\boldsymbol{\varphi}_2 \ \boldsymbol{\varphi}_2] \equiv \begin{bmatrix} \varphi_{11} & \varphi_{12} \\ \varphi_{21} & \varphi_{22} \end{bmatrix} \in \mathbb{R}^{2\times2} \tag{4.2.1}$$

where $\omega_1^2 \leq \omega_2^2$ and $\boldsymbol{\Phi}$ is invertible. As the two generalized coordinates are towards the same direction, the sufficient condition that the r-th mode shape has a node reads

$$\varphi_{1r}\varphi_{2r} \leq 0, \quad r = 1, 2 \tag{4.2.2}$$

The following discussion is about how Eq. (4.2.2) holds true.

Recalling the orthogonal relations of mode shapes with respect to the mass matrix and the stiffness matrix, namely, $\boldsymbol{\Phi}^{\mathrm{T}}\boldsymbol{M}\boldsymbol{\Phi} = \boldsymbol{I}_2$ and $\boldsymbol{\Phi}^{\mathrm{T}}\boldsymbol{K}\boldsymbol{\Phi} = \boldsymbol{\Omega}^2$, leads to

$$\boldsymbol{M} = \boldsymbol{\Phi}^{-\mathrm{T}}\boldsymbol{\Phi}^{-1}, \quad \boldsymbol{K} = \boldsymbol{\Phi}^{-\mathrm{T}}\boldsymbol{\Omega}^2\boldsymbol{\Phi}^{-1} \tag{4.2.3}$$

Their off-diagonal entries are

$$\begin{cases} m_{12} = m_{21} = -\dfrac{1}{(\det \boldsymbol{\Phi})^2}(\varphi_{11}\varphi_{21} + \varphi_{12}\varphi_{22}) \\[4mm] k_{12} = k_{21} = -\dfrac{1}{(\det \boldsymbol{\Phi})^2}(\omega_2^2\varphi_{11}\varphi_{21} + \omega_1^2\varphi_{12}\varphi_{22}) \end{cases} \tag{4.2.4}$$

From Eq. (4.2.4), one arrives at the following relations

$$\begin{cases} k_{12} - m_{12}\omega_1^2 = \dfrac{1}{(\det \boldsymbol{\Phi})^2}(\omega_1^2 - \omega_2^2)\varphi_{11}\varphi_{21} \\[4mm] k_{12} - m_{12}\omega_2^2 = \dfrac{1}{(\det \boldsymbol{\Phi})^2}(\omega_2^2 - \omega_1^2)\varphi_{12}\varphi_{22} \end{cases} \tag{4.2.5}$$

The discussions about Eq. (4.2.5) include the following two cases.

Case (1): Two natural frequencies are different, namely, $\omega_1^2 < \omega_2^2$. According to Eq. (4.2.5) and $\omega_1^2 \leq \omega_2^2$, if $k_{12} - m_{12}\omega_1^2 \geq 0$, then $\varphi_{11}\varphi_{21} \leq 0$ holds such that $\boldsymbol{\varphi}_1$ has a node. If $k_{12} - m_{12}\omega_2^2 \leq 0$, then $\varphi_{12}\varphi_{22} \leq 0$ holds so that $\boldsymbol{\varphi}_2$ has a node. This is the sufficient condition when the two mode shapes have a node. Regarding to the value of m_{12}, this sufficient condition includes two sub-cases as follows.

Sub-case a: $m_{12} < 0$. If $m_{12}\omega_2^2 \leq k_{12}$ holds, then $\boldsymbol{\varphi}_1$ has a node. If $k_{12} \leq m_{12}\omega_1^2$ holds, then $\boldsymbol{\varphi}_2$ has a node. If $m_{12}\omega_2^2 < k_{12} < m_{12}\omega_1^2 < 0$ holds, $\boldsymbol{\varphi}_1$ and $\boldsymbol{\varphi}_2$ may not have any node.

Sub-case b: $m_{12} \geq 0$. If $m_{12}\omega_2^2 \leq k_{12}$, then $\boldsymbol{\varphi}_1$ has a node. If $k_{12} \leq m_{12}\omega_1^2$, then $\boldsymbol{\varphi}_2$ has a node. If $0 < m_{12}\omega_1^2 \leq k_{12} \leq m_{12}\omega_2^2$, both $\boldsymbol{\varphi}_1$ and $\boldsymbol{\varphi}_2$ have a node.

Furthermore, we focus on a special sub-case when $m_{12} = 0$ in Sub-case b. That is, the system does not have any inertial coupling. Now, if $k_{12} < 0$ holds, then $\boldsymbol{\varphi}_2$ has a node. If $k_{12} = 0$ holds, both $\boldsymbol{\varphi}_1$ and $\boldsymbol{\varphi}_2$ have a node. If $k_{12} > 0$ holds, $\boldsymbol{\varphi}_1$ has a node. In this special sub-case, the system has a diagonal mass matrix, the flexibility matrix is an oscillatory matrix if $k_{12} < 0$ holds. Therefore, this special sub-case falls into the studies of Gantmacher and Krein, but the other sub-cases discussed are beyond their scope.

Figure 4.8 presents the above relations of the node number in terms of the inertial coupling, elastic coupling and natural frequencies in a parametric plane of (m_{12}, k_{12}).

Case (2): Two natural frequencies are exactly identical, namely, $\omega_2^2 = \omega_1^2 = \omega_0^2$ holds.

Equation (4.2.3) becomes

$$\boldsymbol{K} = \boldsymbol{\Phi}^{-T}\boldsymbol{\Omega}^2\boldsymbol{\Phi}^{-1} = \omega_0^2\boldsymbol{\Phi}^{-T}\boldsymbol{\Phi}^{-1} = \omega_0^2\boldsymbol{M} \tag{4.2.6}$$

To meet the requirement in Sect. 4.2.3, the displacement impedance matrix is incidentally defined as

(a) Region with vertical lines: φ_1 has a
     node.

(b) Region with horizontal lines: φ_2 has a
     node.

(c) Region with cross lines: both φ_1 and
     φ_2 have a node.

(d) Blank region: φ_1 and φ_2 may not have
     any node.

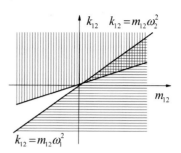

Fig. 4.8 Classification of mode shapes of a 2-DoF system with respect to system parameters

$$\boldsymbol{Z}(\omega) \equiv (\boldsymbol{K} - \omega^2 \boldsymbol{M}) = (\omega_0^2 - \omega^2)\boldsymbol{M} \tag{4.2.7}$$

Now, any two linearly independent non-zero vectors serve as the mode shapes of the system, and so do their linear combinations. Thus, the two mode shapes of a pair of repeated natural frequencies of any 2-DoF system have the following properties:

Property 4.2.1 The two mode shapes have either no node or one node, or one of mode shape has a node.

Property 4.2.2 An arbitrary point on the system can serve as the node of a mode shape.

We prove the above two properties as follows. At first, when the two natural frequencies of the system are identical, any two linearly dependent non-zero vectors serve as the mode shapes of the system. Thus, it is feasible to construct two vectors to satisfy the node number and to reach Property 4.2.1.

To prove Property 4.2.2, select three points A, B and C in the system of concern and establish generalized coordinates u_A, u_B and u_C towards the same direction at the three points. Now we impose a constraint on point C such that $u_C = 0$ holds. Thus, the system degenerates to an SDoF system. Let the SDoF system generate a free harmonic vibration of frequency ω_0 and get $u_A(t) = \bar{u}_A \sin(\omega_0 t)$ and $u_B(t) = \bar{u}_B \sin(\omega_0 t)$. It is easy to understand that the non-zero vector $\boldsymbol{\varphi} = [\bar{u}_A \ \bar{u}_B]^\mathrm{T}$ serves as a mode shape of repeated natural frequencies and has a node at point C, which yields the immovable condition $u_C(t) = 0$.

It is worthy to notice that the studies of Gantmacher and Krein did not touch upon any vibration systems with repeated natural frequencies. As such, the complicated case with repeated natural frequencies here is beyond their studies.

4.2.2.2 Case Studies

Example 4.2.4 Figure 4.9a presents a 2-DoF system, where the uniform rigid bar AB with length L and mass m is symmetrically supported by two massless springs with

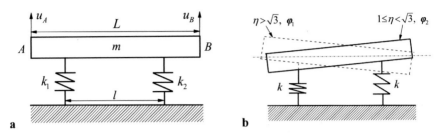

Fig. 4.9 The vibration problem of a 2-DoF system. **a** schematic of system; **b** two mode shapes when two springs are identical

stiffness coefficients k_1 and k_2, respectively. The distance between the two springs is l, which yields $\eta \equiv L/l \geq 1$. Study the node number of two mode shapes of the system for different lengths and spring stiffness coefficients.

As shown in Fig. 4.9a, take the displacements u_A and u_B as the generalized coordinates, and write the kinetic and potential energies of the system as follows

$$
\begin{aligned}
T &= \frac{m}{2}\left[\frac{\dot{u}_A(t)+\dot{u}_B(t)}{2}\right]^2 + \frac{mL^2}{24}\left[\frac{\dot{u}_B(t)-\dot{u}_A(t)}{L}\right]^2 \\
&= \frac{m}{8}[\dot{u}_A(t)+\dot{u}_B(t)]^2 + \frac{m}{24}[\dot{u}_B(t)-\dot{u}_A(t)]^2
\end{aligned}
\tag{a}
$$

$$
\begin{aligned}
V &= \frac{1}{2}k_1\left\{u_A(t) + \frac{[u_B(t)-u_A(t)](\eta-1)}{2\eta}\right\}^2 \\
&\quad + \frac{1}{2}k_2\left\{u_A(t) + \frac{[u_B(t)-u_A(t)](\eta+1)}{2\eta}\right\}^2
\end{aligned}
\tag{b}
$$

Equations (a) and (b) lead to the mass and stiffness matrices of the system, namely,

$$
\begin{cases}
M = \dfrac{m}{6}\begin{bmatrix} 2 & 1 \\ 1 & 2 \end{bmatrix} \\[2mm]
K = \dfrac{1}{4\eta^2}\begin{bmatrix} k_1(\eta+1)^2+k_2(\eta-1)^2 & (k_1+k_2)(\eta^2-1) \\ (k_1+k_2)(\eta^2-1) & k_1(\eta-1)^2+k_2(\eta+1)^2 \end{bmatrix}
\end{cases}
\tag{c}
$$

In what follows, two cases are discussed for the spring stiffness coefficients.

Case (1): When the stiffness coefficients of two springs are identical and denoted as $k_1 = k_2 = k$, Eq. (c) becomes

$$
M = \frac{m}{6}\begin{bmatrix} 2 & 1 \\ 1 & 2 \end{bmatrix}, \quad K = \frac{k}{2\eta^2}\begin{bmatrix} \eta^2+1 & \eta^2-1 \\ \eta^2-1 & \eta^2+1 \end{bmatrix}
\tag{d}
$$

In this case, the system is symmetric about the mid-span section such that it has symmetric mode shape $\boldsymbol{\varphi}_s = [\,1\ 1\,]^T$ and anti-symmetric mode shape $\boldsymbol{\varphi}_a = [\,1\ -1\,]^T$. Substitution of them into Rayleigh quotient leads to the corresponding natural frequencies, namely,

$$\omega_s = \sqrt{\frac{\boldsymbol{\varphi}_s^T \boldsymbol{K} \boldsymbol{\varphi}_s}{\boldsymbol{\varphi}_s^T \boldsymbol{M} \boldsymbol{\varphi}_s}} = \sqrt{\frac{2k}{m}}, \quad \omega_a = \sqrt{\frac{\boldsymbol{\varphi}_a^T \boldsymbol{K} \boldsymbol{\varphi}_a}{\boldsymbol{\varphi}_a^T \boldsymbol{M} \boldsymbol{\varphi}_a}} = \sqrt{\frac{6k}{m\eta^2}} \tag{e}$$

Now, the symmetric mode shape does not have any node, and the anti-symmetric mode shape has a node at the mass center of the bar. The ranking of two natural frequencies depends on dimensionless parameter $\eta = L/l$, that is, the ratio of length L of the bar to distance l between the two springs.

When $1 \leq \eta < \sqrt{3}$, Eq. (e) gives relation $\omega_1 = \omega_s < \omega_a = \omega_2$ such that the second mode shape has a node. When $\eta > \sqrt{3}$, $\omega_1 = \omega_a < \omega_s = \omega_2$ holds and the first mode shape has a node. Figure 4.9b depicts these two sub-cases. It is easy to verify that if $1 \leq \eta < \sqrt{3}$, then $k_{12} < m_{12}\omega_1^2$ holds, while if $\eta > \sqrt{3}$, then $k_{12} > m_{12}\omega_2^2$ holds. Recalling the theoretical analysis, the two sub-cases correspond to the conditions $k_{12} < m_{12}\omega_1^2$ and $k_{12} > m_{12}\omega_2^2$ discussed in sub-case b under case (1) when two natural frequencies are distinct. In Fig. 4.8, the two sub-cases fall into the region with horizontal lines only and the region with vertical lines only, respectively.

When $\eta = \sqrt{3}$, the system has a pair of repeated natural frequencies $\omega_a = \omega_s = \sqrt{2k/m}$. Hence, the mode shapes can be any linearly independent non-zero vectors and the node is quite arbitrary.

Case (2): When the stiffness coefficients of the two springs are different, denote them as $0 < k_1 < k_2$. For simplicity, we focus on the case when $\eta = \sqrt{3}$. Now, the mass matrix and stiffness matrix of the system satisfy

$$\boldsymbol{M} = \frac{m}{6} \begin{bmatrix} 2 & 1 \\ 1 & 2 \end{bmatrix}, \quad \boldsymbol{K} = \frac{1}{6} \begin{bmatrix} k_1(2+\sqrt{3}) + k_2(2-\sqrt{3}) & k_1 + k_2 \\ k_1 + k_2 & k_1(2-\sqrt{3}) + k_2(2+\sqrt{3}) \end{bmatrix} \tag{f}$$

The two natural modes read

$$\begin{cases} \omega_1 = \sqrt{2k_1/m}, \ \boldsymbol{\varphi}_1 = \begin{bmatrix} 1+\sqrt{3} & 1-\sqrt{3} \end{bmatrix}^T \\ \omega_2 = \sqrt{2k_2/m}, \ \boldsymbol{\varphi}_2 = \begin{bmatrix} 1-\sqrt{3} & 1+\sqrt{3} \end{bmatrix}^T \end{cases} \tag{g}$$

Each mode shape in Eq. (g) has a node at the connection point of the rigid bar to the right spring or to the left spring, respectively. As such, the first natural frequency depends on the stiffness coefficient k_1 of the left spring and has nothing to do with the stiffness coefficient k_2 of the right spring. Similarly, the second natural frequency has nothing to do with the left spring. It is easy to verify that $m_{12}\omega_1^2 = k_1/3$, $m_{12}\omega_2^2 = k_2/3$, $k_{12} = (k_1 + k_2)/6$ hold true. Hence, this sub-case corresponds to sub-case b under case (1), that is, $0 < m_{12}\omega_1^2 < k_{12} < m_{12}\omega_2^2$, when the natural frequencies are different. This sub-case falls into the region with cross-lines in Fig. 4.8 and both mode shapes have a node. In addition, this sub-case is beyond the study of Gantmacher and Krein.

Remark 4.2.4 Even though Example 4.2.4 is a simple case study on a 2-DoF system, it provides a useful reference for many engineering problems. In the ground vibration test of a rocket, for instance, the rocket is usually supported on two sets of air springs of lower stiffness such that the rocket seems suspended in air. The stiffness of rocket is relatively much higher than those of air springs. As such, the first two natural mode shapes measured are the rigid-body modes of the rocket, similar to the two mode shapes in Example 4.2.4. If the distance between two sets of air springs is relatively large, the first mode shape does not have any node and the second mode shape has a node, while the third mode shape is close to the first bending mode shape of the rocket at free status and has two nodes. Yet, if the above distance is sufficiently small, the first mode shapes has a node while the second one doesn't.

The analysis and case studies in the subsection indicate that the rules of the node number of a mode shape are much more complicated than available knowledge. When the system has inertial coupling or repeated natural frequencies, the rules are beyond the previous studies of Gantmacher and Krein.

4.2.3 Design Feasibility of Nodes in a Mode Shape

When a system has more than two DoFs, it is almost impossible to use the ideas in Sect. 4.2.2 to check the node number of a mode shape. This subsection presents the study of a kind of MDoF systems to demonstrate the design feasibility of nodes in a mode shape.

Figure 4.10 depicts a composite system $S_c \equiv S_a \cup S_b$, where S_a is a damped n-DOF sub-system with displacement impedance matrix $\mathbf{Z}_n^a(\omega) \equiv [a_{ij}(\omega)]$, S_b is an un-damped 2-DoF sub-system with repeated natural frequencies ω_0 and mass matrix

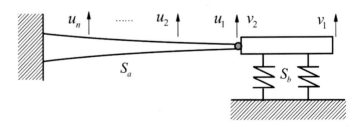

Fig. 4.10 The schematic of a composite system $S_c \equiv S_a \cup S_b$ of $n + 1$ DoFs

$M \equiv [m_{ij}] \in \mathbb{R}^{2 \times 2}$. From Eq. (4.2.7), the displacement impedance matrix of S_b takes the following form

$$Z_2^b(\omega) \equiv [b_{ij}(\omega)] = (\omega_0^2 - \omega^2)M \in \mathbb{R}^{2 \times 2} \tag{4.2.8}$$

In the composite system S_c, let the first coordinate u_1 of sub-system S_a and the second coordinate v_2 of sub-system S_b be connected so that the displacement impedance matrix of the composite system is a square matrix of order $n + 1$ as follows

$$Z_{n+1}^c(\omega) = \begin{bmatrix} b_{11}(\omega) & b_{12}(\omega) & 0 & \cdots & 0 \\ b_{21}(\omega) & a_{11}(\omega) + b_{22}(\omega) & a_{12}(\omega) & \cdots & a_{1n}(\omega) \\ 0 & a_{21}(\omega) & a_{22}(\omega) & \cdots & a_{2n}(\omega) \\ \cdots & \cdots & \cdots & \cdots \\ 0 & a_{n1}(\omega) & a_{n2}(\omega) & \cdots & a_{nn}(\omega) \end{bmatrix} \tag{4.2.9}$$

Applying the Laplace expansion formula to the determinant in Eq. (4.2.9), we derive the characteristic equation governing the natural frequencies of composite system S_c, namely,

$$b_{11}(\omega) \det[Z_n^a(\omega)] + \det[Z_2^b(\omega)] \cdot \det[Z_{n-1}^a(\omega)] = 0 \tag{4.2.10}$$

Substitution of Eq. (4.2.8) into Eq. (4.2.10) yields

$$(\omega_0^2 - \omega^2)\{m_{11} \det[Z_n^a(\omega)] + (\omega_0^2 - \omega^2) \det(M) \cdot \det[Z_{n-1}^a(\omega)]\} = 0 \quad (4.2.11)$$

As shown in Eq. (4.2.11), ω_0 is a natural frequency of the composite system. Corresponding to ω_0, the first row and the first column in impedance matrix $Z_{n+1}^c(\omega_0)$ are zeros such that vector $\varphi_0 = [1 \ 0 \cdots 0]^T \in \mathbb{R}^{n+1}$ is the mode shape corresponding to natural frequency ω_0. As such, the mode shape has n nodes. When the composite system undergoes the natural vibration with frequency ω_0, only the first DoF of sub-system S_b undergoes displacement v_1, while other DoFs of the composite system are at rest.

One may wonder whether this mode shape is the $(n + 1)$-th mode shape or not? The answer is undetermined as follows.

From Eq. (4.2.11), the other n natural frequencies satisfy the following characteristic equation

$$\det[Z_n^a(\omega)] + m_{11}^{-1}(\omega_0^2 - \omega^2) \det(M) \cdot \det[Z_{n-1}^a(\omega)] = 0 \tag{4.2.12}$$

Assuming $m_e \equiv m_{11}^{-1} \det(M)$ and $k_e \equiv m_{11}^{-1}\omega_0^2 \det(M)$, and substituting them into Eq. (4.2.12), we have

$$\det \begin{bmatrix} a_{11}(\omega) + (k_e - m_e\omega^2) & a_{12}(\omega) & a_{13}(\omega) \cdots a_{1n}(\omega) \\ a_{21}(\omega) & a_{22}(\omega) & a_{23}(\omega) \cdots a_{2n}(\omega) \\ a_{31}(\omega) & a_{32}(\omega) & a_{33}(\omega) \cdots a_{3n}(\omega) \\ \cdots & \cdots & \cdots \quad \cdots \quad \cdots \\ a_{n1}(\omega) & a_{n2}(\omega) & a_{n3}(\omega) \cdots a_{nn}(\omega) \end{bmatrix} = 0 \qquad (4.2.13)$$

Equation (4.2.13) governs the natural frequencies of sub-system S'_a, which can be imagined as a sub-system modified from sub-system S_a by attaching a mass m_e to the first DoF in S_a and inserting a spring of stiffness coefficient k_e between this DoF and the fixed boundary. According to the inclusion theorem of eigenvalues[11], an arbitrary natural frequency ω_r of sub-system S_a decreases to $\underset{\sim}{\omega}_r$ after a lumped mass m_e is attached to it and then increases from $\underset{\sim}{\omega}_r$ to $\tilde{\omega}_r$ after a spring with stiffness coefficient k_e is inserted between sub-system S_a and the fixed boundary. Hence, we reach the variation relations of natural frequencies of sub-system S_a before and after the above modification, namely,

$$\begin{cases} 0 \le \underset{\sim}{\omega}_1 \le \omega_1 \le \underset{\sim}{\omega}_2 \le \omega_2 \le \cdots \le \underset{\sim}{\omega}_{n-1} \le \omega_{n-1} \le \underset{\sim}{\omega}_n \le \omega_n \\ 0 \le \underset{\sim}{\omega}_1 \le \tilde{\omega}_1 \le \underset{\sim}{\omega}_2 \le \tilde{\omega}_2 \le \cdots \le \underset{\sim}{\omega}_{n-1} \le \tilde{\omega}_{n-1} \le \underset{\sim}{\omega}_n \le \tilde{\omega}_n \end{cases} \qquad (4.2.14)$$

There follows

$$\tilde{\omega}_1 \le \omega_2, \quad \omega_{r-1} \le \tilde{\omega}_r \le \omega_{r+1}, \quad r = 2, \ldots, n-1 \qquad (4.2.15)$$

Equation (4.2.15) shows that the natural frequencies $\tilde{\omega}_2, \tilde{\omega}_3, \cdots, \tilde{\omega}_{n-1}$ of composite system S_c fall into a frequency band of $[\omega_1, \omega_n]$. Thus, the natural frequency ω_0 of the composite system falls into the above frequency band if the natural frequency ω_0 of the 2-DoF sub-system S_a yields $\omega_1 \le \omega_0 \le \omega_n$. That is, the mode shape φ_0 with n nodes can be one of the mode shapes of composite system S_c from the second order to the n-th order, but rather than the $(n+1)$-th order.

According to the analysis in Sect. 4.2.2, when the 2-DoF sub-system S_b has repeated natural frequencies, there always exists a mode shape with the second coordinate being zero, and the mode shape remains unchanged after sub-system S_b is connected to sub-system S_a. That is, when the natural vibration of composite system takes the natural frequency of sub-system S_b, all DoFs of sub-system S_a keep rest while only the first DoF of sub-system S_b undergoes vibration. In practice, this phenomenon is a local resonance.

Remark 4.2.5 As the above mode shape has nothing to do with sub-system S_a, it is feasible to select proper sub-system S_b and let the natural vibration happen in a desirable frequency band as a local vibration of sub-system S_b. This vibration pattern

---

[11] Courant R, Hilbert D (1989). Methods of Mathematical Physics. Volume I. New York: John Wiley & Sons Inc., 31–34.

is different from the pattern of an anti-resonance, where sub-system S_b works as a dynamic vibration absorber and the connection point of two sub-systems is subject to a harmonic force, but immovable. In the pattern of an anti-resonance, the other DoFs may still undergo vibrations.

Remark 4.2.6 To verify the above results, let the 2-DoF system in Example 4.2.4 be sub-system S_b and take $\eta = \sqrt{3}$ and $k_1 = k_2 = k$, then sub-system S_b has two identical natural frequencies $\omega_0 = \sqrt{2k/m}$. Given an arbitrary n-DoF system S_a, one can assemble S_a and S_b, and get a mode shape φ_0 with n nodes on sub-system S_a.

4.2.4 Concluding Remarks

This section presented the detailed discussions about the node number of a mode shape of discrete chain systems and the solution to Problem 3A proposed in Chap. 1. The major conclusions of the section are as follows.

(1) The mode shape of a discrete system is a geometric shape, but the analysis and computation of the mode shape rely on the frame of coordinates for the system. For a discrete chain system exhibiting one-dimensional vibration, it is feasible to define all coordinates towards the same direction, check the change of signs of two adjacent entries of an eigenvector in view of the study of Gantmacher and Krein, and determine the node number of a mode shape in the frame of coordinates. Given a discrete chain system, it is necessary to check if the node appears at a location away from any origins of coordinates.

(2) The theory of Gantmacher and Krein is not applicable to all vibration systems, even not to all discrete chain systems. For example, when a discrete chain system exhibits any inertial couplings, their theory may not work. When a discrete system has repeated natural frequencies, their theory fails to work.

(3) As shown in the study of an arbitrary 2-DoF system, the two mode shapes of the system may not have any node, or both have a node or one of the mode shapes has a node.

(4) It is feasible to construct a composite system by using an arbitrary n-DOF sub-system and a 2-DoF sub-system with two identical natural frequencies such that a local resonance appears at the 2-DoF sub-system and the n-DOF system keeps rest. This way, one can design a mode shape that has n nodes, but the corresponding frequency order is quite arbitrary.

4.3 Anti-resonances of a Harmonically Excited System

Anti-resonance, observed around the turn between the nineteenth and the twentieth centuries, refers to a phenomenon, where the vibration system to a harmonic force of a specific frequency is locally at rest. The first successful application of anti-resonance of 2-DoF systems was due to the invention of H. Frahm, a German engineer, who utilized two water tanks on a ship to effectively reduce the ship rolling excited by the waves of a rough sea in 1902[12]. The invention became an astonishing issue in both industry and academia. The top journal *Nature* even reported the invention before Frahm received the US patent in 1911. Since then, anti-resonance has witnessed numerous studies and applications in order to improve the dynamic properties of both mechanical and structural systems[13,14]. A recent study of anti-resonance, for instance, is the design for acoustic meta-materials[15].

However, the studies on the mechanisms of anti-resonance of MDoF systems have not been enough. For example, to attenuate the vibration of a damped SDoF system, or called the *primary system*, to a harmonic force, an un-damped SDoF system, or called the *secondary system*, is attached to the primary system such that the steady-state vibration of the primary system becomes zero when the excitation frequency coincides with the natural frequency of the secondary system. It is easy to make the assertion that the work done by the external force vanishes since the primary system keeps rest. However, some books stated that the secondary system absorbs the energy from the primary system input by the external force. That is, the input energy to the primary system transfers to the secondary system. This statement contradicts to the above assertion. In addition, readers may question the relation between the resonance and the node of a mode shape after reading Sect. 4.2.

The author studied the above problems and documented the results as a university report in 1988, but did not publish the results until 2018. Based on the recent paper of the author[16], this section presents the mechanisms behind both direct anti-resonance and cross anti-resonance, and reveals the relation between an anti-resonance and the node of a mode shape so as to address problem 3B proposed in Chap. 1. In order to gain an insight into the nature of anti-resonance, this section begins with the anti-resonance of an arbitrary 2-DoF system, and then Sect. 4.4 will discusses the anti-resonance of a kind of MDoF systems.

---

[12] Den Hartog J P (1984). Mechanical Vibrations. 4th Edition. New York: Dover Publications, 87–113.

[13] Sun J Q, Jolly M R, Norris M A (1995). Passive, adaptive and active tuned vibration absorbers - A survey. The 50th Anniversary Issue of ASME Journal of Mechanical Design and ASME Journal of Vibration and Acoustics 117(B): 234–242.

[14] Wang B P (2016). Eigenvalue problems in forced harmonic response analysis in structural dynamics. Chinese Journal of Computational Mechanics, 33(4): 549–555.

[15] Ma G C, Sheng P (2016). Acoustic metamaterials: from local resonances to broad horizons. Science Advances, 2: e1501595.

[16] Hu H Y (2018). On the anti-resonance problem of a linear system. Journal of Dynamics and Control, 16(5): 385–390 (in Chinese).

4.3.1 Anti-resonances of 2-DoF Systems

Consider the steady-state vibration of a 2-DoF system subject to an external force, which yields the following dynamic equation in a matrix form

$$M\ddot{u}(t) + C\dot{u}(t) + Ku(t) = f(t), \quad t \in [0, +\infty) \tag{4.3.1}$$

where $u(t) \equiv [\, u_1(t) \, u_2(t) \,]^T : t \mapsto \mathbb{R}^2$ represents the system displacement vector, $f \equiv [\, f_1(t) \, f_2(t) \,] : t \mapsto \mathbb{R}^2$ is the external force vector, $M \in \mathbb{R}^{2\times 2}$ is the positive definite symmetric mass matrix, $K \in \mathbb{R}^{2\times 2}$ is the positive definite symmetric stiffness matrix, $C \in \mathbb{R}^{2\times 2}$ is the symmetric damping matrix. Let the three matrices be denoted as

$$M \equiv \begin{bmatrix} m_{11} & m_{12} \\ m_{21} & m_{22} \end{bmatrix}, \quad K \equiv \begin{bmatrix} k_{11} & k_{12} \\ k_{21} & k_{22} \end{bmatrix}, \quad C \equiv \begin{bmatrix} c_{11} & c_{12} \\ c_{21} & c_{22} \end{bmatrix} \tag{4.3.2}$$

Applying Fourier transform to both sides of Eq. (4.3.1) gives

$$Z(\omega)U(\omega) = F(\omega), \quad \omega \in (-\infty, +\infty) \tag{4.3.3}$$

with

$$F(\omega) \equiv \begin{bmatrix} F_1(\omega) \\ F_2(\omega) \end{bmatrix}, \quad U(\omega) \equiv \begin{bmatrix} U_1(\omega) \\ U_2(\omega) \end{bmatrix}, \quad Z(\omega) \equiv K + i\omega C - \omega^2 M \tag{4.3.4}$$

where $i \equiv \sqrt{-1}$, $F : \omega \mapsto \mathbb{C}^2$ is the excitation spectrum vector, $U : \omega \mapsto \mathbb{C}^2$ is the displacement spectrum vector, $Z : \omega \mapsto \mathbb{C}^{2\times 2}$ is the *displacement impedance matrix* of the system. Their entries are complex functions of frequency $\omega \in [0, +\infty)$.

The displacement impedance matrix in Eq. (4.3.4) is also named the *dynamic stiffness matrix* since its entry represents a stiffness coefficient of the system associated with excitation frequency ω. Substitution of Eq. (4.3.2) into Eq. (4.3.4) leads to the displacement impedance matrix as follows

$$Z(\omega) \equiv \begin{bmatrix} Z_{11}(\omega) & Z_{12}(\omega) \\ Z_{21}(\omega) & Z_{22}(\omega) \end{bmatrix} = \begin{bmatrix} k_{11} - m_{11}\omega^2 + ic_{11}\omega & k_{12} - m_{12}\omega^2 + ic_{12}\omega \\ k_{21} - m_{21}\omega^2 + ic_{21}\omega & k_{22} - m_{22}\omega^2 + ic_{22}\omega \end{bmatrix} \tag{4.3.5}$$

It is worthy to notice that the property of dynamic stiffness matrix $Z(\omega)$ is different from static stiffness K for $\omega \neq 0$. As usual, matrix $Z(\omega)$ is symmetric, but may not be positive definite. For example, matrix $Z(\omega)$ is singular at two natural frequencies $\omega_{1,2}$ when $C = 0$, and $Z(\omega)$ is negative definite when $|\omega| \to +\infty$.

From Eq. (4.3.3), the displacement spectrum vector yields

$$U(\omega) = Z^{-1}(\omega)F(\omega) \equiv H(\omega)F(\omega) \qquad (4.3.6)$$

where $H(\omega) \equiv Z^{-1}(\omega)$ is the *frequency response matrix* of the system and also called the *dynamic flexibility matrix* of the system since its entry represents a flexibility coefficient of the system associated with excitation frequency ω. From Eq. (4.3.5), we have

$$
\begin{aligned}
H(\omega) &= \frac{1}{\det[Z(\omega)]} \begin{bmatrix} Z_{22}(\omega) & -Z_{12}(\omega) \\ -Z_{21}(\omega) & Z_{11}(\omega) \end{bmatrix} \\
&= \frac{1}{\det[Z(\omega)]} \begin{bmatrix} k_{22} - m_{22}\omega^2 + ic_{22}\omega & -(k_{12} - m_{12}\omega^2 + ic_{12}\omega) \\ -(k_{21} - m_{21}\omega^2 + ic_{21}\omega) & k_{11} - m_{11}\omega^2 + ic_{11}\omega \end{bmatrix}
\end{aligned} \qquad (4.3.7)
$$

As anti-resonance is a local behavior of the system, the anti-resonant frequency depends on both excitation location and measurement location. As such, denote the anti-resonant frequency as ω_{rs}, $r, s = 1, 2$ when the system is excited at DoF s and measured at DoF r.

Without loss of generality, the study in this section focuses on the case when the system is subject to external force $f_1(t)$ at the first DoF only. That is, $F_1(\omega) \neq 0$ and $F_2(\omega) \equiv 0$. In this case, $H_{11}(\omega_{11}) = 0$ holds, i.e., $U_1(\omega_{11}) = 0$ holds, if there exists a frequency ω_{11} such that $Z_{22}(\omega_{11}) = 0$ and $Z_{21}(\omega_{11}) \neq 0$ hold true. This phenomenon is named an anti-resonance of direct frequency response function $H_{11}(\omega)$, or a *direct anti-resonance* for short. In this case, furthermore, the cross frequency response yields $U_2(\omega_{11}) = H_{21}(\omega_{11})F_1(\omega) = -F_1(\omega)Z_{21}(\omega_{11})/\det[Z(\omega_{11})] \neq 0$.

In addition, if there exists the other frequency $\omega_{21} \neq \omega_{11}$ such that $Z_{21}(\omega_{21}) = 0$ and $Z_{22}(\omega_{21}) \neq 0$ hold true, there follow $H_{21}(\omega_{21}) = 0$ and $U_2(\omega_{21}) = 0$. This phenomenon is named an anti-resonance of cross frequency response function $H_{21}(\omega)$, or a *cross anti-resonance* for short. In this case, the direct frequency response satisfies $U_1(\omega_{21}) = H_{11}(\omega_{21})F_1(\omega) \neq 0$. Accordingly, it is easy to derive the similar results when the system is only subject to external force $f_2(t)$ at the second DoF.

4.3.2   Mechanisms Behind Two Kinds of Anti-Resonances

4.3.2.1   Analysis of Direct Anti-Resonance

The subsection begins with the analysis of a direct anti-resonance phenomenon, where there exists a frequency ω_{11} so that the system frequency responses satisfy $U_1(\omega_{11}) = 0$ and $U_2(\omega_{11}) \neq 0$ when the external forces yield $F_1(\omega) \neq 0$ and $F_2(\omega) \equiv 0$. Many books explained this phenomenon as "energy transfer" such that all of the energy input to the first DoF goes into the second DoF[17,18]. As the secondary system received the name "dynamic vibration absorber"[19], some books claimed that the secondary system absorbs the vibration of the primary system[20]. The following analysis will correct the above assertions.

(1)   Energy analysis for external force

It is easy to see that the external force does not input energy to the system in case of a direct anti-resonance. In fact, the direct anti-resonance means $U_1(\omega_{11}) = 0$ and the work done by the external force $F_1(\omega_{11})$ at the first DoF is zero. More specifically, let $f_1(t) = F_1(\omega_{11})\sin(\omega_{11}t)$ be the external force corresponding to $F_1(\omega_{11})$ in time domain. As the external force is applied to $u_1(t) = 0$, the work done by $f_1(t)$ in an arbitrary interval $[0, \quad T]$ yields

$$W \equiv \int_0^T u_1(t)f_1(t)\,\mathrm{d}t = \int_0^{2\pi/\omega_{11}} u_1(t)F_1(\omega_{11})\sin(\omega_{11}t)\mathrm{d}t = 0 \qquad (4.3.8)$$

That is, the external force $f_1(t)$ does no work and can not input any energy to the system at a direct anti-resonance. Hence, the assertion of energy transfer does not hold true. In other words, the reason why $U_1(\omega_{11}) = 0$ holds is that $Z_{22}(\omega_{11}) = 0$ gives rise to $H_{11}(\omega_{11}) = 0$. That is, the direct dynamic flexibility is zero when $\omega = \omega_{11}$. This also implies that the direct dynamic stiffness is infinitely large so that the external force can not input energy to the system.

(2)   Energy dissipation of cross frequency response

It is easy to prove that the cross frequency response is an energy-conservative harmonic vibration when the direct anti-resonance occurs. In fact, the reason why the direct frequency response exhibits an anti-resonance, that is, $U_1(\omega_{11}) = 0$, is that the internal forces of the system are balanced at the first DoF. From (4.3.7), when $\omega = \omega_{11}$, the elastic force, damping force and inertial force at the second DoF satisfy

$$Z_{22}(\omega_{11}) = k_{22} + ic_{22}\omega_{11} - m_{22}\omega_{11}^2 = 0 \qquad (4.3.9)$$

Equation (4.3.9) is a complex condition and is equivalent to two real ones, namely, $c_{22} = 0$ and $k_{22} - m_{22}\omega_{11}^2 = 0$. The first condition implies that the second DoF does not dissipate energy and the second one indicates that the second DoF undergoes the harmonic vibration of frequency $\omega_{11} = \sqrt{k_{22}/m_{22}}$. It is the harmonic vibration

[17] Balachandran B, Magrab E B. Vibrations (2019). 3rd Edition. Cambridge: Cambridge University Press, 491.

[18] Braun S, Ewins D, Rao S S (2002). Encyclopedia of Vibration. San Diego: Academic Press, 9.

[19] Den Hartog J P (1984). Mechanical Vibrations. 4th Edition. New York: Dover Publications. 87–93.

[20] Zheng Z C (1980). Mechanical Vibration. Volume I. Beijing: Press of Mechanical Industry, 147 (in Chinese).

without energy dissipation that the system keeps a local energy-conservative vibration when the external force does not input any energy to the system.

(3)   Analysis of transient response

We prove that the external force inputs energy to the system before the direct frequency response arrives at an anti-resonance, and the external force keeps the amplitude of cross frequency response when the direct frequency response reaches the anti-resonance. In fact, in the transient stage of the direct frequency response before reaching an anti-resonance, the external force inputs energy to the system. The input energy excites the second DoF to undergo an energy-conservative vibration and generate the elastic restoring force to balance the external force. The work done by the external force approaches zero with decreasing of vibration amplitude of the first DoF. Thus, the cross frequency response yields

$$U_2(\omega_{11}) = -\frac{Z_{21}(\omega_{11})F_1(\omega_{11})}{\det[\mathbf{Z}(\omega_{11})]} = \frac{(k_{21} - m_{21}\omega_{11}^2 + ic_{21}\omega_{11})F_1(\omega_{11})}{\det[\mathbf{Z}(\omega_{11})]} \qquad (4.3.10)$$

As shown in Eq. (4.3.10), the external force keeps the amplitude of cross frequency response. Once the external force decreases, the system undergoes a transient stage and the vibration amplitude of the second DoF decreases and updates the anti-resonance after the dynamic equilibrium. Accordingly, one observes $U_2(\omega_{11}) \to 0$ if $F_1(\omega_{11}) \to 0$.

Example 4.3.1 As shown in Fig. 4.11, a 2-DoF system is initially at rest and subject to a harmonic force, the amplitude of which varies with time. The system parameters and excitation frequency are as follows

$$m = 1 \text{ kg}, \quad k = 10^4 \text{ N/m}, \quad c = 0.3\sqrt{mk} = 30 \text{ N} \cdot \text{s/m}, \quad \omega = \sqrt{\frac{k}{m}} = 100 \text{ rad/s} \tag{a}$$

Study the direct anti-resonance of the system when the excitation amplitude $F_1(t)$ is a piecewise continuous function with unit in N, namely,

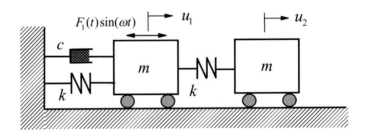

Fig. 4.11 A 2-DoF system under a harmonic force of variable amplitude

$$F_1(t) = \begin{cases} 200, & t \in [0, \ t_1] \\ 200 - 100(t - t_1)/(t_2 - t_1), & t \in (t_1, \ t_2] \\ 100, & t \in (t_2, \ t_3] \end{cases} \tag{b}$$

where $F_1(t)$ takes either a fast or a slow switch pattern as follows

$$\begin{cases} \text{fast switch pattern:} \quad t_1 = 1.5 \text{ s}, \quad t_2 = 1.6 \text{ s}, \quad t_3 = 3.0 \text{ s} \\ \text{slow switch pattern:} \quad t_1 = 1.5 \text{ s}, \quad t_2 = 2.0 \text{ s}, \quad t_3 = 3.0 \text{ s} \end{cases} \tag{c}$$

The dynamic equations of the system are in the form of Eq. (4.3.1), where the mass matrix, stiffness matrix and damping matrix are as follows

$$\boldsymbol{M} \equiv \begin{bmatrix} m & 0 \\ 0 & m \end{bmatrix}, \quad \boldsymbol{K} \equiv \begin{bmatrix} 2k & -k \\ -k & k \end{bmatrix}, \quad \boldsymbol{C} \equiv \begin{bmatrix} c & 0 \\ 0 & 0 \end{bmatrix} \tag{d}$$

Substituting Eqs. (a), (b), (c) and (d) into Eq. (4.3.1) and computing the initial value problem of Eq. (4.3.1) via Runge–Kutta integrator in software Maple, we get the displacement histories of the system shown in Fig. 4.12. It is easy to see that no matter whether the switch pattern of excitation amplitude is fast or slow, the system displacements satisfy the above analysis after a transient stage. That is, the amplitude of $u_1(t)$ approaches zero while the amplitude of $u_2(t)$ is proportional to the excitation amplitude.

Remark 4.3.1 The above analysis indicates that the assertions of "energy transfer" and "vibration absorption" are not appropriate. From the dynamic equilibrium in Eq. (4.3.9), the term of "dynamic cancellation" seems better. Yet, the conventional term of "dynamic vibration absorber" is still used throughout this text.

Remark 4.3.2 The above results are universal since the system of concern is an arbitrary 2-DoF system. For instance, the results are not only valid for the conventional 2-DoF system composed of a primary system and an un-damped secondary system,

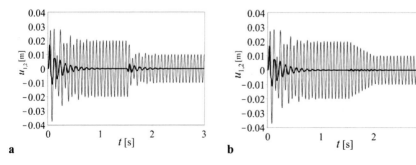

Fig. 4.12 The displacement histories of a 2-DoF system to a harmonic force with variable amplitude (thick solid line: $u_1(t)$, thin solid line: $u_2(t)$). **a** the fast switch pattern; **b** the slow switch pattern

but also for the 2-DoF system composed of a primary system and an un-damped secondary system with an elastic component attached to the rigid ground[21].

4.3.2.2 Analysis of Cross Anti-Resonance

Few studies have dealt with the mechanism of a cross anti-resonance of the displacement spectrum, namely, $U_2(\omega_{21}) = 0$, of a 2-DoF system in Eq. (4.3.6) so far. As a matter of fact, the mechanisms behind a cross anti-resonance and a direct anti-resonance are different. In a cross anti-resonance, the external force inputs energy to the first DoF and drives the system to undergo vibration, but the input energy can not flow to the second DoF.

To be more specific, the cross anti-resonance of a 2-DoF system implies that there exists a frequency ω_{21} for $Z_{21}(\omega_{21}) = 0$ such that the cross frequency response function yields $H_{21}(\omega_{21}) = 0$. In this case, the dynamic stiffness between two DoFs is infinitely large and the energy transfer becomes impossible.

Now, we check the condition of cross anti-resonance, namely,

$$Z_{21}(\omega_{21}) \equiv k_{21} - m_{21}\omega_{21}^2 = 0, \quad c_{21} = 0 \qquad (4.3.11)$$

This condition includes three simple cases as follows.

(1) The 2-DoF system has both inertial coupling term m_{21} and elastic coupling term k_{21} in Eq. (4.3.5). The coupling in Eq. (4.3.11) disappears when the excitation frequency is $\omega = \omega_{21}$. In this case, the external force inputs energy to the first DoF, but the energy does not transfer to the second DoF.

(2) The 2-DoF system has only an inertial coupling term m_{21}. When $\omega_{21} = 0$, this coupling disappears such that the second DoF becomes rest.

(3) The 2-DoF system has no coupling between two DoFs and the external force only drives the vibration of the first DoF.

4.3.3 Design Feasibility of Anti-Resonances

The subsection presents the influence of system parameters on an anti-resonance and explores the design feasibility of the anti-resonance through a 2-DoF system in Fig. 4.13a.

For brevity, let the mass of uniform rigid bar be $m = 6$ and the stiffness coefficients of both springs be $k = 1$, define the ratio of bar length L to the distance l between two springs as $\eta \equiv L/l \geq 1$ and discuss the influence of η on the anti-resonance.

---

[21] Cheung Y L, Wong W O (2011). H-infinity optimization of a variant design of the dynamic vibration absorber: Revisited and new results. Journal of Sound and Vibration, 330(16): 3901–3912.

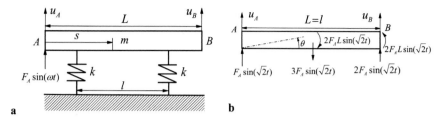

Fig. 4.13 The vibration problem of a 2-DoF system under a harmonic force. **a** the schematic of the system; **b** the force analysis for a direct anti-resonance

We take displacements u_A and u_B at two ends of the bar to describe its motion. From Example 4.2.4, the kinetic and potential energies of the system are

$$\begin{cases} T = \dfrac{3}{4}[\dot{u}_A(t) + \dot{u}_B(t)]^2 + \dfrac{1}{4}[\dot{u}_B(t) - \dot{u}_A(t)]^2 \\[2mm] V = \dfrac{1}{2}\left\{ u_A(t) + \dfrac{[u_B(t) - u_A(t)](\eta - 1)}{2\eta} \right\}^2 + \dfrac{1}{2}\left\{ u_A(t) + \dfrac{[u_B(t) - u_A(t)](\eta + 1)}{2\eta} \right\}^2 \end{cases}$$

(4.3.12)

Equation (4.3.12) leads to the mass matrix and stiffness matrix as follows

$$M = \begin{bmatrix} 2 & 1 \\ 1 & 2 \end{bmatrix}, \quad K = \frac{1}{2\eta^2}\begin{bmatrix} \eta^2 + 1 & \eta^2 - 1 \\ \eta^2 - 1 & \eta^2 + 1 \end{bmatrix}$$

(4.3.13)

For given η, solving the following eigenvalue problem

$$(K - \omega^2 M)\varphi = 0$$

(4.3.14)

we obtain the natural vibrations of the system for three cases as follows.

(1) When $1 \leq \eta < \sqrt{3}$, Example 4.2.4 has given two natural frequencies $\omega_1 = \omega_s = 1/\sqrt{3}$ and $\omega_2 = \omega_a = 1/\eta$. They correspond to a symmetric mode shape $\varphi_1 = [1\ 1]^T$ and an anti-symmetric mode shape $\varphi_2 = [1\ -1]^T$. In this case, the mass center of the bar is the node of the second mode shape.

(2) When $\sqrt{3} < \eta < +\infty$, the natural frequencies are $\omega_1 = \omega_a = 1/\eta$ and $\omega_2 = \omega_s = 1/\sqrt{3}$. They correspond to an anti-symmetric mode shape $\varphi_1 = [1\ -1]^T$ and a symmetric mode shape $\varphi_2 = [1\ 1]^T$, respectively. Now, the mass center of the bar is the node of the first mode shape.

(3) When $\eta = \sqrt{3}$, the system has two repeated natural frequencies $\omega_1 = \omega_2 = 1/\sqrt{3}$, while the mode shapes are two linearly independent vectors, such as a symmetric vector $\varphi_1 = [1\ 1]^T$ and an anti-symmetric vector $\varphi_2 = [1\ -1]^T$.

In what follows, we discuss the design problem of an anti-resonance of the system subject to a harmonic force at the left end, focusing on the following two case studies.

4.3.3.1 Design of Anti-resonances at Two Ends of the Bar

Example 4.3.2 Consider the bar subject to a harmonic force at the left end and design the anti-resonances at two ends of the bar when $m = 6$ and $k = 1$.

From Eq. (4.3.13), we have the following frequency response matrix

$$H(\omega) = (K - \omega^2 M)^{-1}$$

$$= \frac{1}{2(3\omega^2 - 1)(\eta^2\omega^2 - 1)} \begin{bmatrix} \eta^2 + 1 - 4\eta^2\omega^2 & 2\eta^2\omega^2 + 1 - \eta^2 \\ 2\eta^2\omega^2 + 1 - \eta^2 & \eta^2 + 1 - 4\eta^2\omega^2 \end{bmatrix} \quad \text{(a)}$$

Equation (a) leads to the direct anti-resonant frequency and the cross anti-resonant frequency as follows

$$\omega_{AA} = \omega_{BB} = \sqrt{\frac{1}{4} + \frac{1}{4\eta^2}}, \quad \omega_{AB} = \omega_{BA} = \sqrt{\frac{1}{2} - \frac{1}{2\eta^2}} \quad \text{(b)}$$

Recalling the natural frequencies of the system discussed above, we check the variations of natural frequencies and anti-resonant frequencies in Eq. (b) with respect to η and have the following discussions.

When $1 \leq \eta < \sqrt{3}$, as shown in Fig. 4.14, the first natural frequency, i.e., $\omega_1 = \omega_s = 1/\sqrt{3}$, is a constant while the second natural frequency, i.e., $\omega_2 = \omega_a = 1/\eta$, decreases with an increase of η. The natural frequencies and anti-resonant frequencies satisfy $0 \leq \omega_{AB} = \omega_{BA} < \omega_1 < \omega_{AA} = \omega_{BB} < \omega_2$. That is, the direct anti-resonant frequency is between two natural frequencies while the cross anti-resonant frequency is lower than the first natural frequency.

Fig. 4.14 The natural frequencies and anti-resonant frequencies with respect to η (thick solid line: ω_s, thin solid line: ω_a, dashed line: ω_{AA}, ω_{BB}, dot line: ω_{AB}, ω_{BA})

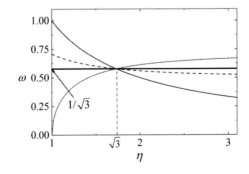

When $\sqrt{3} < \eta < +\infty$, as shown in Fig. 4.14, the first natural frequency is $\omega_1 = \omega_a = 1/\eta$ and decreases with increasing of η, while the second natural frequency is $\omega_2 = \omega_s = 1/\sqrt{3}$ and remains constant. The natural frequencies and anti-resonant frequencies satisfy $0 < \omega_1 < \omega_{AA} = \omega_{BB} < \omega_2 < \omega_{AB} = \omega_{BA}$. That is, the direct anti-resonant frequency is still between two natural frequencies while the cross anti-resonant frequency is higher than the second natural frequency.

When $\eta = \sqrt{3}$, both natural frequencies and both anti-resonant frequencies are identical such that $\omega_1 = \omega_2 = \omega_{AA} = \omega_{BB} = \omega_{AB} = \omega_{BA} = 1/\sqrt{3}$ holds. In Fig. 4.14, the four curves intersect at point $(\sqrt{3}, \ 1/\sqrt{3})$, exhibiting a degeneration. In this case, the frequency response function has a zero and a pole at $\omega = 1/\sqrt{3}$. Given $\omega \neq 1/\sqrt{3}$, substituting $\eta = \sqrt{3}$ into Eq. (a) leads to the frequency response matrix of the system as follows

$$\boldsymbol{H}(\omega) = \frac{1}{2(3\omega^2 - 1)^2} \begin{bmatrix} 4 - 12\omega^2 & 6\omega^2 - 2 \\ 6\omega^2 - 2 & 4 - 12\omega^2 \end{bmatrix} = \frac{1}{3\omega^2 - 1} \begin{bmatrix} -2 & 1 \\ 1 & -2 \end{bmatrix} \quad \text{(c)}$$

In this case, there is no anti-resonance for other excitation frequencies.

As discussed above, Fig. 4.14 enables one to select the ratio η of the bar length to the distance between two springs in order to put the anti-resonant frequencies in a desirable frequency band. The dashed line shows that the variation of direct anti-resonant frequencies is relatively small, while the dot line shows a relatively large variation of cross anti-resonant frequencies with the lower bound being zero. Hence, the design of cross anti-resonant frequencies is more flexible.

4.3.3.2 Design of Anti-Resonance at an Arbitrary Position of the Bar

Example 4.3.3 Study how to make an arbitrary point on the bar undergo an anti-resonance when the harmonic force is applied to the left end of the bar.

From Eqs. (4.3.6) and (a) in Example 4.3.2, we get the frequency responses at two ends of the bar when the left end of the bar is subject to a harmonic force $f_A(t) = F_A \sin(\omega t)$, namely,

$$\begin{bmatrix} U_A(\omega) \\ U_B(\omega) \end{bmatrix} = \frac{1}{2(3\omega^2 - 1)(\eta^2 \omega^2 - 1)} \begin{bmatrix} \eta^2 + 1 - 4\eta^2\omega^2 & 2\eta^2\omega^2 + 1 - \eta^2 \\ 2\eta^2\omega^2 + 1 - \eta^2 & \eta^2 + 1 - 4\eta^2\omega^2 \end{bmatrix} \begin{bmatrix} F_A \\ 0 \end{bmatrix}$$

$$= \frac{F_A}{2(3\omega^2 - 1)(\omega^2 - 1)} \begin{bmatrix} \eta^2 + 1 - 4\eta^2\omega^2 \\ 2\eta^2\omega^2 + 1 - \eta^2 \end{bmatrix} \quad \text{(a)}$$

In Fig. 4.13a, let an arbitrary point be measured as s from the left end and study its frequency response. Denote $\xi = s/(\eta L) \in [0, \ 1]$ as the dimensionless distance, then we have the frequency response of point s as follows

$$U_s(\omega) = (1 - \xi)U_A(\omega) + \xi U_B(\omega)$$

$$= \frac{(1 - \xi)(\eta^2 + 1 - 4\eta^2\omega^2) + \xi(2\eta^2\omega^2 + 1 - \eta^2)}{2(3\omega^2 - 1)(\omega^2 - 1)} F_A \qquad (b)$$

The nominator of the right-hand side of Eq. (b) gives the following condition of an anti-resonance of $U_s(\omega)$

$$(1 - \xi)(\eta^2 + 1 - 4\eta^2\omega^2) + \xi(2\eta^2\omega^2 + 1 - \eta^2) = 0 \qquad (c)$$

Solving Eq. (c) for ω leads to the anti-resonant frequency ω_{sA} for a dimensionless distance ξ, namely,

$$\omega_{sA} = \sqrt{\frac{\eta^2(1 - 2\xi) + 1}{2\eta^2(2 - 3\xi)}} \qquad (d)$$

In a parameter region of $(\eta, \xi) \in [0, 4] \otimes [1, 0]$, Fig. 4.15 depicts the three-dimensional contour lines of Eq. (d), which enables one to select parametric combination of (η, ξ) in order to get a desirable anti-resonant frequency ω_{sA}. As the surface in Fig. 4.15 looks a little bit complicated, Fig. 4.16 demonstrates two typical profiles of the surface. Figure 4.16a presents the case when $\eta = 1$, showing the monotonic increase of ω_{sA} with an increase of ξ, while Fig. 4.16b is the case when $\eta = 2$, exhibiting two monotonic decreasing branches. When $\eta = \sqrt{3}$, Eq. (b) degenerates to $\omega_{sA} = 1/\sqrt{3} \approx 0.577$, which corresponds to the horizontal solid line in Fig. 4.16 and the degenerate case (3) in Example 4.3.2.

In view of better static stability of the system, take $\eta = 1$ and substitute it into Eq. (d), then we have the following anti-resonant frequency

Fig. 4.15 The anti-resonant frequency ω_{sA} of the bar versus parametric combination (η, ξ)

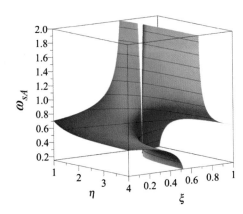

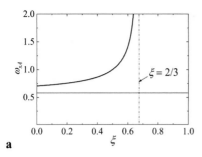 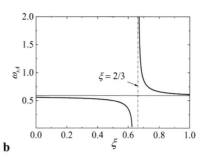

Fig. 4.16 The anti-resonant frequency ω_{sA} at the position $\xi = s/\eta L$ on the bar. **a** $\eta = 1 < \sqrt{3}$; **b** $\eta = 2 > \sqrt{3}$

$$\omega_{sA} = \sqrt{\frac{1 - \xi}{2 - 3\xi}} \qquad\qquad (e)$$

In what follows, the aenomena in statics, is discussed for three cases.

(1) When $\xi = 0$, $U_s(\omega) = U_A(\omega) = H_{AA}(\omega)F_A$ is the direct frequency response and its anti-resonant frequency $\omega_{AA} = 1/\sqrt{2}$ is between two natural frequencies $\omega_1 = 1/\sqrt{3}$ and $\omega_2 = 1$. In the direct anti-resonance, $U_A(\omega_{AA}) = 0$ and $U_B(\omega_{AA}) = -2$ hold. That is, $u_A(t) = 0$ and $u_B(t) = -2F_A \sin(\sqrt{1/2}\, t)$. Now, we analyze the forces applied to the rigid bar from Fig. 4.13b. The external force to the left end of the bar is $f_A(t) = F_A \sin(\sqrt{1/2}\, t)$, the elastic restoring force to the right end of the bar is $-ku_B(t) = 2F_A \sin(\sqrt{1/2}\, t)$, the inertial force to the mass center of the bar is $-m\ddot{u}_B(t)/2 = -3F_A \sin(\sqrt{1/2}\, t)$, and the superposition of these forces is zero. In addition, if the counter clockwise direction is positive, the torque of the elastic restoring force to the right end of the bar with respect to the left end of the bar is $-kLu_B(t) = 2F_AL \sin(\sqrt{1/2}\, t)$, the torque of distributed inertial forces with respect to the left end of the bar is $-(mL^2/3)[\ddot{u}_B(t)/L] = -2F_AL \sin(\sqrt{1/2}\, t)$, and their summation is also zero. As a consequence, the left end of the bar remains rest such that the left spring has no deformation at all even though the left end of the bar is subject to a harmonic force. In this case, the bar undergoes a rotational harmonic vibration with respect to the left end of the bar and the vibration amplitude is $2F_A$.

(2) When $\xi > 0$, $U_s(\omega)$ is a cross frequency response and Eq. (e) gives its anti-resonant frequency. In particular, the anti-frequency is $\omega_{sA} = \omega_2 = 1$ when $\xi = 1/2$, which is not only the mass center of the bar, but also the node of the second mode shape. In this case, the bar exhibits the second resonance under harmonic force $f_A(t) = F_A \sin t$ to the left end of the bar, rotating around the mass center of the bar. In general, when a cross anti-resonant frequency coincides with a natural frequency, the node of corresponding mode shape serves as the location of the cross anti-resonance.

(3)   Figure 4.16a shows that if the external force to the left end of the bar is a sweeping harmonic force in frequency domain, the left end of the bar keeps immovable when the excitation frequency is $\omega = 1/\sqrt{2}$. This immovable point shifts towards the right direction and yields $s = (2\omega^2 - 1)L/(3\omega^2 - 1)$ with an increase of excitation frequency ω. Finally, the immovable point approaches $s = 2L/3$. That is, the design range for the anti-resonance on the bar harmonically excited at the left end is $s \in [0, \ 2L/3)$ when $\eta = 1$.

4.3.4   Concluding Remarks

This section dealt with the anti-resonance problems of a 2-DoF damped system to a harmonic force, revealed the mechanisms behind both direct and cross anti-resonances and demonstrated the design feasibility of anti-resonances. The section has solved Problem 3B proposed in Chap. 1 and led to the major conclusions as follows.

(1)   During a direct anti-resonance of a 2-DoF system, the harmonic excitation does not input energy to the system and the cross frequency response of the system is energy-conservative harmonic vibration, the amplitude of which is maintained by the harmonic excitation. During a cross anti-resonance of the system, the harmonic excitation inputs energy to the system, but the dynamic impedance between two DoFs is infinitely large so that the energy can not flow between two DoFs. For the mechanisms of direct anti-resonance, the previous viewpoints of "dynamic absorption" and "energy transfer" are not appropriate. The viewpoint of "dynamic cancellation" based on the dynamic equilibrium of all forces seems better.

(2)   The steady-state vibration of a rigid bar supported by two identical springs and harmonically excited at the left end of the bar well demonstrates interesting anti-resonances and their designs. For instance, the vibration at the right end of the bar has a cross anti-resonant frequency, which is either lower than the first natural frequency or higher than the second natural frequency. In addition, the cross anti-resonance may appear from the left end of the bar to 2/3 of full length of the bar with an increase of excitation frequency. When a cross anti-resonant frequency coincides with a natural frequency, a cross anti-resonance occurs at the node of the corresponding mode shape.

4.4   Dynamic Modifications of a System

In practice, a system designed or manufactured may not meet the dynamic requirements and needs to be modified. This task, named the *dynamic modification* of a system, usually begins with an available dynamic model, either a continuous one or a discrete one. Given the dynamic inputs and desirable outputs of a vibration system, the dynamic modification has not only a great variety of options, but also a promising integration with new design approaches, such as topological optimizations.

This section briefly presents the basic ideas of local dynamic modification of an MDoF system. Let the MDoF system of concern be the *primary system* and an auxiliary system attached to the primary system be the *secondary system* for dynamic modification. The section uses the dynamic sub-structuring method to establish the frequency response matrix of the composite system composed of the primary system and the secondary system, and then discusses how to design the secondary system in order to generate an anti-resonance or adjust resonances of the primary system.

It is worthy to point out that the dynamic modification based on the frequency response matrix is applicable to any continuous systems when their frequency response function is available. In fact, the examples in this section will be the dynamic modification of a hinged-hinged beam. The frequency response function of a continuous system computed or measured in a finite frequency band implies the modal truncation out of the frequency band.

4.4.1 Frequency Response of a Composite System

Figure 4.17 shows a composite system composed of a primary system denoted by P and a secondary system denoted by S. Let the DoFs of the primary system be divided into two sets such that set A includes all interfacial DoFs to the secondary system and set B includes the other DoFs of the primary system. Similarly, let the DoFs of the secondary system be also divided into two sets. The DoFs in set C connect to the primary system and those in set D are all other DoFs of the secondary system. Accordingly, sets B and D may be empty.

At first, consider the dynamic response of primary system P to external forces in frequency domain $\omega \in [0, +\infty)$ and express it as

$$U_P(\omega) = H_{PP}(\omega) F_P(\omega) \tag{4.4.1}$$

where $U_P : \omega \mapsto \mathbb{C}^P$ is the frequency response vector of the primary system, $F_P : \omega \mapsto \mathbb{C}^P$ is the external force vector applied to the primary system, and $H_{PP} : \omega \mapsto \mathbb{C}^{P \times P}$ is the frequency response matrix of the primary system. For brevity, the dimension of a matrix or a vector, as well as frequency ω, does not

Fig. 4.17 The schematic of a composite system composed of a primary system and a secondary system

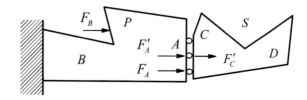

explicitly appear if not necessary. From the division of above DoFs, Eq. (4.4.1) can be recast as the following form of block matrices

$$U_P(\omega) \equiv \begin{bmatrix} U_A \\ U_B \end{bmatrix} = \begin{bmatrix} H_{AA} & H_{AB} \\ H_{BA} & H_{BB} \end{bmatrix} \begin{bmatrix} F_A \\ F_B \end{bmatrix} \equiv H_{PP}(\omega) F_P(\omega) \tag{4.4.2}$$

Second, we study the frequency response of the secondary system connected to the primary system. In the practice of dynamic modification, the secondary system is artificially attached to the primary system and not subject to external forces. That is, the secondary system is only subject to an interfacial force vector F'_C at set C of DoFs. Thus, the frequency response vector U_S of the secondary system yields

$$U_S \equiv \begin{bmatrix} U'_C \\ U'_D \end{bmatrix} = \begin{bmatrix} H_{CC} & H_{CD} \\ H_{DC} & H_{DD} \end{bmatrix} \begin{bmatrix} F'_C \\ 0 \end{bmatrix} \tag{4.4.3}$$

When the primary system and the secondary system are connected each other, set A of DoFs of the primary system is subject to an interfacial force vector F'_A. Then, the frequency response vector U_P of the primary system becomes vector U'_P, which is the superposition of the original frequency response vector U_P and an incremental vector due to interfacial force vector F'_A, namely,

$$U'_P \equiv \begin{bmatrix} U'_A \\ U'_B \end{bmatrix} = \begin{bmatrix} H_{AA} & H_{AB} \\ H_{BA} & H_{BB} \end{bmatrix} \left\{ \begin{bmatrix} F_A \\ F_B \end{bmatrix} + \begin{bmatrix} F'_A \\ 0 \end{bmatrix} \right\} \tag{4.4.4}$$

The compatibility conditions at the interfacial DoFs give

$$U'_A - U'_C = 0, \quad F'_A + F'_C = 0 \tag{4.4.5}$$

Substitution of Eq. (4.4.5) into the first row in Eq. (4.4.3) leads to

$$U'_A = U'_C = H_{CC} F'_C = -H_{CC} F'_A \tag{4.4.6}$$

Then, substituting Eq. (4.4.6) into the first row in Eq. (4.4.4) gives

$$(H_{AA} + H_{CC}) F'_A = -(H_{AA} F_A + H_{AB} F_B) \tag{4.4.7}$$

Now, we claim that matrix $H_{AA} + H_{CC}$ is invertible. If the claim is not true, there exists a frequency $\tilde{\omega}$ so that $\det[H_{AA}(\tilde{\omega}) + H_{CC}(\tilde{\omega})] = 0$ holds. When $F_A(\tilde{\omega}) = 0$ and $F_B(\tilde{\omega}) = 0$ hold, Eq. (4.4.7), as a set of linear homogeneous equations, has a set of non-trivial solutions, that is, $F'_A(\tilde{\omega}) \neq 0$. Substitution of $F'_A(\tilde{\omega}) \neq 0$ into Eq. (4.4.6) gives $U'_A(\tilde{\omega}) \neq 0$. This implies that the primary system free of external forces undergoes a vibration when it is connected with a secondary system free of any external forces. The above contradiction indicates that matrix $H_{AA} + H_{CC}$ is invertible.

As matrix $H_{AA} + H_{CC}$ is invertible, solving Eq. (4.4.7) gives

$$F'_A = -F'_C = -(H_{AA} + H_{CC})^{-1}(H_{AA} F_A + H_{AB} F_B) \qquad (4.4.8)$$

Substituting Eq. (4.4.8) into the combination of Eqs. (4.4.3) and (4.4.4), we get the frequency response vector of the composite system, namely,

$$\begin{bmatrix} U'_A \\ U'_B \\ U'_D \end{bmatrix} = \left\{ \begin{bmatrix} H_{AA} & H_{AB} & 0 \\ H_{BA} & H_{BB} & 0 \\ 0 & 0 & 0 \end{bmatrix} - \begin{bmatrix} H_{AA} \\ H_{BA} \\ -H_{DC} \end{bmatrix} (H_{AA} + H_{CC})^{-1} \begin{bmatrix} H_{AA} & H_{AB} & 0 \end{bmatrix} \right\} \begin{bmatrix} F_A \\ F_B \\ 0 \end{bmatrix} \qquad (4.4.9)$$

It is easy to see that the frequency response matrix of the composite system reads

$$H \equiv \begin{bmatrix} H_{AA} & H_{AB} & 0 \\ H_{BA} & H_{BB} & 0 \\ 0 & 0 & 0 \end{bmatrix} - \begin{bmatrix} H_{AA} \\ H_{BA} \\ -H_{DC} \end{bmatrix} (H_{AA} + H_{CC})^{-1} \begin{bmatrix} H_{AA} & H_{AB} & 0 \end{bmatrix} \qquad (4.4.10)$$

As set A of the primary system is connected to set C of the secondary system, set C does not show up in Eqs. (4.4.9) and (4.4.10).

In the dynamic modification for the primary system, it is reasonable for engineers to pay much attention to the relation between the external forces and the system responses. For example, they ought to study the dynamic response of set I of DoFs when the primary system is subject to external forces at set J of DoFs. In this case, let H_{IJ} be the frequency response matrix from set J to set I, H_{IA} and H_{JA} be the frequency response matrices from set A to set I and set J, respectively. After assembly of the primary system and the secondary system, the frequency response matrix of the modified system from set J to set I in view of Eq. (4.4.10) becomes available, namely,

$$H'_{IJ} = H_{IJ} - H_{IA}(H_{AA} + H_{CC})^{-1} H_{AJ} \qquad (4.4.11)$$

This is an important formula in the vibration analysis and dynamic modification for the primary system in frequency domain.

Remark 4.4.1 In practice, it is very useful to replace the frequency response matrix H_{CC} at interfacial DoFs in Eq. (4.4.11) with corresponding displacement impedance matrix $Z_{CC} = H_{CC}^{-1}$ and then derive the following two expressions

$$H'_{IJ} = H_{IJ} - H_{IA}[H_{CC}(Z_{CC}H_{AA} + I)]^{-1}H_{AJ}$$
$$= H_{IJ} - H_{IA}(I + Z_{CC}H_{AA})^{-1}Z_{CC}H_{AJ} \qquad (4.4.12a)$$

$$H'_{IJ} = H_{IJ} - H_{IA}[(H_{AA}Z_{CC} + I)H_{CC}]^{-1}H_{AJ}$$
$$= H_{IJ} - H_{IA}Z_{CC}(I + H_{AA}Z_{CC})^{-1}H_{AJ} \qquad (4.4.12b)$$

Section 4.4.3 will present the use of above expressions when a secondary system is inserted into the primary system.

Remark 4.4.2 The other expression of Eq. (4.4.9) reads

$$\begin{bmatrix} U'_A \\ U'_B \\ U'_D \end{bmatrix} = \begin{bmatrix} U_A \\ U_B \\ 0 \end{bmatrix} - \left\{ \begin{bmatrix} H_{AA} \\ H_{BA} \\ -H_{DC} \end{bmatrix} (H_{AA} + H_{CC})^{-1} \begin{bmatrix} H_{AA} & H_{AB} & 0 \end{bmatrix} \right\} \begin{bmatrix} F_A \\ F_B \\ 0 \end{bmatrix}$$
$$(4.4.13)$$

where the first term in the right-hand side includes the original frequency response vector U_P of the primary system in view of Eq. (4.4.2), and sub-matrix H_{BB} in Eq. (4.4) does not show up. In general, set A involves a few of DoFs, but set B may have many DoFs so that it is difficult to get sub-matrix H_{BB}. It is beneficial, hence, to use Eq. (4.4.13) to predict the dynamic response of the composite system since it is based on the original frequency response vector U_P of the primary system[22]. In practice, the original frequency response of the primary system measured is often contaminated with noise. In this case, Eq. (4.4.13) provides a rough estimation for the dynamic modification.

4.4.2 Adjusting an Anti-Resonance of a Primary System

Consider the local dynamic modification of the primary system at one DoF $a \in A$. In this case, the secondary system has only one DoF $c \in C$ for connection so that Eq. (4.4.11) can be simplified as

$$H'_{IJ}(\omega) = H_{IJ}(\omega) - \frac{1}{H_{aa}(\omega) + H_{cc}(\omega)} H_{Ia}(\omega)H_{aJ}(\omega) \qquad (4.4.14)$$

From Eq. (4.4.14), the frequency response matrix of the modified system from set J of excitation DoFs to the connection DoF a reads

---

[22] Li Y F, Hu H Y (1988). Vibration attenuation design of complicated structures based on their original responses. Journal of Nanjing Aeronautical Institute, 5(1): 133–142.

Fig. 4.18 The schematic of an SDoF secondary system

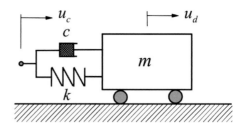

$$H'_{aJ}(\omega) = H_{aJ}(\omega) - \frac{H_{aa}(\omega)}{H_{aa}(\omega) + H_{cc}(\omega)} H_{aJ}(\omega)$$

$$= \frac{H_{cc}(\omega)}{H_{aa}(\omega) + H_{cc}(\omega)} H_{aJ}(\omega) \qquad (4.4.15)$$

When the secondary system in Fig. 4.18 is used for the dynamic modification, the dynamic equation of the secondary system in time domain reads

$$\begin{bmatrix} 0 & 0 \\ 0 & m \end{bmatrix}\begin{bmatrix} 0 \\ \ddot{u}_d(t) \end{bmatrix} + \begin{bmatrix} c & -c \\ -c & c \end{bmatrix}\begin{bmatrix} \dot{u}_c(t) \\ \dot{u}_d(t) \end{bmatrix} + \begin{bmatrix} k & -k \\ -k & k \end{bmatrix}\begin{bmatrix} u_c \\ u_d \end{bmatrix} = \begin{bmatrix} 0 \\ 0 \end{bmatrix} \qquad (4.4.16)$$

Applying Fourier transform to both sides of Eq. (4.4.16), we get the displacement impedance matrix and the frequency response matrix of the secondary system as follows

$$Z_{SS}(\omega) \equiv \begin{bmatrix} k + ic\omega & -k - ic\omega \\ -k - ic\omega & k + ic\omega - m\omega^2 \end{bmatrix} \qquad (4.4.17)$$

$$H_{SS}(\omega) = -\frac{1}{m\omega^2(k + ic\omega)}\begin{bmatrix} k + ic\omega - m\omega^2 & k + ic\omega \\ k + ic\omega & k + ic\omega \end{bmatrix} \qquad (4.4.18)$$

In particular, the frequency response function at the connection point is

$$H_{cc}(\omega) = -\frac{k + ic\omega - m\omega^2}{m\omega^2(k + ic\omega)} \qquad (4.4.19)$$

According to the conditions of anti-resonance in Sect. 4.3, let $c = 0$ in Eq. (4.4.19) and get $H_{cc}(\omega_{aa}) = 0$ when $\omega_{aa} = \sqrt{k/m}$. Substitution of $H_{cc}(\omega_{aa}) = 0$ into Eq. (4.4.15) leads to

$$H'_{aJ}(\omega_{aa}) = 0 \qquad (4.4.20)$$

Compared with the problem of anti-resonance of an SDoF primary system in Sect. 4.3, Eq. (4.4.20) gives an important result for a primary MDoS system.

Remark 4.4.3 Equation (4.4.20) implies that the row matrix $H'_{aJ}(\omega)$ for the frequency responses of the primary system exhibits anti-resonance at frequency $\omega = \omega_{aa}$. That is, the primary system does not only exhibit a direct anti-resonance at the connection DoF a, bus also a cross anti-resonance between the connection DoF a and an arbitrary set J of DoFs. This result enables one to understand why the dynamic vibration absorber attached to the top floor of Building 101 in Taipei, China is able to reduce the vibration of the top floor driven by the distributed wind flow to the entire building.

Remark 4.4.4 The symmetry of frequency response matrix gives

$$H'_{Ja}(\omega_{aa}) = H'^T_{aJ}(\omega_{aa}) = 0 \tag{4.4.21}$$

This implies that once the primary system is subject to an external force at the connection DoF a, the vibration of entire primary system disappear at frequency ω_{aa}. That is, the un-damped secondary system makes the displacement impedance of connection DoF be infinitely large at frequency ω_{aa}.

According to Eq. (4.4.14), we get the frequency response matrix of other DoFs ($a \notin I$, $a \notin J$) of the primary system at anti-frequency ω_{aa}, namely,

$$H'_{IJ}(\omega_{aa}) = H_{IJ}(\omega_{aa}) - \frac{1}{H_{aa}(\omega_{aa})} H_{Ia}(\omega_{aa}) H_{aJ}(\omega_{aa}) \tag{4.4.22}$$

Equation (4.4.22) shows that the external forces at set J of DoFs still cause the vibration of the primary system at other DoFs. That is, the anti-resonance is a local behavior of an MDoF system.

From Eq. (4.4.10), we get the frequency response function of the primary system at DoF d due to the external force at the connection DoF a, namely,

$$H_{da}(\omega) = \frac{H_{dc}(\omega) H_{aa}(\omega)}{H_{aa}(\omega) + H_{cc}(\omega)} = \frac{k H_{aa}(\omega)}{k - m\omega^2 - km\omega^2 H_{aa}(\omega)} \tag{4.4.23}$$

At the direct anti-resonant frequency $\omega_{aa} = \sqrt{k/m}$, Eq. (4.4.23) can be simplified as

$$H_{da}(\omega_{aa}) = -\frac{1}{m\omega_{aa}^2} = -\frac{1}{k} \tag{4.4.24}$$

This is just the condition under which the elastic restoring force of the secondary system balances the external force.

4.4.3 Adjusting Resonances of a Primary System

This subsection presents how to use a secondary system to adjust the resonances of a primary system, including shifting the resonant frequencies and suppressing resonant peaks. The secondary system used can be a lumped mass, a massless spring and a massless dashpot, respectively.

4.4.3.1 A Lumped Mass Attached

Let $c \to +\infty$ and $k \to +\infty$ in Fig. 4.18, then the secondary system connected to the primary system becomes a lumped mass m attached to the connection DoF a of the primary system. In this case, Eq. (4.4.19) becomes

$$H_{cc}(\omega) = -\frac{1}{m\omega^2} \tag{4.4.25}$$

Substitution of Eq. (4.4.25) into Eq. (4.4.15) leads to

$$H'_{aJ}(\omega) = \frac{1}{1 - m\omega^2 H_{aa}(\omega)} H_{aJ}(\omega) \tag{4.4.26}$$

According to the denominator in Eq. (4.4.26), it is easy to study the influence of lumped mass m on the poles of the frequency response matrix $H'_{aJ}(\omega)$ and adjust the resonant frequencies of the primary system. The following example demonstrates the detailed procedure of both computation and modal test.

Example 4.4.1 The primary system of concern is a hinged-hinged beam of length L, study the dynamic modification to reduce the first three resonant frequencies when a lumped mass is attached to the beam at a quarter span measured from the left end.

We first take the position of $L/4$ measured from the left end of the beam as the connection point. In view of modal expression of frequency response function, the direct frequency response function of an un-damped beam with the first three modes taken into account reads

$$H_{aa}(\omega) = \sum_{r=1}^{3} \frac{\varphi_r(L/4)\varphi_r(L/4)}{M_r(\omega_r^2 - \omega^2)} \tag{a}$$

In the dynamic modification of a lumped mass, it is more convenient to use the frequency response function of acceleration, which is expressed, from Eq. (a), as

$$G_{aa}(\omega) \equiv -\omega^2 H_{aa}(\omega) = \sum_{r=1}^{3} \frac{\omega^2 \varphi_r(L/4)\varphi_r(L/4)}{M_r(\omega^2 - \omega_r^2)} \tag{b}$$

Now, we attach the lumped mass m to the beam. Let $J = a$ and both sides of Eq. (4.4.26) be multiplied by $-\omega^2$, then we get the direct frequency response function of acceleration of the modified beam, namely,

$$G'_{aa}(\omega) = \frac{G_{aa}(\omega)}{1 + mG_{aa}(\omega)} \tag{c}$$

where the zeros $\omega'_r, r = 1, 2, 3$ of the denominator in Eq. (c) are the resonant frequencies, i.e., the natural frequencies, of the modified beam.

From Table 2.1, the natural frequencies, mode shapes and modal masses of a hinged-hinged beam in Eq. (b) are as follows

$$\omega_r = \frac{r^2 \pi^2}{L^2} \sqrt{\frac{EI}{\rho A}}, \quad \varphi_r(x) = \sin\left(\frac{r\pi x}{L}\right),$$

$$M_r = \int_0^L \rho A \varphi_r^2(x)\,dx = \frac{\rho AL}{2}, \quad r = 1, 2, 3 \tag{d}$$

We take the parameters of a hinged-hinged steel beam that the author used in BIT School of Aerospace Engineering to demonstrate the dynamics modification, namely,

$$L = 2\,\text{m}, \ A = 6 \times 10^{-4}\,\text{m}^2, \ I = 5 \times 10^{-5}\text{m}^4, \ \rho = 7850\,\text{kg/m}^3, \ E = 210\,\text{GPa} \tag{e}$$

Substitution of Eq. (e) into Eq. (d) leads to the first three natural frequencies ω_r, which can be converted, from $f_r = \omega_r/(2\pi)$, as

$$f_1 = 5.86\,\text{Hz}, \quad f_2 = 23.45\,\text{Hz}, \quad f_3 = 52.77\,\text{Hz} \tag{f}$$

In view of Eq. (b), we compute the frequency response function of acceleration $G_{aa}(f)$ of the beam at the connection point as shown in Fig. 4.19a, where $f = \omega/(2\pi)$. As the damping is negligible, $G_{aa}(f)$ approaches infinity at the first three resonant frequencies. In the modal test, of course, the measured frequency response function exhibited finite peaks.

Now, we use a mass block with $m = 4.866$ kg and plot a horizontal dashed line $G(f) = -1/m \approx -0.206$ kg^{-1} in Fig. 4.19a. From the intersection points of the dashed line with curve $G_{aa}(f)$, the vertical lines to the abscissa predict the resonant frequencies of modified beam as shown in Table 4.1. Figure 4.19b gives a comparison between the frequency response functions of the beam before and after the dynamic modification.

Table 4.1 also presents the resonant frequencies computed by using Software ABAQUS of finite elements and measured via a modal test for the beam. The computational results of the two methods got very good agreement, but the experimental

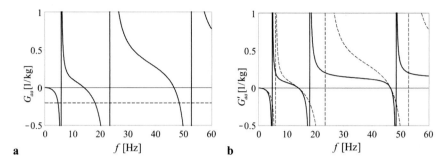

Fig. 4.19 The frequency response functions of acceleration of a hinged-hinged beam before and after mass modification. **a** solid line: the frequency response function before modification, dashed line: $G(f) = -1/m$; **b** solid line: the frequency response function after modification, dashed line: the frequency response function before modification

Table 4.1 A comparison of resonant frequencies of a modified beam via different methods	Methods	1st natural frequency	2nd natural frequency	3rd natural frequency
	Method in this section	4.72	18.07	48.58
	Finite element method	4.72	18.02	48.36
	Modal test	4.99	18.26	48.40

results were slightly different from computational ones. There are several reasons for the deviations. One is that the hinged boundaries of the beam in test were not perfect. The other is that the mass block attached to the beam was not a lumped mass, but exhibited a rotational inertia.

Furthermore, it is easy, from Fig. 4.19a, to see the following rules.

(1) When a lumped mass is attached to a beam, all resonant frequencies of the beam decrease. The larger the lumped mass, the more the resonant frequencies of beam decrease. Given a lumped mass, the resonant frequencies of lower order of the beam decrease slightly and those of higher order of the beam decrease significantly.

(2) To design the attached mass for the beam modification, one simply selects a desired resonant frequency f'_r of the modified beam, plots a vertical line starting from f'_r at the abscissa to the frequency response function of acceleration $G_{aa}(f)$ of the beam and gets the intersection point denoted by $G_{aa}(f'_r)$. Then, $m = -1/G_{aa}(f'_r)$ is the mass to be attached. This procedure is also applicable to the mass modification of other structures.

4.4.3.2   A Massless Spring Inserted

The local stiffness modification based on a massless spring for the primary system has two ways. One is to insert a spring between the primary system and the rigid ground. The other is to insert a spring between two points of the primary system.

To demonstrate the first way, let $m \to +\infty$ in Fig. 4.18, the secondary system connected to the primary system becomes a spring between the primary system and the rigid ground. Equation (4.4.19) gives

$$H_{cc}(\omega) = \frac{1}{k} \qquad (4.4.27)$$

Substitution of Eq. (4.4.27) into Eq. (4.4.15) leads to

$$H'_{aJ}(\omega) = \frac{1}{1 + k H_{aa}(\omega)} H_{aJ}(\omega) \qquad (4.4.28)$$

From the denominator in Eq. (4.4.28), it is easy to discuss the influence of the stiffness coefficient of a spring on the poles of the frequency response matrix $H'_{aJ}(\omega)$ and adjust the resonant frequencies. The following example presents the detailed procedure.

Example 4.4.2 Let the hinged-hinged beam be the primary system and insert a spring between the beam at a quarter span measured from the left end and the rigid ground. Study the influence of the local stiffness modification on the first three resonant frequencies.

From Eq. (a) in Example 4.4.1, we have the direct frequency response function of an un-damped beam with the first three modes taken into account, namely,

$$H_{aa}(\omega) = \sum_{r=1}^{3} \frac{\varphi_r(L/4)\varphi_r(L/4)}{M_r(\omega_r^2 - \omega^2)} \qquad (a)$$

To insert the spring of stiffness coefficient k, let $J = a$ in Eq. (4.4.28) and get the frequency response function of the modified beam as follows

$$H'_{aa}(\omega) = \frac{H_{aa}(\omega)}{1 + k H_{aa}(\omega)} \qquad (b)$$

The poles, that is, the zeros of the denominator, of Eq. (b) are the resonant frequencies of the modified beam. They are also the natural frequencies to be adjusted.

Let the parameters in Example 4.4.1 be substituted into Eq. (a) and compute the direct frequency response function $H_{aa}(f)$ of the beam before modification shown in Fig. 4.20a, where $f = \omega/(2\pi)$. To determine the poles of Eq. (b), plot a

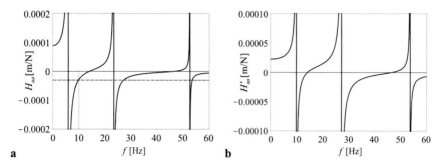

Fig. 4.20 The frequency response functions of a hinged-hinged beam before and after a spring is inserted between the beam and the rigid ground. **a** solid line: the frequency response function before modification, dashed line: $H(f) = -1/k$; **b** the frequency response function of after modification

horizontal dashed line $H(f) = -1/k$ in Fig. 4.20a, and then the vertical lines from the intersection points with $H_{aa}(f)$ to the abscissa give the poles of Eq. (b), that is, the resonant frequencies of the modified beam.

According to Fig. 4.20, it is clear to see the following rules:

(1) After the spring is inserted between the beam and the rigid ground, all resonant frequencies of the beam increase. The larger the stiffness coefficient of the spring is, the more the resonant frequencies of the beam increase. Given the stiffness coefficient of the spring, the increase of resonant frequencies of lower order of the beam are large, while those of higher orders are small.

(2) To design the stiffness coefficient of the spring, one simply selects a desired resonant frequency f'_r of the beam, plots a vertical line starting from f'_r at the abscissa to the frequency response function $H_{aa}(f)$, and gets an intersection point denoted by $H_{aa}(f'_r)$. Then, $k = -1/H_{aa}(f'_r)$ is the stiffness coefficient of the spring to be inserted. This procedure is also applicable to the first way of local stiffness modification for other structures.

For instance, let $f'_1 = 10$ Hz be the first resonant frequency of the modified beam in Example 4.4.2, plot a vertical dot line from $f'_1 = 10$ Hz at the abscissa to the frequency response $H_{aa}(f)$ in Fig. 4.20a. Then, the horizontal dashed line passing through the intersection point leads to value $H'_{aa}(10)$ at the ordinate, and $k = -1/H'_{aa}(10) \approx 33$ kN/m is the stiffness coefficient of the spring to be designed. Figure 4.20b presents the frequency response function of the modified beam and the first resonant frequency of the modified beam appears at 10 Hz exactly.

Now, we turn to the second way of the local stiffness modification based on a massless spring for the primary system. The displacement impedance of the spring reads

$$\mathbf{Z}_{CC}(\omega) = \begin{bmatrix} k & -k \\ -k & k \end{bmatrix} \tag{4.4.29}$$

As matrix $Z_{CC}(\omega)$ is singular, we use Eq. (4.4.12) to perform the dynamic modification. To get the variation of resonant frequencies in the modification, we only need to use the frequency response sub-matrix at the interfacial DoFs and simplify Eq. (4.4.12a) as

$$H'_{AA} = H_{AA} - H_{AA}(I + Z_{CC}H_{AA})^{-1}Z_{CC}H_{AA}$$
$$= H_{AA}(I + Z_{CC}H_{AA})^{-1} \tag{4.4.30}$$

where $H_{AA}(\omega)$ is expressed as

$$H_{AA}(\omega) = \begin{bmatrix} H_{11}(\omega) & H_{12}(\omega) \\ H_{21}(\omega) & H_{22}(\omega) \end{bmatrix} \tag{4.4.31}$$

Substituting Eqs. (4.4.29) and (4.4.31) into Eq. (4.4.30), we derive the frequency response sub-matrix $H'_{AA}(\omega)$ in the interfacial DoFs for the primary system after the dynamic modification and their entries satisfy

$$\begin{cases} H'_{11}(\omega) = \dfrac{k[H_{11}(\omega) - H_{12}(\omega)]^2}{\Delta(\omega)}, & H'_{22}(\omega) = \dfrac{k[H_{21}(\omega) - H_{22}(\omega)]^2}{\Delta(\omega)} \\ H'_{12}(\omega) = H'_{21}(\omega) = \dfrac{k[H_{11}(\omega) - H_{12}(\omega)][H_{12}(\omega) - H_{22}(\omega)]}{\Delta(\omega)} \\ \Delta(\omega) = 1 + k[H_{11}(\omega) + H_{22}(\omega) - H_{12}(\omega) - H_{21}(\omega)] \end{cases} \tag{4.4.32}$$

Equation (4.4.32) enables one to predict the dynamic effect of a spring inserted between two DoFs of the primary system. Here, the zeros of denominator $\Delta(\omega)$ are the resonant frequencies of the primary system after the modification. One can design the stiffness coefficient of the spring by using a graphic method similar to that in Example 4.4.2 so as to adjust the resonant frequencies of the primary system.

4.4.3.3　A Massless Dashpot Inserted

The local damping modification of the primary system is quite similar to the local stiffness modification via a massless spring. One is to insert a massless dashpot between the primary system and the rigid ground, and the other is to insert a massless dashpot between two DoFs of the primary system. Because of similarity, only the first way is discussed as follows.

Let $m \to +\infty$ and $k = 0$ in Fig. 4.18, then secondary system connected to the primary system degenerates to a massless dashpot between the primary system and the rigid ground, and Eq. (4.4.19) becomes

$$H_{cc}(\omega) = \frac{1}{\mathrm{i}c\omega} \tag{4.4.33}$$

Substitution of Eq. (4.4.33) into Eq. (4.4.15) leads to

$$\boldsymbol{H}'_{aJ}(\omega) = \frac{1}{1 + \mathrm{i}c\omega H_{aa}(\omega)} \boldsymbol{H}_{aJ}(\omega) \tag{4.4.34}$$

In practice, the primary system under damping modification is usually an un-damped system or a very weakly-damped system. Consider, as an example, an un-damped system, which behaves at the resonant frequencies as follows

$$\lim_{\omega \to \omega_r} |\boldsymbol{H}_{aJ}(\omega)| \to +\infty, \quad r = 1, 2, \ldots \tag{4.4.35}$$

After the dashpot is inserted, the frequency response function of the primary system yields

$$\left| H'_{aj}(\omega_r) \right| = \left| \frac{H_{aj}(\omega_r)}{1 + \mathrm{i}c\omega_r H_{aa}(\omega_r)} \right| \approx \left| \frac{1}{c\omega_r} \right|, \quad j \in J, \quad r = 1, 2, \ldots \tag{4.4.36}$$

That is, the infinitely large resonant peaks become finite.

When s independent massless dashpots with damping coefficients c_j, $j = 1, 2, \ldots, s$ are inserted between the primary system and the rigid ground, it is beneficial to collect these dashpots into a single secondary system. From Eq. (4.4.33), the displacement impedance sub-matrix of the secondary system at the interfacial DoFs yields

$$\boldsymbol{Z}_{CC}(\omega) = \mathrm{i}\omega\boldsymbol{C}, \quad \boldsymbol{C} \equiv \mathop{\mathrm{diag}}_{1 \le j \le s} [c_j] \tag{4.4.37}$$

Substitution of Eq. (4.4.37) into Eq. (4.4.12b) leads to the frequency response sub-matrix of the primary system between sets J and I of DoFs, namely,

$$\boldsymbol{H}'_{IJ}(\omega) = \boldsymbol{H}_{IJ}(\omega) - \mathrm{i}\omega\boldsymbol{H}_{IA}(\omega)\boldsymbol{C}\left[\boldsymbol{I} + \mathrm{i}\omega\boldsymbol{H}_{AA}(\omega)\boldsymbol{C}\right]^{-1}\boldsymbol{H}_{AJ}(\omega) \tag{4.4.38}$$

Equation (4.4.38) is the same as the result in previous study[23], but the deduction procedure is much simpler and exhibits a clear picture in physics.

Remark 4.4.5 The local dynamic modification based on frequency response matrix features the direct parametric design for the secondary system, but has some shortcomings. For instance, both Eqs. (4.4.11) and (4.4.12) require the computation of

---

[23] Zhang W (1980). The forced response of the damped dynamic systems. Applied Mathematics and Mechanics, 1(3): 431–440.

inverse matrix, which is suitable for a few of interfacial DoFs. If the frequency response matrix $\boldsymbol{H}_{CC}(\omega)$ of the secondary system is the summation of s matrices of rank one[24], namely,

$$
\begin{cases}
\boldsymbol{H}_{CC}(\omega) = \sum_{j=1}^{s} \boldsymbol{a}_j(\omega)\boldsymbol{b}_j^{\mathrm{T}}(\omega), \quad \boldsymbol{a}_j : \omega \mapsto \mathbb{C}^C, \quad \boldsymbol{b}_j : \omega \mapsto \mathbb{C}^C \\
j = 1, \ldots, s \ll P + D
\end{cases}
\tag{4.4.39}
$$

the computation of inverse matrix can be simplified. In addition, any measured frequency response function has noise and needs to be converted to theoretical one by fitting measured data to eliminate the ill-conditions in computation.

4.4.4   Concluding Remarks

This section presented how to adjust the dynamic properties of a primary system, based on the description of frequency response matrix, by using some local dynamic modifications. A part of contents associated with anti-resonances in the section is a further study for Problem 3B proposed in Chap. 1. The major conclusions of the section are as follows.

(1)   It is easy to establish the frequency response matrix of a composite system composed of a primary system and a second system so as to analyze the dynamic influence of the secondary system on the primary system. In addition, the results cover some previous studies, but look much simpler than previous ones.

(2)   The above description of frequency response matrix of a composite system clearly shows the dynamic influence of an un-damped dynamic vibration absorber attached to the primary system and gives an extension of the study in Sect. 4.3. That is, the un-damped dynamic vibration absorber fully removes the vibration of the primary system between the connection DoF and arbitrary DoFs. In addition, the description provides a simple way for designing the parameters of a lumped mass, a spring and a dashpot attached to the primary system in order to adjust the resonant peaks.

(3)   The above approaches feature that the design of dynamic modification is feasible based on the original frequency response of the primary system when the full frequency response matrix is not available.

---

[24] Zheng G T (2016). Sequel of Structural Dynamics: Applications to Flight Vehicles. Beijing: Science Press, 300–303 (in Chinese).

Fig. 4.21 The schematic of a dynamic anti-resonant vibration isolation system

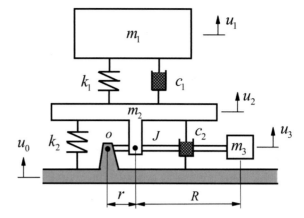

4.5 Further Reading and Thinking

(1) Read the reference[25] and make a comparison between the modal analysis in Sect. 4.1 and the complex modal analysis in the reference. Find out their unity from the beauty of science, point out their difference and explain the possible mechanisms.

(2) Read the references[26,27] and prove that the r mode shape of the tether pendulum in Example 3.1.6 has $r - 1$ nodes.

(3) Figure 4.21 presents the schematic of a dynamic anti-resonant vibration isolation system invented by W. G. Flannelly in the USA. In the vibration isolation system of double decks shown in Fig. 4.21, a rigid connecting bar, a part of which is overshadowed by a dashpot, with an auxiliary mass m_3 is mounted through two pivots on the two decks, respectively. This device is able to produce an anti-resonance in order to reduce the vibration transmissibility from the base excitation u_0 to the equipment displacement u_1. Read the reference[28] and study the influence of auxiliary mass m_3, length r and length R of the connecting bar on the vibration transmissibility.

(4) Based on the ideas of dynamic modification, take the dynamic anti-resonant vibration isolation system with two decks in Fig. 4.21 as the primary system and the connecting bar and the auxiliary mass as the secondary system, and study the influence of the secondary system on the vibration transmissibility of the primary system in frequency domain.

[25] Jinsberg J H (2001). Mechanical and Structural Vibrations: Theory and Applications. Baffins Lane: John Wiley & Sons Inc., 565–646.

[26] Gantmacher F R, Krein M G (2002). Oscillation Matrices and Kernels and Small Vibrations of Mechanical Systems. 2nd Edition. Rhode Island: AMS Chelsea Publishing, 113–179.

[27] Wang D J, Wang Q S, He B C (2019). Qualitative Theory in Structural Mechanics. Singapore: Springer Nature, 87–118.

[28] Flannelly W G (1969). The dynamic anti-resonant vibration isolator. US Patent 3445080, 1–10.

(5)   Read the reference[29] and discuss the applications of anti-resonance to the dynamic modeling and analysis of engineering structures.

---

[29] Hanson D, Watersb T P, Thompson D J (2006). The role of anti-resonance frequencies from operational modal analysis in finite element model updating. Mechanical System and Signal Processing, 21(1): 74–97.

Chapter 5
Natural Vibrations of One-Dimensional Structures

The advances in computational mechanics and the popularization of associated software have greatly changed the transference and development of vibration mechanics over the past four decades. Both academia and industry have been widely using the computational software based on the finite element methods to establish the dynamic model of a mechanical/structural system, compute and analyze vibration problems. Following this trend, most elementary courses of vibration mechanics for undergraduate students today only deal with the natural vibrations of a uniform rod and a uniform slender beam. As a consequence, the students can not well understand the vibrations of a continuous system, especially the mechanisms behind them. In practice, the young generation of engineers is facing difficulties when they solve the problems proposed in Chap. 1 or explain the reasons why some engineering systems behave so.

This chapter, therefore, presents the detailed studies on the natural vibrations of some one-dimensional structures, including a tether pendulum under gravity, the dual problems of rods and beams in natural vibrations. The objective of the chapter is to assist readers to understand the mechanisms behind the above problems and enter into a new academic realm, where they can not reach through both finite element methods and their software only.

5.1 Natural Vibrations of a Tether Pendulum

Section 1.2.1 presented the requirements for the preliminary study of a tethered satellite and proposed Problem 1A and Problem 1B associated with the dynamic modeling of a tether pendulum, that is, a thin tether fixed at one end and attached with a lumped mass at the other end. In addition, Sect. 1.2.1 also outlined the basic ideas to study those two problems.

Many scientists have studied the *tether pendulum* and named it a *massive cord* for the case when the tip mass vanishes so as to emphasize the effect of distributed inertia

© Science Press 2022
H. Hu, *Vibration Mechanics*,
https://doi.org/10.1007/978-981-16-5457-2_5

Fig. 5.1 The small planer
vibration of a tether
pendulum under gravity

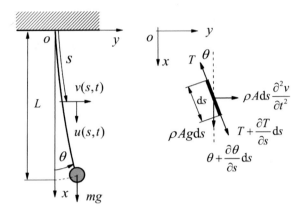

of the tether/cord under gravity[1,2]. These studies, however, mainly focused on some special cases, such as the fundamental natural frequency of a tether pendulum, and the natural vibrations of a massive cord. No completed studies have been available for the natural vibrations of a tether pendulum under gravity.

Section 3.1.3 completed the first step of the study outlined in Sect. 1.2.1, including the dynamic modeling via a lumped parameter method and the basic analysis of the relation among natural vibrations and system parameters for the tether pendulum. This section presents the second and third steps, including the dynamic model of a continuous system and the exact solutions for the natural vibrations of the tether pendulum, as well as a comparison between the continuous model and two discrete models. The section will complete the studies on Problem 1A and Problem 1B proposed in Chap. 1.

5.1.1 Dynamic Equation of a Tether Pendulum

This subsection deals with the planar vibration of a tether pendulum under gravity as shown in Fig. 5.1, where L is the tether length keeping constant, A is the cross-sectional area, ρ is the material density. The lumped mass m attached at the bottom tip of the tether may represent a satellite in the study of tethered satellite system. In addition, $\eta \equiv m/(\rho AL) > 0$ represents the ratio of tip mass to the tether mass, or called the *tip mass ratio* for short.

In the fixed frame of coordinates oxy in Fig. 5.1, let s be the arc-length coordinate of tether measured from the origin, $u(s,t)$ and $v(s,t)$ be the dynamic displacements

[1] Armstrong H L (1976). Effect of the mass of the cord on the period of a simple pendulum. American Journal of Physics, 44(6): 564–566.

[2] Epstein S T, Olsson M G (1977). Comment on "effect of the mass of the cord on the period of a simple pendulum". American Journal of Physics, 45(7): 671–672.

of a tether section with arc-length s in directions x and y, and θ be the rotation angle in the counter-clockwise direction.

We begin with the static problem of the tether pendulum and determine the static tension $T(s)$ at the section with arc-length s. As the tether is assumed not stretchable, the displacement along direction x is much less than that along direction y for small vibration. By neglecting the dynamic effect in direction x, we establish the static equation of the tether as follows

$$\left[T(s) + \frac{\mathrm{d}T(s)}{\mathrm{d}s}\mathrm{d}s\right]\cos\left(\theta + \frac{\partial\theta}{\partial s}\mathrm{d}s\right) - T(s)\cos(\theta) + \rho Ag\mathrm{d}s = 0 \qquad (5.1.1)$$

The static equilibrium equation gives $T(L) = mg$ at the bottom tip of the tether.

When the tether pendulum undergoes a small vibration, the rotation angle θ is so small that $\cos(\theta) \approx 1$ and $\cos(\theta + (\partial\theta/\partial s)\mathrm{d}s) \approx 1$ hold true. Substituting them into Eq. (5.1.1) and using the static relation of $T(L) = mg$, we derive the following boundary value problem of an ordinary differential equation

$$\begin{cases} \dfrac{\mathrm{d}T(s)}{\mathrm{d}s} + \rho Ag = 0, \quad s \in [0, \ L] \\ T(L) = mg \end{cases} \qquad (5.1.2)$$

Integration of Eq. (5.1.2) leads to the linear relation of static tension and arc-length

$$T(s) = mg + \rho Ag(L - s) = \rho Ag(L + l - s) = \rho Ag(L' - s) \qquad (5.1.3)$$

where $L' \equiv L + l$, and $l \equiv m/(\rho A) > 0$ is the ratio of tip mass to linear density of tether and has the dimension of length.

By analyzing the forces applied to an infinitesimal tether shown in Fig. 5.1 and using the static tension to approximate the dynamic tension, we establish the dynamic equation of the tether, namely,

$$\rho A\mathrm{d}s\frac{\partial^2 v(s, t)}{\partial t^2}$$
$$= \left[T(s) + \frac{\mathrm{d}T(s)}{\mathrm{d}s}\mathrm{d}s\right]\left\{\sin[\theta(s, t)] + \frac{\partial\sin[\theta(s, t)]}{\partial s}\mathrm{d}s\right\} - T(s)\sin[\theta(s, t)]$$
$$\approx \left\{\frac{\mathrm{d}T(s)}{\mathrm{d}s}\sin[\theta(s, t)] + T(s)\frac{\partial\sin[\theta(s, t)]}{\partial s}\right\}\mathrm{d}s = \frac{\partial}{\partial s}\{T(s)\sin[\theta(s, t)]\}\mathrm{d}s$$
$$(5.1.4)$$

Substituting Eq. (5.1.3) into Eq. (5.1.4) and using the geometric relations

$$\begin{cases} \sin[\theta(s, t)] \approx \dfrac{\partial v(s, t)}{\partial s} \\ \dfrac{\partial\sin[\theta(s, t)]}{\partial s} = \cos[\theta(s, t)]\dfrac{\partial\theta(s, t)}{\partial s} \approx \dfrac{\partial^2 v(s, t)}{\partial s^2} \end{cases} \qquad (5.1.5)$$

we simplify Eq. (5.1.4) as a partial differential equation with a variable coefficient as follows

$$\frac{\partial^2 v(s, t)}{\partial t^2} + g\frac{\partial v(s, t)}{\partial s} - g(L' - s)\frac{\partial^2 v(s, t)}{\partial s^2} = 0, \quad s \in [0, L] \qquad (5.1.6a)$$

The boundary conditions at two ends of the tether satisfy

$$v(0, t) = 0, \quad gv_s(L, t) + v_{tt}(L, t) = 0 \qquad (5.1.6b)$$

where the bottom tip boundary comes from the dynamic equation of the tip mass as follows

$$m\frac{\partial^2 v(L, t)}{\partial t^2} = -T(L)\theta(L, t) = -mg\frac{\partial v(s, t)}{\partial s}\bigg|_{s=L} \qquad (5.1.7)$$

Remark 5.1.1 If the tip mass vanishes, i.e., $m = 0$, then $T(L) = 0$. This makes Eq. (5.1.7) be an identity, which does not offer any boundary information. In this case, the tension at the bottom tip of the tether disappears such that the bending stiffness of the tether at the bottom tip approaches zero, and the horizontal deformation goes to infinity. This degenerated case will be discussed in Sect. 5.1.2.

5.1.2   Analysis of Natural Vibrations

5.1.2.1   Exact Solution of Natural Vibrations

Let the solution of Eq. (5.1.6a) be in a form of separated variables as follows

$$v(s, t) = \varphi(s)q(t) \qquad (5.1.8)$$

Substitution of Eq. (5.1.8) into Eq. (5.1.6a) leads to

$$\frac{1}{q(t)}\frac{d^2 q(t)}{dt^2} = \frac{1}{\varphi(s)}\left[g(L' - s)\frac{d^2\varphi(s)}{ds^2} - g\frac{d\varphi(s)}{ds}\right] \qquad (5.1.9)$$

As the left-hand side of Eq. (5.1.9) depends on time t and the right-hand side depends on arc-length s respectively, they must be a constant γ such that Eq. (5.1.9) becomes

$$\begin{cases} g(L' - s)\dfrac{d^2\varphi(s)}{ds^2} - g\dfrac{d\varphi(s)}{ds} - \gamma\varphi(s) = 0 \\ \dfrac{d^2 q(t)}{dt^2} - \gamma q(t) = 0 \end{cases} \qquad (5.1.10)$$

If $\gamma \geq 0$, the solutions $q(t)$ of the second equation in Eq. (5.1.10) satisfy

$$\begin{cases} q(t) = a_1 \exp(\sqrt{\gamma}t) + a_2 \exp(-\sqrt{\gamma}t), & \gamma > 0 \\ q(t) = a_1 t + a_2, & \gamma = 0 \end{cases} \qquad (5.1.11)$$

They both are divergent with time and not the meaningful solutions required. Thus, the only realistic case is $\gamma = -\omega^2 < 0$ such that the solution of the second equation in Eq. (5.1.10) reads

$$q(t) = a \cos(\omega t + \theta) \qquad (5.1.12)$$

In this case, the first equation in Eq. (5.1.10) becomes

$$g(L' - s)\frac{d^2\varphi(s)}{ds^2} - g\frac{d\varphi(s)}{ds} + \omega^2\varphi(s) = 0. \qquad (5.1.13a)$$

Substituting Eqs. (5.1.8) and (5.1.12) into Eq. (5.1.7) leads to the boundary conditions of $\varphi(s)$ as follows

$$\varphi(0) = 0, \quad g\varphi_s(L) - \omega^2\varphi(L) = 0 \qquad (5.1.13b)$$

Now, we solve the boundary value problem of Eq. (5.1.13), which includes an ordinary differential equation with variable coefficient $g(L' - s)$. It is worthy to point out that software Maple is able to transform Eq. (5.1.13a) into an ordinary differential equation of Bessel type and provides an exact general solution. To help readers understand the solution procedure, the detailed deduction is described as following.

Define a new variable corresponding to the arc-length s as follows

$$z = 2\omega\sqrt{\frac{L' - s}{g}}, \quad s \in [0, \ L] \qquad (5.1.14a)$$

or equivalently,

$$s = L' - g(z/2\omega)^2, \quad z \in [2\omega\sqrt{l/g}, \ 2\omega\sqrt{L'/g}] \qquad (5.1.14b)$$

In view of Eq. (5.1.14a), we derive

$$\begin{cases} \dfrac{d\varphi(s)}{ds} = \dfrac{d\varphi(z)}{dz}\dfrac{dz}{ds} = -\dfrac{\omega}{\sqrt{g(L' - s)}}\dfrac{d\varphi(z)}{dz} \\ \dfrac{d^2\varphi(s)}{ds^2} = \dfrac{d}{dz}\left[-\dfrac{\omega}{\sqrt{g(L' - s)}}\dfrac{d\varphi(z)}{dz}\right]\dfrac{dz(s)}{ds} = \dfrac{\omega}{\sqrt{g(L' - s)}}\dfrac{d}{dz}\left[\dfrac{2\omega^2}{gz}\dfrac{d\varphi(z)}{dz}\right] \\ \quad = \dfrac{\omega^2}{g(L' - s)}\dfrac{d^2\varphi(z)}{dz^2} - \dfrac{\omega}{2\sqrt{g(L' - s)^3}}\dfrac{d\varphi(z)}{dz} \end{cases}$$

$$(5.1.15)$$

Substitution of Eqs. (5.1.14) and (5.1.15) into Eq. (5.1.13a) leads to

$$g(L' - s) \left[\frac{\omega^2}{g(L' - s)} \frac{d^2\varphi(z)}{dz^2} - \frac{\omega}{2\sqrt{g(L' - s)^{3/2}}} \frac{d\varphi(z)}{dz} \right]$$
$$+ \frac{g\omega}{\sqrt{g(L' - s)}} \frac{d\varphi(z)}{dz} + \omega^2 \varphi(z) = 0 \qquad (5.1.16)$$

Equation (5.1.16) can be recast, with the help of Eq. (5.1.14a), as

$$z^2 \frac{d^2\varphi(z)}{dz^2} + z \frac{d\varphi(z)}{dz} + z^2 \varphi(z) = 0 \qquad (5.1.17)$$

This is an ordinary differential equation of Bessel type of zero order, or called *Bessel equation* of order zero for short. The Bessel equation of order zero has the following general solution in terms of *Bessel functions*[3]

$$\varphi(z) = c_1 J_0(z) + c_2 Y_0(z), \quad z \in [2\omega\sqrt{l/g}, \ 2\omega\sqrt{L'/g}] \qquad (5.1.18)$$

where c_1 and c_2 are integration constants, $J_0(z)$ is the first kind of Bessel function of order zero, and $Y_0(z)$ is the second kind of Bessel function of order zero, respectively. Their derivatives satisfy

$$\frac{dJ_0(z)}{dz} = -J_1(z), \quad \frac{dY_0(z)}{dz} = -Y_1(z) \qquad (5.1.19)$$

where $J_1(z)$ is the first kind of Bessel function of order one, while $Y_1(z)$ is the second kind of Bessel function of order one. Figure 5.2 presents the Bessel functions and their derivatives with an increase of z, showing their oscillations.

Remark 5.1.2 To understand Eq. (5.1.18), it is helpful to recall the general solution of a second-order ordinary differential equation with invariant coefficients. In the case of two distinct eigenvalues, the base functions of the general solution are triangle functions and hyperbolic-triangle functions. As shown in Fig. 5.2, the Bessel functions are oscillatory and serve as the base functions for the general solution of Eq. (5.1.17) similarly. An important feature of the Bessel functions is that the first kind of Bessel functions and their derivatives are bounded, but the second kind of them approach infinity when $z \to 0$.

From Eq. (5.1.14) and (5.1.19), we have

$$\frac{d\varphi(s)}{ds} = \frac{d\varphi(z)}{dz} \frac{dz}{ds} = \frac{c_1\omega}{\sqrt{g(L' - s)}} J_1(z) + \frac{c_2\omega}{\sqrt{g(L' - s)}} Y_1(z) \qquad (5.1.20)$$

---

[3] Courant R, Hilbert D (1989). Methods of Mathematical Physics. Volume I. New York: John Wiley & Sons Inc., 466–501.

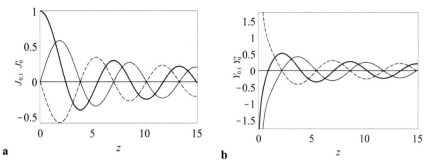

Fig. 5.2 The Bessel functions and their derivatives. **a** the first kind of Bessel functions and one of their derivatives (thick solid line: $J_0(z)$, thin solid line: $J_1(z)$, dashed line: $J_0'(z)$); **b** the second kind of Bessel functions and one of their derivatives (thick solid line: $Y_0(z)$, thin solid line: $Y_1(z)$, dashed line: $Y_0'(z)$)

Substitution of Eqs. (5.1.18) and (5.1.20) into Eq. (5.1.13b) leads to

$$\begin{cases} c_1 J_0\left(2\omega\sqrt{\dfrac{L'}{g}}\right) + c_2 Y_0\left(2\omega\sqrt{\dfrac{L'}{g}}\right) = 0 \\ \omega\sqrt{\dfrac{g}{l}}\left[c_1 J_1\left(2\omega\sqrt{\dfrac{l}{g}}\right) + c_2 Y_1\left(2\omega\sqrt{\dfrac{l}{g}}\right)\right] - \omega^2\left[c_1 J_0\left(2\omega\sqrt{\dfrac{l}{g}}\right) + c_2 Y_0\left(2\omega\sqrt{\dfrac{l}{g}}\right)\right] = 0 \end{cases}$$
$$(5.1.21)$$

This is a set of linear homogeneous equations for unknowns c_1 and c_2, and has non-trivial solutions if and only if the following condition holds true

$$J_0\left(2\omega\sqrt{\frac{L'}{g}}\right)\left[Y_1\left(2\omega\sqrt{\frac{l}{g}}\right) - \omega\sqrt{\frac{l}{g}}Y_0\left(2\omega\sqrt{\frac{l}{g}}\right)\right]$$
$$- Y_0\left(2\omega\sqrt{\frac{L'}{g}}\right)\left[J_1\left(2\omega\sqrt{\frac{l}{g}}\right) - \omega\sqrt{\frac{l}{g}}J_0\left(2\omega\sqrt{\frac{l}{g}}\right)\right] = 0 \qquad (5.1.22)$$

Equation (5.1.22) is the eigenvalue equation of the free vibration of the system. Given a set of system parameters, the solutions of Eq. (5.1.22) give natural frequencies ω_r, $r = 1, 2, \ldots$. Substituting a natural frequency ω_r and $c_1 = 1$ into the first equation in Eq. (5.1.21), we get c_2 such that the corresponding mode shape is

$$\varphi_r(s) = J_0\left(2\omega_r\sqrt{\frac{L'-s}{g}}\right) - \frac{J_0(2\omega_r\sqrt{L'/g})}{Y_0(2\omega_r\sqrt{L'/g})}Y_0\left(2\omega_r\sqrt{\frac{L'-s}{g}}\right), \quad r = 1, 2, \ldots$$
$$(5.1.23)$$

Remark 5.1.3 As stated in Remark 5.1.1, the tip boundary condition does not work if the tip mass in the tether pendulum vanishes. This case requires an extra discussion. Noting that $s = L$ corresponds to $z = 0$, it is easy to see, from Eq. (5.1.18), that if $c_2 \neq 0$ holds, the following singularity comes

$$\lim_{s \to L^-} \varphi(s) = \lim_{z \to 0} \varphi(z) = c_1 J_0(0) + c_2 Y_0(0) = \infty \qquad (5.1.24)$$

There follows $c_2 = 0$ according to the fact that the motion of the tether tip is always bounded. In solving equations of mathematical physics, the requirement imposed on a solution from physics like this is called a *natural boundary condition*.

5.1.2.2   Discussions About Natural Vibrations

(1)   The case for positive tip mass

Now, define a variable transform

$$\lambda \equiv \frac{2\omega}{\omega_0} = 2\omega \sqrt{\frac{L}{g}} \qquad (5.1.25)$$

where $\omega_0 \equiv \sqrt{g/L}$ is the natural frequency of a simple pendulum with the same tether length as the tether pendulum. Substituting Eq. (5.1.25) into Eq. (5.1.22) and utilizing the relations $l/L = m/(\rho AL) = \eta$ and $L'/L = \eta + 1$, we derive the simplified eigenvalue equation as follows

$$J_0(\sqrt{\eta + 1}\,\lambda)[Y_1(\sqrt{\eta}\,\lambda) - \frac{\sqrt{\eta}\,\lambda}{2}Y_0(\sqrt{\eta}\,\lambda)]$$
$$- Y_0(\sqrt{\eta + 1}\,\lambda)[J_1(\sqrt{\eta}\,\lambda) - \frac{\sqrt{\eta}\,\lambda}{2}J_0(\sqrt{\eta}\,\lambda)] = 0 \qquad (5.1.26)$$

The eigenvalue λ_r of Eq. (5.1.26) only depends on the tip mass ratio η, while the natural frequencies are proportional to that of the simple pendulum, namely,

$$\omega_r = \frac{\omega_0 \lambda_r}{2} = \frac{\lambda_r}{2}\sqrt{\frac{g}{L}}, \quad r = 1, 2, \ldots \qquad (5.1.27)$$

Substitution of Eq. (5.1.27) into Eq. (5.1.23) leads to the mode shapes depending on the tip mass ratio η as follows

$$\varphi_r(s) = J_0\left(\lambda_r\sqrt{\eta + 1 - \frac{s}{L}}\right) - \frac{J_0(\sqrt{\eta + 1}\,\lambda_r)}{Y_0(\sqrt{\eta + 1}\,\lambda_r)}Y_0\left(\lambda_r\sqrt{\eta + 1 - \frac{s}{L}}\right), \quad r = 1, 2, \ldots$$
$$(5.1.28)$$

As analyzed above, the tip mass ratio η plays an essential role in the natural vibrations of the tether pendulum. The mode shapes only rely on the tip mass ratio η, while the natural frequencies depend on the tip mass ratio η and are proportional to $\sqrt{g/L}$. The above analysis verifies the assertion drawn from a simplified model of lumped parameters in Sect. 5.3.1.3.

(2) **The case for no tip mass**

In this case, $c_2 = 0$ holds and Eq. (5.1.22) becomes

$$J_0(\lambda) = 0 \tag{5.1.29}$$

Solving Eq. (5.1.29) for eigenvalues λ_r, $r = 1, 2, \ldots$ gives natural frequencies in Eq. (5.1.27). As shown in Fig. 5.2a, the zeros λ_r, $r = 1, 2, \ldots$ of Bessel function $J_0(\lambda)$ are uniformly distributed, and so are the natural frequencies ω_r, $r = 1, 2, ..$ in frequency domain.

From $\lambda_1 \approx 2.404$, the first natural frequency of a massive cord yields

$$\omega_1 \approx 1.202\sqrt{g/L} = 1.202\omega_0 \tag{5.1.30}$$

Now, the mode shape in Eq. (5.1.23) takes a simple form as follows

$$\varphi_r(s) = J_0\left(\lambda_r\sqrt{1 - \frac{s}{L}}\right), \quad r = 1, 2, \ldots \tag{5.1.31}$$

That is, the mode shape is the same as the shape of Bessel function $J_0(z)$. Let $s_{r,j}$ be the position of the j-th node of the r-th mode shape, then Eq. (5.1.31) leads to

$$s_{r,j} = L\left[1 - \left(\frac{z_j}{\lambda_r}\right)^2\right], \quad 1 \leq j \leq r - 1, \quad r = 2, 3, \ldots, \tag{5.1.32}$$

where z_j is the j-th zero of $J_0(z)$. Because $(z_j/\lambda_r)^2 < 1$ holds, there must be $s_{r,j} \to L$ when the mode order r increases. As such, most nodes of a mode shape of higher order are accumulated around the bottom tip of the tether. This phenomenon is reasonable since both tension and bending stiffness of the tether vanishes at the bottom tip of the tether.

5.1.2.3 Numerical Results and Discussions

Example 5.1.1 Given parameters $g = 9.8$ m/s^2, $L = 9.8$ m and four tip mass ratios $\eta = 10.0, \quad 1.0, \quad 0.1, \quad 0.0$, compute and discuss the natural vibrations of the tether pendulum.

According to whether the tip mass ratio is positive or zero, we first solve Eq. (5.1.26) or Eq. (5.1.29) for the first four natural frequencies of the tether pendulum

Table 5.1 The first four natural frequencies (Hz) of the tether pendulum versus tip mass ratio η

Frequency order	$\eta = 10.0$	$\eta = 1.0$	$\eta = 0.1$	$\eta = 0.0$
1	0.1604	0.1681	0.1842	0.1914
2	1.6355	0.6496	0.4482	0.4393
3	3.2474	1.2315	0.7503	0.6886
4	4.8645	1.8271	1.0720	0.9383

shown in Table 5.1, and then compute the mode shapes of the tether pendulum shown in Fig. 5.3 from Eq. (5.1.28) or Eq. (5.1.31).

As shown in Table 5.1, the first natural frequency of the tether pendulum slightly increases and the other natural frequencies obviously decrease when the tip mass ratio goes down. Given the tip mass ratio η, the distribution of natural frequencies higher than the second order is relatively uniform. In particular, the results verify the phenomenon when the tip mass ratio $\eta \to 0$. When $\eta = 10.0$, the tip mass is so large that the bottom tip of the tether is almost immovable. That is, the tether pendulum looks like a fixed-fixed string and exhibits the uniform distribution of natural frequencies.

As shown in Fig. 5.3, the bottom tip of the tether is almost fixed when the tip mass ratio is $\eta = 10.0$ and the first mode shape looks like a straight line, which is the mode shape of a simple pendulum. The mode shapes, from the second order, of the tether pendulum, are close to the sinusoidal functions which describe the mode shapes of a fixed-fixed string. When the tip mass ratio is $\eta = 0.1$ or $\eta = 0.0$, the

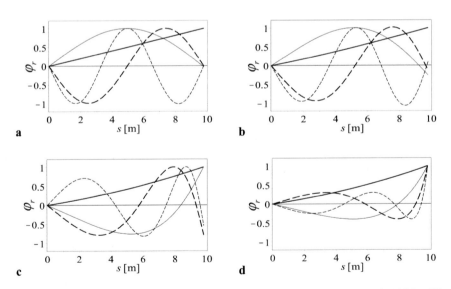

Fig. 5.3 The first four mode shapes of the tether pendulum for different tip mass ratio (thick solid line: 1st order, thin solid line: 2nd order, thick dashed line: 3rd order, thin dashed line: 4th order). **a** $\eta = 10.0$; **b** $\eta = 1.0$; **c** $\eta = 0.1$; **d** $\eta = 0.0$

mode shapes look quite different from the above case. As theoretically analyzed, the bottom tip of the tether undergoes the large vibration and the nodes of a mode shape of higher order accumulate to the bottom tip of the tether.

In what follows, the natural vibrations of the tether pendulum are discussed for two extreme cases.

When $\eta = 0.0$, Eq. (5.1.30) gives $\omega_1 \approx 1.202\omega_0$. The natural frequency of the simple pendulum is $f_0 \equiv \omega_0/(2\pi) \approx 0.1592$ Hz. The first natural frequency of the tether without tip mass in Table 5.1 is $f_1 \approx 0.1914$ Hz and the ratio is just $f_1/f_0 \approx 1.202$.

When $\eta = 10.0$, the first natural frequency of the tether pendulum in Table 5.1 is $f_1 \approx 0.1604$ Hz $\approx 1.0075 f_0$, which is very close to that of the simple pendulum. As shown in Table 2.1, the natural frequencies of a fixed-fixed string satisfy

$$\tilde{\omega}_r = \frac{r\pi}{L}\sqrt{\frac{\overline{T}}{\rho A}}, \quad r = 1, 2, \ldots \tag{a}$$

where $\overline{T}$ is the constant tension in the string. If the averaged tension of the top end and bottom end is denoted as $\overline{T}$, there follows

$$\tilde{\omega}_r = \frac{r\pi}{L}\sqrt{\frac{T(0) + T(L)}{2\rho A}} = \frac{r\pi}{L}\sqrt{\frac{(1 + 2\eta)\rho ALg}{2\rho A}} = r\pi\omega_0\sqrt{\frac{(1 + 2\eta)}{2}}, \quad r = 1, 2, \ldots \tag{b}$$

Substitution of $\eta = 10.0$ and $\omega_0 = 1$ rad/s into Eq. (b) leads to

$$\tilde{f}_r = 1.6202 \text{ Hz}, \quad 3.2404 \text{ Hz}, \quad 4.8606 \text{ Hz}, \quad \cdots \tag{c}$$

Compared with the case $\eta = 10.0$ in Table 5.1, the natural frequencies, higher than the second order, of a fixed-fixed string well approximate the natural frequencies of the tether pendulum and serve as the lower bounds for the natural frequencies of the tether pendulum.

5.1.3 Comparison Between Continuous Model and Discrete Models

Based on the exact solutions of natural frequency in Sect. 5.1.2, this subsection demonstrates the efficacy of the discrete models established in Sect. 3.1.3. In view of the analysis in Sect. 5.1.2, it is sufficient to establish a discrete model of a few DoFs when the tip mass ratio η is relatively large and the mode order is relatively low. Otherwise, it is necessary to establish a discrete model of more DoFs. Now we discuss the computational problems of the first three natural modes of the tether pendulum for two tip mass ratios.

5.1.3.1   Case Study of a Heavy Tip Mass

Example 5.1.2 Let the tip mass ratio be $\eta = 10.0$ and the DoFs of the discrete model be $N = 6$. We use discrete models A and B established in Sect. 3.1.1, compute the first three natural modes of the tether pendulum and make comparisons as follows.

Table 5.2 presents a comparison of the continuous model, discrete models A and B for the first three natural frequencies of the tether pendulum. As shown in the table, both discrete modes provide very good results of the first natural frequency with the first four decimal places being identical. For the second and the third natural frequencies, both discrete models offer good results with relative errors around 0.79 to 4.55%. In comparison, discrete model B is better than discrete model A. That is, when the distributed mass of the tether is lumped to the bottom end of each segment in a discrete model, the approximate natural frequencies higher than the second order slightly increase and are more close to the exact results.

In this case, the first mode shape of the continuous model is close to a straight line, and the results of both discrete modes and the continuous model coincide very well. Figure 5.4 shows a comparison of the continuous model and the discrete model

Table 5.2 A comparison of the first three natural frequencies (Hz) from continuous model and two discrete models when $\eta = 10.0$	Mode order	Continuous model	Discrete model A (relative error %)	Discrete model B (relative error %)
	1	0.1604	0.1604 (0.00)	0.1604 (0.00)
	2	1.6355	1.6162 (1.18)	1.6225 (0.79)
	3	3.2474	3.0996 (4.55)	3.1119 (4.17)

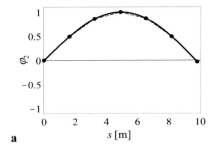
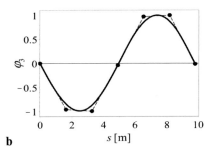

Fig. 5.4 A comparison of the mode shapes from continuous model and discrete model A when $\eta = 10.0$ (solid line: continuous model, dashed line with circles: discrete model A). **a** the 2nd mode shape; **b** the 3rd mode shape

A for the second and third mode shapes. As shown in the figure, even the discrete model A with less accuracy in natural frequencies computed provides good results.

5.1.3.2 Case Study of a Light Tip Mass

Example 5.1.3 Let the tip mass ratio be $\eta = 0.1$, and the DoFs of the discrete model be $N = 6$. We use discrete models A and B to compute the first three natural frequencies of the tether pendulum and make comparisons as below.

Table 5.3 presents a comparison among the continuous model and discrete models A and B for the first three natural frequencies. In this case, no discrete model shows superiority for the natural frequencies of different orders, but both discrete models can meet the computational requirements in engineering. From the computational results, discrete model B allocates more distributed mass to the bottom tip of the tether, decreases the first natural frequency and increases the second natural frequency. Such a mass allocation observes the rule of natural frequencies predicted by the continuous model.

Figure 5.5 shows a comparison between the continuous model and discrete model A for the second and third mode shapes. As shown in the figure, the accuracy of the third mode shape becomes poor since the nodes moves towards the bottom tip in this case, but the discrete segments with equal length are selected for the discrete model.

Table 5.3 A comparison of the first three natural frequencies (Hz) from continuous model and two discrete models when $\eta = 0.1$

Mode order	Continuous model	Discrete model A (relative error %)	Discrete model B (relative error %)
1	0.1842	0.1832 (0.54)	0.1796 (2.49)
2	0.4482	0.4299 (0.38)	0.4534 (1.16)
3	0.7503	0.6889 (8.18)	0.7539 (0.48)

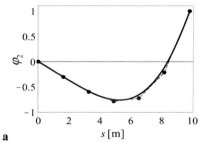
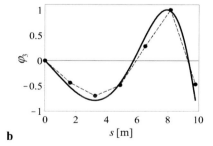

Fig. 5.5 A comparison of the mode shapes from continuous model and discrete model A when $\eta = 0.1$ (solid line: continuous model, dashed line with circles: discrete model A). **a** the 2nd mode shape; **b** the 3rd mode shape

Remark 5.1.4 To improve the accuracy of discrete models, there have been several available approaches, such as using shorter discrete segments of equal length and using denser and denser discretization for the bottom part of the tether. The later approach is the idea of the so-called adaptive meshing in finite element methods.

5.1.4 Concluding Remarks

This section presented the exact solutions of the planar natural vibration of a tether pendulum under gravity in order to meet the requirements of the preliminary study of a tethered satellite system. The section also made comparisons among the continuous model and two discrete models established in Sect. 3.1.3. These studies, together with the study in Sect. 3.1.3, have solved Problems 1A and 1B proposed in Chap. 1, and led to the major conclusions as follows.

(1)   The mode shapes of a tether pendulum only rely on the tip mass ratio η, while the corresponding natural frequencies not only depend on the tip mass ratio η, but are also proportional to $\omega_0 = \sqrt{g/L}$, which is the natural frequency of the simple pendulum with the same tether length. To study the natural vibration of a tethered satellite, one only needs to adjust the proportion factor $\sqrt{g/L}$ if the gravitational acceleration g or the tether length L changes.

(2)   When the tip mass ratio is large enough, say, $\eta = 10.0$, the natural vibrations of a tether pendulum are relatively simple. The natural vibration of the first order looks like that of a simple pendulum, the natural vibrations higher than the second order look like those of a fixed-fixed string. As the mass of a tethered satellite is usually much larger than the mass of a tether, the above results greatly simplify the study of natural vibrations of a tethered satellite system.

(3)   When the tip mass ratio approaches zero, both tension and bending stiffness near the bottom end of the tether approach zero. In this case, the bottom end of the tether undergoes large natural vibrations and the nodes of mode shapes of higher order accumulate to the bottom part of the tether. In the space mission of debris removal, for example, if the tethered tug captures a light space debris, the small tip mass ratio induces the large vibration at the tethered tug with captured debris. This case needs to be avoided or well treated.

(4)   In the computation of natural vibrations of a tether pendulum under gravity, the lumped parameter model is much simpler than the continuous model, but well predicts the relation between the system parameters and the natural vibrations. Both lumped parameter models established in Sect. 3.1.3 work well when the tip mass ratio is relatively large. When the tip mass ratio approaches zero, the two discrete models do not show any superiority, but can offer reasonable results for engineers.

5.2 Duality Analysis of Rods in Natural Vibrations

The straight rods made of linear homogeneous elastic materials, or called rods for short, are essential components in engineering. They also serve as the simplified models in the analysis and design of engineering systems. The studies on the dynamics of rods have been mature. The typical studies include the longitudinal dynamics of a rod based on the assumption of one-dimensional deformation[4], the similar dynamics of a non-uniform rod, a non-local rod and a rod with transverse inertia taken into account, as well as the three-dimensional elasto-dynamics of a rod of circular cross-section without the assumption of deformation[5]. Yet, some dynamic problems are still open.

An interesting observation more than one century ago is that the natural frequencies of a rod may be exactly the same as those of another different rod under certain conditions. For example, the air columns in a pair of horns may have identical natural frequencies when the product of their cross-sectional areas is a constant and the small ends of both horns are closed[6]. To be specific, let horn **a** and horn **b** have the same length L, their small ends be closed and large ends be open, and $A(x)$ and $\widetilde{B}(x)$ be the cross-sectional areas at x measured from the small end. When $A(x)\widetilde{B}(L-x)$ is a constant, then the air columns in these horns have identical natural frequencies.

The subsequent studies indicated that the above phenomenon is the dynamic property of a pair of non-uniform fixed-free rods and has drawn attention from academia. For example, Y. M. Ram and S. Elhay in Australia used the finite difference method to discretize the dynamic model of non-uniform rods and studied the above property[7]. They referred to such a pair of fixed-free rods as a *dual* when the rods have different cross-sectional areas, but identical natural frequencies. Some books of vibration mechanics have mentioned that a uniform free-free rod and a fixed-fixed rod have identical natural frequencies, but not yet analyzed the mechanism behind the phenomenon.

The studies on structural vibrations have dealt with several concepts of duals, such as the dual between an elastic structure and a viscoelastic structure[8], and the symplectic dual for vibration analysis[9], but not well addressed the dual of rods in natural vibrations.

---

[4] Hagedorn P, Das Gupta A (2007). Vibrations and Waves in Continuous Mechanical Systems. Chichester: John Wiley & Sons Inc., 1–112.

[5] Achenbach J D (1973). Wave Propagation in Elastic Solids. Amsterdam: North-Holland Publishing Company, 202–261.

[6] Benade A H (1976). Fundamentals of Musical Acoustics. Oxford: Oxford University Press, 140.

[7] Ram Y M, Elhay S (1995). Dualities in vibrating rods and beams: continuous and discrete models. Journal of Sound and Vibration, 184(5): 648–655.

[8] Chen Q, Zhu D M (1990). Vibrational analysis theory and applications to elastic–viscoelastic composite structure. Computers and Structures, 37(4): 585–595.

[9] Li X J, Xu F Y, Zhang Z (2018). Symplectic method for natural modes of beams resting on elastic foundations. ASCE Journal of Engineering Mechanics, 144(4): 04018009.

This section will deal with a pair of rods as a dual when they have identical natural frequencies and reveal the rule of their cross-sectional areas and their boundary conditions in order to solve Problem 4B proposed in Chap. 1. The section begins with the condition of cross-sectional areas and homogeneous boundary conditions for a dual of non-uniform rods, and then focuses on the condition when a pair of rods has the same cross-sectional area and the mechanism behind such a pair of uniform rods. Finally, the section presents how to extend the above studies to a pair of rods when their materials and cross-sectional areas vary axially. The major contents come from the recent paper of the author[10].

5.2.1 A Dual of Different Cross-Sections

This subsection is about the study of a dual problem of rod **a** and rod **b** in natural vibrations. The two rods are made of the same linear homogeneous elastic material, but have the cross-sectional areas and homogeneous boundaries to be determined.

5.2.1.1   Dual Conditions of Two Rods and Their Boundaries

To begin with, consider the dynamic equation of the longitudinal displacement $u(x, t)$, or named the displacement for short, of rod **a** as follows

$$\rho A(x) \frac{\partial^2 u(x, t)}{\partial t^2} - \frac{\partial}{\partial x}\left[EA(x) \frac{\partial u(x, t)}{\partial x} \right] = 0 \qquad (5.2.1a)$$

where L is the length of the rod, ρ is the material density, E is Young's modulus of material, $x \in [0, L]$ is the coordinate measured from the left end of the rod, $A(x) > 0$ is the cross-sectional area of the rod, which has the second-order smoothness for $x \in [0, L]$.

The homogeneous boundary conditions of displacement and internal force of rod **a** at two ends can be expressed as a unified form as follows

$$u(x_B, t) = 0, \quad u_x(x_B, t) \equiv \left. \frac{\partial u(x, t)}{\partial x} \right|_{x=x_B} = 0, \quad x_B \in \{0, \ L\} \qquad (5.2.1b)$$

In Eq. (5.2.1b), the two kinds of boundary conditions and the coordinate $x_B \in \{0, \ L\}$ form a *boundary condition set*, including four kinds of boundaries, namely, fixed-fixed, free-free, fixed-free and free-fixed boundaries. Hereinafter, once $x_B \in \{0, \ L\}$ appears for boundary conditions, it implies the boundary condition set, rather than a

---

[10] Hu H Y (2020) Duality relations of rods in natural vibrations. Journal of Dynamics and Control, 18(2): 1–8 (in Chinese).

single boundary condition. As such, Eqs. (5.2.1a) and (5.2.1b) describe the dynamic boundary value problem of rod **a** in terms of displacement $u(x, t)$.

In Eq. (5.2.1a), the term in brackets is the internal force of rod **a** and yields

$$N(x, t) \equiv EA(x)\frac{\partial u(x, t)}{\partial x} \tag{5.2.2}$$

Substituting Eq. (5.2.2) into Eq. (5.2.1a) leads to

$$\frac{\partial^2 u(x, t)}{\partial t^2} = \frac{1}{\rho A(x)}\frac{\partial N(x, t)}{\partial x} \tag{5.2.3}$$

Taking the time derivative of Eq. (5.2.2) twice, exchanging the order of partial derivatives and utilizing Eq. (5.2.3), we have

$$\frac{\partial^2 N(x, t)}{\partial t^2} = EA(x)\frac{\partial}{\partial x}\frac{\partial^2 u(x, t)}{\partial t^2} = EA(x)\frac{\partial}{\partial x}\left[\frac{1}{\rho A(x)}\frac{\partial N(x, t)}{\partial x}\right] \tag{5.2.4}$$

Thus, Eq. (5.2.4) serves as the dynamic equation of rod **a** in terms of internal force $N(x, t)$ as follows

$$\rho \widetilde{A}(x)\frac{\partial^2 N(x, t)}{\partial t^2} = \frac{\partial}{\partial x}\left[E\widetilde{A}(x)\frac{\partial N(x, t)}{\partial x}\right], \quad \widetilde{A}(x) \equiv \frac{1}{\rho EA(x)} \tag{5.2.5a}$$

where $\widetilde{A}(x)$ plays a role similar to cross-sectional area $A(x)$, but has the unit in $m^2 \cdot s^2 \cdot kg^{-2}$. From the displacement boundary condition set of rod **a** in Eq. (5.2.1b), it is easy to derive the boundary condition set in terms of internal force

$$\begin{cases} N_x(x_B, t) = \frac{\partial}{\partial x}\left[EA(x)\frac{\partial u(x, t)}{\partial x}\right]\bigg|_{x=x_B} = \rho A(x_B)\frac{\partial^2 u(x_B, t)}{\partial t^2} = 0 \\ N(x_B, t) = EA(x_B)u_x(x_B, t) = 0, \quad x_B \in \{0, L\} \end{cases} \tag{5.2.5b}$$

Equations (5.2.5a) and (5.2.5b) describe the dynamic boundary value problem of rod **a** in terms of internal force $N(x, t)$.

Next, we consider rod **b**, which has the same parameters as rod **a**, but different cross-sectional area $\widetilde{B}(x)$ with the second-order smoothness for $x \in [0, L]$. Furthermore, we exchange the boundary conditions of rod **a** with rod **b**. Then, the dynamic boundary value problem of rod **b** in terms of displacement $\tilde{u}(x, t)$ reads

$$\begin{cases} \rho\widetilde{B}(x)\frac{\partial \tilde{u}^2(x, t)}{\partial t^2} = \frac{\partial}{\partial x}\left[E\widetilde{B}(x)\frac{\partial \tilde{u}(x, t)}{\partial x}\right] \\ \tilde{u}_x(x_B, t) = 0, \quad \tilde{u}(x_B, t) = 0, \quad x_B \in \{0, L\} \end{cases} \tag{5.2.6}$$

Define the internal force of rod **b** as

$$\widetilde{N}(x, t) = E\widetilde{B}(x)\frac{\partial \tilde{u}(x, t)}{\partial x} \tag{5.2.7}$$

Similar to the analysis for rod **a**, we get the dynamic boundary value problem of rod **b** in terms of internal force $\widetilde{N}(x, t)$, namely,

$$\begin{cases} \rho B(x)\dfrac{\partial^2 \widetilde{N}(x, t)}{\partial t^2} = \dfrac{\partial}{\partial x}\left[EB(x)\dfrac{\partial \widetilde{N}(x, t)}{\partial x}\right] \\ \widetilde{N}(x_B, t) = 0, \quad \widetilde{N}_x(x_B, t) = 0, \quad x_B \in \{0, \ L\}, \quad B(x) \equiv \dfrac{1}{\rho E\widetilde{B}(x)} \end{cases} \tag{5.2.8}$$

Now, we take the cross-sectional area and a similar function of rod **b** as

$$\widetilde{B}(x) = \gamma \widetilde{A}(x) = \frac{\gamma}{\rho EA(x)}, \quad B(x) = \frac{A(x)}{\gamma}, \quad \gamma \equiv 1 \ \text{kg}^2 \cdot \text{s}^{-2} \tag{5.2.9}$$

Substituting Eq. (5.2.9) into Eqs. (5.2.6) and (5.2.8), we find that Eqs. (5.2.6) and (5.2.5) have the same form, and so do Eqs. (5.2.8) and (5.2.1). Hence, rod **a** and rod **b** have identical natural frequencies. In general, $\widetilde{B}(x) \neq A(x)$ holds, such a pair of rods is referred to as a *dual of different cross-sections*.

Furthermore, no matter whether a comparison between Eqs. (5.2.1b) and (5.2.5b) or a comparison between the boundary condition sets in Eqs. (5.2.6) and (5.2.8) is made, the boundary conditions in terms of displacement and those in terms of internal force form a *dual of boundary conditions*. That is, the fixed boundary condition of displacement corresponds to that of the gradient of internal force being zero, and the free boundary condition of displacement corresponds to that of internal force being zero.

5.2.1.2 Dual of Two Rods with Different Cross-Sections and Classifications

From the above concept of a dual of different cross-sections and a dual of boundary conditions, it is easy to make the following assertions.

(1) The dual of different cross-sections includes four pairs. That is, fixed-fixed rod **a** and free-free rob **b**, free-free rod **a** and fixed-fixed rod **b**, fixed-free rod **a** and free-fixed rod **b**, and free-fixed rod **a** and fixed-free rod **b**.

(2) The two rods in a dual of different cross-sections have identical natural frequencies, and their mode shapes of displacement and internal force are in the same form. For instance, given the r-th displacement mode shape $u_r(x)$ of rod **a**, the corresponding internal force mode shape, i.e., $N_r(x) = EA(x)[\partial u_r(x)/\partial x]$, is available from Eq. (5.2.2). Then, the r-th displacement mode shape of rod **b** can be taken as $\tilde{u}_r(x) = N_r(x)$. On the contrary, given the r-th displacement mode shape $\tilde{u}_r(x)$ of rod **b**, it is easy to get the corresponding internal force mode shape, i.e., $\widetilde{N}_r(x) = \gamma E\widetilde{A}(x)[\partial \tilde{u}_r(x)/\partial x]$, from Eqs. (5.2.7) and (5.2.9). Hence, the r-th displacement mode shape of rod **a** can be taken as $u_r(x) = \widetilde{N}_r(x)$.

Example 5.2.1 Given a fixed-fixed rod **a** with cross-sectional area function $A(x) = A_0(1 + \alpha x)^2$, where $A_0 > 0$, $\alpha > -1/L$ are two constants, find the other rod **b** in a dual of different cross-sections for rod **a**.

From Eq. (5.2.1), the dynamic boundary value problem of rod **a** in terms of displacement yields

$$
\begin{cases}
\rho A_0(1 + \alpha x)^2 \dfrac{\partial^2 u(x, t)}{\partial t^2} - \dfrac{\partial}{\partial x}\left[EA_0(1 + \alpha x)^2 \dfrac{\partial u(x, t)}{\partial x} \right] = 0 \\
u(0, t) = 0, \quad u(L, t) = 0
\end{cases}
\tag{a}
$$

It is easy to recast Eq. (a), via a new function $v(x, t) \equiv (1 + \alpha x)u(x, t)$[11], as

$$
\begin{cases}
\rho \dfrac{\partial^2 v(x, t)}{\partial t^2} - E \dfrac{\partial^2 v(x, t)}{\partial x^2} = 0 \\
v(0, t) = 0, \quad v(L, t) = 0
\end{cases}
\tag{b}
$$

This is the boundary value problem of a uniform fixed-fixed rod and Table 2.1 presents the natural frequencies and the displacement mode shapes as follows

$$
\omega_r = \kappa_r c_0, \quad v_r(x) = a_r \sin(\kappa_r x), \quad \kappa_r = \frac{r\pi}{L}, \quad r = 1, 2, \ldots
\tag{c}
$$

where $c_0 \equiv \sqrt{E/\rho}$ is the longitudinal wave speed of a uniform rod, and it happens to be the longitudinal wave speed of this non-uniform rod[12]. The non-uniform fixed-fixed rod **a** has the same natural frequencies in Eq. (c), while the corresponding displacement mode shapes satisfy

$$
u_r(x) = \frac{v_r(x)}{1 + \alpha x} = \frac{a_r \sin(\kappa_r x)}{1 + \alpha x}, \quad r = 1, 2, \ldots
\tag{d}
$$

Consider a non-uniform free-free rod **b** with the following cross-sectional area

$$
\widetilde{B}(x) = \frac{\gamma}{\rho E A(x)} = \frac{\gamma}{\rho E A_0(1 + \alpha x)^2}
\tag{e}
$$

From the condition of a dual of different cross-sections, such a free-free rod **b** and the fixed-fixed rod **a** have identical natural frequencies, while the displacement mode shapes of rod **b** can be taken as

[11] Abrate S (1995). Vibration of non-uniform rods and beams. Journal of Sound and Vibration, 185(4): 703–716.

[12] Guo S Q, Yang S P (2012). Wave motions in non-uniform one-dimensional waveguides. Journal of Vibration and Control, 18(1): 92–100.

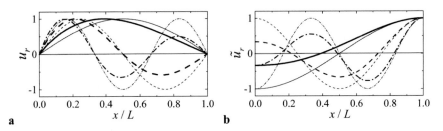

Fig. 5.6 The first three normalized mode shapes of a dual of different cross-sections (thick line: $\alpha = 2/L$, thin line: $\alpha = 0$; solid line: 1st order, dashed line: 2nd order, dot line: 3rd order). **a** fixed-fixed rod **a**; **b** free-free rod **b**

$$\tilde{u}_r(x) = EA(x)\frac{\partial u_r(x)}{\partial x} = b_r[\kappa_r(1 + \alpha x)\cos(\kappa_r x) - \alpha \sin(\kappa_r x)], \quad b_r \equiv a_r EA_0$$
$$(f)$$

To verify the above results, one can substitute Eq. (f) into the dynamic boundary value problem of non-uniform free-free rod **b** and confirm that Eq. (f) gives the elastic mode shapes of rod **b** when the rigid-body mode shape of the rod is neglected.

The thick curves in Fig. 5.6 show the first three normalized mode shapes with respect to the maximal displacement for $\alpha = 2/L$. Here, the cross-sectional area of rod **a** gets the maximum at the right end so that the large amplitude of all mode shapes appear at the left part of the rod, while rod **b** behaves oppositely since its cross-sectional area gets the maximum at the left end. As a reference, the thin curves show the first three normalized mode shapes for $\alpha = 0$ and exhibit symmetry or anti-symmetry because of the constant cross-sectional area.

5.2.1.3 Trivial Duals and Non-Trivial Duals

Besides a dual with different cross-sections, rod **a** and rod **b** have identical natural frequencies in the following two cases.

(1) Mirror: Rod **a** and rod **b**, together with their boundaries, are symmetric with respect to the mid-span section at $x = L/2$. In this case, the two rods have identical natural frequencies while their mode shapes keep unchanged under a mirror reflection.

(2) Similarity: Rod **a** and rod **b** have the same boundaries, and their cross-sectional areas are $A(x)$ and $\widetilde{B}(x) = \beta A(x)$, where $\beta > 0$. Substitution of $\widetilde{B}(x)$ into the partial differential equation in Eq. (5.2.6) leads to the same form of Eq. (5.2.1a). Hence, the two rods have identical natural frequencies and mode shapes. These two rods are said to be the same rods in duality studies.

Remark 5.2.1 The cases of mirror and similarity have the following features. At first, they are trivial and do not require any analysis for getting to know the above vibration behaviors. Second, the two rods in either a mirror case or a similarity case have identical natural frequencies and fall into a dual defined. Third, the two rods do not form a dual of different cross-sections. Fourth, the rod in a pair of mirror or similarity can transfer dual information to each other. That is, if one rod in the pair finds a rod as its dual, the other rod in the pair falls into the dual, too. As such, the above pair of rods is named a *trivial dual*, while the dual of different cross-sections is a *non-trivial dual*.

5.2.2 A Dual of Identical Cross-Sections

One may wonder whether two rods under a dual of boundaries have identical natural frequencies when they have identical cross-sectional areas. This subsection presents a case study first, and then extends the result of the case study to the general case.

5.2.2.1 A Dual of Two Rods with Exponential Cross-Sections

Example 5.2.2 A fixed-fixed rod **a** in Fig. 5.7 has the cross-sectional area $A(x) = A_0 \exp(\alpha x)$ with $A_0 > 0$ and $\alpha \geq 0$ being two constants, and the cross-sectional areas at two ends are $A(0) = A_0$ and $A(L) = A_0 \exp(\alpha L)$, respectively. Find a free-free rod **b** with the same cross-sectional area as a dual of rod **a**.

We first consider a free-free rod **b** in Fig. 5.7, where the cross-sectional area $\widetilde{B}(x)$ yields Eq. (5.2.9), namely,

$$\widetilde{B}(x) = \gamma \widetilde{A}(x) = \frac{\gamma}{\rho EA(x)} = \frac{\gamma}{\rho EA_0} \exp(-\alpha x) \tag{a}$$

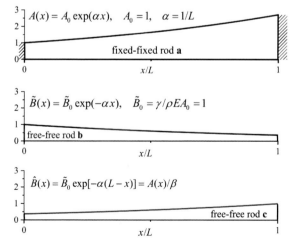

Fig. 5.7 The cross-sectional areas of fixed-fixed rod **a**, free-free rod **b** and its mirror rod **c** under the dual condition of identical cross-sections

Thus, free-free rod **b** and fixed-fixed rod **a** form a dual of different cross-sections.

Let the mirror of rod **b** be free-free rod **c**, in Fig. 5.7, which carries the dual information of rod **b** such that free-free rod **c** and fixed-fixed rod **a** form a dual of different cross-sections. The cross-sectional area of rod **c** in Fig. 5.7 yields

$$\widehat{B}(x) \equiv \widetilde{B}(L - x) = \frac{\gamma}{\rho E A(L)} \exp(\alpha x) = \frac{A(x)}{\beta}, \quad \beta \equiv \frac{\rho E A_0 A(L)}{\gamma} > 0 \quad \text{(b)}$$

That is, rod **c** and rod **a** have the similar cross-sectional areas and their only difference is a scaling factor denoted as $1/\beta$.

Finally, let the cross-sectional area of a free-free rod **d** be equal to the cross-sectional area of rod **c** multiplied by β, then rod **d** and rod **c** also have identical natural frequencies. Now, fixed-fixed rod **a** and free-free rod **d** can be named a *dual of identical cross-sections* since they have the same cross-sectional area and identical natural frequencies.

To be specific, the previous study of S. Q. Guo and S. P. Yang in China gave the natural frequencies and displacement mode shapes of the above fixed-fixed rod **a** with an exponential cross-sectional area as following[13]

$$\begin{cases} \omega_r = c_0 \sqrt{\lambda_r^2 + \left(\frac{a}{2}\right)^2}, \quad \lambda_r = \frac{r\pi}{L}, \quad c_0 \equiv \sqrt{\frac{E}{\rho}} \\ u_r(x) = a_r \exp\left(-\frac{\alpha x}{2}\right) \sin(\lambda_r x), \quad r = 1, 2, \ldots \end{cases} \quad \text{(c)}$$

As will be discussed in Sect. 7.2.1, c_0 has the dimension of velocity, but is not the longitudinal wave speed of non-uniform rod **a**. From the displacement mode shapes in Eq. (c), the internal force mode shapes of rod **a** satisfy

$$\begin{cases} N_r(x) = EA(x)\frac{\partial u_r(x)}{\partial x} = b_r \exp\left(\frac{\alpha x}{2}\right)\left[\lambda_r \cos(\lambda_r x) - \frac{\alpha}{2}\sin(\lambda_r x)\right] \\ b_r \equiv a_r EA_0, \quad r = 1, 2, \ldots \end{cases} \quad \text{(d)}$$

As fixed-fixed rod **a** and free-free rod **b** form a dual of different cross-sections, the natural frequencies of free-free rod **b** are ω_r, $r = 1, 2, \ldots$ in Eq. (c). Let internal force mode shapes $N_r(x)$ of rod **a** be the displacement mode shapes $\tilde{u}_r(x)$ of rod **b**, we get the displacement mode shapes of rod **c** via a mirror transform $x \to L - x$ or $\alpha \to -\alpha$, namely,

$$\hat{u}_r(x) = b_r \exp\left(-\frac{\alpha x}{2}\right)\left[\lambda_r \cos(\lambda_r x) + \frac{\alpha}{2}\sin(\lambda_r x)\right], \quad r = 1, 2, .. \quad \text{(e)}$$

They are just the displacement mode shapes of free-free rod **a** in the previous study (see footnote 13).

---

[13] Guo S Q, Yang S P (2010). Free axial vibration of a non-uniform rod. Journal of Shijiazhuang Tiedao University, 23(2): 59–63 (in Chinese).

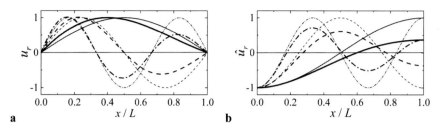

Fig. 5.8 The first three normalized displacement mode shapes of a dual of identical cross-sections (thick line: $\alpha = 2/L$, thin line: $\alpha = 0$; solid line: 1st order, dashed line: 2nd order, dot line: 3rd order). **a** fixed-fixed rod **a**; **b** free-free rod **a**

In Fig. 5.8, the thick curves represent the first three normalized displacement mode shapes with respect to the maximal displacement for $\alpha = 2/L$, where the large amplitude of all mode shapes appears at the left part of the rods since the cross-sectional area gets the maximum at the right end. As a reference, the thin curves in Fig. 5.8 show the first three normalized mode shapes for $\alpha = 0$.

5.2.2.2 Dual Condition for Identical Cross-Sections

It is easy to extend the ideas in Example 5.2.2 and get the dual conditions for two rods with identical cross-sectional areas. That is, there exists a constant $\beta > 0$ such that the cross-sectional area $A(x)$ yields

$$A(x) = \beta \widehat{B}(x) = \beta \widetilde{B}(L - x) = \frac{\beta \gamma}{\rho E A(L - x)} \tag{5.2.10}$$

Let Eq. (5.2.10) be recast as

$$\frac{A(x)}{\sqrt{\beta \gamma / \rho E}} \frac{A(L - x)}{\sqrt{\beta \gamma / \rho E}} = 1 \tag{5.2.11}$$

or an equivalent form as follows

$$\ln\left[\frac{A(x)}{\sqrt{\beta \gamma / \rho E}}\right] + \ln\left[\frac{A(L - x)}{\sqrt{\beta \gamma / \rho E}}\right] = 0 \tag{5.2.12}$$

Define the following function in terms of a new variable y

$$f(y) \equiv \ln\left[\frac{A(y + L/2)}{\sqrt{\beta \gamma / \rho E}}\right], \quad y \equiv x - \frac{L}{2} \in \left[-\frac{L}{2}, \frac{L}{2}\right] \tag{5.2.13}$$

Using $f(y)$ to express Eq. (5.2.12), we find that $f(y)$ satisfies the condition of an odd function, namely,

$$f(y) + f(-y) = 0 \tag{5.2.14}$$

From Eqs. (5.2.13) and (5.2.14), it is easy to derive

$$A(x) = \sqrt{\frac{\beta\gamma}{\rho E}} \exp\left[f\left(x - \frac{L}{2}\right)\right], \quad x \in [0, L] \tag{5.2.15}$$

Given an arbitrary smooth odd function $f(y)$, $y \in [-L/2, \quad L/2]$, the cross-sectional area $A(x)$ governed by Eq. (5.2.15) yields the dual condition of identical cross-sections in Eq. (5.2.10). It is easy to verify that the cross-sectional area in Example 5.2.2 has the simplest form satisfying Eq. (5.2.15).

5.2.2.3 A Dual of Two Rods with Identical Cross-Sections

Now, consider the dual problem of a fixed-fixed rod and a free-free rod, the cross-sectional areas of which satisfy Eq. (5.2.15). To demonstrate the function of Eq. (5.2.15), take the following smooth odd function

$$g(y) \equiv \alpha \sin\left(\frac{3\pi y}{L}\right), \quad y \in \left[-\frac{L}{2}, \frac{L}{2}\right], \quad \alpha > 0, \quad \beta \equiv \frac{\rho E A_0^2}{\gamma} \tag{5.2.16}$$

Substitution of Eq. (5.2.16) into Eq. (5.2.15) leads to the cross-sectional area of rod **a**, namely,

$$A(x) = A_0 \exp\left[\alpha \cos\left(\frac{3\pi x}{L}\right)\right], \quad A_0 > 0, \quad \alpha > 0 \tag{5.2.17}$$

It is easy to verify that Eq. (5.2.17) yields the condition in Eq. (5.2.10).

Figure 5.9 presents the cross-sectional areas of the three rods, that is, $A(x)$ of fixed-fixed rod **a**, $\widetilde{B}(x) = \gamma\widetilde{A}(x)$ of free-free rod **b**, and $\widehat{B}(x) = A(x)/\beta$ of rod **c** mirrored from rod **b**. Let the cross-sectional area of rod **c** multiplied by β and allocate the result $A(x)$ to a free-free rod **d**. Thus, rod **a** and rod **d** are a pair of dual and have identical cross-sectional areas. Hence, a fixed-fixed rod and a free-free rod form a dual if their cross-section areas are identical and yield Eq. (5.2.17).

5.2.2.4 Rods Beyond the Dual of Identical Cross-Sections

According to whether the cross-sectional area yields the dual condition of identical cross-sectional areas, the following two cases are discussed.

(1) For fixed-free rod **a** with cross-sectional area satisfying Eq. (5.2.15), the dual of different cross-sectional areas is free-fixed rod **b**, and the mirror of rod **b** is free-fixed rod **c**. It is easy to see that the difference between fixed-free rod **c** and

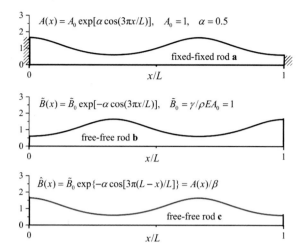

Fig. 5.9 The cross-sectional areas of fixed-fixed rod **a**, free-free rod **b** and its mirror rod **c** under the dual condition of identical cross-sections

fixed-free rod **a** is the cross-sectional areas at their ends. From the discussions about the trivial dual, rod **a** and rod **c** are similar and said to be the same rod. Thus, the fixed-free rod can only take itself as a dual of identical cross-section, and such a dual is trivial. Similarly, the free-fixed rod also has a trivial dual only.

(2)   The cross-sectional areas of many rods do not yield Eq. (5.2.15). For instance, rod **a** in Example 5.2.1 has the cross-sectional area of a quadratic function, which does not satisfy Eq. (5.2.15). Thus, rod **a** can not find any rod to form a dual of identical cross-sections. For example, the previous study of S. Abrate in the USA[14] has well supported the above assertion.

5.2.3   A Dual of Two Uniform Rods

When rod **a** has a uniform cross-sectional area denoted by $A(x) = A_0 > 0$, it is right to eliminate A_0 from Eq. (5.2.1a) and recast the dynamic boundary value problem of rod **a** in terms of displacement in Eq. (5.2.1) and that in terms of internal force in Eq. (5.2.5), respectively, as follows

$$\begin{cases} \dfrac{\partial^2 u(x,t)}{\partial t^2} - c_0^2 \dfrac{\partial^2 u(x,t)}{\partial x^2} = 0 \\ u(x_B, t) = 0, \quad u_x(x_B, t) = 0, \quad x_B \in \{0, L\} \end{cases} \tag{5.2.18}$$

---

[14] Abrate S (1995). Vibration of non-uniform rods and beams. Journal of Sound and Vibration, 185(4): 703–716.

$$
\begin{cases}
\dfrac{\partial^2 N(x,t)}{\partial t^2} - c_0^2 \dfrac{\partial^2 N(x,t)}{\partial x^2} = 0 \\[3mm]
N_x(x_B, t) = 0, \quad N(x_B, t) = 0, \quad x_B \in \{0,\ L\}
\end{cases}
\tag{5.2.19}
$$

where $c_0 \equiv \sqrt{E/\rho}$ is the longitudinal wave speed of rod **a**.

In view of Eq. (5.2.9), the dynamic boundary value problems of rod **b** in Eqs. (5.2.6) and (5.2.8) become

$$
\begin{cases}
\dfrac{\partial^2 \tilde{u}(x,t)}{\partial t^2} - c_0^2 \dfrac{\partial^2 \tilde{u}(x,t)}{\partial x^2} = 0 \\[3mm]
\tilde{u}_x(x_B, t) = 0, \quad \tilde{u}(x_B, t) = 0, \quad x_B \in \{0,\ L\}
\end{cases}
\tag{5.2.20}
$$

$$
\begin{cases}
\dfrac{\partial^2 \tilde{N}(x,t)}{\partial t^2} - c_0^2 \dfrac{\partial^2 \tilde{N}(x,t)}{\partial x^2} = 0 \\[3mm]
\tilde{N}(x_B, t) = 0, \quad \tilde{N}_x(x_B, t) = 0, \quad x_B \in \{0,\ L\}
\end{cases}
\tag{5.2.21}
$$

Obviously, Eqs. (5.2.18) and (5.2.21) have the same form, and so do Eqs. (5.2.19) and (5.2.20). As the above dynamic boundary value problems do not involve the cross-sectional area, it is reasonable to take rod **a** and rod **b** as the same rod and get the dual of identical cross-sections by changing boundary conditions. Hence, one arrives at the following assertions.

(1)   A fixed-fixed uniform rod and a free-free uniform rod form a dual, and have identical natural frequencies $\omega_r = \kappa_r c_0$, $\quad \kappa_r = r\pi/L$, $\quad r = 1, 2, \ldots$ if the rigid-body motion of the free-free rod is neglected. The displacement mode shapes of the fixed-fixed uniform rod are $u_r(x) = a_r \sin(\kappa_r x)$, $\quad r = 1, 2, \ldots$, and the corresponding internal force mode shapes satisfy $N_r(x) = b_r \cos(\kappa_r x)$, $\quad r = 1, 2, \ldots$. They serve as the displacement mode shapes of a free-free uniform rod, while the internal force mode shapes of the free-free rod are proportional to $u_r(x) = a_r \sin(\kappa_r x)$, $r = 1, 2, \ldots$. The thin curves in Fig. 5.8 depict the first three displacement mode shapes, respectively.

(2)   Both fixed-free uniform rod and free-fixed uniform rod satisfy the dual condition of identical cross-sections, but they are a pair of mirrors and a trivial dual.

A free-free rod and a fixed-fixed rod have different boundaries, and exhibit distinct mode shapes. One may doubt whether they have identical natural frequencies.

Example 5.2.3 Compute the natural frequencies of a dual of uniform rods via Rayleigh quotient.

It is easy to derive the referenced kinetic energy and potential energy for the r-th natural vibration of a fixed-fixed uniform rod as follows

$$
\begin{cases}
T_{\text{ref}} = \dfrac{\rho A}{2} \displaystyle\int_0^L u_r^2(x)\,dx = \dfrac{\rho A b_r^2}{2} \int_0^L \sin^2(\kappa_r x)\,dx = \dfrac{\rho A L b_r^2}{4} \\[4mm]
V = \dfrac{EA}{2} \displaystyle\int_0^L \left[\dfrac{du_r(x)}{dx}\right]^2 dx = \dfrac{E A b_r^2 \kappa_r^2}{2} \int_0^L \cos^2(\kappa_r x)\,dx = \dfrac{E A L b_r^2 \kappa_r^2}{4}
\end{cases}
\tag{a}
$$

Similarly, the referenced kinetic energy and potential energy for the r-th natural vibration of a free-free uniform rod are

$$\begin{cases} T_{ref} = \dfrac{\rho A}{2} \displaystyle\int_0^L \tilde{u}_r^2(x)dx = \dfrac{\rho A b_r^2}{2} \displaystyle\int_0^L \cos^2(\kappa_r x)dx = \dfrac{\rho A L b_r^2}{4} \\[4mm] V = \dfrac{EA}{2} \displaystyle\int_0^L \left[\dfrac{d\tilde{u}_r(x)}{dx}\right]^2 dx = \dfrac{EAb_r^2\kappa_r^2}{2} \displaystyle\int_0^L \sin^2(\kappa_r x)dx = \dfrac{EALb_r^2\kappa_r^2}{4} \end{cases} \quad \text{(b)}$$

Equations (a) and (b) are identical such that the Rayleigh quotients of the two rods, as well as the natural frequencies of the two rods, are identical, namely,

$$\omega_r = \sqrt{\frac{V}{T_{ref}}} = \kappa_r c_0, \quad r = 1, 2, \ldots. \tag{c}$$

Hence, the ratio of potential energy to referenced kinetic energy keeps the same for the two rods even though they have different mode shapes.

Remark 5.2.2 If the above problem is discussed from the viewpoint of symplectic mechanics, the displacement and internal force of a rod can be taken as a dual of variables. That is, the displacement boundary conditions of a fixed-fixed rod and the internal force boundary conditions of a free-free rod form a dual. A pair of rods subject to a dual of constraints exhibits identical natural frequencies.

5.2.4 A Dual of Two Rods with Axially Varying Material Properties

This subsection extends the studies in Sects. 5.2.1 and 5.2.2 to the non-uniform rods with axially varying material properties, and demonstrates the feasibility via an example.

5.2.4.1 Dynamic Boundary Value Problem of Rod

Consider a non-uniform rod with axially varying material density $\rho(x) > 0$, Young's modulus $E(x) > 0$ and cross-sectional area $A(x) > 0$. The free vibration of such a rod yields

$$\rho(x)A(x)\frac{\partial^2 u(x, t)}{\partial t^2} - \frac{\partial}{\partial x}\left[E(x)A(x)\frac{\partial u(x, t)}{\partial x}\right] = 0 \tag{5.2.22a}$$

The boundary condition set of the rod reads

$$u(x_B, t) = 0, \quad u_x(x_B, t) = 0, \quad x_B \in \{0, L\} \tag{5.2.22b}$$

From the discussion about the dynamic modeling of a non-uniform rod in Sect. 3.1.1, we define the following coordinate transform

$$y \equiv \int_0^x \sqrt{\frac{\rho(s)}{E(s)}} \, ds \tag{5.2.23}$$

As $dy/dx = \sqrt{\rho(x)/E(x)} > 0$ holds, Eq. (5.2.23) monotonically varies with respect to x such that there exists an inverse transform $x \equiv x(y)$. Substitution of Eq. (5.2.23) into Eq. (5.2.22a) leads to

$$\sqrt{\frac{\rho(x)}{E(x)}} \left\{ A(x)\sqrt{\rho(x)E(x)}\frac{\partial^2 u(y, t)}{\partial t^2} - \frac{\partial}{\partial y}\left[A(x)\sqrt{\rho(x)E(x)}\frac{\partial u(y, t)}{\partial y} \right] \right\} = 0 \tag{5.2.24}$$

Using the inverse transform $x \equiv x(y)$, we recast Eq. (5.2.24) as

$$\begin{cases} \overline{A}(y)\dfrac{\partial^2 u(y, t)}{\partial t^2} - \dfrac{\partial}{\partial y}\left[\overline{A}(y)\dfrac{\partial u(y, t)}{\partial y} \right] = 0 \\ \overline{A}(y) \equiv A[x(y)]\sqrt{\rho[x(y)]E[x(y)]} \end{cases} \tag{5.2.25a}$$

where $\overline{A}(y)$ is named the *cross-sectional property function* with the unit in kg · s^{-1}. Substituting Eq. (5.2.23) into Eq. (5.2.22b), it is easy to see the corresponding boundary condition set as follows

$$u(y_B, t) = 0, \quad u_y(y_B, t) = 0, \quad y_B \in \{0, \overline{L}\}, \quad \overline{L} \equiv \int_0^L \sqrt{\frac{\rho(x)}{E(x)}} \, dx \tag{5.2.25b}$$

As Eqs. (5.2.25) and (5.2.1) have the same form, it is easy to follow the studies in Sects. 5.2.1 and 5.2.2 to discuss the dual problem of two rods with axial varying material properties and cross-sectional areas. From agreement of homogeneous boundary condition sets in Eqs. (5.2.22b) and (5.2.25b), one should only study the dual condition of the cross-sectional property function.

5.2.4.2   Dual Condition

Let $\rho_a(x)$, $E_a(x)$ and $A(x)$ be the material density, Young's modulus and cross-sectional area of rod **a**, and $\rho_b(x)$, $E_b(x)$ and $\widetilde{B}(x)$ be the corresponding functions of rod **b**. Hence, their cross-sectional property functions satisfy

$$\overline{A}(y) \equiv A[x(y)]\sqrt{\rho_a[x(y)]E_a[x(y)]}, \quad \overline{B}(y) \equiv \widetilde{B}[x(y)]\sqrt{\rho_b[x(y)]E_b[x(y)]} \tag{5.2.26}$$

According to Eq. (5.2.9), we have the dual condition of different cross-sections for rods **a** and **b**, namely,

$$\overline{A}(y)\overline{B}(y) = \gamma \tag{5.2.27}$$

From Eq. (5.2.15), their dual condition of identical cross-sections is as follows

$$\overline{A}(y) = \sqrt{\beta\gamma}\,\exp\left[f\left(y - \frac{\overline{L}}{2}\right)\right], \quad y \in [0, \overline{L}] \tag{5.2.28}$$

The following example demonstrates the feasibility of a dual of identical cross-sections for the two rods.

Example 5.2.4 Let the cross-sectional area and the material density of a rod be constant values, and Young's modulus of the rod be axially varying as follows

$$A(x) \equiv A_0 > 0, \quad \rho(x) \equiv \rho_0 > 0, \quad E(x) \equiv E_0(1 + \eta x)^2, \quad E_0 > 0, \quad \eta > 0 \tag{a}$$

Study the dual condition of identical cross-sections for the rod.

Substitution of Eq. (a) into Eq. (5.2.23) leads to

$$y = \frac{1}{c_0}\int_0^x \frac{1}{1 + \eta s}\,ds = \frac{\ln(1 + \eta x)}{c_0\eta} \in [0, \overline{L}], \quad \overline{L} \equiv \frac{\ln(1 + \eta L)}{c_0\eta}, \quad c_0 \equiv \sqrt{\frac{E_0}{\rho_0}} \tag{b}$$

Now, c_0 has the dimension of velocity, but is not the longitudinal wave speed. From the first equation in Eq. (b), it is easy to derive

$$x = \frac{\exp(c_0\eta y) - 1}{\eta} \tag{c}$$

There follows the cross-sectional property function

$$\overline{A}(y) \equiv A_0\sqrt{E_0\rho_0[1 + \eta x(y)]^2} = A_0\sqrt{\rho_0 E_0}\,\exp(c_0\eta y) \tag{d}$$

It is easy to verify that Eq. (d) satisfies the dual condition of identical cross-sections in Eq. (5.2.28). As a matter of fact, the cross-sectional property function and the cross-sectional area in Example 5.2.2 have the same form. Hence, the fixed-fixed non-uniform rod and the free-free non-uniform rod have identical natural frequencies.

Remark 5.2.3 In general, it is not so easy, like this example, to get the analytic expression of $x(y)$ for a rod with material properties and cross-sectional area varied along x. Thus, numerical methods need to be developed to determine the dual conditions.

5.2.5 Concluding Remarks

This section studied how two rods, which look different in cross-sectional areas or boundaries, have identical frequencies. The section has solved Problem 4B and arrived at the following major conclusions.

(1) Given the cross-sectional area and homogeneous boundaries for a non-uniform rod made of homogeneous elastic materials, it is feasible to find the other non-uniform rod with homogeneous boundaries such that the two rods become a dual of different cross-sectional areas. That is, the two rods have different cross-sectional areas but identical natural frequencies, whereas the mode shapes of the two rods are the derivatives of each other. In this case, a fixed-fixed rod and a free-free rod form a dual, and so do a fixed-free rod and a free-fixed rod.

(2) The two rods in a dual of different cross-sectional areas may become a dual of identical sections if their cross-sectional areas satisfy a specific condition. That is, they have the same cross-sectional area and identical natural frequencies, but different boundaries. In this case, a fixed-fixed rod and a free-free rod are a dual, while a fixed-free rod and a free-fixed rod degenerate to a mirror. The uniform rods are a kind of special case of them.

(3) For a rod with axially varying material properties and cross-sectional area, it is possible to transform the dynamic boundary value problem of the rod into the dynamic boundary value problem of the uniform rod made of a homogeneous material so as to study the dual problem by using the above results.

(4) As the torsional vibration of a shaft and the longitudinal vibration of a rod have the same form of dynamic equation and boundaries, the studies and conclusions in this section are available for the duality of shafts in torsional natural vibrations.

5.3 Duality Analysis of Beams in Natural Vibrations

This section presents the dual analysis of the Euler–Bernoulli beams made of linear homogeneous elastic materials, or called E-B beams or beams for short. Beams serve as important components in engineering systems, and they are also useful models for both analysis and design of engineering systems. Recent studies of vibrating beams have emphasized on the dynamic design of a great variety of beam problems. For example, acoustic "black-hole", a new concept for designing non-uniform beams for vibration attenuation, has drawn much attention and seen applications to vibration reduction[15,16]. The other example is the topological optimizations of

[15] Krylov V V, Tilman F J B S (2004). Acoustic "black holes" for flexural waves as effective vibration dampers. Journal of Sound and Vibration, 274(3–5): 605–619.

[16] Gao N S, Wei Z Y, Zhang R H, et al. (2019). Low-frequency elastic wave attenuation in a composite acoustic black hole beam. Applied Acoustics, 154: 68–76.

the cross-sections and boundaries for beams[17]. In the above studies, it is a popular practice to solve an inverse problem of structural dynamics. That is, one ought to determine the cross-sections and boundaries of a beam under given dynamic properties of the beam. The existence, uniqueness and solvability of the above inverse problems greatly depend on a proper understanding of the vibration properties of beams.

Recalling the inclusion theorem of eigenvalues for the constraints of a beam, for example, one knows that a constraint imposed on the beam increases (does not decrease at least) all natural frequencies of the beam, and each natural frequency of the modified beam falls into an interval of two natural frequencies of the original beam. Thus, engineers usually believe that the beam with a constraint imposed gets higher natural frequencies, and do not pay much attention to the rigid-body motion if the original beam is not fully constrained.

The author pointed out, in an elementary textbook, that a free-free uniform beam and a clamped-clamped uniform beam have identical natural frequencies when the rigid-body motion of the free-free beam is not taken into consideration. This phenomenon also appears for a hinged-free uniform beam and a hinged-clamped uniform beam[18]. Many other books also have mentioned the phenomena[19], but not yet presented any studies. Although these phenomena do not violate the inclusion theorem of eigenvalues, one may wonder why the natural frequencies of a free-free beam keep unchanged when the clamped-clamped boundaries are imposed on it. One may further question whether this is a coincidence by chance or there exists an intrinsic relation between a pair of beams, and whether this is true for a pair of non-uniform beams.

In 1967, Karnopp in Canada studied the natural vibrations of non-uniform beams by using the dual variational principle. He referred to two beams under different boundaries as a *dual* when the two beams have identical natural frequencies[20]. Subsequently, Y. M. Ram and S. Elhay in Australia used the difference method to discretize the differential equation of a non-uniform beam and studied the dual problem of two beams in natural vibrations[21]. Wang and his coworkers in China introduced the concept of conjugate beams in mechanics of material to the structural vibration, studied the relation between beams with under-constraints and those with over-constraints, and discussed the rules of the node number of a mode shape[22]. Yet,

[17] Zargham S, Ward T A, Ramli R, et al. (2016). Topology optimization: A review for structural designs under vibration problems. Structural and Multidisciplinary Optimization, 53(6): 1157–1177.

[18] Hu H Y (1998). Mechanical Vibrations and Shocks. Beijing: Press of Aviation Industry, 159–160 (in Chinese).

[19] Rao S S (2007). Vibration of Continuous Systems. Hoboken: John Wiley & Sons Inc., 328–331.

[20] Karnopp B H (1967). Duality relations in the analysis of beam oscillations. Zeitschrift für Angewandte Mathematik und Physik ZAMP, 18(4): 575–580.

[21] Ram Y M, Elhay S (1995). Dualities in vibrating rods and beams: continuous and discrete models. Journal of Sound and Vibration, 184(5): 648–655.

[22] Wang D J, Wang Q C, He B C (2019). Qualitative Theory of Structural Mechanics. Singapore: Springer Nature, 129–132.

Fig. 5.10 The schematic of
a non-uniform beam

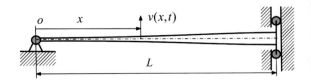

the systematic studies on the duality of non-uniform beams in natural frequencies
have not been available.

Based on the recent paper of the author[23], this section will present the detailed
study of dual analysis of two beams in natural vibrations and solve Problem 4A
proposed in Chap. 1. The study begins with the dual analysis of two beams with
different cross-sectional areas and second moments of cross-sectional areas. Then,
the study turns to the dual analysis of two beams with identical cross-sections. Finally,
the section discusses the dual of two uniform beams and reveals a new kind of dual.
The ideas in this subsection are similar to those in Sect. 5.2, but the complicated
boundaries of beams require a detailed analysis.

5.3.1   A Dual of Different Cross-Sections

Given a non-uniform beam with homogeneous boundaries, this subsection presents
how to find the other non-uniform beam and the corresponding boundaries such that
the two beams have identical natural frequencies. That is, they form a dual. The
subsection also gives the classifications of all duals of beams and points out the
relation among mirror, similarity and duals.

5.3.1.1   Displacement and Bending Descriptions for a Beam

We begin with the non-uniform beam **a** of length L as shown in Fig. 5.10, where the
coordinate $x \in [0, L]$ is measured from the left end of the beam. Let $v(x, t)$ be the
dynamic transverse displacement, or named the displacement for short, to describe
the free vibration of the beam, which yields

$$\rho A(x)\frac{\partial^2 v(x, t)}{\partial t^2} + \frac{\partial^2}{\partial x^2}\left[EI(x)\frac{\partial^2 v(x, t)}{\partial x^2}\right] = 0 \qquad (5.3.1)$$

where ρ is material density, E is Young's modulus of material, $A(x) > 0$ and
$I(x) > 0$ are respectively the cross-sectional area of the beam and the second moment
of cross-sectional area of the beam, and they both have the second-order smoothness
for $x \in [0, L]$.

---

[23] Hu H Y (2020). Duality relations of beams in natural vibrations. Chinese Journal of Theoretical
and Applied Mechanics, 52(1): 139–149 (in Chinese).

Noting that the term in the pair of brackets in Eq. (5.3.1) is the bending moment of beam **a** during vibration, denote it as

$$M(x, t) \equiv EI(x) \frac{\partial^2 v(x, t)}{\partial x^2} \qquad (5.3.2)$$

Applying Eq. (5.3.2) to Eq. (5.3.1) leads to

$$\frac{\partial^2 v(x, t)}{\partial t^2} = -\frac{1}{\rho A(x)} \frac{\partial^2 M(x, t)}{\partial x^2} \qquad (5.3.3)$$

Taking the time derivative of Eq. (5.3.2) twice, exchanging the order of partial derivatives and utilizing Eq. (5.3.3), we derive

$$\frac{\partial^2 M(x, t)}{\partial t^2} = EI(x) \frac{\partial^2}{\partial x^2} \left[\frac{\partial^2 v(x, t)}{\partial t^2} \right] = -EI(x) \frac{\partial^2}{\partial x^2} \left[\frac{1}{\rho A(x)} \frac{\partial^2 M(x, t)}{\partial x^2} \right] \qquad (5.3.4)$$

It is easy to recast Eq. (5.3.4) as the dynamic equation of beam **a** in terms of bending moment $M(x, t)$, namely,

$$\begin{cases} \rho \widetilde{A}(x) \dfrac{\partial^2 M(x, t)}{\partial t^2} + \dfrac{\partial^2}{\partial x^2} \left[E\widetilde{I}(x) \dfrac{\partial^2 M(x, t)}{\partial x^2} \right] = 0 \\[2mm] \widetilde{A}(x) \equiv \dfrac{1}{\rho EI(x)}, \quad \widetilde{I}(x) \equiv \dfrac{1}{\rho EA(x)} \end{cases} \qquad (5.3.5)$$

Equations (5.3.5) and (5.3.1) have the same form, but function $\widetilde{A}(x)$ similar to the cross-sectional area $A(x)$ has the unit in $s^2 \cdot kg^{-2}$, while function $\widetilde{I}(x)$ similar to the second moment of cross-sectional area $I(x)$ has the unit in $m^2 \cdot s^2 \cdot kg^{-2}$.

Next, we consider a non-uniform beam **b** made of the same material of beam **a**. Beam **b** also has length L, but cross-sectional area $\widetilde{B}(x) > 0$ and the second moment of cross-sectional area $\widetilde{J}(x) > 0$. They also have the second-order smoothness for $x \in [0, L]$. Let $\tilde{v}(x, t)$ be the displacement of beam **b**, which yields

$$\rho \widetilde{B}(x) \frac{\partial^2 \tilde{v}(x, t)}{\partial t^2} + \frac{\partial^2}{\partial x^2} \left[E\widetilde{J}(x) \frac{\partial^2 \tilde{v}(x, t)}{\partial x^2} \right] = 0 \qquad (5.3.6)$$

Define the term in the pair of brackets in Eq. (5.3.6) as the bending moment of beam **b** as follows

$$\widetilde{M}(x, t) \equiv E\widetilde{J}(x) \frac{\partial^2 \tilde{v}(x, t)}{\partial x^2} \qquad (5.3.7)$$

Similar to the analysis for beam **a**, the dynamic equation of beam **b** in terms of bending moment $\widetilde{M}(x, t)$ reads

$$\begin{cases} \rho B(x)\dfrac{\partial^2 \widetilde{M}(x,t)}{\partial t^2} + \dfrac{\partial^2}{\partial x^2}\left[EJ(x)\dfrac{\partial^2 \widetilde{M}(x,t)}{\partial x^2}\right] = 0 \\ B(x) \equiv \dfrac{1}{\rho E\widetilde{J}(x)}, \quad J(x) \equiv \dfrac{1}{\rho E\widetilde{B}(x)} \end{cases} \tag{5.3.8}$$

Let the cross-sectional area and the second moment of cross-sectional area of beam **b** satisfy

$$\begin{cases} \widetilde{B}(x) = \widetilde{\gamma}\widetilde{A}(x) = \dfrac{\widetilde{\gamma}}{\rho E\widetilde{I}(x)} \\ \widetilde{J}(x) = \widetilde{\gamma}\widetilde{I}(x) = \dfrac{\widetilde{\gamma}}{\rho E\widetilde{A}(x)}, \quad \widetilde{\gamma} \equiv 1\ m^2 \cdot \text{kg}^2 \cdot \text{s}^{-2} \end{cases} \tag{5.3.9}$$

Substituting Eq. (5.3.9) into Eqs. (5.3.6) and (5.3.8), we find that the partial differential equations in Eqs. (5.3.6) and (5.3.5) are identical, and so are those in Eqs. (5.3.8) and (5.3.1). As such, beam **a** and beam **b** are referred to as a *dual of beams* of different cross-sections when their cross-sectional properties satisfy Eq. (5.3.9). That is, the dynamic equation of displacement for beam **a** and the dynamic equation of bending moment for beam **b** have the same form, while the displacement of beam **b** and the bending moment of beam **a** yield the dynamic equations of the same form.

Whether the above beams **a** and **b** have identical natural frequencies depends on if their boundaries are a dual. Thus, we turn to the study of *homogeneous boundaries*, including a clamped boundary, a hinged boundary, a free boundary and a sliding boundary as shown in the right end of the beam in Fig. 5.10. The sliding boundary makes the rational angle of the cross-section be zero and the shear force be zero at the beam end, but is movable on the horizontal plane to release the axial deformation and axial force.

Let $x_B \in \{0,\ L\}$ be the coordinates at two beam ends, consider the four boundaries in terms of displacement and bending moment, and list the results in Table 5.4. When the boundary conditions of $v(x_B, t) = 0$ and $v_x(x_B, t) = 0$ described by the displacement and slope of a beam are converted into those described by the bending moment, one needs to use Eq. (5.3.3) and conduct the following deductions

Table 5.4 The homogeneous boundaries described by displacement and bending moment

Boundary type	Description of displacement	Description of bending moment
F: Free	$EI(x_B)v_{xx}(x_B, t) = 0$ $[EI(x_B)v_{xx}(x_B, t)]_x = 0$	$M(x_B, t) = 0$ $M_x(x_B, t) = 0$
S: Sliding	$v_x(x_B, t) = 0$ $[EI(x_B)v_{xx}(x_B, t)]_x = 0$	$[E\widetilde{I}(x_B)M_{xx}(x_B, t)]_x = 0$ $M_x(x_B, t) = 0$
H: Hinged	$v(x_B, t) = 0$ $EI(x_B)v_{xx}(x_B, t) = 0$	$M_{xx}(x_B, t) = 0$ $M(x_B, t) = 0$
C: Clamped	$v(x_B, t) = 0$ $v_x(x_B, t) = 0$	$M_{xx}(x_B, t) = 0$ $[E\widetilde{I}(x_B)M_x(x_B, t)]_x = 0$

$$v(x_B, t) = 0 \quad \Rightarrow \quad \frac{\partial^2 v(x_B, t)}{\partial t^2} = 0 \quad \Rightarrow \quad \left[\frac{1}{\rho A(x)} \frac{\partial^2 M(x, t)}{\partial x^2}\right]\Bigg|_{x=x_B} = 0$$

$$\Rightarrow \quad M_{xx}(x_B, t) = 0 \tag{5.3.10}$$

$$v_x(x_B, t) = 0 \quad \Rightarrow \quad \frac{\partial^2 v_x(x_B, t)}{\partial t^2} = 0 \quad \Rightarrow \quad \frac{\partial}{\partial x}\left[\frac{1}{\rho A(x)} \frac{\partial^2 M(x, t)}{\partial x^2}\right]\Bigg|_{x=x_B} = 0$$

$$\Rightarrow \quad \left[\widetilde{EI}(x_B) M_{xx}(x_B, t)\right]_x = 0 \tag{5.3.11}$$

where replacement of $1/[\rho A(x)]$ by $\widetilde{EI}(x)$ leads to the corresponding conditions when the shear force is zero.

In Table 5.4, a comparison between the second column and the third column shows that the free boundary of displacement is equivalent to the clamped boundary of bending moment, the sliding boundary of displacement is equivalent to the sliding boundary of bending moment, the hinged boundary of displacement is equivalent to the hinged boundary of bending moment, and the clamped boundary of displacement is equivalent to the free boundary of bending moment. Each pair of them is named a *dual of boundaries*.

5.3.1.2   Dual of Two Beams with Different Cross-Sections and Classification

When beam **a** and beam **b** have the same length and yield the dual condition of different cross-sections and a dual of boundaries, their dynamic boundary value problems in terms of displacement and bending moment have the same form. Hence, they have identical natural frequencies. In this case, beams **a** and **b** are called a *dual of different cross-sections*. If the four boundaries allocated to each end of a beam, there are 16 types of beams in total as shown in Table 5.5, where T-1 stands for Type 1 and so on.

According the above concept of a dual of different cross-sections, the following assertions can be confirmed.

(1)   In Table 5.5, the beams in a dual of different cross-sections can be classified as 7 types, and Type 2, Type 3 and Type 4 can be further divided into sub-types A and B. However, if the cross-sectional areas and the second moments of cross-sectional area of beams **a** and **b** are symmetric with respect to the mid-span

Table 5.5   The beams with homogeneous boundaries (F: Free, S: Sliding, H: Hinged, C: Clamped)

Boundary type	Right end: F		Right end: S		Right end: H		Right end: C	
Left end: F	T-1	F-F	T-2B	F-S	T-3B	F-H	*T-7*	F-C
Left end: S	T-2A	S-F	**T-5**	S-S	**T-4B**	S-H	T-2A	S-C
Left end: H	T-3A	H-F	**T-4A**	H-S	**T-6**	H-H	T-3A	H-C
Left end: C	*T-7*	C-F	T-2B	C-S	T-3B	C-H	T-1	C-C

section $x = L/2$ (i.e., mirror symmetric for short), the sub-types A and B are symmetric to each other and do not require to be identified.

(2) In the above 7 types of dual, a dual of two beams in Types 1, 2 and 3 have different boundaries, a dual of two beams in Types 4, 5 and 6 have the same boundaries, a dual of two beams in Type 7 have two boundaries mirrored to each other. Hence, Types 4, 5, 6 are denoted by boldface fonts, and Type 7 by italic boldface fonts.

(3) In the above 7 types of dual, each type includes two beams with identical natural frequencies, and the displacement mode shapes of one beam are proportional to the bending moment mode shapes of the other beam. For example, given the r-th displacement mode shape $v_r(x)$ of beam **a**, the r-th bending moment mode shape of beam **a** yields $M_r(x) = EI(x)[\partial^2 v_r(x)/\partial x^2]$ such that the r-th displacement mode shape of beam **b** is $\tilde{v}_r(x) = M_r(x)$. On the contrary, given the r-th displacement mode shape $\tilde{v}_r(x)$ of beam **b**, the r-th displacement mode shape of beam **a** yields $v_r(x) = \widetilde{EI}(x)[\partial^2 \tilde{v}_r(x)/\partial x^2]$.

Example 5.3.1 Given clamped-clamped beam **a**, which has cross-sectional area $A(x) = A_0(1 + \alpha x)^4$ with $A_0 > 0$ and $\alpha > -1/L$, and the second moment of cross-sectional area $I(x) = A_0 r_g^2 (1 + \alpha x)^4$, where $r_g > 0$ is the *radius of gyration* of the cross-section. Find beam **b** in a dual of beam **a**, such that they have different cross-sections but identical natural frequencies.

According to Eq. (5.3.1), the dynamic boundary value problem of beam **a** reads

$$\begin{cases} \rho(1+\alpha x)^4 \dfrac{\partial^2 v(x,t)}{\partial t^2} + Er_g^2 \dfrac{\partial^2}{\partial x^2}\left[(1+\alpha x)^4 \dfrac{\partial v^2(x,t)}{\partial x}\right] = 0 \\ v(0,t) = 0, \quad v_x(0,t) = 0, \quad v(L,t) = 0, \quad v_x(L,t) = 0 \end{cases} \quad \text{(a)}$$

Define a new function $w(x,t) \equiv (1 + \alpha x)^2 v(x,t)$ and transform Eq. (a) as[24]

$$\begin{cases} \dfrac{\partial^2 w(x,t)}{\partial t^2} + c_0^2 r_g^2 \dfrac{\partial^4 w(x,t)}{\partial x^4} = 0, \quad c_0 \equiv \sqrt{\dfrac{E}{\rho}} > 0 \\ w(0,t) = 0, \quad w_x(0,t) = 0, \quad w(L,t) = 0, \quad w_x(L,t) = 0 \end{cases} \quad \text{(b)}$$

where c_0 has the dimension of velocity, but is not the longitudinal wave speed of the non-uniform beam[25]. Equation (b) describes the dynamic boundary value problem of a clamped-clamped uniform beam, which has the natural frequencies and the corresponding displacement mode shapes, from Table 2.1, as follows

$$\omega_r = \kappa_r^2 r_g c_0, \quad \kappa_1 \approx \frac{4.730}{L}, \quad \kappa_2 \approx \frac{7.853}{L}, \quad \kappa_r \approx \frac{(2r+1)\pi}{2L}, \quad r \geq 3 \quad \text{(c)}$$

---

[24] Abrate S (1995). Vibration of non-uniform rods and beams. Journal of Sound and Vibration, 185(4): 703–716.

[25] Guo S Q, Yang S P (2012). Wave motions in non-uniform one-dimensional waveguides. Journal of Vibration and Control, 18(1): 92–100.

$$\begin{cases} w_r(x) = a_r[\cosh(\kappa_r x) - \cos(\kappa_r x) - d_r \sinh(\kappa_r x) + d_r \sin(\kappa_r x)] \\ d_r \equiv \dfrac{\cosh(\kappa_r L) - \cos(\kappa_r L)}{\sinh(\kappa_r L) - \sin(\kappa_r L)}, \quad r = 1, 2, \ldots \end{cases} \tag{d}$$

The clamped-clamped non-uniform beam **a** shares the above natural frequencies in Eq. (c), and has the corresponding displacement mode shapes as follows

$$\begin{aligned} v_r(x) &= \frac{w_r(x, t)}{(1 + \alpha x)^2} \\ &= \frac{a_r}{(1 + \alpha x)^2}[\cosh(\kappa_r x) - \cos(\kappa_r x) - d_r \sinh(\kappa_r x) + d_r \sin(\kappa_r x)], \\ r &= 1, 2, \ldots \end{aligned} \tag{e}$$

Following the dual relation of different cross-sections in Eq. (5.3.9), we take the cross-sectional area and the second moment of cross-sectional area of non-uniform beam **b** as follows

$$\begin{cases} \widetilde{B}(x) = \dfrac{\widetilde{\gamma}}{\rho EI(x)} = \dfrac{\widetilde{\gamma}}{\rho EA_0 r_g^2 (1 + \alpha x)^4} \\ \widetilde{J}(x) = \dfrac{\widetilde{\gamma}}{\rho EA(x)} = \dfrac{\widetilde{\gamma}}{\rho EA_0 (1 + \alpha x)^4} \end{cases} \tag{f}$$

Thus, the natural frequencies of free-free non-uniform beam **b** are the same as those of clamed-clamped non-uniform beam **a**, and the corresponding displacement mode shapes are

$$\begin{aligned} \tilde{v}_r(x) &= EI(x)\frac{\partial^2 v_r(x)}{\partial x^2} \\ &= a_r EA_0 r_g^2 (1 + ax)^4 \cdot \\ &\quad \left\{ \frac{\kappa_r^2[\cosh(\kappa_r x) + \cos(\kappa_r x) - d_r \sinh(\kappa_r x) - d_r \sin(\kappa_r x)]}{(1 + \alpha x)^2} \right. \\ &\quad + \frac{4\alpha\kappa_r[d_r \cosh(\kappa_r x) - d_r \cos(\kappa_r x) - \sinh(\kappa_r x) - \sin(\kappa_r x)]}{(1 + \alpha x)^3} \\ &\quad \left. + \frac{6\alpha^2[\cosh(\kappa_r x) - \cos(\kappa_r x) - d_r \sinh(\kappa_r x) + d_r \sin(\kappa_r x)]}{(1 + \alpha x)^4} \right\}, \\ r &= 1, 2, \ldots \end{aligned} \tag{g}$$

It is easy to verify the correctness of the above results by substituting Eq. (g) into the dynamic boundary value problem of free-free non-uniform beam **b**.

Figure 5.11 presents the first three normalized displacement mode shapes with respect to the maximal displacement of the two beams in a dual when $\alpha = 2/L$. As the cross-sectional area of beam **a** reaches the maximum at the right end, the

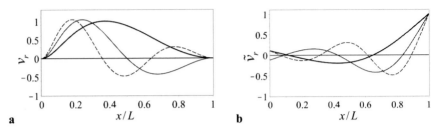

Fig. 5.11 The first three normalized displacement mode shapes of two beams in a dual of different cross-sections for $\alpha = 2/L$ (thick solid line: 1st order, thin solid line: 2nd order, dashed line: 3rd order). **a** clamped-clamped beam **a**; **b** free-free beam **b**

mode shapes exhibit smaller values at the right part of the beam. On the contrary, the mode shapes of beam **b** get larger values at the right part of the beam since the cross-sectional area has the minimum at the right end.

Remark 5.3.1 As shown in the above analysis and example, the displacement and bending moment of a non-uniform beam are a dual of variables such that the dynamic boundary value problems of two non-uniform beams with different boundaries have the same form and share the identical natural frequencies when the beams and their boundaries satisfy the dual condition. It will be possible to understand the mechanics behind the above results from the discussions about the reflection of bending waves of an E-B beam at different boundaries in Sect. 7.4.1.

5.3.1.3 Trivial Duals and Non-trivial Duals

Besides a dual with different cross-sections, beam **a** and beam **b** have identical natural frequencies in the following two cases.

(1) Mirror: Beam **a** and beam **b**, together with their boundaries, are symmetric with respect to the mid-span section at $x = L/2$. In this case, the two beams have identical natural frequencies while their mode shapes keep the same under a mirror reflection.

(2) Similarity: Beam **a** and beam **b** have the same boundaries, their cross-sectional areas are $A(x)$ and $\beta A(x)$, and their second moments of cross-sectional area are $I(x)$ and $\beta I(x)$, respectively. Here $\beta > 0$ should make beam **b** still be an E-B beam. According to the dynamic equation of an E-B beam, $\beta > 0$ has no influence on the natural vibrations of the beam. Hence, the two beams have identical natural frequencies and mode shapes. The above two beams are said to be the same beam in duality studies.

Remark 5.3.2 The mirror and similarity of beams have the same features of rods as follows. At first, they are trivial and do not require any analysis to get to know the above vibration behaviors. Second, the two beams in either a mirror case or a similarity case have identical natural frequencies and fall into a dual defined. Third, the two beams do not form a dual of different cross-sections. Fourth, the beam in a pair of mirror or similarity can transfer dual information to each other. That is, if one beam in the pair finds a beam as its dual, the other beam in the pair falls into the dual, too. As such, the above pair of beams is called *a trivial dual*, while the dual of different cross-sections is a *non-trivial dual*.

5.3.2 A Dual of Identical Cross-Sections

This subsection discusses whether two beams with identical cross-sections have the same natural frequencies. Similar to the study on rods in Sect. 5.2.2, the subsection begins with a case study, and then turns to more general cases.

5.3.2.1 A Dual of Two Beams with Exponential Cross-Sections

Example 5.3.2 The clamed-clamed beam **a** has a rectangular cross-section of a constant height and an exponentially variable width. That is, the cross-sectional area of the beam yields $A(x) = A_0 \exp(\alpha x)$ with $A_0 > 0$, and the second moment of cross-sectional area of the beam is $I(x) = A_0 r_g^2 \exp(\alpha x)$, where $r_g > 0$ is the gyration radius of cross-section. Discuss the conditions under which beam **a** with different boundaries has the same natural frequencies.

We begin with a free-free beam **b** with the following cross-sectional area and the second moment of cross-sectional area

$$
\begin{cases}
\widetilde{B}(x) = \dfrac{\widetilde{\gamma}}{\rho EI(x)} = \dfrac{\widetilde{\gamma}}{\rho E A_0 r_g^2} \exp(-\alpha x), \\[3mm]
\widetilde{J}(x) = \dfrac{\widetilde{\gamma}}{\rho EA(x)} = \dfrac{\widetilde{\gamma}}{\rho EA_0} \exp(-\alpha x).
\end{cases}
\tag{a}
$$

Hence, free-free beam **b** and clamped-clamped beam **a** form a dual of different cross-sections.

Define beam **c** as the mirror of beam **b** such that the cross-sectional area and the second moment of cross-sectional area satisfy

$$
\begin{cases}
\widehat{B}(x) = \dfrac{\widetilde{\gamma}}{\rho EA_0 r_g^2} \exp[\alpha(x-L)] = \dfrac{A(x)}{\beta} \\[3mm]
\widehat{J}(x) = \dfrac{\widetilde{\gamma}}{\rho EA_0} \exp[\alpha(x-L)] = \dfrac{I(x)}{\beta}, \quad \beta \equiv \dfrac{\rho EA_0 A(L) r_g^2}{\widetilde{\gamma}}
\end{cases}
\tag{b}
$$

As beam **c** carries the dual information of beam **b**, then free-free beam **c** and clamped-clamped beam **a** form a dual of different cross-sections. Let $\widehat{B}(x)$ and $\widehat{J}(x)$ in Eq.

(b) be multiplied by β, and allocate the results to beam **d**. Thus, beam **d** and beam **c** are a pair of similar beams and have identical natural frequencies.

In a logic way, clamped-clamed beam **a** and free-free beam **d** form a dual of different cross-sections. They have identical natural frequencies while the displacement mode shapes of one beam are the bending moment mode shapes of the other beam, and vice versa. Yet, beam **d** and beam **a** have the same cross-sectional area and the same second moment of cross-sectional area. Hence, they are named a *dual of identical cross-sections*.

To well understand the dual of identical cross-sections, we solve the natural vibrations of clamped-clamped beam **a**, and then get the natural vibrations of free-free beam **d** in the dual from duality result.

Substituting Eq. (a) into Eq. (5.3.1) and eliminating $A_0 \exp(\alpha x) > 0$, we get a partial differential equation with invariant coefficients, namely,

$$\frac{\partial^2 v(x,t)}{\partial t^2} + c_0^2 r_g^2 \left[\frac{\partial^4 v(x,t)}{\partial x^4} + 2\alpha \frac{\partial^3 v(x,t)}{\partial x^3} + \alpha^2 \frac{\partial^2 v(x,t)}{\partial x^2} \right] = 0, \quad c_0 \equiv \sqrt{\frac{E}{\rho}} \tag{c}$$

As will be seen in Sect. 7.2.1, c_0 has the dimension of velocity, but is not the longitudinal wave speed of the beam. Let $v(x,t) = v(x) \sin(\omega t)$ be the solution candidate of Eq. (c) and substitute it into Eq. (c), then we get an ordinary differential equation as follows

$$\frac{d^4 v(x)}{dx^4} + 2\alpha \frac{d^3 v(x)}{dx^3} + \alpha^2 \frac{d^2 v(x)}{dx^2} - \kappa^4 v(x) = 0, \quad \kappa \equiv \sqrt{\frac{\omega}{c_0 r_g}} \tag{d}$$

Equation (d) has the following characteristic equation

$$\lambda^4 + 2\alpha \lambda^3 + \alpha^2 \lambda^2 - \kappa^4 = 0 \tag{e}$$

As the exact solutions of Eq. (e) look complicated, we turn to the approximate solutions when $0 < \alpha \ll 1$ holds. Let the approximate solution of Eq. (e) be expressed as $\lambda \approx \eta_0 + \eta_1 \alpha$. Substituting it into Eq. (e) and equating the same order of α, we get $\eta_0 \in \{\pm i\kappa, \pm \kappa\}$ and $\eta_1 = -1/2$ such that the approximate eigenvalues are

$$\lambda_{1,2} \approx -\frac{\alpha}{2} \pm i\kappa, \quad \lambda_{3,4} \approx -\frac{\alpha}{2} \pm \kappa \tag{f}$$

Hence, the solution of Eq. (d) takes the following form

$$v(x) = \exp\left(-\frac{\alpha x}{2}\right) [c_1 \cos(\kappa x) + c_2 \sin(\kappa x) + c_3 \cosh(\kappa x) + c_4 \sinh(\kappa x)] \tag{g}$$

Substitution of Eq. (g) into the two clamped boundary conditions of beam **a** gives

$$
\begin{cases}
c_1[\cos(\kappa L) - \cosh(\kappa L)] + c_2[\sin(\kappa L) - \sinh(\kappa L)] = 0 \\[2mm]
c_1\left\{\dfrac{\alpha}{2}[\cosh(\kappa L) - \cos(\kappa L)] - \kappa[\sin(\kappa L) + \sinh(\kappa L)]\right\} \\[2mm]
+ c_2\left\{\dfrac{\alpha}{2}[\sinh(\kappa L) - \sin(\kappa L)] + \kappa[\cos(\kappa L) - \cosh(\kappa L)]\right\} = 0
\end{cases}
\tag{h}
$$

From the sufficient and necessary condition of non-trivial solution of Eq. (h), we find that the characteristic equation of beam **a** is as same as the characteristic equation of a clamped-clamped uniform beam[26]. Thus, the natural frequencies of beam **a**, from Table 2.1, satisfy

$$
\omega_r = \kappa_r^2 r_g c_0, \quad \kappa_1 \approx \frac{4.730}{L}, \quad \kappa_2 \approx \frac{7.853}{L}, \quad \kappa_r \approx \frac{(2r+1)\pi}{2L}, \quad r \geq 3 \tag{i}
$$

Substituting wave number κ_r in Eq. (i) and the non-trivial solution of Eq. (h) into Eq. (g), we derive the displacement mode shapes of beam **a**, namely,

$$
\begin{cases}
v_r(x) = c_1 \exp\left(-\dfrac{\alpha x}{2}\right)[\cosh(\kappa_r x) - \cos(\kappa_r x) - d_r \sinh(\kappa_r x) + d_r \sin(\kappa_r x)] \\[3mm]
d_r \equiv \dfrac{\cosh(\kappa_r L) - \cos(\kappa_r L)}{\sinh(\kappa_r L) - \sin(\kappa_r L)}, \quad r = 1, 2, \ldots
\end{cases}
\tag{j}
$$

Now, we turn to free-free beam **b**, which satisfies the dual condition in Eq. (a). In this case, the natural frequencies of beam **b** are the same as those of beam **a** in Eq. (i), while the displacement mode shapes of beam **b** can be taken as the bending moment mode shapes of beam **a** as follows

$$
\begin{cases}
\tilde{v}_r(x) = EI(x)\dfrac{\partial^2 v_r(x)}{\partial x^2} = b_r \exp\left(\dfrac{\alpha x}{2}\right)\big[(\alpha^2 + 4\alpha d_r \kappa_r + 4\kappa_r^2)\cosh(\kappa_r x) \\[2mm]
\quad - (\alpha^2 - 4\alpha d_r \kappa_r - 4\kappa_r^2)\cos(\kappa_r x) - (\alpha^2 d_r + 4\alpha \kappa_r + 4 d_r \kappa_r^2)\sinh(\kappa_r x) \\[2mm]
\quad + (\alpha^2 d_r - 4\alpha \kappa_r - 4 d_r \kappa_r^2)\sin(\kappa_r x)\big], \quad b_r \equiv \dfrac{c_1 EA_0 r_g^2}{4}, \quad r = 1, 2, \ldots
\end{cases}
\tag{k}
$$

By replacing the coordinate x in Eq. (k) with $L - x$, or changing the sign of α, we obtain the displacement mode shapes $\hat{v}_r(x)$, $r = 1, 2, \ldots$ of beam **c**. They are also the displacement mode shapes of beam **d** or beam **a** under free-free boundaries.

Figure 5.12 shows the first three normalized displacement mode shapes with respect to the maximal displacement of clamped-clamped beam and free-free beam, respectively, when $\alpha = 1/(2L)$. As both cross-sectional area and second moment of cross-sectional area get the maximal values at the right end of the beams, the mode shapes exhibit smaller values at the right part of the beams.

---

[26] Timoshenko S, Young D H, Weaver Jr W (1974). Vibration Problems in Engineering. 4th Edition. New York: John Wiley and Sons Inc., 424–425.

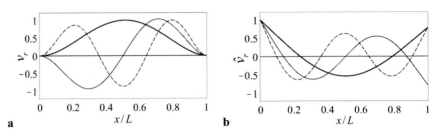

Fig. 5.12 The first three normalized displacement mode shapes of two beams in a dual of identical cross-sections for $\alpha = 1/(2L)$ (thick solid line: 1st order, thin solid line: 2nd order, dashed line: 3rd order). **a** clamped-clamped beam **a**; **b** free-free beam **a**

5.3.2.2   Dual Conditions of Identical Cross-Sections

It is easy to extend the ideas in Example 5.3.2 and get the following dual conditions of two beams with identical cross-sections. That is, there exists a constant $\beta > 0$ such that the cross-sectional area $A(x)$ and the second moment of cross-sectional area $I(x)$ of a beam satisfy

$$A(x) = \beta \widehat{B}(x) = \frac{\beta \widetilde{\gamma}}{\rho EI(L-x)}, \quad I(x) = \beta \widehat{J}(x) = \frac{\beta \widetilde{\gamma}}{\rho EA(L-x)} \qquad (5.3.12)$$

As the two conditions in Eq. (5.3.12) are equivalent, we study the first one and recast it as

$$\frac{A(x)}{\beta} \frac{I(L-x)}{\widetilde{\gamma}/\rho E} = 1 \qquad (5.3.13)$$

Equation (5.3.13) has an equivalent form as follows

$$\ln\left[\frac{A(x)}{\beta}\right] + \ln\left[\frac{I(L-x)}{\widetilde{\gamma}/\rho E}\right] = 0 \qquad (5.3.14)$$

If a new variable y and the following two functions are defined

$$\begin{cases} y \equiv x - L/2 \in [-L/2, \ L/2], \\ f(y) \equiv \ln\left[\frac{A(L/2+y)}{\beta}\right], \quad g(-y) \equiv \ln\left[\frac{I(L/2-y)}{\widetilde{\gamma}/\rho E}\right] \end{cases} \qquad (5.3.15)$$

Equation (5.3.14) can be recast as

$$f(y) + g(-y) = 0. \qquad (5.3.16)$$

From Eqs. (5.3.15) and (5.3.16), it is easy to derive

$$A(x) = \beta \exp\left[f\left(x - \frac{L}{2}\right)\right], \quad I(x) = \frac{\tilde{\gamma}}{\rho E} \exp\left[-f\left(\frac{L}{2} - x\right)\right] \quad (5.3.17)$$

Once a smooth function $f(y)$, $y \in [-L/2, L/2]$ is assumed, it is easy to derive $A(x)$ and $I(x)$ from Eq. (5.3.17) such that they yield Eq. (5.3.12). Even though the cross-sectional properties of the beam have to yield Eq. (5.3.17), the smooth function $f(y)$ is quite arbitrary and gives enough design options for a dual of identical cross-sections. Here are the discussions about the following two cases.

(1) Case for $f(y)$ being an odd function

In this case, $A(x)$ and $I(x)$ are anti-symmetric about the mid-span section $x = L/2$, and have the following relations

$$I(x) = \frac{\tilde{\gamma}}{\beta \rho E} A(x) = r_g^2 A(x), \quad \beta = \frac{\tilde{\gamma}}{\rho E r_g^2} \quad (5.3.18)$$

where the radius of gyration of cross-section r_g is a constant. One can easily verify that the cross-sectional area and the second moment of cross-sectional area of the beam in Example 5.3.2 satisfy Eqs. (5.3.17) and (5.3.18) such that this case is a dual of identical cross-sections.

Now, we select a more complicated odd function $f(y) = -\alpha \sin(\pi y/L)$ with $\alpha > 0$. Substitution of this function into Eq. (5.3.17) leads to

$$\begin{cases} A(x) = A_0 \exp\left[\alpha \cos\left(\frac{\pi x}{L}\right)\right], & A_0 \equiv \beta \\[4mm] I(x) = I_0 \exp\left[\alpha \cos\left(\frac{\pi x}{L}\right)\right], & I_0 \equiv A_0 r_g^2 \end{cases} \quad (5.3.19)$$

It is easy to verify that Eq. (5.3.19) yields the dual conditions in Eq. (5.3.12).

Let the cross-sectional area and the second moment of cross-sectional area in Eq. (5.3.19) allocate to clamped-clamped beam **a**, and then discuss the dual of identical cross-sections. Figure 5.13 shows the cross-sectional areas of clamped-clamped beam **a**, free-free beam **b** in a dual of different cross-sections, free-free beam **c** mirrored from beam **b**, from the top to the bottom. Thus, the cross-sectional area and the second moment of cross-sectional area of beam **c** are $\hat{B}(x) = A(x)/\beta$ and $\hat{J}(x) = I(x)/\beta$, respectively. Let them multiplied by β and allocate the results to beam **d**. As such, the cross-sectional area and the second moment of cross-sectional area of free-free beam **d** are the same as those of clamped-clamped beam **a**, and they have identical natural frequencies. In other words, once Eq. (5.3.19) holds, when the clamed-clamped boundaries of beam **a** are replaced with free-free boundaries, the natural frequencies of the beam remain unchanged.

(2) Case for $f(y)$ being an even function

In this case, both $A(x)$ and $I(x)$ are symmetric with respect to the mid-span section $x = L/2$, while the radius of gyration of cross-section varies as follows

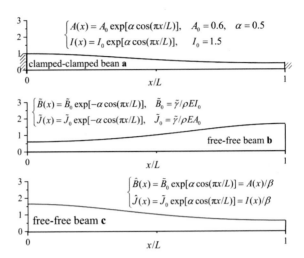

Fig. 5.13 The cross-sectional areas of clamed-clamed beam **a**, free-free beam **b** in a dual, and beam **c** mirrored from beam **b**

$$r_g^2 = \frac{I(x)}{A(x)} = \frac{\tilde{\gamma}}{\beta \rho E} \exp\left[-2f\left(x - \frac{L}{2}\right)\right] \tag{5.3.20}$$

To be more specific, taking an even function $f(y) = \alpha y^2$ with $\alpha > 0$ and substituting it into Eq. (5.3.17) lead to

$$\begin{cases} A(x) = A_0 \exp\left[\alpha(x - L/2)^2\right], & A_0 \equiv \beta \\ I(x) = I_0 \exp\left[-\alpha(x - L/2)^2\right], & I_0 \equiv \dfrac{\tilde{\gamma}}{\rho E} \end{cases} \tag{5.3.21}$$

It is easy to see that Eq. (5.3.21) satisfies Eq. (5.3.12). Thus, we can allocate the cross-sectional properties in Eq. (5.3.21) to clamped-clamped beam **a**, and discuss the dual problem. Figure 5.14 presents the cross-sectional areas of clamped-clamed beam **a**, free-free beam **b** in a dual of beam **a**, free-free beam **c** mirrored from beam **b**, from the top to the bottom. Similar to the discussions in case (1), once Eq. (5.3.21) holds, when the clamed-clamped boundaries of beam **a** are replaced with free-free boundaries, the natural frequencies of the beam remain unchanged.

5.3.2.3 Classification of Beams in a Dual of Identical Cross-Sections

Now, we check the boundary conditions under which two beams satisfying Eq. (5.3.12) fall into a dual of identical cross-sections according to Table 5.5. As studied in Figs. 5.13 and 5.14, a clamped-clamped beam in Type 1 and a free-free beam in Type 1 form a dual of identical cross-sections. It is similar to prove that the beams in each of Type 2, Type 3 and Type 4 form a dual of identical cross-sections in the same type.

Fig. 5.14 The cross-sectional areas of clamed-clamed beam **a**, free-free beam **b** in a dual, and beam **c** mirrored from beam **b**

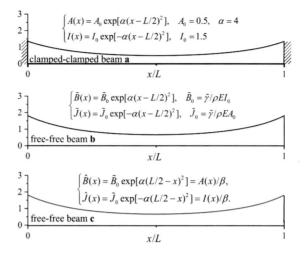

Consider, as an example, a sliding-free beam **a** in Type 2A. Noting that a sliding-clamped beam **b** and sliding-free beam **a** are in a dual of different cross-sections, while a clamped-sliding beam **c** is the mirror of beam **b**, we get a clamped-sliding beam **d**, which is similar to beam **c** and has the same cross-sectional area and the same second moment of cross-sectional area with beam **a**. Hence, a sliding-free beam in Type 2A and a clamped-sliding beam in Type 2B are in a dual of identical cross-sections.

It is similar to prove the following four types of dual: a sliding-clamped beam in Type 2A and a free-sliding beam in Type 2B, a hinged-free beam in Type 3A and a clamed-hinged beam in Type 3B, a hinged-clamped beam in Type 3A and a free-hinged beam in Type 3B, and a hinged-sliding beam in Type 4A and a sliding-hinged beam in Type 4B.

5.3.2.4 Beams Beyond a Dual of Identical Cross-Sections

As the dual conditions of identical cross-section are stricter than the dual conditions of different cross-sections, the type number of the dual of identical cross-sections is less than that of the dual of different cross-sections in Table 5.5.

Consider, as an example, a clamped-free beam **a** in Type 7. Noting that a free-clamped beam **b** and the clamped-free beam **a** are in a dual of different cross-sections, while a clamped-free beam **c** is the mirror of beam **b**. From the discussion about the case of similarity, beam **a** and beam **c** are the same beam and fall into a trivial dual. In other words, all clamped-free beams in Type 7 are in a trivial dual, and so are all free-clamed beams in the same type. In addition, the above clamped-free beam **a** and free-clamped beam **b** are a pair of mirror, which is also a trivial dual. Similarly, it is easy to prove that the beams in Type 5 and Type 6 are also in a trivial dual only.

From previous studies, it is also possible to list many non-uniform beams, which do not satisfy the dual condition of identical cross-sections in Eq. (5.3.12). For instance, the second moment of cross-sectional area of beam **a** in Example 5.3.1 is $I(x) = r_g^2 A(x)$, but the cross-sectional area $A(x)$ does not have the form in Eq. (5.3.17). Thus, beam **a** fails to have a dual of identical cross-sections. As studied by S. Abrate[27], the natural frequencies of such a beam with clamped-clamped boundaries are different from those of a beam with free-free boundaries.

5.3.3 A Dual of Two Uniform Beams

5.3.3.1 Degeneration from Non-uniform Beams

To begin with, consider a uniform beam **a** with cross-sectional area $A(x) \equiv A_0 > 0$ and the second moment of cross-sectional area $I(x) \equiv A_0 r_g^2 > 0$, and denote the longitudinal wave speed of beam as $c_0 \equiv \sqrt{E/\rho}$. From Eqs. (5.3.1) and (5.3.5), we get the dynamic boundary value problems of beam **a** in terms of displacement and bending moment respectively as follows

$$\begin{cases} \dfrac{\partial^2 v(x,t)}{\partial t^2} + c_0^2 r_g^2 \dfrac{\partial^4 v(x,t)}{\partial x^4} = 0 \\ v(x_B, t) = 0, \quad v_x(x_B, t) = 0, \\ v_{xx}(x_B, t) = 0, \quad v_{xxx}(x_B, t) = 0, \quad x_B \in \{0, L\} \end{cases} \tag{5.3.22}$$

$$\begin{cases} \dfrac{\partial^2 M(x,t)}{\partial t^2} + c_0^2 r_g^2 \dfrac{\partial^2 M(x,t)}{\partial x^2} = 0 \\ M_{xx}(x_B, t) = 0, \quad M_{xxx}(x_B, t) = 0, \\ M(x_B, t) = 0, \quad M_x(x_B, t) = 0, \quad x_B \in \{0, L\} \end{cases} \tag{5.3.23}$$

In Eqs. (5.3.22) and (5.3.23), all possible boundary condition sets are listed according to the rule of a dual of boundaries. It is easy to take the combination of boundaries of a specific beam from Table 5.4.

Now, we select beam **b** according to Eq. (5.3.9) and get the dynamic boundary value problems of beam **b** in terms of displacement and bending moment respectively as follows

$$\begin{cases} \dfrac{\partial^2 \tilde{v}(x,t)}{\partial t^2} + c_0^2 r_g^2 \dfrac{\partial^2 \tilde{v}(x,t)}{\partial x^2} = 0 \\ \tilde{v}_{xx}(x_B, t) = 0, \quad \tilde{v}_{xxx}(x_B, t) = 0, \\ \tilde{v}(x_B, t) = 0, \quad \tilde{v}_x(x_B, t) = 0, \quad x_B \in \{0, L\} \end{cases} \tag{5.3.24}$$

[27] Abrate S (1995). Vibration of non-uniform rods and beams. Journal of Sound and Vibration, 185(4): 703–716.

$$\begin{cases} \dfrac{\partial^2 \widetilde{M}(x,t)}{\partial t^2} + c_0^2 r_g^2 \dfrac{\partial^2 \widetilde{M}(x,t)}{\partial x^2} = 0 \\ \widetilde{M}(x_B, t) = 0, \quad \widetilde{M}_x(x_B, t) = 0, \\ \widetilde{M}_{xx}(x_B, t) = 0, \quad \widetilde{M}_{xxx}(x_B, t) = 0, \quad x_B \in \{0, L\} \end{cases} \tag{5.3.25}$$

As the above four equations do not have variable cross-sectional functions, one can treat beams **a** and **b** as identical beams and only select proper boundaries to form a dual of identical cross-section. Now, Eqs. (5.3.22) and (5.3.25) share the same form of boundary conditions, and so do Eqs. (5.3.23) and (5.3.24). Hence, it is easy to draw the following assertions.

(1) The first three types of uniform beams in Table 5.5 keep the dual of identical cross-sections. Type 1 includes a free-free beam and a clamped-clamped beam. Type 2 covers a sliding-free beam and a fixed-sliding beam, and the mirror of the dual. Type 3 covers a hinged-free beam and a clamped-hinged beam, and the mirror of the dual. Each of the three duals includes two kinds of beam. One is not fully constrained and may undergo rigid-body motions, while the other is not only fully constrained, but also statically undetermined. Regardless of the rigid-body motions, the two beams in a dual have identical natural frequencies.

(2) The two beams in Type 4 are a pair of mirrors, and so are the two beams in Type 7. The two beams in Type 5 are the same beam, and so are the two beams in Type 6. They all are trivial duals.

5.3.3.2 An Extended Dual of Two Uniform Beams

Under the descriptions of displacement and bending moment, a sliding-sliding beam and a hinged-hinged beam do not have any dual relations. However, the sliding-sliding beam may undergo a rigid-body motion, but the hinged-hinged beam doesn't. Such a pair of beams is very close to the first three types of dual of identical cross-sections from the consideration of constraints. Hence, it is reasonable to guess whether the two uniform beams form a dual under some other descriptions. Here is a short, but interesting discussion.

From Eq. (5.3.22), the dynamic boundary value problem of a sliding-sliding uniform beam in terms of displacement reads

$$\begin{cases} \dfrac{\partial^2 v(x,t)}{\partial t^2} + c_0^2 r_g^2 \dfrac{\partial^4 v(x,t)}{\partial x^4} = 0 \\ v_x(0,t) = 0, \quad v_{xxx}(0,t) = 0, \quad v_x(L,t) = 0, \quad v_{xxx}(L,t) = 0 \end{cases} \tag{5.3.26}$$

Differentiating the partial differential equation in Eq. (5.3.26) with respect to x and letting $\theta(x,t) \equiv v_x(x,t)$, we get the dynamic boundary value problem of a sliding-sliding uniform beam in terms of slope, namely,

$$\begin{cases} \dfrac{\partial^2 \theta(x,t)}{\partial t^2} + c_0^2 r_g^2 \dfrac{\partial^4 \theta(x,t)}{\partial x^4} = 0 \\ \theta(0,t) = 0, \quad \theta_{xx}(0,t) = 0, \quad \theta(L,t) = 0, \quad \theta_{xx}(L,t) = 0 \end{cases} \tag{5.3.27}$$

It is interesting to see that Eq. (5.3.27) looks the same as the dynamic boundary value problem of a hinged-hinged uniform beam in terms of displacement. Thus, a sliding-sling uniform beam and a hinged-hinged uniform beam form a dual under the description of displacement and slope. Regardless of the rigid-body motion of the sliding-sliding uniform beam normal to the beam axis, two beams have identical natural frequencies as follows

$$\omega_r = \kappa_r^2 r_g c_0, \quad \kappa_r = \frac{r\pi}{L}, \quad r = 1, 2, \dots. \tag{5.3.28}$$

In view of the displacement mode shapes of a hinged-hinged uniform beam, the slope mode shapes of the sliding-sliding beam yields

$$\theta_r(x) = a_r \sin(\kappa_r x), \quad r = 1, 2, \dots. \tag{5.3.29}$$

Integrating Eq. (5.3.29) with respect to x gives the displacement mode shape of the sliding-sliding uniform beam with no rigid-body motion taken into account, namely,

$$v_r(x) = b_r \cos(\kappa_r x), \quad b_r \equiv -\frac{a_r}{\kappa_r}, \quad r = 1, 2, \dots \tag{5.3.30}$$

Remark 5.3.3 In the current frame of mechanics, there is no dual relation under the description of displacement and slope. Hence, this is a new relation of dual. As will be studied in Sect. 7.4.1, the hinged boundary and sliding boundary for a uniform beam share a feature. That is, the reflection of an incident traveling wave at these two boundaries gives only a traveling wave reflected, and does not produce any evanescent wave, which decays away from a clamped boundary or a free boundary.

5.3.3.3   Rigid-Body Motions

Finally, we consider, as an example, uniform beams and show that in a non-trivial dual, one of two beams undergoes rigid-body motions. In a non-trivial dual, we first select a beam, which does not undergo any rigid-body motion and denote the wave number of vibration as $\kappa > 0$. Then, the slope and the curvature of the beam are as follows

$$\tilde{v}_x(x, t) = [c_1 \cos(\kappa x) + c_2 \sin(\kappa x) + c_3 \cosh(\kappa x) + c_4 \sinh(\kappa x)]q(t) \tag{5.3.31a}$$

$$\tilde{v}_{xx}(x, t) \equiv \frac{\widetilde{M}(x)}{EI}$$
$$= [d_1 \cos(\kappa x) + d_2 \sin(\kappa x) + d_3 \cosh(\kappa x) + d_4 \sinh(\kappa x)]q(t) \tag{5.3.31b}$$

Integrating Eqs. (5.3.31a) and (5.3.31b) with respect to x once and twice respectively, leads to

$$v(x, t) = c_5(t)$$
$$+ \frac{1}{\kappa}[c_4 \cosh(\kappa x) + c_3 \sinh(\kappa x) - c_2 \cos(\kappa x) + c_1 \sin(\kappa x)]q(t) \quad (5.3.32a)$$

$$v(x, t) = d_6(t) + x d_5(t)$$
$$+ \frac{1}{\kappa^2 EI}[d_4 \sinh(\kappa x) + d_3 \cosh(\kappa x) - d_2 \sin(\kappa x) - d_1 \cos(\kappa x)]q(t)$$
$$(5.3.32b)$$

where $c_5(t)$, $d_5(t)$ and $d_6(t)$ are functions in time and associated with rigid-body motions.

Equation (5.3.32a) describes the dynamics of a sliding-sliding beam or a sliding-free beam, both undergoing a rigid-body motion described by $c_5(t)$ perpendicular to the un-deformed beam axis. Equation (5.3.32b) describes the dynamics of a free-free beam undergoing a planar rigid-body motion described by $d_6(t) + x d_5(t)$. If $d_6(t) = 0$ holds true, Eq. (5.3.32b) corresponds to the dynamics of a hinged-free beam, which undergoes rotation $x d_5(t)$ about the left end of bam. These four kinds of beams have insufficient constraints to eliminate $c_5(t)$, $d_5(t)$ and $d_6(t)$ such that they must undergo rigid-body motions.

Remark 5.3.4 The above analysis begins with the elastic vibration of a beam around the equilibrium position, derives the rigid-body motion by integration, and superposes both rigid-body motion and elastic vibration. Thus, Eq. (5.3.32) can only describe the rectilinear rigid-body motion of a sliding-sliding beam or a sliding-free beam near the equilibrium position, and the rigid-body rotation of a hinged-free beam or a free-free beam about the left end of beam at a lower angular velocity. If a beam rotates with higher angular velocity, it is necessary to study the nonlinear coupling between the rigid-body motion and the elastic vibration of the beam from the viewpoint of the dynamics of flexible multi-body systems[28,29].

5.3.4 Concluding Remarks

This section presented detailed studies on the duality of Euler–Bernoulli beams made of a linear homogeneous elastic material, or called E-B beams or beams for short, in natural vibrations. The studies have solved Problem 4A proposed in Chap. 1, enabled

---

[28] Bauchau O A (2011). Flexible Multibody Dynamics. New York: Springer-Verlag, 569–637.

[29] Shabana A A (2005). Dynamics of Multibody Systems. 3rd Edition. New York: Cambridge University Press, 188–343.

one to well understand the natural vibrations of a non-uniform beam under various conditions, and laid a theoretical foundation in both cross-sectional properties and boundaries in the dynamic design of non-uniform beams. The main conclusions of the section are as follows.

(1) For non-uniform beams under homogeneous boundaries, the descriptions of displacement and bending moment enable one to study two beams in a dual of different cross-sections. That is, they have different cross-sections, but identical natural frequencies. The study showed that the non-uniform beams have 7 types of dual of different cross-sections. Type 1 includes a free-free beam and a clamped-clamped beam, Type 2 includes a sliding-free beam and a sliding-clamped beam (and its mirror), Type 3 covers a hinged-free beam and a hinged-clamped beam (and its mirror), Type 4 covers a hinged-sliding beam and a sliding-hinged beam (and its mirror), Type 5 covers a sliding-sliding beam and a sliding-sliding beam, Type 6 includes a hinged-hinged beam and a hinged-hinged beam, and Type 7 covers a clamped-free beam and a free-clamped beam.

(2) The two beams with the same cross-section in the above dual are named a dual of identical cross-sections. The study showed that two beams in such a dual must have specific cross-sectional area and the second moment of cross-sectional area. They yield the dual condition described by an exponential function, but still have enough design options. Once the cross-sectional area and the second moment of cross-sectional area satisfy the dual condition, the first four types of dual of different cross-sections become the duals of identical cross-sections. The other three types are trivial duals because two beams in a dual of identical cross-sections are a pair of mirrors.

(3) For uniform beams, the first three types of dual of identical cross-sections keep the dual relation, but Type 4 degenerates to a pair of mirrors. In this case, the descriptions of displacement and slope of two beams enable one to find a new dual, which includes a sliding-sliding beam and a hinged-hinged beam. In the above four types of dual, one beam may undergo rigid-body motions while the other beam is fully constrained.

The above conclusions can be summarized in Table 5.6, where different homogeneous boundaries give different types of dual when the cross-sectional area and the second moment of cross-sectional area satisfy the dual conditions in the first row. Here symbol $\Leftrightarrow$ represents a non-trivial dual, and symbol $\rightleftarrows$ means a trivial dual. Similar to Table 5.5, the boldface fonts imply that the two beams have the same boundaries while the italic boldface fonts imply that the boundaries of two beams are a pair of mirrors.

Table 5.6 The dual conditions of beams and classifications for homogeneous boundaries (F: Free, S: Sliding, H: Hinged, C: Clamped)

Type	Different sections $\widetilde{B}(x) = \tilde{\gamma}/\rho EI(x),$ $\widetilde{J}(x) = \tilde{\gamma}/\rho EA(x)$	Identical sections $A(x)I(L-x) = \beta\tilde{\gamma}/\rho E$	Uniform sections $A(x) = A_0,$ $I(x) = A_0 r_g^2$
1	F-F $\Leftrightarrow$ C-C	F-F $\Leftrightarrow$ C-C	F-F $\Leftrightarrow$ C-C
2	S-F $\Leftrightarrow$ S-C	S-F $\Leftrightarrow$ C-S	S-F $\Leftrightarrow$ C-S
3	H-F $\Leftrightarrow$ H-C	H-F $\Leftrightarrow$ C-H	H-F $\Leftrightarrow$ C-H
4	**H-S $\Leftrightarrow$ H-S**	***H-S $\Leftrightarrow$ S-H***	***H-S $\rightleftharpoons$ S-H***
5	**S-S $\Leftrightarrow$ S-S**	**S-S $\rightleftharpoons$ S-S**	
6	**H-H $\Leftrightarrow$ H-H**	**H-H $\rightleftharpoons$ H-H**	S-S $\Leftrightarrow$ H-H
7	***C-F $\Leftrightarrow$ F-C***	***C-F $\rightleftharpoons$ F-C***	***C-F $\rightleftharpoons$ F-C***

5.4 Further Reading and Thinking

(1) In the ground vibration test of a rocket, the rocket is suspended through several robber cables. Figure 5.15 depicts a model of the ground vibration test, where the rocket is simplified as a uniform rod with length L, linear density ρA, tension stiffness EA, and the tension stiffness coefficient of a single rubber cable is k. Study the longitudinal vibration of the rocket under gravitational acceleration g and discuss the influences of hanging location l and angle θ on the natural frequencies and mode shapes of the rocket. In addition, when the rocket is simplified as a rod with axially varying density $\rho(x)A(x)$ and tensional stiffness $E(x)A(x)$, read Sect. 5.2.4 and the reference[30], and then discuss how to analyze the natural vibrations of the rocket and the influences of hanging location l and angle θ on the natural frequencies and mode shapes.

Fig. 5.15 The schematic of the ground vibration test of a rocket

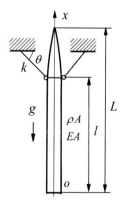

---

[30] Bapat C N (1995). Vibration of rods with uniformly tapered sections. Journal of Sound and Vibration, 185(1): 185–189.

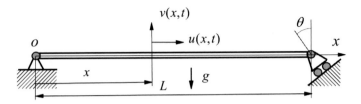

Fig. 5.16 The schematic of a beam with an inclined support

(2) Figure 5.16 shows a slender structure with an inclined support in structural engineering. The slender structure can be simplified as an E-B beam of length L, with a rectangular cross-section of height h and width b, material density ρ and Young's modulus E. Study the decoupling condition for the longitudinal vibration and transverse vibration of the beam under gravitational acceleration g, and discuss the influence of angle θ for the inclined support on the natural frequencies of bending vibration of the beam.

(3) Read the references[31,32,33], take the beam in Fig. 5.16 as a Timoshenko beam and study the influence of angle θ for the inclined support on the first natural frequency of bending vibration of the beam.

(4) Read the references[34,35,36], and give a different explanation to the mechanism of duality of E-B beams in natural vibrations. According the simplicity of vibration mechanics, discuss the dual problem of a pair of E-B beams under hinged-hinged boundaries and sliding-sliding boundaries.

(5) Discuss whether two Timoshenko beams under clamped-clamed boundaries and free-free boundary conditions respectively can be a dual in natural vibrations, and explain the mechanism behind the phenomenon.

[31] Jinsberg J H (2001). Mechanical and Structural Vibrations: Theory and Applications. Baffins Lane: John Wiley & Sons Inc., 325–498.

[32] Rao S S (2007). Vibration of Continuous Systems. Hoboken: John Wiley & Sons Inc., 317–389.

[33] Hagedorn P, Das Gupta A (2007). Vibrations and Waves in Continuous Mechanical Systems. Chichester: John Wiley & Sons Inc., 113–178.

[34] Ram Y M, Elhay S (1995). Dualities in vibrating rods and beams: Continuous and discrete models. Journal of Sound and Vibration, 184(5): 648–655.

[35] Abrate S (1995). Vibration of non-uniform rods and beams. Journal of Sound and Vibration, 185(4): 703–716.

[36] Wang D J, Wang Q C, He B C (2019). Qualitative Theory of Structural Mechanics. Singapore: Springer Nature, 215–269.

Chapter 6
Natural Vibrations of Symmetric Structures

Studies on symmetric engineering systems stemmed from the early works for buildings and bridges with mirror-symmetry, wheels and circular shafts with axial symmetry, and windmills and hydro-turbines with cyclosymmetry. In the beginning stage of vibration mechanics, the studies associated with symmetry only dealt with the mirror-symmetric and axial-symmetric systems, such as a hinged-hinged beam and a circular plate, and were far from modern science.

Two milestones of modern science came from the studies on symmetry. One is the group theory established by E. Galois, a great French mathematician, in 1832. The other is the correspondence between a differential symmetry and a conservation law in physics, proved by E. Noether, a great German mathematician, in 1918. Following their seminal works, physicists and chemists made great achievements, such as quantum mechanics and theory of molecular vibrations, after the 1930s. Their publications written in abstract language of mathematics, however, did not draw much attention from engineering circle until the 1960s.

In 1964, engineers in NASA began to compute the natural vibrations of Saturn-I launch vehicle. To simplify the computation, they utilized the symmetry of the first-stage booster, which was basically a central cylinder surrounded by eight symmetrically placed outer propellant tanks. Their success aroused an interest of engineering circle in studying the dynamics of symmetrical systems[1]. Since then, several computational methods have been available to model and compute the dynamics of important symmetric systems, such as an aero-engine, a hydro-turbine, a cooling tower of power plant, and a vacuum vessel of fusion reactor[2]. Some recent studies have

---

[1] Evensen D A (1976). Vibration analysis of multi-symmetric structures. AIAA Journal, 14(4): 446–453.

[2] Rong B, Lu K, Ni X J (2020). Hybrid finite element transfer matrix method and its parallel solution for fast calculation of large-scale structural eigenproblem. Applied Mathematical Modeling, 77(1): 169–181.

© Science Press 2022
H. Hu, *Vibration Mechanics*,
https://doi.org/10.1007/978-981-16-5457-2_6

focused on the qualitative theory of symmetric structures from the viewpoint of modern science[3].

To study the dynamic problems of mechanical and structural systems, the capability of an engineer may have two levels. The elementrary level is to utilize the mirror-symmetry or axial-symmetry of a system to reduce the computational cost and to understand the influence of symmetry on the system dynamics. The advanced level is to utilize group theory to analyze the dynamics of more complicated symmetric systems and to well understand the influence of symmetry and symmetry-breaking on the system dynamics.

The objective of this chapter is to address Problems 5A and 5B proposed in Chap. 1. The chapter presents the dynamic analysis of mirror-symmetric structures and cyclosymmetric structures since these two kinds of structures are popular and can compose more complicated structures. The chapter begins with the natural vibrations of mirror-symmetric structures, and focuses on their repeated natural frequencies and close frequencies due to symmetry-breaking. Then, the chapter turns to the studies on the natural vibrations of cyclosymmetric systems by using group theory, including the high-efficient computation and qualitative analysis of modal properties.

6.1   Natural Vibrations of Mirror-Symmetric Structures

The *mirror-symmetry* of a structure means that the structure keeps identical after the reflection about a plane. For example, the free-free uniform rod and the hinged-hinged uniform beam discussed in elementary textbooks exhibit mirror-symmetry about their mid-span cross-sections. The other mirror-symmetric structures include, but not limited to, bicycles, ground vehicles and airplanes in engineering.

This section begins with the decoupling problem of a mirror-symmetric structure in natural vibrations, and deals with how to decouple a mirror-symmetric structure by using the methods of mechanics and mathematics, respectively. Then, the section presents detailed studies on the natural vibrations of a thin rectangular plate with hinged boundaries, including the rigorous deduction of exact solutions of natural vibrations, the natural vibrations of repeated frequencies, the natural modes with very close frequencies and their uncertainties in dynamic tests in order to address Problem 5A proposed in Chap. 1.

6.1.1   Decoupling of Mirror-Symmetric Structures

In studying the vibration problems of a complicated system, the routine work today is to establish the dynamic model of finite elements for the system first and then

---

[3] Wang D J, Wang Q S, He B C (2019). Qualitative Theory in Structural Mechanics. Singapore: Springer Nature, 271–326.

compute the natural vibrations of the system as the start of further works. Thus, one needs to solve the following eigenvalue problem

$$(\boldsymbol{K} - \omega^2 \boldsymbol{M})\boldsymbol{\varphi} = \boldsymbol{0} \qquad (6.1.1)$$

Here, both mass matrix $\boldsymbol{M} \in \mathbb{R}^{n \times n}$ and stiffness matrix $\boldsymbol{K} \in \mathbb{R}^{n \times n}$ may have a very high order. If the vibration system includes non-holonomic constraints discussed in Sect. 3.1, the eigenvalue problem becomes asymmetric and requires even higher cost of computations.

When the system of concern is mirror-symmetric, the natural vibrations are either symmetric or anti-symmetric. As discussed for a hinged-hinged beam with mirror-symmetry in Example 2.2.2, the symmetric loads do not excite any anti-symmetric vibrations of a symmetric system, and the anti-symmetric loads do not excite any symmetric vibrations of the symmetric system. In other words, the symmetric vibrations and anti-symmetric vibrations of a mirror-symmetric system do not couple with each other. Hence, it is possible to decouple the eigenvalue problem in Eq. (6.1.1).

This subsection begins with the vibration description of a mirror-symmetric system, and then presents how to decouple the corresponding eigenvalue problem in Eq. (6.1.1) via methods of mechanics and mathematics, respectively.

6.1.1.1 Vibration Description of a Mirror-Symmetric Structure

In Fig. 6.1, the dashed line depicts the structure S with mirror-symmetry about plane P, which is called a *symmetry plane*. Plane P divides structure S into two identical sub-structures denoted by S_L and S_R. We first establish the local frame of coordinates $ox_R y_R z_R$ in Fig. 6.1 to describe the vibration of right sub-structure S_R, and then get the local frame of coordinates $ox_L y_L z_L$ through the mirror reflection about plane P so as to describe the vibration of left sub-structure S_L. Accordingly, when the local frame $ox_R y_R z_R$ is right-handed, the local frame $ox_L y_L z_L$ is left-handed due to the mirror reflection.

Fig. 6.1 The schematic of a mirror-symmetric structure and two local frames of coordinates for right and left sub-structures

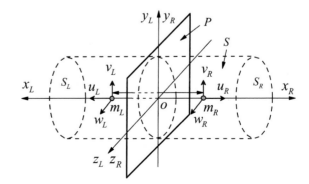

Now, consider an arbitrary mass point m_R in S_R. The mirror reflection of m_R gives mass point m_L in S_L. Let vectors $[\,u_R\ v_R\ w_R\,]^T$ and $[\,u_L\ v_L\ w_L\,]^T$ describe the displacement components of m_R and m_L, respectively. The vibration of structure S is said to be the *symmetric vibration* about plane P if $u_R = u_L,\quad v_R = v_L,\quad w_R = w_L$ hold true, or the *anti-symmetric vibration* about plane P if $u_R = -u_L,\quad v_R = -v_L,\quad w_R = -w_L$ hold.

Remark 6.1.1 The concept of mirror-symmetry is introduced in a global sense, but the symmetric and anti-symmetric vibrations are defined in the local frames. Hence, it is necessary to check these vibrations defined in two local frames $o_{x_R y_R z_R}$ and $o_{x_L y_L z_L}$ in a global sense as follows.

(1) When structure S is a spatial beam, for example, the symmetric bending vibration defined in local frames $o_{x_R y_R z_R}$ and $o_{x_L y_L z_L}$ gives a picture that the motions of S_L and S_R are identical in two sides of plane P, while an anti-symmetric bending vibration exhibits that the motions of S_L and S_R are in opposite directions in two sides of plane P such that any mass point of structure S in plane P is immovable. It is easy to understand the above assertion from the symmetric and anti-symmetric mode shapes of a beam with identical boundaries at two ends in Sect. 5.3.
(2) When structure S is a rod, for instance, the symmetric axial vibration defined in local frames $o_{x_R y_R z_R}$ and $o_{x_L y_L z_L}$ behaves such that the motions of S_L and S_R are in opposite directions in two sides of plane P and any mass point of structure S in plane P is immovable, while the anti-symmetric axial vibration exhibits that the motions of both S_L and S_R are identical in two sides of plane P.

6.1.1.2 A Case Study of Decoupling Procedure

Example 6.1.1 Consider a 4-DoF system with mirror-symmetry about plane P in Fig. 6.2a and study the symmetric and anti-symmetric natural vibrations from the viewpoint of mechanics and mathematics.

(1) Analysis from viewpoint of mechanics

The analysis begins with the symmetric vibrations of the system. In this case, as discussed about the axial vibration in Remark 6.1.1, the mass point of the system in plane P is immovable such that a constraint can be imposed on plane P and the 4-DoF system is divided into two 2-DoF systems S and S' as shown in Fig. 6.2b. The constraint makes the middle spring be two springs with a half length. Thus, each of them has doubled stiffness coefficient $2k$.

Now, we study the natural vibrations of sub-system S in the local frame of coordinates and focus on the dynamic equation of sub-system S as follows

$$M_s \ddot{u}_s(t) + K_s u_s(t) = 0 \tag{a}$$

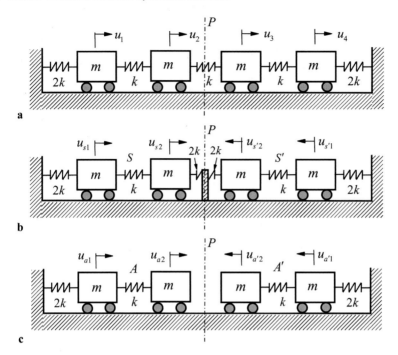

Fig. 6.2 The vibration analysis of a mirror-symmetric system. **a** A mirror-symmetric system and the global frame of coordinates; **b** A simplified model of symmetric vibrations and the local frame of coordinates; **c** A simplified model of anti-symmetric vibrations and the local frame of coordinates

where

$$M_s = m \begin{bmatrix} 1 & 0 \\ 0 & 1 \end{bmatrix}, \quad K_s = k \begin{bmatrix} 3 & -1 \\ -1 & 3 \end{bmatrix}, \quad u_s = \begin{bmatrix} u_{s1} \\ u_{s2} \end{bmatrix} \tag{b}$$

Solving the eigenvalue problem of Eq. (b) leads to a pair of natural modes as follows

$$\begin{cases} \omega_{s1} = \sqrt{2k/m}, \ \varphi_{s1} = \begin{bmatrix} 1 & 1 \end{bmatrix}^T \\ \omega_{s2} = \sqrt{4k/m}, \ \varphi_{s2} = \begin{bmatrix} 1 & -1 \end{bmatrix}^T \end{cases} \tag{c}$$

From the definition of symmetric vibration of a mirror-symmetric structure, subsystem S' has the same natural modes described in the local frame of coordinates, namely,

$$\varphi_{s'1} = \varphi_{s1}, \quad \varphi_{s'2} = \varphi_{s2} \tag{d}$$

If one prefers the description of the symmetric mode shapes in the global frame of coordinates in Fig. 6.2a, it is easy to use the transform of the two frames and express the symmetric mode shapes as

$$\begin{bmatrix} u_1 \\ u_2 \end{bmatrix} = \begin{bmatrix} u_{s1} \\ u_{s2} \end{bmatrix} = \boldsymbol{u}_s, \quad \begin{bmatrix} u_3 \\ u_4 \end{bmatrix} = -\begin{bmatrix} u_{s'2} \\ u_{s'1} \end{bmatrix} = -\begin{bmatrix} u_{s2} \\ u_{s1} \end{bmatrix} = -\tilde{\boldsymbol{I}}_2 \boldsymbol{u}_s, \quad \tilde{\boldsymbol{I}}_2 \equiv \begin{bmatrix} 0 & 1 \\ 1 & 0 \end{bmatrix}$$

$$(e)$$

There follow the two symmetric mode shapes in the global frame of coordinates

$$\begin{cases} \boldsymbol{\psi}_{s1} = \begin{bmatrix} \boldsymbol{\varphi}_{s1}^{\mathrm{T}} & -\boldsymbol{\varphi}_{s1}^{\mathrm{T}} \tilde{\boldsymbol{I}}_2 \end{bmatrix}^{\mathrm{T}} = \begin{bmatrix} 1 & 1 & -1 & -1 \end{bmatrix}^{\mathrm{T}} \\ \boldsymbol{\psi}_{s2} = \begin{bmatrix} \boldsymbol{\varphi}_{s2}^{\mathrm{T}} & -\boldsymbol{\varphi}_{s2}^{\mathrm{T}} \tilde{\boldsymbol{I}}_2 \end{bmatrix}^{\mathrm{T}} = \begin{bmatrix} 1 & -1 & 1 & -1 \end{bmatrix}^{\mathrm{T}} \end{cases}$$

$$(f)$$

Here, $\boldsymbol{\psi}_{s1}$ has one node and is the second mode shape of the 4-DoF system, while $\boldsymbol{\psi}_{s2}$ has three nodes, and is the fourth mode shape of the system.

Next, we study the anti-symmetric vibration. From the discussion about the axial vibration in Remark 6.1.1, the adjacent lumped masses in two sides of plane P undergo the same motion and the spring between them has no deformation and can be dropped off. Now the 4-DoF system is simplified as two 2-DoF systems A and A' in Fig. 6.2c.

The natural vibration of sub-system A in the local frame of coordinates yields

$$\boldsymbol{M}_a \ddot{\boldsymbol{u}}_a(t) + \boldsymbol{K}_a \boldsymbol{u}_a(t) = \boldsymbol{0}$$

$$(g)$$

where

$$\boldsymbol{M}_a = m \begin{bmatrix} 1 & 0 \\ 0 & 1 \end{bmatrix}, \quad \boldsymbol{K}_a = k \begin{bmatrix} 3 & -1 \\ -1 & 1 \end{bmatrix}, \quad \boldsymbol{u}_a = \begin{bmatrix} u_{a1} \\ u_{a2} \end{bmatrix}$$

$$(h)$$

Solving the eigenvalue problem of Eq. (h) leads to a pair of natural modes as follows

$$\begin{cases} \omega_{a1} = \sqrt{(2 - \sqrt{2})k/m}, \ \boldsymbol{\varphi}_{a1} = \begin{bmatrix} 1 & 1 + \sqrt{2} \end{bmatrix}^{\mathrm{T}} \\ \omega_{a2} = \sqrt{(2 + \sqrt{2})k/m}, \ \boldsymbol{\varphi}_{a2} = \begin{bmatrix} 1 & 1 - \sqrt{2} \end{bmatrix}^{\mathrm{T}} \end{cases}$$

$$(i)$$

From the definition of anti-symmetric axial vibration of a mirror-symmetric system, the mode shapes of sub-system A' and those of sub-system A have opposite signs in their local frame of coordinates, that is,

$$\boldsymbol{\varphi}_{a'1} = -\boldsymbol{\varphi}_{a1}, \quad \boldsymbol{\varphi}_{a'2} = -\boldsymbol{\varphi}_{a2}$$

$$(j)$$

In the global frame of coordinates, the anti-symmetric mode shapes satisfy

$$\begin{bmatrix} u_1 \\ u_2 \end{bmatrix} = \begin{bmatrix} u_{a1} \\ u_{a2} \end{bmatrix} = \boldsymbol{u}_a, \quad \begin{bmatrix} u_3 \\ u_4 \end{bmatrix} = -\begin{bmatrix} u_{a'2} \\ u_{a'1} \end{bmatrix} = \begin{bmatrix} u_{a2} \\ u_{a1} \end{bmatrix} = \tilde{\boldsymbol{I}}_2 \boldsymbol{u}_a \tag{k}$$

There follow the anti-symmetric mode shapes in the global frame of coordinates, namely,

$$\begin{cases} \boldsymbol{\psi}_{a1} = \begin{bmatrix} \boldsymbol{\varphi}_{a1}^{\mathrm{T}} & \boldsymbol{\varphi}_{a1}^{\mathrm{T}}\tilde{\boldsymbol{I}}_2 \end{bmatrix}^{\mathrm{T}} = \begin{bmatrix} 1 & 1+\sqrt{2} & 1+\sqrt{2} & 1 \end{bmatrix}^{\mathrm{T}} \\ \boldsymbol{\psi}_{a2} = \begin{bmatrix} \boldsymbol{\varphi}_{a2}^{\mathrm{T}} & \boldsymbol{\varphi}_{a2}^{\mathrm{T}}\tilde{\boldsymbol{I}}_2 \end{bmatrix}^{\mathrm{T}} = \begin{bmatrix} 1 & 1-\sqrt{2} & 1-\sqrt{2} & 1 \end{bmatrix}^{\mathrm{T}} \end{cases} \tag{l}$$

They have no node and two nodes, respectively, and are the first and the third mode shapes accordingly.

To this end, we have completed the natural vibration analysis of the 4-DoF system in Fig. 6.2a. Equations (f) and (l) show that the symmetric mode shapes in geometry have anti-symmetry in algebraic expressions, while the anti-symmetric mode shapes in geometry have symmetric expressions. This is due to the mirro reflection between the frames of corrdinates for two sub-systems.

(2) Analysis from the viewpoint of mathematics

In the global frame of coordinates in Fig. 6.2a, the dynamic equations of the 4-DoF system in a matrix form are

$$\boldsymbol{M}\ddot{\boldsymbol{u}}(t) + \boldsymbol{K}\boldsymbol{u}(t) = \boldsymbol{0} \tag{m}$$

where

$$\boldsymbol{M} = m \begin{bmatrix} 1 & 0 & 0 & 0 \\ 0 & 1 & 0 & 0 \\ 0 & 0 & 1 & 0 \\ 0 & 0 & 0 & 1 \end{bmatrix}, \quad \boldsymbol{K} = k \begin{bmatrix} 3 & -1 & 0 & 0 \\ -1 & 2 & -1 & 0 \\ 0 & -1 & 2 & -1 \\ 0 & 0 & -1 & 3 \end{bmatrix} \tag{n}$$

As well known, any motion of a system can be decomposed into a symmetric part and an anti-symmetric part. From the expressions of a symmetric vibration and an anti-symmetric vibration in the global frame coordinates, i.e., Eqs. (e) and (k), we have

$$\boldsymbol{u} = \begin{bmatrix} \boldsymbol{I}_2 \\ -\tilde{\boldsymbol{I}}_2 \end{bmatrix} \boldsymbol{u}_s + \begin{bmatrix} \boldsymbol{I}_2 \\ \tilde{\boldsymbol{I}}_2 \end{bmatrix} \boldsymbol{u}_a = \begin{bmatrix} \boldsymbol{I}_2 & \boldsymbol{I}_2 \\ -\tilde{\boldsymbol{I}}_2 & \tilde{\boldsymbol{I}}_2 \end{bmatrix} \begin{bmatrix} \boldsymbol{u}_s \\ \boldsymbol{u}_a \end{bmatrix} = \boldsymbol{P} \begin{bmatrix} \boldsymbol{u}_s \\ \boldsymbol{u}_a \end{bmatrix} \tag{o}$$

where

$$\boldsymbol{P} \equiv \begin{bmatrix} \boldsymbol{I}_2 & \boldsymbol{I}_2 \\ -\tilde{\boldsymbol{I}}_2 & \tilde{\boldsymbol{I}}_2 \end{bmatrix} \tag{p}$$

Equation (p) serves as a transform matrix from a symmetric vibration and an anti-symmetric vibration to an arbitrary vibration. Let Eq. (m) be substituted by Eq. (o) and pre-multiplied by P^{T}, we get the decoupled dynamic equations of the system as follows

$$\begin{bmatrix} M_{ss} & 0 \\ 0 & M_{aa} \end{bmatrix}\begin{bmatrix} \ddot{u}_s(t) \\ \ddot{u}_a(t) \end{bmatrix} + \begin{bmatrix} K_{ss} & 0 \\ 0 & K_{aa} \end{bmatrix}\begin{bmatrix} u_s(t) \\ u_a(t) \end{bmatrix} = 0 \qquad (q)$$

where

$$M_{ss} = M_{aa} = m\begin{bmatrix} 2 & 0 \\ 0 & 2 \end{bmatrix}, \quad K_{ss} = k\begin{bmatrix} 6 & -2 \\ -2 & 6 \end{bmatrix}, \quad K_{aa} = k\begin{bmatrix} 6 & -2 \\ -2 & 2 \end{bmatrix} \qquad (r)$$

To this end, the dynamic equations of the 4-DoF system in Eq. (m) are decoupled as Eq. (q), which includes two sets of differential equations governing symmetric vibrations and anti-symmetric vibrations, respectively. After solving their eigenvalue problems for natural frequencies, as well as symmetric and anti-symmetric eigenvectors, we substitute the symmetric and anti-symmetric eigenvectors into Eq. (o) and get the mode shapes of the 4-DoF system. It is easy to verify the agreement of the above results with those obtained from the viewpoint of mechanics.

Remark 6.1.2 It is interesting to find that the decoupled mass matrix and stiffness matrix via the above methods look different. The reasons are as follows. In the analysis from the viewpoint of mechanics, we either imposed a constraint on the middle spring or removed the middle spring to realize decoupling in view of the deformation of the middle spring such that Eqs. (b) and (h) describe the mass matrix and stiffness matrix of sub-system S or A, which can only result in the energy of sub-systems S or A. In the analysis from the viewpoint of mathematics, however, we used the decomposition of an arbitrary motion and introduced a transform such that Eqs. (q) and (r) describe the mass matrix and stiffness matrix of entire system decoupled. These two matrices result in the energy of the entire system and are equal to the doubled mass and stiffness matrices of sub-system S or sub-system A.

6.1.1.3 General Analysis

When a vibration system has a single symmetry plane, the above two methods have no major difference. Once a vibration system has two or even more symmetry planes, the analysis from the viewpoint of mathematics provides benefits for programmable analysis. As such, we extend the analysis from viewpoint of mathematics to a general case.

According to Fig. 6.1, use symmetry plane P to divide a mirror-symmetric structure into sub-structures S_L and S_R, and then establish the finite element model of n-DoFs for sub-structure S_R in local frame of coordinates $ox_R y_R z_R$. There follows the finite element model of n-DoFs for S_L via mirror reflection. As such, we have the finite element model of $2n$-DoFs for the entire structure.

Now, take the local frame $ox_R y_R z_R$ of sub-structure S_R as the global frame of the entire system, and derive the dynamic equations of the system as follows

$$\boldsymbol{M}\ddot{\boldsymbol{u}}(t) + \boldsymbol{K}\boldsymbol{u}(t) = \boldsymbol{0} \tag{6.1.2}$$

where the mass matrix and stiffness matrix are in the following block form

$$\boldsymbol{M} \equiv \begin{bmatrix} \boldsymbol{M}_{RR} & \boldsymbol{M}_{RL} \\ \boldsymbol{M}_{LR} & \boldsymbol{M}_{LL} \end{bmatrix} \in \mathbb{R}^{2n \times 2n}, \quad \boldsymbol{K} \equiv \begin{bmatrix} \boldsymbol{K}_{RR} & \boldsymbol{K}_{RL} \\ \boldsymbol{K}_{LR} & \boldsymbol{K}_{LL} \end{bmatrix} \in \mathbb{R}^{2n \times 2n} \tag{6.1.3}$$

In the global frame, the finite element model of sub-structure S_L is that of sub-structure S_R such that the sub-matrices in Eq. (6.1.3) satisfy the following relations

$$\begin{cases} \tilde{\boldsymbol{I}}_n \boldsymbol{M}_{LL} \tilde{\boldsymbol{I}}_n = \boldsymbol{M}_{RR}, & \tilde{\boldsymbol{I}}_n \boldsymbol{M}_{LR} = \boldsymbol{M}_{RL} \tilde{\boldsymbol{I}}_n \\ \tilde{\boldsymbol{I}}_n \boldsymbol{K}_{LL} \tilde{\boldsymbol{I}}_n = \boldsymbol{K}_{RR}, & \tilde{\boldsymbol{I}}_n \boldsymbol{K}_{LR} = \boldsymbol{K}_{RL} \tilde{\boldsymbol{I}}_n \end{cases} \tag{6.1.4}$$

It is easy to prove, as an example, the first equation in Eq. (6.1.4). The mirror reflection makes submatrices $\boldsymbol{M}_{RR}$ and $\boldsymbol{M}_{LL}$ be in the following form

$$\boldsymbol{M}_{RR} = \begin{bmatrix} m_{11} & m_{12} & \cdots & m_{1n} \\ m_{21} & m_{22} & \cdots & m_{21} \\ \vdots & \vdots & \ddots & \vdots \\ m_{n1} & m_{n2} & \cdots & m_{nn} \end{bmatrix}, \quad \boldsymbol{M}_{LL} = \begin{bmatrix} m_{nn} & m_{n(n-1)} & \cdots & m_{n1} \\ m_{(n-1)n} & m_{(n-1)(n-1)} & \cdots & m_{(n-1)1} \\ \vdots & \vdots & \ddots & \vdots \\ m_{1n} & m_{1(n-1)} & \cdots & m_{11} \end{bmatrix} \tag{6.1.5}$$

That is, the multiplication of $\boldsymbol{M}_{LL}$ by $\tilde{\boldsymbol{I}}_n$ rearranges the columns of sub-matrix $\boldsymbol{M}_{LL}$ in a reversed order, sub-matrix $\tilde{\boldsymbol{I}}_n(\boldsymbol{M}_{LL}\tilde{\boldsymbol{I}}_n)$ makes the rows of sub-matrix $\boldsymbol{M}_{LL}\tilde{\boldsymbol{I}}_n$ be in a reversed order, and finally $\tilde{\boldsymbol{I}}_n \boldsymbol{M}_{LL} \tilde{\boldsymbol{I}}_n = \boldsymbol{M}_{RR}$ holds true.

For the axial vibration of the structure in Fig. 6.1, Eq. (o) in Example 6.1.1 suggests the following transform

$$\boldsymbol{u} = \boldsymbol{P} \begin{bmatrix} \boldsymbol{u}_s \\ \boldsymbol{u}_a \end{bmatrix} \equiv \begin{bmatrix} \boldsymbol{I}_n & \boldsymbol{I}_n \\ -\tilde{\boldsymbol{I}}_n & \tilde{\boldsymbol{I}}_n \end{bmatrix} \begin{bmatrix} \boldsymbol{u}_s \\ \boldsymbol{u}_a \end{bmatrix} \tag{6.1.6a}$$

For the bending vibration of the structure, the direction of local frame of coordinates requires that the transform yields

$$\boldsymbol{u} = \boldsymbol{P} \begin{bmatrix} \boldsymbol{u}_s \\ \boldsymbol{u}_a \end{bmatrix} \equiv \begin{bmatrix} \boldsymbol{I}_n & \boldsymbol{I}_n \\ \tilde{\boldsymbol{I}}_n & -\tilde{\boldsymbol{I}}_n \end{bmatrix} \begin{bmatrix} \boldsymbol{u}_s \\ \boldsymbol{u}_a \end{bmatrix} \tag{6.1.6b}$$

Let Eq. (6.1.2) be substituted by Eq. (6.1.6) and pre-multiplied by P^T, we derive, with the help of Eq. (6.1.4), the dynamic equations of the system in the following matrix form

$$\tilde{M}\begin{bmatrix} \ddot{u}_s(t) \\ \ddot{u}_a(t) \end{bmatrix} + \tilde{K}\begin{bmatrix} u_s(t) \\ u_a(t) \end{bmatrix} = 0 \tag{6.1.7}$$

where $\tilde{M}$ and $\tilde{K}$ are the diagonal block matrices as following

$$\begin{cases} \tilde{M} = \begin{bmatrix} I_n & I_n \\ \mp\tilde{I}_n & \pm\tilde{I}_n \end{bmatrix}^T \begin{bmatrix} M_{RR} & M_{RL} \\ M_{LR} & M_{LL} \end{bmatrix} \begin{bmatrix} I_n & I_n \\ \mp\tilde{I}_n & \pm\tilde{I}_n \end{bmatrix} \\ \quad = \begin{bmatrix} 2M_{RR} \mp M_{RL}\tilde{I}_n \mp \tilde{I}_n M_{LR} & 0 \\ 0 & 2M_{RR} \pm M_{RL}\tilde{I}_n \pm \tilde{I}_n M_{LR} \end{bmatrix} \\ \tilde{K} = \begin{bmatrix} I_n & I_n \\ \mp\tilde{I}_n & \pm\tilde{I}_n \end{bmatrix}^T \begin{bmatrix} K_{RR} & K_{RL} \\ K_{LR} & K_{LL} \end{bmatrix} \begin{bmatrix} I_n & I_n \\ \mp\tilde{I}_n & \pm\tilde{I}_n \end{bmatrix} \\ \quad = \begin{bmatrix} 2K_{RR} \mp K_{RL}\tilde{I}_n \mp \tilde{I}_n K_{LR} & 0 \\ 0 & 2K_{RR} \pm K_{RL}\tilde{I}_n \pm \tilde{I}_n K_{LR} \end{bmatrix} \end{cases} \tag{6.1.8}$$

To this end, the eigenvalue problem in Eq. (6.1.7) takes the following diagonal block matrix form

$$\tilde{Z}(\omega)\varphi = \begin{bmatrix} Z_{ss}(\omega) & 0 \\ 0 & Z_{aa}(\omega) \end{bmatrix} \begin{bmatrix} \varphi_s \\ \varphi_a \end{bmatrix} = 0 \tag{6.1.9}$$

where

$$\begin{cases} Z_{ss}(\omega) \equiv 2K_{RR} \mp K_{RL}\tilde{I}_n \mp \tilde{I}_n K_{LR} - \omega^2(2M_{RR} \mp M_{RL}\tilde{I}_n \mp \tilde{I}_n M_{LR}) \\ Z_{aa}(\omega) \equiv 2K_{RR} \pm K_{RL}\tilde{I}_n \pm \tilde{I}_n K_{LR} - \omega^2(2M_{RR} \pm M_{RL}\tilde{I}_n \pm \tilde{I}_n M_{LR}) \end{cases} \tag{6.1.10}$$

In Eq. (6.1.10), $Z_{ss}(\omega)$ and $Z_{aa}(\omega)$ are the displacement impedance sub-matrices associated with symmetric vibrations and anti-symmetric vibrations, respectively.

When a structure has two symmetry planes which are perpendicular to each other, it is easy to extend the transform in Eq. (6.1.6) and decouple the eigenvalue problem of the natural vibrations of the entire structure to a 4 × 4 diagonal block form as follows

$$Z(\omega)\varphi = \begin{bmatrix} Z_{ss}(\omega) & 0 & 0 & 0 \\ 0 & Z_{sa}(\omega) & 0 & 0 \\ 0 & 0 & Z_{as}(\omega) & 0 \\ 0 & 0 & 0 & Z_{aa}(\omega) \end{bmatrix} \begin{bmatrix} \varphi_{ss} \\ \varphi_{sa} \\ \varphi_{as} \\ \varphi_{aa} \end{bmatrix} = 0 \tag{6.1.11}$$

Fig. 6.3 The symmetry utilization of an equilateral triangle plate. **a** Using 1/2 triangle from the mirror symmetry; **b** Using 1/6 triangle from group theory

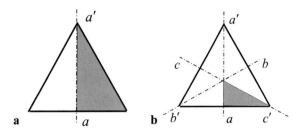

where subscript ss represents the symmetry with respect to two symmetry planes, subscripts sa and as mean the symmetry about one plane and the anti-symmetry about the other plane, and subscript aa means anti-symmetry about two planes. Accordingly, one only needs to establish the finite element modeling for 1/4 structure.

It is similar to construct the transform to decouple the dynamic equations of a structure with three perpendicular symmetry planes of mirror-symmetry. This way, one has an eigenvalue problem in the form of 8×8 diagonal block matrix, and the computation cost reduces to 1/8 compared with the analysis of entire structures.

Remark 6.1.3 The analysis becomes complicated when the structure has several symmetry planes not perpendicular to each other. For instance, the equilateral triangle in Fig. 6.3a has three symmetry planes, which are not perpendicular to each other. If one uses the above transform to simplify the computation of natural vibrations, only one of symmetry planes is available. Figure 6.3a shows, as an example, the choice of plane aa'. In this case, one can take the shaded region as a sub-structure and reduces the computational cost by 1/2. Yet, group theory in mathematics enables one to use a shaded region in Fig. 6.3b as a sub-structure and reduce the computational cost by 1/6.

6.1.2 Free Vibrations of Thin Rectangular Plates

Many elementary textbooks presented the natural vibrations of a thin rectangular plate with hinged boundaries, but some of them had problems in their solution procedure. This subsection assists readers to understand the solution procedure in detail.

6.1.2.1 Dynamic Equation

Figure 6.4 depicts a thin rectangular plate made of a linear homogeneous elastic material, or called a thin plate for short. The thin plate has a uniform thickness h, and the two adjacent edges of the plate has lengths a and b.

Now, we refer to a plane with equal distances to two surfaces of the plate as a *neutral plane*, and take one corner of the neutral plane as an origin to establish the

Fig. 6.4 The schematic of a
thin rectangular plate

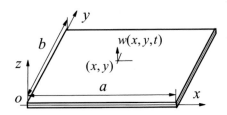

frame of coordinates $oxyz$ shown in Fig. 6.4. The study of linear vibrations of a thin plate is usually based on four assumptions proposed by G. Kirchihoff, a German physicist, to simplify the deformation description as follows.

(1)   The thickness variation of the plate is negligible during the plate deformation. This assumption leads to the neglection of normal strain of the plate, i.e., $\varepsilon_z = 0$ holds.

(2)   The bending deformation of the plate is much less than the plate thickness such that the neutral plane before and after deformation keeps plane and there is no tension and compression strains in the neutral plane.

(3)   A line perpendicular to the neutral plane before deformation keeps perpendicular to the deformed neutral plane during deformation. This assumption leads to negligible transverse shear strains, i.e., $\gamma_{xz} = \gamma_{yz} = 0$ holds true.

(4)   In vibration modeling, the rotary inertia of distributed mass of the plate is negligible.

Remark 6.1.4 In theory of elastic plates, a thin plate is called the *Kirchhoff plate* if it satisfies the above assumptions made by Kirchhoff. A comparison between the Kirchhoff assumptions of a thin plate and the Euler–Bernoulli assumptions for a slender beam indicates that they are quite similar and only applicable to the vibrations with small wave numbers. The major difference between a thin plate and a slender beam is that the deformations along side directions of the beam are not subject to any boundaries, while all edges of the plate are subject to boundary conditions. As such, their expressions of bending stiffness look similar, but have different meanings and units.

Under the assumptions of Kirchhoff, the dynamic deformation $w(x, y, t)$ of the neutral plane is used to describe the dynamics of a thin plate and to analyze the forces applied to an arbitrary rectangular infinitesimal element. Thus, we derive the dynamic equation of the thin plate as follows

$$\rho h \frac{\partial^2 w(x, y, t)}{\partial t^2} + D \nabla^4 w(x, y, t) = 0 \tag{6.1.12}$$

where ρ is the material density, ∇^4 is the *double Laplace operator* in the frame of Cartesian coordinates and D is the *bending stiffness of the plate*. They satisfy

$$\nabla^4 \equiv \nabla^2\nabla^2 \equiv \left(\frac{\partial^2}{\partial x^2} + \frac{\partial^2}{\partial y^2}\right)\left(\frac{\partial^2}{\partial x^2} + \frac{\partial^2}{\partial y^2}\right) = \frac{\partial^4}{\partial x^4} + 2\frac{\partial^4}{\partial x^2 \partial y^2} + \frac{\partial^4}{\partial y^4}$$

$$\tag{6.1.13}$$

$$D \equiv \frac{Eh^3}{12(1 - v^2)} \tag{6.1.14}$$

where E is Young's modulus and v is Poisson's ratio. For brevity, the term "thin" is not mentioned hereinafter since it is a fundamental requirement for the plate of concern.

6.1.2.2 General Solution of Free Vibration

Most elementary textbooks presented either the solution candidate guessed or an assumed solution for the natural vibrations of a rectangular plate. To assist readers to understand the study procedure, we discuss how to derive the solution of free vibration of a rectangular plate. Similar to the vibration solution of an E-B beam, let the solution candidate be in the following form of separated variables

$$w(x, y, t) = \varphi(x, y)\sin(\omega t + \theta) \tag{6.1.15}$$

Substituting Eq. (6.1.15) into Eq. (6.1.12) leads to a partial differential equation governing vibration shape $\varphi(x, y)$ as follows

$$\nabla^4\varphi(x, y) - \lambda^4\varphi(x, y) = 0, \quad \lambda \equiv \sqrt[4]{\frac{\rho h \omega^2}{D}} \tag{6.1.16}$$

Equation (6.1.13) enables one to recast the partial differential equation in Eq. (6.1.16) as

$$[\nabla^2\varphi(x, y) + \lambda^2\varphi(x, y)][\nabla^2\varphi(x, y) - \lambda^2\varphi(x, y)] = 0 \tag{6.1.17}$$

There follow two partial differential equations

$$\begin{cases} \nabla^2\varphi(x, y) + \lambda^2\varphi(x, y) = 0 \\ \nabla^2\varphi(x, y) + (i\lambda)^2\varphi(x, y) = 0 \end{cases} \tag{6.1.18}$$

Remark 6.1.5 At present, the general solution of Eq. (6.1.17) has not been available. Thus, we look for a kind of general solutions in terms of separated variables, that is, a linear combination of the solutions of two equations in Eq. (6.1.18), respectively.

Consider, as an example, the first equation in Eq. (6.1.18) and let its solution be in a form of separated variables, namely,

$$\tilde{\varphi}(x, y) = f(x)g(y) \tag{6.1.19}$$

Substituting Eq. (6.1.19) into the first equation in Eq. (6.1.18) leads to

$$\frac{1}{f(x)}\frac{d^2 f(x)}{dx^2} + \frac{1}{g(y)}\frac{d^2 g(x)}{dy^2} + \lambda^2 = 0 \tag{6.1.20}$$

Assume that there are two constants $\alpha > 0$ and $\beta > 0$ satisfying

$$\frac{1}{f(x)}\frac{d^2 f(x)}{dx^2} = -\alpha^2, \quad \frac{1}{g(y)}\frac{d^2 g(x)}{dy^2} = -\beta^2, \quad \alpha^2 + \beta^2 = \lambda^2 \tag{6.1.21}$$

Substituting the solutions of Eq. (6.1.21) into Eq. (6.1.18) leads to a solution of the first equation in Eq. (6.1.18), but the contrary assertion may not be true.

The two ordinary differential equations in Eq. (6.1.21) have the following general solutions

$$f(x) = b_1 \sin(\alpha x) + b_2 \cos(\alpha x), \quad g(y) = b_3 \sin(\beta y) + b_4 \cos(\beta y) \tag{6.1.22}$$

Then, the first equation in Eq. (6.1.18) has a kind of general solutions in terms of separated variables, namely,

$$\begin{aligned}
\tilde{\varphi}(x, y) &= f(x)g(y) = [b_1 \sin(\alpha x) + b_2 \cos(\alpha x)][b_3 \sin(\beta y) + b_4 \cos(\beta y)] \\
&= a_1 \sin(\alpha x) \sin(\beta y) + a_2 \sin(\alpha x) \cos(\beta y) \\
&\quad + a_3 \cos(\alpha x) \sin(\beta y) + a_4 \cos(\alpha x) \cos(\beta y)
\end{aligned} \tag{6.1.23}$$

Similarly, we derive a kind of general solutions for the second equation in Eq. (6.1.18) as follows

$$\begin{aligned}
\hat{\varphi}(x, y) &= a_5 \sinh(\overline{\alpha}x) \sinh(\overline{\beta}y) + a_6 \sinh(\overline{\alpha}x) \cosh(\overline{\beta}y) \\
&\quad + a_7 \cosh(\overline{\alpha}x) \sinh(\overline{\beta}y) + a_8 \cosh(\overline{\alpha}x) \cosh(\overline{\beta}y)
\end{aligned} \tag{6.1.24}$$

where $\overline{\alpha}^2 + \overline{\beta}^2 = \lambda^2$. Therefore, a kind of general solution, of Eq. (6.1.16), in terms of separated variables reads

$$\begin{aligned}
\varphi(x, y) &= \tilde{\varphi}(x, y) + \hat{\varphi}(x, y) \\
&= a_1 \sin(\alpha x) \sin(\beta y) + a_2 \sin(\alpha x) \cos(\beta y) \\
&\quad + a_3 \cos(\alpha x) \sin(\beta y) + a_4 \cos(\alpha x) \cos(\beta y) \\
&\quad + a_5 \sinh(\overline{\alpha}x) \sinh(\overline{\beta}y) + a_6 \sinh(\overline{\alpha}x) \cosh(\overline{\beta}y) \\
&\quad + a_7 \cosh(\overline{\alpha}x) \sinh(\overline{\beta}y) + a_8 \cosh(\overline{\alpha}x) \cosh(\overline{\beta}y)
\end{aligned} \tag{6.1.25}$$

Substituting Eq. (6.1.25) into Eq. (6.1.15) leads to a kind of general solution of free vibrations of a rectangular plate.

Remark 6.1.6 It is also possible to assume that the solution candidate of Eq. (6.1.18) is the summation of exponential functions. As such, the general solution assumed has 16 terms. Apart from eight terms in Eq. (6.1.25), each of other eight terms is the product of a triangle function and a hyperbolic-triangle function and can describe the vibration of a rectangular plate with a non-hinged boundary.

6.1.2.3 Natural Vibrations of a Rectangular Plate with Hinged Boundaries

Now, we use subscripts x and y to denote the corresponding partial derivatives and write the boundary conditions of the four hinged edges of the plate as follows

$$\begin{cases} w(0, y, t) = 0, & w_{xx}(0, y, t) = 0, & w(a, y, t) = 0, & w_{xx}(a, y, t) = 0, & y \in [0, b] \\ w(x, 0, t) = 0, & w_{yy}(x, 0, t) = 0, & w(x, b, t) = 0, & w_{yy}(x, b, t) = 0, & x \in [0, a] \end{cases}$$
(6.1.26)

There follow the boundary conditions for Eq. (6.1.25)

$$\begin{cases} \varphi(0, y) = 0, & \varphi_{xx}(0, y) = 0, & \varphi(a, y) = 0, & \varphi_{xx}(a, y) = 0, & y \in [0, b] \\ \varphi(x, 0) = 0, & \varphi_{yy}(x, 0) = 0, & \varphi(x, b) = 0, & \varphi_{yy}(x, b) = 0, & x \in [0, a] \end{cases}$$
(6.1.27)

We determine the constants in Eq. (6.1.25) and get an eigenvalue problem in the following four steps.

(1) Conditions $\varphi(0, 0) = 0$ and $\varphi_{xx}(0, 0) = 0$ lead to $a_4 = 0$ and $a_8 = 0$.
(2) Conditions $\varphi(0, y) = 0$ and $\varphi_{xx}(0, y) = 0$, $y \in [0, b]$ give $a_3 = 0$ and $a_7 = 0$.
(3) Conditions $\varphi(x, 0) = 0$ and $\varphi_{yy}(x, 0) = 0$, $x \in [0, a]$ give $a_2 = 0$ and $a_6 = 0$.
(4) Conditions $\varphi(a, b) = 0$ and $\varphi_{xx}(a, b) = 0$ give rise to $a_5 = 0$, and the characteristic equation for the non-trivial solutions as follows

$$\varphi(a, b) = a_1 \sin(\alpha a) \sin(\beta b) = 0$$
(6.1.28)

Solving Eq. (6.1.28) for eigenvalues leads to

$$\alpha_r = \frac{r\pi}{a}, \quad \beta_s = \frac{s\pi}{b}, \quad \alpha_r^2 + \beta_s^2 = \lambda_{rs}^2, \quad r, s = 1, 2, \ldots \tag{6.1.29}$$

From the definition of λ in Eq. (6.1.16), we derive the natural frequencies and the corresponding mode shapes of the rectangular plate with hinged boundaries as follows

$$\omega_{rs} = \lambda_{rs}^2 \sqrt{\frac{D}{\rho h}} = \pi^2 \sqrt{\frac{D}{\rho h}} \left(\frac{r^2}{a^2} + \frac{s^2}{b^2} \right), \quad r, s = 1, 2, \ldots \tag{6.1.30}$$

$$\varphi_{rs}(x, y) = \sin(\alpha_r x) \sin(\beta_s y) = \sin\left(\frac{r\pi x}{a} \right) \sin\left(\frac{s\pi y}{b} \right), \quad r, s = 1, 2, \ldots \tag{6.1.31}$$

The mode shapes here have mirror-symmetry about two symmetry planes $x = a/2$ and $y = b/2$, respectively.

Equation (6.1.29) indicates that the mode shape of the plate has two wave numbers α_r and β_s in directions x and y, respectively. The ranking of natural frequencies in Eq. (6.1.30), hence, requires two indices. If $a > b$ holds, Eq. (6.1.30) yields

$$\omega_{rs} = \frac{\pi^2}{b^2} \sqrt{\frac{D}{\rho h}} \left(\frac{b^2}{a^2} r^2 + s^2 \right), \quad r, s = 1, 2, \ldots \tag{6.1.32}$$

In this case, the contribution of index r to natural frequency ω_{rs} is less than that of index s. Hence, when the corresponding mode shape has two or more nodal lines in direction x, the first nodal line in direction y shows up. The contrary phenomenon appears if $a < b$ holds. When $a = b$, the rectangular plate becomes a square plate and the mode shapes exhibit more complicated rules, which will be discussed in next subsection.

6.1.3 Repeated Natural Frequencies of a Thin Rectangular Plate

To understand the modal properties with a pair of repeated frequencies step by step, this subsection begins with the natural vibrations of a square plate and then discusses the natural vibrations of a rectangular, but not square, plate.

6.1.3.1 Natural Modes of a Square Plate with Repeated Frequencies

For a square plate with hinged boundaries, substituting $b = a$ into Eqs. (6.1.30) and (6.1.31) gives the natural frequencies and modes shapes as follows

$$\omega_{rs} = \frac{\pi^2}{a^2} \sqrt{\frac{D}{\rho h}} (r^2 + s^2), \quad r, s = 1, 2, \dots \tag{6.1.33}$$

$$\varphi_{rs}(x, y) = \sin\left(\frac{r\pi x}{a}\right) \sin\left(\frac{s\pi y}{a}\right), \quad r, s = 1, 2, \dots \tag{6.1.34}$$

They indicate the following mode properties of a square plate with hinged boundary. That is, when $r \neq s$, a pair of repeated natural frequencies $\omega_{rs} = \omega_{sr}$ appears in Eq. (6.1.33) and there follows a pair mode shapes

$$\begin{cases} \varphi_{rs}(x, y) = \sin\left(\frac{r\pi x}{a}\right) \sin\left(\frac{s\pi y}{a}\right) \\ \\ \varphi_{sr}(x, y) = \sin\left(\frac{s\pi x}{a}\right) \sin\left(\frac{r\pi y}{a}\right), \quad r \neq s, \quad r, s = 1, 2, \dots \end{cases} \tag{6.1.35}$$

As $r \neq s$ holds, $\varphi_{rs}(x, y)$ and $\varphi_{sr}(x, y)$ have different wave numbers in directions x and y such that they are two linearly independent triangle functions. In view of linear algebra, the subspace of the two mode shapes is two-dimensional, and $\varphi_{rs}(x, y)$ and $\varphi_{sr}(x, y)$ serve as two base functions for the subspace. As such, define a linear combination of $\varphi_{rs}(x, y)$ and $\varphi_{sr}(x, y)$ as

$$\psi(x, y) = c_1 \varphi_{rs}(x, y) + c_2 \varphi_{sr}(x, y) \tag{6.1.36}$$

It is also a mode shape for repeated frequencies $\omega_{rs} = \omega_{sr}$.

Given $a = 1$, $r = 1$ and $s = 2$, Fig. 6.5 presents the three-dimensional contours of the two mode shapes in Eq. (6.1.35). The mode shape in Fig. 6.5a has a nodal line $y = a/2 = 0.5$, while the mode shape in Fig. 6.5b has a nodal line $x = a/2 = 0.5$. That is, Fig. 6.5a after a rotation by $\pi/2$ is identical to Fig. 6.5b.

Figure 6.6 shows two linear combinations of the mode shapes with repeated frequencies. Now, the nodal line in Fig. 6.6a is the diagonal $y = 1 - x$ of the square plate while the nodal line in Fig. 6.6b is not a straight line. The above linear combinations have an infinite number of options. Thus, the nodal lines of a mode shape with repeated frequencies do not have a unique pattern. To clearly demonstrate the nodal lines of a mode shape, Fig. 6.7 is used to show the two mode shapes in

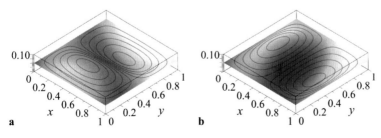

Fig. 6.5 Two mode shapes with repeated frequencies of a square plate with hinged boundaries ($a = 1$, $r = 1$, $s = 2$). **a** $\varphi_{12}(x, y) = \sin(\pi x/a) \sin(2\pi y/a)$; **b** $\varphi_{21}(x, y) = \sin(2\pi x/a) \sin(\pi y/a)$

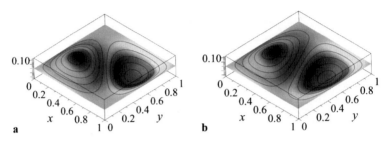

Fig. 6.6 Two linear combinations of mode shapes of repeated frequencies of a square plate with hinged boundaries ($a = 1$, $r = 1$, $s = 2$). **a** $c_1 = 1.0$, $c_2 = 1.0$; **b** $c_1 = 1.0$, $c_2 = 0.5$

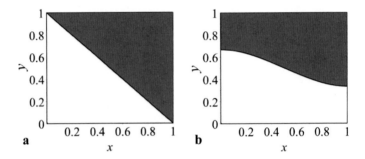

Fig. 6.7 Two linear combinations of mode shapes of repeated frequencies of a square plate with hinged boundaries ($a = 1$, $r = 1$, $s = 2$). **a** $c_1 = 1.0$, $c_2 = 1.0$; **b** $c_1 = 1.0$, $c_2 = 0.5$

Fig. 6.6. The gray region represents negative amplitude and the white region means positive amplitude, while their interface gives a nodal line. In practice, Fig. 6.7 is called the *figure of nodal lines*.

The nodal lines of mode shapes in Eq. (6.1.35) are parallel to the edges of the square plate and look mirror-symmetric, but those of the linear combination of two mode shapes may not be so.

To check the linear combinations of mode shapes with repeated frequencies, we take $c_1 = 1.0$ and some values for $c_2 \in [-1, 1]$. Figure 6.8 shows the nodal lines of six linear combinations of mode shapes in Eq. (6.1.36) when $r = 1$ and $s = 3$. Each sub-figure in Fig. 6.8 exhibits the mirror-symmetry about $x = a/2$ and $y = b/2$.

Figure 6.9 gives the nodal line figures of six linear combinations of mode shapes in Eq. (6.1.36) for $r = 2$ and $s = 3$. Now, the nodal lines do not show the mirror-symmetry about $x = a/2$ and $y = b/2$. It is interesting to observe that the second row of sub-figures, after a rotation by an angle $\pi/2$, is identical to the first row of sub-figures. That is, these mode shapes have rotation-symmetry.

To understand the symmetric properties of the nodal lines in Figs. 6.8 and 6.9, it is better to notice that a square plate has two important kinds of symmetry as follows.

(1) Mirror-symmetry: A square plate satisfies four operations of mirror-symmetry in total. The first is the mirror-symmetric reflection about section $x = a/2$, the

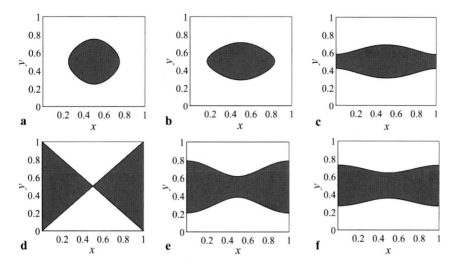

Fig. 6.8 Six linear combinations of two mode shapes with repeated frequencies of a square plate ($a = 1$, $r = 1$, $s = 3$). **a** $c_1 = 1.0$, $c_2 = 1.0$; **b** $c_1 = 1.0$, $c_2 = 0.5$; **c** $c_1 = 1.0$, $c_2 = 0.25$; **d** $c_1 = 1.0$, $c_2 = -1.0$; **e** $c_1 = 1.0$, $c_2 = -0.5$; **f** $c_1 = 1.0$, $c_2 = -0.25$

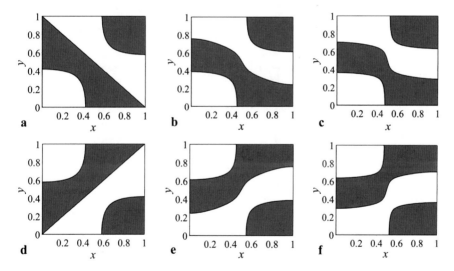

Fig. 6.9 Six linear combinations of two mode shapes with repeated frequencies of a square plate ($a = 1$, $r = 2$, $s = 3$). **a** $c_1 = 1.0$, $c_2 = 1.0$; **b** $c_1 = 1.0$, $c_2 = 0.5$; **c** $c_1 = 1.0$, $c_2 = 0.25$; **d** $c_1 = 1.0$, $c_2 = -1.0$; **e** $c_1 = 1.0$, $c_2 = -0.5$; **f** $c_1 = 1.0$, $c_2 = -0.25$

second is about section $y = b/2 = a/2$, the third is about diagonal section $y = x$, and the forth is about diagonal section $y = a - x$.

(2) Cyclosymmetry: To describe the symmetry, a straight line, defined as the *central axis*, is assumed to pass through the center of the plate and be parallel to axis z in Fig. 6.4. Then, a square plate yields four operations of cyclosymmetry. That is, the plate is exactly identical via the counter-clockwise rotations by angles $\pi/2$, π, $3\pi/2$ and 2π, respectively.

This subsection only deals with the mirror-symmetry, while Sects. 6.2 and 6.3 will present studies on the vibrations of cyclosymmetric structures.

6.1.3.2 Natural Modes of a Rectangular Plate with Repeated Frequencies

The present books of vibration mechanics and theory of plates have not dealt with the natural vibrations of a rectangular plate with repeated frequencies. In fact, the rectangular plate also exhibits abundant natural modes with repeated frequencies.

Now, we consider a rectangular plate under the condition $0 < b < a$. If a natural vibration of the rectangular plate has a pair of repeated frequencies $\omega_{rs} = \omega_{ij}$, thus, Eq. (6.1.30) leads to

$$\omega_{rs} = \pi^2 \sqrt{\frac{D}{\rho h}} \left(\frac{r^2}{a^2} + \frac{s^2}{b^2} \right) = \pi^2 \sqrt{\frac{D}{\rho h}} \left(\frac{i^2}{a^2} + \frac{j^2}{b^2} \right) = \omega_{ij} \tag{6.1.37}$$

Equation (6.1.37) is equivalent to that the following algebraic equation under condition $0 < b < a$ has positive integer solutions (r, s) and (i, j), namely,

$$\frac{r^2}{a^2} + \frac{s^2}{b^2} = \frac{i^2}{a^2} + \frac{j^2}{b^2} \tag{6.1.38}$$

In mathematics, this is a complicated problem of integer solutions to an undetermined equation. When the low-order natural vibrations of a rectangular plate are concerned with, the problem becomes simple. The following discussions focus on two case studies.

(1) From repeated frequencies to length ratio

In this case study, one looks for the length ratio a/b from a pair of repeated natural frequencies $\omega_{rs} = \omega_{ij}$. For this purpose, one recasts Eq. (6.1.38) as

$$\frac{a}{b} = \sqrt{\frac{i^2 - r^2}{s^2 - j^2}} \tag{6.1.39}$$

From condition of $0 < b < a$, the frequency order in Eq. (6.1.39) yields

$$s \neq j, \quad r \neq i, \quad s^2 - j^2 < i^2 - r^2 \tag{6.1.40}$$

Equation (6.1.40) enables one to select frequency order and determine ratio a/b from Eq. (6.1.39). Given the natural frequencies and $\sqrt{D/(\rho h)}$, one can also determine lengths a and b for the plate from Eq. (6.1.37).

Example 6.1.2 Given $\sqrt{D/(\rho h)}$ for a rectangular plate with hinged boundaries, find the ratio a/b such that the repeated natural frequencies have the minimal wave number.

In view of the minimal wave number, select $j = 1$ first. Now, $s = 2$ is the minimal s satisfying the first inequality in Eq. (6.1.40). From the second and third inequalities in Eq. (6.1.40), $r = 1$ and $i = 3$ are the minimal solutions. Thus, a pair of repeated natural frequencies with the minimal wave number are $\omega_{12} = \omega_{31}$. Substituting the above frequency orders into Eq. (6.1.39) leads to

$$\frac{a}{b} = \sqrt{\frac{8}{3}} \tag{a}$$

To be more specific, consider a rectangular steel plate with following parameters

$$b = 1.0 \text{ m}, \quad h = 0.005 \text{ m}, \quad \rho = 7800.0 \text{ kg/m}, \quad E = 210.0 \text{ GPa}, \quad v = 0.28. \tag{b}$$

Substituting Eqs. (a), (b) and the above frequency orders into Eq. (6.1.37), we get a pair of repeated natural frequencies as follows

$$f_{12} = f_{31} \equiv \frac{\omega_{31}}{2\pi} = \frac{35\pi}{16} \sqrt{\frac{D}{\rho h}} \approx 53.61 \text{ Hz} \tag{c}$$

From Eq. (6.1.31), the corresponding mode shapes are

$$\varphi_{12}(x, y) = \sin\left(\frac{\pi x}{\sqrt{8/3}}\right) \sin(2\pi y), \quad \varphi_{31}(x, y) = \sin\left(\frac{3\pi x}{\sqrt{8/3}}\right) \sin(\pi y) \tag{d}$$

In the subspace expanded by the above two mode shapes, any mode shape yields the following linear combination

$$\psi(x, y) = c_1 \varphi_{12}(x, y) + c_2 \varphi_{31}(x, y) \tag{e}$$

Figure 6.10 presents the mode shapes with a pair of repeated frequencies for $c_1 = 1.0$ and $c_2 \in [0, +\infty)$. Here, Fig. 6.10f is equivalent to the case when $c_1 = 0.0$ and $c_2 = 1$. Figure 6.10 shows how mode shape $\psi(x, y)$ evolutes from $\varphi_{12}(x, y)$ to $\varphi_{31}(x, y)$, exhibiting one nodal line parallel to the long plate edges to two nodal lines parallel to the short plate edges.

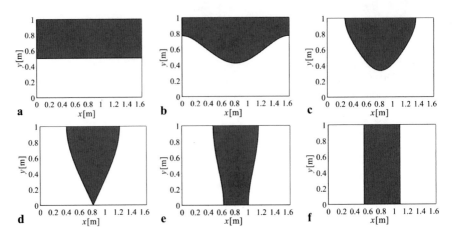

Fig. 6.10 Six mode shapes with a pair of repeated frequencies and the minimal wave number of a rectangular plate ($a/b = \sqrt{8/3}$, $\omega_{12} = \omega_{31}$). **a** $c_1 = 1.0$, $c_2 = 0.0$; **b** $c_1 = 1.0$, $c_2 = 0.5$; **c** $c_1 = 1.0$, $c_2 = 1.0$; **d** $c_1 = 1.0$, $c_2 = 2.0$; **e** $c_1 = 1.0$, $c_2 = 4.0$; **f** $c_1 = 1.0$, $c_2 \to +\infty$

Remark 6.1.7 The author pointed out in an elementary textbook that when the ratio a/b is a rational number, the rectangular plate with hinged boundaries exhibits repeated natural frequencies[4]. Example 6.1.2 demonstrates that this condition is not necessary. Although it is impossible to make a plate with an irrational ratio a/b, one can replace the irrational ratio $a/b = \sqrt{8/3} \approx 1.6322993$ with a feasible ratio $a/b = 1.632$ and makes a rectangular plate with a pair of close natural frequencies $f_{12} = 53.62$ Hz and $f_{31} = 53.66$ Hz. The computation results show that the corresponding mode shapes almost coincide with those in Fig. 6.10. Such a rectangular plate under a small disturbance also exhibits complicated evolution results. Section 6.1.4 will discuss this case in detail.

(2)   From length ratio to repeated frequencies

In this case study, a rational length ratio a/b is given, where a and b are integers, one determines the natural frequency orders in a lower frequency band such that $\omega_{rs} = \omega_{ij}$ holds. For this purpose, one first recasts Eq. (6.1.38) as

$$r^2 = i^2 + \frac{a^2}{b^2}(j^2 - s^2) = i^2 + \frac{a^2}{b^2}(j + s)(j - s) \qquad (6.1.41)$$

In order that the natural frequency orders are low enough, it is reasonable to take a factor $(j + s) = b^2$ in Eq. (6.1.41) and derive

$$r^2 = i^2 + a^2(j - s) \qquad (6.1.42)$$

_____
[4] Hu H Y. Mechanical Vibrations and Shocks. Beijing: Press of Aviation Industry, 1998: 178 (in Chinese).

Under the condition of $(j + s) = b^2$, one can select $(j - s)$ and calculate $r = \sqrt{i^2 + a^2(j - s)}$ for a given i in a certain range, then the integer results of r serve as the natural frequency orders expected. Of course, it is also feasible to take a factor $j^2 - s^2 = lb^2$, where l is a positive integrer, and get more integer solutions of Eq. (6.1.41), but the solution procedure is usually complicated.

Example 6.1.3 Given the length ratio $a/b = 4/3$ for a rectangular plate, determine the natural frequencies with both wave numbers in directions x and y less than 10.

Substituting ratio $a/b = 4/3$ into Eq. (6.1.41) leads to

$$r^2 = i^2 + \frac{16}{9}(j + s)(j - s) \tag{a}$$

At first, we select $j + s = 9$ and get eight pairs of candidate solutions $(s, j) = (1, 8)$, $(2, 7)$, $\ldots$, $(8, 1)$, where the last four pairs are the same as the first four pairs after an exchange of subscripts s and j. Now, we check the first four pairs and calculate the following results

$$\sqrt{i^2 + 16(j - s)}, \quad i = 1, 2, \ldots 10, \quad (s, j) = (1, 8), (2, 7), (3, 6), (4, 5) \tag{b}$$

This procedure leads to four pairs of natural frequency orders, which have the wave numbers in both directions less than 10, as follows

$$\begin{cases} (s, j) = (2, 7): & (r, i) = (9, 1) \\ (s, j) = (3, 6): & (r, i) = (7, 1), \quad (8, 4) \\ (s, j) = (4, 5): & (r, i) = (5, 3) \end{cases} \tag{c}$$

In view of the order of row and column in Eq. (c), we have the following four pairs of repeated natural frequencies

$$\omega_{92} = \omega_{17}, \quad \omega_{73} = \omega_{16}, \quad \omega_{83} = \omega_{46}, \quad \omega_{54} = \omega_{35} \tag{d}$$

Next, if one selects the other option, such as $j - s = 9$ or $j + s = 18$, the integer solution yields $r > 10$. It is interesting that if one takes $j^2 - s^2 = 45$, it is feasible to get the following integer solution and the repeated natural frequencies with both wave numbers less than 10 as follows

$$(s, j) = (6, 9): \quad (r, i) = (9, 1) \quad \Rightarrow \quad \omega_{96} = \omega_{19} \tag{e}$$

However, the repeated natural frequencies are higher than those in Eq. (d).

Now, we focus on the four pairs of repeated natural frequencies in Eq. (d) and check the linear combination of two mode shapes with ascending values of natural frequency as follows

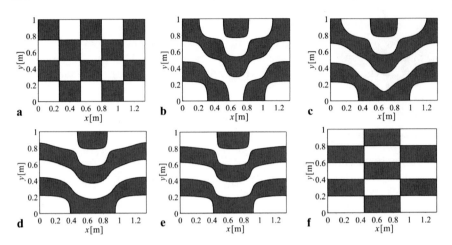

Fig. 6.11 Six mode shapes with a pair of repeated frequencies of a rectangular plate ($a/b = 4/3$, $f_{54} = f_{35} = 368.40$ Hz). **a** $c_1 = 1.0$, $c_2 = 0.0$; **b** $c_1 = 1.0$, $c_2 = 0.5$; **c** $c_1 = 1.0$, $c_2 = 1.0$; **d** $c_1 = 1.0$, $c_2 = 2.0$; **e** $c_1 = 1.0$, $c_2 = 4.0$; **f** $c_1 = 1.0$, $c_2 \to +\infty$

$$\psi(x, y) = c_1 \varphi_{rs}(x, y) + c_2 \varphi_{ij}(x, y) \tag{f}$$

We begin with the mode shapes corresponding to repeated natural frequencies $\omega_{54} = \omega_{35}$, or equivalently, $f_{54} = f_{35} = 368.40$ Hz. Figure 6.11 depicts the evolution of $\psi(x, y)$ from $\varphi_{54}(x, y)$ to $\varphi_{35}(x, y)$. The number of nodal lines in Fig. 6.11a is seven, and reduces to six in Fig. 6.11c, and keeps as six till Fig. 6.11f.

Next, we examine the mode shapes of $\omega_{73} = \omega_{16}$, or equivalently $f_{73} = f_{16} = 448.05$ Hz. Figure 6.12 presents the evolution of $\psi(x, y)$ from $\varphi_{73}(x, y)$ to $\varphi_{16}(x, y)$, exhibiting the number of nodal lines from eight in Fig. 6.12a to five in Fig. 6.12c, and then keeping five till Fig. 6.12f.

Afterwards, we study the mode shapes of $\omega_{83} = \omega_{46}$, i.e., $f_{83} = f_{46} = 551.45$ Hz. Figure 6.13 shows the evolution of $\psi(x, y)$ from $\varphi_{83}(x, y)$ to $\varphi_{46}(x, y)$, exhibiting the number of nodal lines from nine in Fig. 6.13a to eleven in Fig. 6.13c, and then to eight from Fig. 6.13d to Fig. 6.13f.

Finally, we check the mode shapes of $\omega_{92} = \omega_{17}$, i.e., $f_{92} = f_{17} = 607.36$ Hz. Figure 6.14 presents the evolution of $\psi(x, y)$ from $\varphi_{92}(x, y)$ to $\varphi_{17}(x, y)$, showing the number of nodal lines from nine in Fig. 6.14a to four in Fig. 6.14b, and then to six in Fig. 6.14c, and back to four in Fig. 6.14d, and finally to six in Fig. 6.14f.

The further examination of repeated natural frequencies in Eq. (e) exhibits more complicated mode shapes. Nevertheless, the dynamic model of Kirchhoff plate does not work well for a natural vibration with high wave number and such a natural vibration is usually not important enough in engineering.

Remark 6.1.8 Compared with a square plate, the rectangular plate has the sparse distribution of repeated natural frequencies in frequency domain. The reason is that the rectangular plate has only two planes $x = a/2$ and $y = b/2$ for mirror-symmetry,

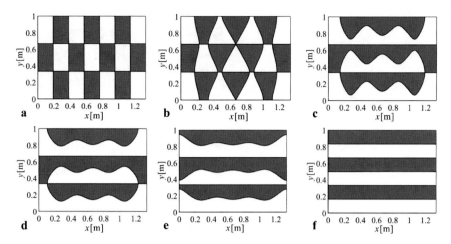

Fig. 6.12 Six mode shapes with a pair of repeated frequencies of a rectangular plate ($a/b =$ 4/3, $f_{73} = f_{16} = 448.05$ Hz). **a** $c_1 = 1.0$, $c_2 = 0.0$; **b** $c_1 = 1.0$, $c_2 = 0.5$; **c** $c_1 = 1.0$, $c_2 =$ 1.0; **d** $c_1 = 1.0$, $c_2 = 2.0$; **e** $c_1 = 1.0$, $c_2 = 4.0$; **f** $c_1 = 1.0$, $c_2 \to +\infty$

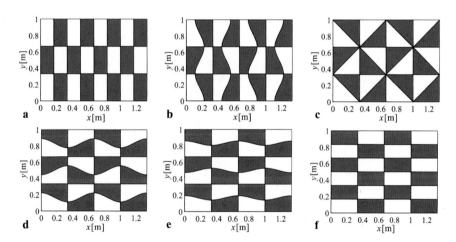

Fig. 6.13 Six mode shapes with a pair of repeated frequencies of a rectangular plate ($a/b =$ 4/3, $f_{83} = f_{46} = 551.45$ Hz). **a** $c_1 = 1.0$, $c_2 = 0.0$; **b** $c_1 = 1.0$, $c_2 = 0.5$; **c** $c_1 = 1.0$, $c_2 =$ 1.0; **d** $c_1 = 1.0$, $c_2 = 2.0$; **e** $c_1 = 1.0$, $c_2 = 4.0$; **f** $c_1 = 1.0$, $c_2 \to +\infty$

and two rotation angles π and 2π for cyclosymmetry. In mathematics, the group of rectangular plates is a subgroup of square plates.

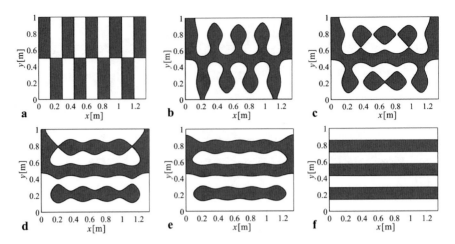

Fig. 6.14 Six mode shapes with a pair of repeated frequencies of a rectangular plate ($a/b = 4/3$, $f_{92} = f_{17} = 607.36$ Hz). **a** $c_1 = 1.0$, $c_2 = 0.0$; **b** $c_1 = 1.0$, $c_2 = 0.5$; **c** $c_1 = 1.0$, $c_2 = 1.0$; **d** $c_1 = 1.0$, $c_2 = 2.0$; **e** $c_1 = 1.0$, $c_2 = 4.0$; **f** $c_1 = 1.0$, $c_2 \to +\infty$

6.1.4 Close Natural Frequencies of a Thin Rectangular Plate

Section 2.2.5 briefly discussed the influence of a structural disturbance, that is, symmetry-breaking, on the natural vibrations of repeated frequencies of a simple symmetric structure. This subsection examines the modal behaviors of a pair of close natural frequencies of a rectangular plate due to some structural disturbances, such as the plate lengh and an attached lumped mass, in order to address Problem 5A in part.

6.1.4.1 Disturbance of Edge Length of a Square Plate

At first, we study a square plate under a small structural disturbance due to measurement and manufacturing errors and call it *a quasi-square plate* for short. Let the quasi-square plate be a rectangular plate with two pairs of slightly different lengths, say, $b = a(1 + \varepsilon)$, where $0 \le |\varepsilon| << 1$ is a small dimensionless parameter. Thus, Eq. (6.1.30) leads to two dimensionless frequencies as follows

$$\overline{\omega}_{rs} \equiv \frac{\omega_{rs}a^2}{\pi^2\sqrt{D/\rho h}} = r^2 + \frac{s^2}{(1+\varepsilon)^2}, \quad \overline{\omega}_{sr} \equiv \frac{\omega_{sr}a^2}{\pi^2\sqrt{D/\rho h}} = s^2 + \frac{r^2}{(1+\varepsilon)^2} \tag{6.1.43}$$

They decrease with an increase of ε. For a sufficiently small ε, Eq. (6.1.43) can be recast as

$$\overline{\omega}_{rs} \approx r^2 + s^2 - 2s^2\varepsilon, \quad \overline{\omega}_{sr} \approx s^2 + r^2 - 2r^2\varepsilon \tag{6.1.44}$$

with

$$\overline{\omega}_{rs} - \overline{\omega}_{sr} \approx 2(r^2 - s^2)\varepsilon \qquad (6.1.45)$$

As $r \neq s$ holds true, Eq. (6.1.44) gives the decreasing rate of $\overline{\omega}_{rs}$ and $\overline{\omega}_{sr}$ with an increase of ε. When $r < s$ holds, $\overline{\omega}_{rs}$ decreases faster than $\overline{\omega}_{sr}$. Otherwise, $\overline{\omega}_{rs}$ goes down more slowly than $\overline{\omega}_{sr}$. Equation (6.1.45) indicates that the two frequencies exchange their rankings with a positive or negative variation of ε.

Next, we check the mode shapes corresponding to the above natural frequencies. Given $|\varepsilon| << 1/(s\pi)$, the two mode shapes can be approximated as

$$\begin{cases} \varphi_{rs}(x, y) = \sin\left(\dfrac{r\pi x}{a}\right)\sin\left[\dfrac{s\pi y}{a(1+\varepsilon)}\right] \approx \sin\left(\dfrac{r\pi x}{a}\right)\sin\left[\dfrac{(1-\varepsilon)s\pi y}{a}\right] \\[2mm] \qquad \approx \sin\left(\dfrac{r\pi x}{a}\right)\sin\left(\dfrac{s\pi y}{a}\right)\left[1 - \dfrac{\varepsilon s\pi y}{a}\cot\left(\dfrac{s\pi y}{a}\right)\right] \\[2mm] \varphi_{sr}(x, y) \approx \sin\left(\dfrac{s\pi x}{a}\right)\sin\left(\dfrac{r\pi y}{a}\right)\left[1 - \dfrac{\varepsilon r\pi y}{a}\cot\left(\dfrac{r\pi y}{a}\right)\right] \end{cases} \qquad (6.1.46)$$

where the terms in the last two pairs of brackets oscillate with a variation of y, and their quantities are $s\pi|\varepsilon|$ and $r\pi|\varepsilon|$, respectively. Hence, a small length variation of a pair of edges of a square plate has a relatively small influence on the natural frequencies of low order, but a large influence on those of high order.

Example 6.1.4 Given a square plate with length a, let the length of a pair of edges in direction y change to $a(1+\varepsilon)$. In the range of $|\varepsilon| \leq 0.01$, discuss the variation of natural vibrations for $r = 1$ and $s = 2$.

Following Eq. (6.1.43), Fig. 6.15a presents the dimensionless natural frequencies with respect to ε. As $s = 2 > 1 = r$ holds, one can predict from Eq. (6.1.44) that $\overline{\omega}_{12}$ goes down fast and $\overline{\omega}_{21}$ goes down slowly. Given $|\varepsilon| \leq 0.01$, the ranking of $\overline{\omega}_{12}$

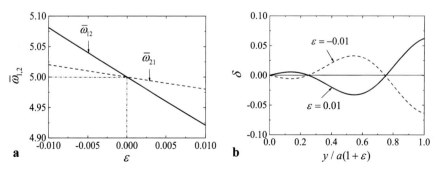

Fig. 6.15 The natural modes of a pair of repeated frequencies of a square plate subject to a length variation of a pair of edges ($r = 1$, $s = 2$, $a = 1$). **a** The variation of dimensionless frequencies; **b** The variation of mode shape difference

and $\overline{\omega}_{21}$ changes in a neighborhood of $\varepsilon = 0$ as shown in Fig. 6.15a, exhibiting a modal bifurcation discussed in Sect. 2.2.5.

Equation (6.1.46) depicts the difference between the mode shape $\varphi_{12}(x, y)$ in Eq. (6.1.35) and the mode shape $\varphi_{12}(x, y)$ before approximation in direction y in Fig. 6.15b for $|\varepsilon| \leq 0.01$. As shown in the figure, their difference is at the quantity level of $2\pi|\varepsilon| \approx 0.06$. As expected in Eq. (6.1.46), the length variation of a square plate has a small influence on the mode shapes of low order.

6.1.4.2 Disturbance of a Lumped Mass to a Square Plate

Now we study the vibration test of a square plate to address Problem 5A. In a vibration test, it is often necessary to install an accelerometer or a connection mount of exciter on the square plate such that a *local lumped mass* is attached to the plate and produces a mass disturbance to the plate. Let the attached mass be much smaller than the mass of the plate, that is, $m << \rho a^2 h$, and the attachment coordinates be (x_0, y_0). In this case, Ritz method serves as a useful tool to check the influence of a mass disturbance on the natural modes with repeated frequencies of a square plate.

As the attached mass is very small, the mode shapes of repeated frequencies are used as two base functions and express the dynamic deformation of the plate subject to the mass disturbance as

$$w(x, y, t) \approx \varphi_{rs}(x, y)q_{rs}(t) + \varphi_{sr}(x, y)q_{sr}(t) \qquad (6.1.47)$$

Remark 6.1.9 will give the possibility if one intends to account for the contributions of other modes.

From Eq. (6.1.47), the kinetic energy of the plate under the mass disturbance reads

$$\begin{aligned} T &= \frac{1}{2}\int_0^a\int_0^a \rho h\left[\frac{\partial w(x, y, t)}{\partial t}\right]^2 dxdy + \frac{m}{2}\left[\frac{\partial w(x_0, y_0, t)}{\partial t}\right]^2 \\ &= \frac{\rho h}{2}\int_0^a\int_0^a [\varphi_{rs}(x, y)\dot{q}_{rs}(t) + \varphi_{sr}(x, y)\dot{q}_{sr}(t)]^2 dxdy \\ &\quad + \frac{m}{2}[\varphi_{rs}(x_0, y_0)\dot{q}_{rs}(t) + \varphi_{sr}(x_0, y_0)\dot{q}_{sr}(t)]^2 \end{aligned} \qquad (6.1.48)$$

From the orthogonality of the mode shapes, it is easy to simplify Eq. (6.1.48) as

$$T = \frac{M_{rs}}{2}\dot{q}_{rs}^2(t) + \frac{M_{sr}}{2}\dot{q}_{sr}^2(t) + \frac{m}{2}[\varphi_{rs}(x_0, y_0)\dot{q}_{rs}(t) + \varphi_{sr}(x_0, y_0)\dot{q}_{sr}(t)]^2 \qquad (6.1.49)$$

where M_{rs} and M_{sr} are the modal mass coefficients associated with the above mode shapes of repeated frequencies, namely,

$$M_{rs} \equiv \rho h \int_0^a \int_0^a \varphi_{rs}^2(x, y) dx dy, \quad M_{sr} \equiv \rho h \int_0^a \int_0^a \varphi_{sr}^2(x, y) dx dy \quad (6.1.50)$$

Substituting Eq. (6.1.47) into the expression of the potential energy of the square plate[5] and using the orthogonal relation of the mode shapes again, we derive the potential energy of the square plate as follows

$$V = \frac{D}{2} \int_0^a \int_0^a \left[\frac{\partial^2 w(x, t)}{\partial x^2} + \frac{\partial^2 w(x, t)}{\partial y^2} \right]^2 dx dy$$
$$+ \frac{(1 - v)D}{2} \int_0^a \int_0^a \left[\frac{\partial^2 w(x, t)}{\partial x \partial y} \frac{\partial^2 w(x, t)}{\partial x \partial y} - \frac{\partial^2 w(x, t)}{\partial x^2} \frac{\partial^2 w(x, t)}{\partial y^2} \right] dx dy$$
$$= \frac{K_{rs}}{2} q_{rs}^2(t) + \frac{K_{sr}}{2} q_{sr}^2(t) \quad (6.1.51)$$

where K_{rs} and K_{sr} are the modal stiffness coefficients associated with the above mode shapes. Though the detailed expression in Eq. (6.1.51) looks complicated, it is an easy work to get it on software Maple. Besides, the computation on Maple does not need to check the modal orthogonality and the ratio of a cross entry to a direct entry arrives at the quantity level of 10^{-14}.

Substituting the above kinetic energy and potential energy into Lagrange's equation of the second kind leads to the approximate dynamic equations of the square plate under the mass disturbance in a matrix form, namely,

$$(M + \Delta M)\ddot{q}(t) + Kq(t) = 0 \quad (6.1.52)$$

where

$$\begin{cases} K \equiv \begin{bmatrix} K_{rs} & 0 \\ 0 & K_{sr} \end{bmatrix}, \quad M \equiv \begin{bmatrix} M_{rs} & 0 \\ 0 & M_{sr} \end{bmatrix} \\ \Delta M \equiv m \begin{bmatrix} \varphi_{rs}^2(x_0, y_0) & \varphi_{rs}(x_0, y_0)\varphi_{sr}(x_0, y_0) \\ \varphi_{rs}(x_0, y_0)\varphi_{sr}(x_0, y_0) & \varphi_{sr}^2(x_0, y_0) \end{bmatrix} \end{cases} \quad (6.1.53)$$

Solving the following eigenvalue problem

$$\left[K - \tilde{\omega}^2(M + \Delta M) \right] \tilde{q} = 0 \quad (6.1.54)$$

we obtain two natural frequencies $\tilde{\omega}_{rs}$ and $\tilde{\omega}_{sr}$ for the square plate under the mass disturbance and the corresponding eigenvectors

$$\tilde{q}_{rs} \equiv \begin{bmatrix} \tilde{q}_{11} & \tilde{q}_{21} \end{bmatrix}^T, \quad \tilde{q}_{sr} \equiv \begin{bmatrix} \tilde{q}_{12} & \tilde{q}_{22} \end{bmatrix}^T \quad (6.1.55)$$

---

[5] Rao S S (2007). Vibration of Continuous Systems. Hoboken: John Wiley & Sons Inc., 459.

From Eq. (6.1.47), the approximate mode shapes of the square plate under the mass disturbance are as following

$$\begin{cases} \tilde{\varphi}_{rs}(x, y) = \tilde{q}_{11}\varphi_{rs}(x, y) + \tilde{q}_{21}\varphi_{sr}(x, y) \\ \tilde{\varphi}_{sr}(x, y) = \tilde{q}_{12}\varphi_{rs}(x, y) + \tilde{q}_{22}\varphi_{sr}(x, y) \end{cases} \tag{6.1.56}$$

Remark 6.1.9 The modal mass matrix and modal stiffness matrix of the square plate before the mass disturbance are diagonal. The mass disturbance makes the modal mass matrix have inertial coupling in matrix ΔM, which clearly exhibits the influence of attached mass and attachment location on the problem. To account for the effects of more natural modes, one only requires to extend the modal mass matrix and modal stiffness matrix, while an entry in ΔM is proportional to the product of the values of two mode shapes at the attachment location.

Example 6.1.5 Given a square plate with the following parameters

$$a = 1.0\,\text{m}, \quad h = 0.005\,\text{m}, \quad \rho = 7800.0\,\text{kg/m}, \quad E = 210.0\,\text{GPa}, \quad \nu = 0.28 \tag{a}$$

when $r = 2$, $s = 3$ and $r = 3$, $s = 2$, the plate has a pair of natural modes with repeated frequencies. Discuss the variation of the natural modes due to the disturbance of an attached mass.

From Eqs. (6.1.33) and (6.1.34), we get the repeated natural frequencies and their mode shapes as follows

$$f_{23} = f_{32} = \frac{13}{2\pi a^2}\sqrt{\frac{D}{\rho h}} = 159.31\,\text{Hz} \tag{b}$$

$$\varphi_{23}(x, y) = \sin\left(\frac{2\pi x}{a}\right)\sin\left(\frac{3\pi y}{a}\right), \quad \varphi_{32}(x, y) = \sin\left(\frac{3\pi x}{a}\right)\sin\left(\frac{2\pi y}{a}\right) \tag{c}$$

Using sofeware Maple, we obtain the modal mass and stiffness matrices as following

$$\begin{cases} M = \begin{bmatrix} M_{12} & 0 \\ 0 & M_{21} \end{bmatrix} = \begin{bmatrix} 9.7500 & 0.0000 \\ 0.0000 & 9.7500 \end{bmatrix}\,\text{kg} \\ K = \begin{bmatrix} K_{12} & 0 \\ 0 & K_{21} \end{bmatrix} = \begin{bmatrix} 9.7686 & 0.0000 \\ 0.0000 & 9.7686 \end{bmatrix} \times 10^3\,\text{kN/m} \end{cases} \tag{d}$$

Now, let the attached mass and attachment location be

$$m = 0.01\,\text{kg}, \quad x_0 = 0.20\,\text{m}, \quad y_0 \in \{0.2\,\text{m}, \quad 0.25\,\text{m}, \quad 0.5\,\text{m}\} \tag{e}$$

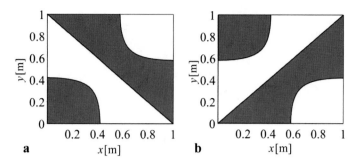

Fig. 6.16 Two natural modes of the square plate with a 0.01 kg mass attached at (0.2 m, 0.2 m). **a** $\tilde{f}_{23} = 159.26$ Hz, $\tilde{\varphi}_{23}(x, y)$; **b** $\tilde{f}_{32} = 159.31$ Hz, $\tilde{\varphi}_{32}(x, y)$

The mass of the plate is $m_p = \rho h a^2 = 39$ kg and the ratio $m/m_p \approx 2.56 \times 10^{-4}$ is very small. Thus, this problem of a mass disturbance falls into the scope of the above theoretical study. If $(x_0, y_0) = (0.2$ m, 0.2 m) is taken as an example, the disturbed mass matrix is

$$\Delta M = \begin{bmatrix} 0.002795 & 0.002795 \\ 0.002795 & 0.002795 \end{bmatrix} \text{kg} \qquad (f)$$

In what follows, we discuss three cases.

(1)   Given $(x_0, y_0) = (0.2$ m, 0.2 m), we solve eigenvalue problem and get the natural frequencies $\tilde{f}_{23} = 159.26$ Hz and $\tilde{f}_{32} = 159.31$ Hz for the disturbed plate. In this case, $\tilde{f}_{23}$ is less than f_{23} by 0.05 Hz, and $\tilde{f}_{32}$ and f_{32} are identical for the first five decimal places. Yet, the two mode shapes get great changes as shown in Fig. 6.16. As the attached mass is at the diagonal $y = x$ of the square plate, the disturbed plate keeps the mirror-symmetry about $y = x$, and so does the two mode shapes.

(2)   Given $(x_0, y_0) = (0.2$ m, 0.25 m), we obtain two natural frequencies $\tilde{f}_{23} = 159.25$ Hz and $\tilde{f}_{32} = 159.31$ Hz. Now $\tilde{f}_{23}$ is smaller than f_{23} by 0.06 Hz, and $\tilde{f}_{32}$ and f_{32} keep the same for the first five decimal places. As predicted, the mode shapes undergo great changes as shown in Fig. 6.17.

(3)   Given $(x_0, y_0) = (0.2$ m, 0.5 m), we obtain $\tilde{f}_{23} = 159.24$ Hz and $\tilde{f}_{32} = 159.33$ Hz. Now, $\tilde{f}_{23}$ is smaller than f_{23} by 0.07 Hz, $\tilde{f}_{32}$ is larger than f_{32} by 0.02 Hz, and the mode shapes are also subject to great variations shown in Fig. 6.18.

Figures 6.17 and 6.18 show that the mode shapes are not symmetric about the diagonal section $y = x$ since the lumped mass attached is not at the diagonal line of the plate.

Remark 6.1.10 Example 6.1.5 indicates that the mode shapes of repeated frequencies of a perfect square plate are extremely sensitive to the attached mass. For a square

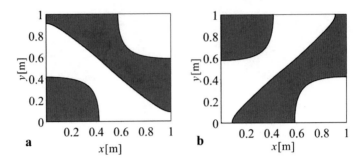

Fig. 6.17 Two natural modes of the square plate with a 0.01 kg mass attached at (0.2m, 0.25m). **a** $\tilde{f}_{23} = 159.25$ Hz, $\tilde{\varphi}_{23}(x, y)$; **b** $\tilde{f}_{32} = 159.31$ Hz, $\tilde{\varphi}_{32}(x, y)$

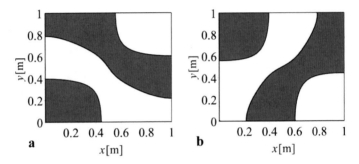

Fig. 6.18 Two natural modes of the square plate with a 0.01 kg mass attached at (0.2m, 0.5m). **a** $\tilde{f}_{23} = 159.24$ Hz, $\tilde{\varphi}_{23}(x, y)$; **b** $\tilde{f}_{32} = 159.33$ Hz, $\tilde{\varphi}_{32}(x, y)$

plate of 39 kg, a lumped mass of 0.01 kg attached causes the small relative variation of a pair of repeated frequencies $f_{23} = f_{32}$ less than 0.1 %, but the mode shapes undergo very large changes. Hence, it is understandable that the mass disturbance due to accelerometers and connection mounts attached to an almost perfect structure with symmetry often greatly reduces the repeatability of modal tests for the mode shapes with a pair of repeated frequencies. In this case, it is desirable to use both contactless measurement and excitation techniques.

6.1.4.3 Disturbance of a Lumped Mass to a Quasi-Square Plate

A square plate in engineering is never a perfect square plate in mathematical sense, but a quasi-square plate with errors of measurement and manufacturing. Now, we take a quasi-square plate discussed in Example 6.1.4 as an example and check the modal sensitivity of the plate to an attached mass in order to further address Problem 5A proposed in Chap. 1.

Example 6.1.6 Let the lengths of the quasi-square plate be $a = 1.0$ m and $b = 1.01a$, and other parameters be the same as those in Example 6.1.4. Study the variation of two mode shapes with a pair of close frequencies of the quasi-square plate due to an attached mass 0.01 kg at $(x_0, y_0) = (0.2$ m, 0.2 m$)$.

From Eq. (6.1.30), the quasi-square plate has a pair of natural frequencies and the corresponding mode shapes as follows

$$f_{23} = 157.13 \text{ Hz}, \quad f_{32} = 158.34 \text{ Hz} \tag{a}$$

$$\varphi_{23}(x, y) = \sin\left(\frac{2\pi x}{a}\right)\sin\left(\frac{3\pi y}{1.01a}\right), \quad \varphi_{32}(x, y) = \sin\left(\frac{3\pi x}{a}\right)\sin\left(\frac{2\pi y}{1.01a}\right) \tag{b}$$

Similar to the procedure in Example 6.1.5, we use Eq. (b) as two base functions and get the natural frequencies of the plate subject to the mass disturbance as follows

$$\tilde{f}_{23} = 157.11 \text{ Hz}, \quad \tilde{f}_{32} = 158.32 \text{ Hz} \tag{c}$$

A comparison between Eqs. (a) and (c) shows that the two natural frequencies decrease about 0.02 Hz due to the attached mass.

Figure 6.19 presents the two mode shapes of the quasi-square plate under the mass disturbance. They look slightly different from the mode shapes of a perfect square plate, but the variations due to the mass disturbance are much smaller than those in Fig. 6.16 in Example 6.1.5.

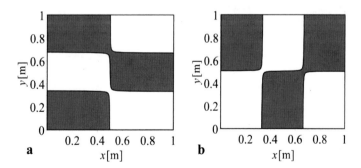

Fig. 6.19 Two natural modes of the quasi-square plate with a 0.01 kg mass attached at $(0.2\text{m}, 0.2\text{m})$. **a** $\tilde{f}_{23} = 157.11$ Hz, $\tilde{\varphi}_{23}(x, y)$; **b** $\tilde{f}_{32} = 158.32$ Hz, $\tilde{\varphi}_{32}(x, y)$

6.1.4.4 Disturbance of a Lumped Mass to a Rectangular Plate

Now, we study the influence of a mass disturbance on the natural modes of a rectangular plate with close frequencies in order to complete the study on Problem 5A proposed in Chap. 1.

Example 6.1.7 As studied in Example 6.1.2, the rectangular plate with hinged boundaries exhibits a pair of repeated natural frequencies when the length ratio is $a/b = \sqrt{8/3} \approx 1.632993$. We study the influence of a mass disturbance to a rectangular plate with hinged boundaries and the following parameters

$$\begin{cases} a = 1.63 \text{ m}, \quad b = 1.0 \text{ m}, \quad h = 0.005 \text{ m} \\ \rho = 7800.0 \text{ kg/m}, \quad E = 210.0 \text{ GPa}, \quad \nu = 0.28 \end{cases} \tag{a}$$

The length ratio $a/b = 1.63$ makes the rectangular plate have a pair of natural modes with very close frequencies as follows

$$\begin{cases} f_{12} = 53.630 \text{ Hz}, \quad \varphi_{12}(x, y) = \sin\left(\dfrac{\pi x}{1.63}\right) \sin(2\pi y) \\ f_{31} = 53.765 \text{ Hz}, \quad \varphi_{31}(x, y) = \sin\left(\dfrac{3\pi x}{1.63}\right) \sin(\pi y) \end{cases} \tag{b}$$

Let a lumped mass $m = 0.1$ kg be attached to the rectangular plate at $(x_0, y_0) = (0.4 \text{ m}, \ 0.25 \text{ m})$. The plate mass is $m_p = \rho abh = 63.65$ kg so that the mass ratio is only $m/m_p = 0.00157$. Following the procedure in Example 6.1.5, we use the mode shapes in Eq. (b) as base functions, compute the approximate natural modes of the disturbed plate and obtain $\tilde{f}_{12} = 53.579$ Hz and $\tilde{f}_{31} = 53.727$ Hz. In this case, the natural frequencies $\tilde{f}_{12}$ and $\tilde{f}_{31}$ only decrease by 0.06 Hz and 0.04 Hz respectively, but their mode shapes in Fig. 6.20 undergo significant changes due to a small disturbance with mass ratio less than 0.16%.

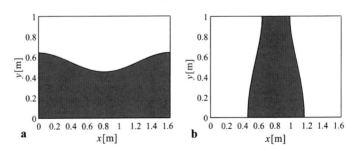

Fig. 6.20 Two natural modes of the rectangular plate with a 0.1 kg mass attached at (0.4m, 0.25m). **a** $\tilde{f}_{12} = 53.579$ Hz, $\tilde{\varphi}_{12}(x, y)$; **b** $\tilde{f}_{31} = 53.727$ Hz, $\tilde{\varphi}_{31}(x, y)$

6.1.5 Concluding Remarks

This section presented the detailed studies on the natural vibrations of mirror-symmetric structures, focusing on the natural modes with a pair of repeated frequencies or close frequencies, in order to address Problem 5A proposed in Chap. 1. The major conclusions are as follows.

(1) A mirror-symmetric structure can be divided to two identical sub-structures such that the natural vibrations of the structure are decoupled as those of two sub-structures with a half of total DoFs from the viewpoint of mechanics or mathematics. The method of mechanics imposes a constraint on the interface of two sub-structures or removes the interface when dealing with the symmetric or anti-symmetric vibrations. The method of mathematics introduces a transform to decouple the mass matrix and stiffness matrix of the entire structure. The method of mechanics looks direct, but the method of mathematics is universal to more complicated structures with multiple mirror-symmetries.

(2) The section presented the possible solution candidate for the free vibrations of a rectangular plate and the natural vibrations of a rectangular plate with hinged boundaries, and indicated that the full general solutions of the free vibrations of a rectangular plate with arbitrary conditions are still under study.

(3) The symmetry of a rectangular plate with hinged boundaries leads to the natural modes with a pair of repeated frequencies such that the natural modes have a two-dimensional subspace. In this case, any linear combination of two independent mode shapes falls into this subspace and exhibits abundant dynamics. Among the rectangular plates with hinged boundary, the square plate has four operations of mirror-symmetry and four operations of cyclosymmetry, and exhibits more natural modes with repeated frequencies. The rectangular, but not square, plate exhibits the natural modes with repeated frequencies relatively sparse in frequency domain, but their mode shapes look more complicated.

(4) When a rectangular plate with hinged boundaries is subject to a structural disturbance, the above symmetry is breaking such that the natural modes with a pair of repeated frequencies split into those with a pair of very close frequencies. The disturbance of a lumped mass, about 0.1 % of the plate mass, causes the great variations in two mode shapes even though the pair of repeated frequencies changes a little. As such, the modal tests of a symmetric structure usually have poor repeatability.

6.2 Vibration Computations of Cyclosymmetric Structures

Sections 1.2.5 and 2.2.4 briefly introduced the concept of cyclosymmetric structures, such as a bladed disc for aero-engines, a rotor for helicopters, a water wheel for hydro-turbines and an antenna of radio-telescope for satellites. The cyclosymmetric structure consists of n identical sectors around a central axis and remains unchanged when the structure rotates through an angle $\theta \equiv 2\pi/n$ about the central axis. The cyclosymmetry has naturally called attention to the relation of mechanics between a single sector and the entire structure. For example, is it possible to follow the decoupling ideas for mirror-symmetric structures in Sect. 6.1.1 so as to decouple the dynamic computations of a cyclosymmetric structure from the cyclosymmetric property of the structure?

The studies on the vibration computations of cyclosymmetric structures have had a long history, but the early study, limited to poor computational facilities, dealt with very simple problems only. The decades of the 1970s and 1980s witnessed the fast developments of several decoupling methods integrated with finite element methods and their successful applications to engineering structures. The basic ideas of these decoupling methods stemmed from different theoretical backgrounds, such as wave propagation, complex constraints, cyclic matrices and theory of cyclic group, but enjoy the same computational efficiency almost. That is, one only needs to mesh one of n sectors of the structure and can derive a set of decoupled eigenvalue problems with their orders reduced to no more than $2/n$ of the original one. As $n >> 10$ holds for many cyclosymmetric structures, such as a bladed disc for aero-engines, these decoupling methods are very attractive in engineering.

Among the decoupling methods, the method of wave propagation utilized the property of standing waves in an annular structure with cyclosymmetry and showed a clear picture in physics[6]. The method of complex constraints expressed the standing waves as complex mode shapes and imposed complex constraints on some blocks in stiffness matrix so as to decouple the eigenvalue problem of the entire structure[7]. The method of cyclic matrix made use of the periodicity of blocks in stiffness matrix to decouple the eigenvalue problem[8]. The method of group theory introduced the generalized displacement vector in each irreducible representative subspace of a cyclic group to describe the the cyclosymmetric structure and then decoupled the dynamic equations of the structure in those subspaces[9]. In addition, those methods have been combined with various sub-structuring methods and parallel computation algorithms in order to further increase the ability of solving tough problems[10].

In China, the designs for hydro-turbines and television towers stimulated the computational studies on the natural vibrations of cyclosymmetric structures on the basis of limited computational facilities in the 1970s and 1980s. One attempt in those studies was based on the theory of cyclic group and the author was one of the team members engaged in the study. The experience of the author confirmed that

[6] Thomas D L (1974). Standing waves in rotationally periodic structures. Journal of Sound and Vibration, 37(2): 288–290.

[7] Thomas D L (1979). Dynamics of rotationally periodic structures. International Journal of Numerical Methods in Engineering, 14(1): 81–102.

[8] Olson B J, Shaw S W, Shi C Z, et al. (2014). Circulant matrices and their application to vibration analysis. Applied Mechanics Review, 66(4): 040803.

[9] Hu H Y, Cheng D L (1986). Generalized mode synthesis of cyclosymmetrical structures. Vibration and Shock, 5(4): 1–7 (in Chinese).

[10] Tran D M (2014). Reduced models of multi-stage cyclic structures using cyclic symmetry reduction and component mode synthesis. Journal of Sound and Vibration, 333(21): 5443–5463.

Fig. 6.21 The schematic of a C_n structure clamped at the central hollow ($n = 4$)

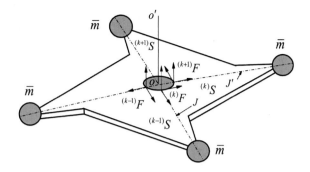

the computations via the cyclic group are complicated in theory, but universal in applications, exhibiting a charming integration of mechanics and mathematics. As such, this section presents the vibration analysis of cyclosymmetric structures based on the theory of cyclic group. The major content comes from the early studies of the author (see footnote 9), but presents a detailed deduction procedure and several demonstrative examples to enhance readability.

Figure 6.21 shows the schematic of a cyclosymmetric structure with four identical sectors. The structure has *cyclosymmetry* about the central axis. That is, the structure remains unchanged when it rotates about axis oo' by an angle among $\pi/2$, π, $3\pi/2$ and 2π. The minimum of those angles is usually denoted as $\theta \equiv 2\pi/n$. Here $n = 4$ in Fig. 6.21. In group theory, all objects and operations with cyclic periodicity form an abstract set, which is named a *cyclic group C_n*. The basic operation in group C_n is the counter-clockwise rotation through an angle θ about axis oo'. Thus, the structure with n identical sectors around a central axis is referred to as a structure on group C_n or a C_n *structure* for short. This section only deals with the case when $n > 2$ since the case of $n = 1$ implies no cyclosymmetry and the case of $n = 2$ is a trivial one, which does not need to use group theory for vibration analysis.

It is worthy to point out that there are two other kinds of structures similar to C_n structures. The first kind is axial-symmetric structures, such as a circular plate and a circular cylinder. An axial-symmetric structure remains unchanged after the rotation through any angles about a central axis. The second kind is a part of cyclosymmetric structures, each sector of which has mirror-symmetry. The examples of those structures include, but not limited to, an equilateral triangle plate in Fig. 6.3 and a square plate with all edges hinged discussed in Sect. 6.1.3. In group theory, the second kind of structures is named the structure on group C_{nv} or a C_{nv} *structure* for short. Here, v in subscripts means the reflection symmetry about a vertical plane. Accordingly, group C_{nv} has all properties of group C_n, which is a *subgroup* of group C_{nv}. From the viewpoint of engineers, the symmetry of a group C_{nv} is the combination of both cyclosymmetry and mirror-symmetry. Of course, all properties of a C_n structure studied in this section are valid for a C_{nv} structure.

To reduce difficulty for readers, Sect. 6.2.1 deals with the annular C_n structures, which are clamped or hollowed out at the central axis oo', where no DoFs have to be taken into account. The annular C_n structures are referred to as the C_n structures without a central axis. Section 6.2.2 will deal with the natural vibrations of the C_n structures with DoFs on the central axis, or named *central DoFs* for short, and solve Problem 5B proposed in Chap. 1.

6.2.1 Decoupling a Cyclosymmetric Structure Without a Central Axis

6.2.1.1 Basic Descriptions

As shown in Fig. 6.21, the C_n structure S is divided into n identical sub-structures $^{(k)}S$ along the circumferential direction such that $^{(k+n)}S = {}^{(k)}S$, $k \in I_n$ holds, where $I_n \equiv \{0, 1, \ldots, n-1\}$ is a set of integer indices.

Let $^{(0)}S$ be the basic sub-structure and establish the local frame $^{(0)}F$ of Cartesian coordinates or cylinder coordinates, where the third axis z is parallel to the central axis. In the above local frame of coordinates, we mesh the sub-structure $^{(0)}S$ via appropriate finite elements.

It is straightforward to get the local frame of coordinates $^{(k)}F$ for sub-structure $^{(k)}S$ and the corresponding finite element model of $^{(k)}S$ through k operations of group C_n, i.e., a counter-clockwise rotation through an angle of $k\theta$ about the central axis, on $^{(0)}F$ and $^{(0)}S$, respectively.

Now, let $^{(k)}u$ be the nodal displacement vector, or called the displacement vector for short, for $^{(k)}S$ in local frame $^{(k)}F$ in a block matrix form as follows

$$^{(k)}u \equiv \left[{}^{(k)}u_I^{\mathrm{T}} \quad {}^{(k)}u_J^{\mathrm{T}} \quad {}^{(k)}u_{J'}^{\mathrm{T}} \right]^{\mathrm{T}}, \quad k \in I_n \tag{6.2.1}$$

where $^{(k)}u_I \in \mathbb{R}^I$ is the internal displacement vector of $^{(k)}S$, $^{(k)}u_J \in \mathbb{R}^J$ is the interfacial displacement vector for $J \equiv {}^{(k)}S \cap {}^{(k-1)}S$, and $^{(k)}u_{J'} \in \mathbb{R}^J$ is the interfacial displacement vector for $J' \equiv {}^{(k)}S \cap {}^{(k+1)}S$. In Eq. (6.2.1) time t does not show up since notations hereinafter look quite complicated. Furthermore, if two adjacent substructures have no interface, the C_n structure is a decoupled one. This special case does not fall into the study.

In general case, the number of DoFs of sub-structure $^{(k)}S$ is

$$m' \equiv I + 2J \tag{6.2.2}$$

That is, the finite element model of $^{(k)}S$ includes two interfaces J and J'. Thus, sub-structure $^{(k)}S$ has mass matrix $M \in \mathbb{R}^{m' \times m'}$ and stiffness matrix $K \in \mathbb{R}^{m' \times m'}$. As will be studied later, only the DoFs of one interface remain due to the compatibility conditions of two interfaces. Thus, the effective DoFs of $^{(k)}S$ reduces to m and the total DoFs of the entire C_n structure is N, namely,

$$m \equiv I + J, \quad N \equiv n \times m = n \times (I + J) \tag{6.2.3}$$

In what follows, we utilize the cyclosymmetry of a C_n structure to introduce the generalized displacement vectors, eliminate redundant DoFs in view of interface compatibility and derive a set of decoupled dynamic equations, which have only m or $2m$ unknowns to be solved.

6.2.1.2 Transform of Generalized Displacements

As studied in classical dynamics, one can use an orthogonal matrix of order three to describe the rotation of a rigid-body through any angles about an axis in a three-dimensional space. The operation of group C_n has a similar representation. Specifically, the k-th operation of group C_n is the rotation of a rigid-body about the central axis by an angle $k\theta$, where $\theta = 2\pi/n$, $k \in I_n$. Thus, group C_n has n different operations and any further operations, i.e., rotations, repeat periodically. In an algebraic language, let $\mathbb{C}^n$ be the space of complex vectors of n dimensions, the above n operations are equivalent to the following n vectors

$$\boldsymbol{e}_j \equiv \frac{1}{\sqrt{n}} \mathrm{col}_k \left[\exp(ijk\theta) \right] \in \mathbb{C}^n, \quad j \in I_n \tag{6.2.4}$$

where $i \equiv \sqrt{-1}$, $\mathrm{col}_k[\cdot]$ is a column vector formed according to index $k \in I_n$.

It is easy to prove that the above vectors are a set of normalized orthogonal base vectors of space $\mathbb{C}^n$, and satisfy the following conjugation relations

$$\boldsymbol{e}_{n-j} \equiv \frac{1}{\sqrt{n}} \mathrm{col}_k \left[\exp[i(n-j)k\theta] \right] = \frac{1}{\sqrt{n}} \mathrm{col}_k \left[\exp(-ijk\theta) \right] = \bar{\boldsymbol{e}}_j \quad j \in \tilde{I}_{n-1} \tag{6.2.5}$$

where the index set $\tilde{I}_{n-1} \equiv \{1, 2, \ldots, n-1\}$ is different from I_n.

Example 6.2.1 Derive the representation of group C_4 in reference to the C_4 structure in Fig. 6.21.

The independent operations of group C_4 include 4 counter-clockwise rotations about the central axis by angles of $k\theta$, $k = 0, 1, 2, 3$, where $\theta = \pi/2$. Their representations correspond to the following normalized orthogonal base vectors

$$\boldsymbol{e}_0 = \frac{1}{2} \begin{bmatrix} 1 \\ 1 \\ 1 \\ 1 \end{bmatrix}, \quad \boldsymbol{e}_1 = \frac{1}{2} \begin{bmatrix} 1 \\ i \\ -1 \\ -i \end{bmatrix}, \quad \boldsymbol{e}_2 = \frac{1}{2} \begin{bmatrix} 1 \\ -1 \\ 1 \\ -1 \end{bmatrix}, \quad \boldsymbol{e}_3 = \frac{1}{2} \begin{bmatrix} 1 \\ -i \\ -1 \\ i \end{bmatrix} \tag{a}$$

If one checks the variation of each entry of the four vectors on a complex plane, the entries in e_0 form a straight line, the entries in e_1 undergo a counter-clockwise rotation by an angel $\theta = \pi/2$ one by one, those in e_2 undergo a counter-clockwise rotation by an angel $2\theta = \pi$ one by one, and those in e_3 undergo a counter-clockwise rotation by an angel $3\theta = 3\pi/2$ one by one. The four vectors completely describe all operations of group C_4. For example, one can verify the orthogonality of two base vectors and the normalization of a base vector as follows

$$e_1^H e_2 = \frac{1}{4}\begin{bmatrix} 1 & -i & -1 & i \end{bmatrix}\begin{bmatrix} 1 & -1 & 1 & -1 \end{bmatrix}^T = 0, \quad e_2^H e_2 = 1 \qquad \text{(b)}$$

where the superscript H stands for the conjugate transpose for a vector or a matrix. For example, Eq. (6.2.5) gives

$$e_3 = \frac{1}{2}\begin{bmatrix} 1 & -i & -1 & i \end{bmatrix}^T = \frac{1}{2}\begin{bmatrix} \bar{1} & \bar{i} & -\bar{1} & -\bar{i} \end{bmatrix}^T = \bar{e}_1 \qquad \text{(c)}$$

Now, we introduce a new description for the displacement vectors of C_n structure. Take the l-th displacement component $^{(k)}u_l$ from displacement vectors $^{(k)}u$, $k \in I_n$ and define a set of new vectors

$$u_l \equiv \mathrm{col}_k\big[^{(k)}u_l\big] \in \mathbb{R}^n \subset \mathbb{C}^n, \quad l \in \tilde{I}_{m'} \equiv \{1, 2, \ldots, m'\} \qquad (6.2.6)$$

As vector u_l falls into the n-dimensional complex vector space $\mathbb{C}^n$, it can be linearly expressed in terms of the normalized orthogonal base vectors as follows

$$u_l = \sum_{j=0}^{n-1} e_j q_{jl} = \frac{1}{\sqrt{n}}\sum_{j=0}^{n-1} \mathrm{col}_k\big[\exp(ijk\theta)\big]q_{jl} \quad l \in \tilde{I}_{m'} \qquad (6.2.7)$$

where the complex coefficient q_{jl} is the inner product of vectors e_j and u_l, namely,

$$q_{jl} \equiv e_j^H u_l = \frac{1}{\sqrt{n}}\sum_{k=0}^{n-1} {}^{(k)}u_l \exp(-ijk\theta) \quad j \in I_n, \quad l \in \tilde{I}_{m'} \qquad (6.2.8)$$

Here, the superscript H for conjugate transpose makes the exponential functions in Eqs. (6.2.7) and (6.2.8) be a conjugate pair.

To get the description for displacement vectors $^{(k)}u$, $k \in I_n$, we use the above m' vectors u_l in Eq. (6.2.7) to form a matrix

$$\begin{bmatrix} u_1 & \cdots & u_{m'} \end{bmatrix} = \big[^{(k)}u_l\big] = \frac{1}{\sqrt{n}}\big[\exp(ijk\theta)\big]\big[q_{jl}\big] \qquad (6.2.9)$$

The corresponding transpose matrix reads

$$\left[^{(k)}u\right]^{\mathrm{T}} = \frac{1}{\sqrt{n}}[q_{jl}]^{\mathrm{T}}[\exp(\mathrm{i}jk\theta)]^{\mathrm{T}} = \frac{1}{\sqrt{n}}[q_{jl}]^{\mathrm{T}}[\exp(\mathrm{i}jk\theta)] \qquad (6.2.10)$$

where the k-th column in the matrix of left-hand side is displacement vector $^{(k)}u$. The j-th column in matrix $[q_{jl}]^{\mathrm{T}}$ in right-hand side of Eq. (6.2.10) can be defined as

$$q_j \equiv \mathrm{col}_l[q_{jl}] \in \mathbb{C}^{m'}, \quad j \in I_n \qquad (6.2.11)$$

Thus, the k-th column in two sides of Eq. (6.2.10) yields

$$^{(k)}u = \frac{1}{\sqrt{n}}\mathrm{row}_j[q_j]\mathrm{col}_j\big[\exp(\mathrm{i}jk\theta)\big] = \frac{1}{\sqrt{n}}\sum_{j=0}^{n-1}\exp(\mathrm{i}jk\theta)q_j, \quad k \in I_n \quad (6.2.12)$$

where $\mathrm{row}_j[\,\cdot\,]$ is a row vector formed according to index j. In Eq. (6.2.12), vector q_j is referred to as the *generalized displacement vector* of the C_n structure in an *irreducible representative subspace* S_j, which is spanned by base vector e_j of group C_n, while Eq. (6.2.12) is a linear transform from vectors q_j, $j \in I_n$ to vectors $^{(k)}u$, $k \in I_n$. Hereinafter, q_j is also called the displacement vector in subspace S_j for short.

It is worthy to notice that vectors q_j, $j \in I_n$ are not linearly independent since Eq. (6.2.5) leads to

$$q_{n-j} \equiv \mathrm{col}_l\big[e^{\mathrm{H}}_{n-j}u_l\big] = \mathrm{col}_l\big[e^{\mathrm{T}}_j u_l\big] = \overline{q}_j, \quad j \in \tilde{I}_{n-1} \qquad (6.2.13)$$

To use a set of linearly independent generalized displacement vectors in the linear transform, we define

$$a_j \equiv \begin{cases} 1, \ j = 0 \\ 2, \ j = 1, 2, \ldots \\ 1, \ j = n/2 = [n/2] \end{cases} \qquad (6.2.14)$$

where $[n/2]$ stands for the maximal integer no more than $n/2$. Using Eq. (6.2.12), we get the displacement vectors $^{(k)}u$, $k \in I_n$ expressed in terms of independent generalized displacement vectors q_j, $j \in I_{[n/2]}$ as follows

$$^{(k)}u = \frac{1}{\sqrt{n}}\sum_{j=0}^{n-1}\exp(\mathrm{i}jk\theta)q_j = \frac{1}{\sqrt{n}}\big\{q_0 + \exp(\mathrm{i}k\theta)q_1 + \cdots + \exp[\mathrm{i}(n-1)k\theta]q_{n-1}\big\}$$

$$= \frac{1}{\sqrt{n}} \left[q_0 + \exp(ik\theta)q_1 + \cdots + \overline{\exp(ik\theta)q_1} \right]$$

$$= \frac{1}{\sqrt{n}} \sum_{j=0}^{[n/2]} a_j \text{Re}[\exp(ijk\theta)q_j], \quad k \in I_n \tag{6.2.15}$$

Example 6.2.2 Given the C_4 structure in Fig. 6.21, study the representation of group C_4 for the vertical displacements of four lumped masses mounted at the corners of a thin massless plate, which provides bending stiffness.

To avoid non-inertial DoFs, we divide the C_4 structure into four sub-structures shown in Fig. 6.21. Thus, sub-structure $^{(k)}S$ has no internal DoF, but one DoF in interfaces J and J', respectively. Each DoF describes the vertical motion of a lumped mass $\overline{m}/2$. Thus, the displacement vector for $^{(k)}S$ reads

$$^{(k)}u \equiv \left[^{(k)}u_J \ ^{(k)}u_{J'} \right]^{\text{T}} \equiv \left[^{(k)}u_1 \ ^{(k)}u_2 \right]^{\text{T}}, \quad k \in I_4 \tag{a}$$

Following the definition in Eq. (6.2.6), we construct $u_1 \in \mathbb{C}^4$ and $u_2 \in \mathbb{C}^4$. Substituting the four orthogonal base vectors derived in Example 6.2.1 into Eq. (6.2.9) leads to a linear expression of u_1 and u_2 as follows

$$[u_1 \ u_2] = \begin{bmatrix} ^{(0)}u_1 & ^{(0)}u_2 \\ ^{(1)}u_1 & ^{(1)}u_2 \\ ^{(2)}u_1 & ^{(2)}u_2 \\ ^{(3)}u_1 & ^{(3)}u_2 \end{bmatrix} = \frac{1}{2} \begin{bmatrix} 1 & 1 & 1 & 1 \\ 1 & i & -1 & -i \\ 1 & -1 & 1 & -1 \\ 1 & -i & -1 & i \end{bmatrix} \begin{bmatrix} q_{01} & q_{02} \\ q_{11} & q_{12} \\ q_{21} & q_{22} \\ q_{31} & q_{32} \end{bmatrix} \tag{b}$$

Transposing Eq. (b) gives

$$\begin{bmatrix} ^{(0)}u_1 & ^{(1)}u_1 & ^{(2)}u_1 & ^{(3)}u_1 \\ ^{(0)}u_2 & ^{(1)}u_2 & ^{(2)}u_2 & ^{(3)}u_2 \end{bmatrix} = \frac{1}{2} \begin{bmatrix} q_{01} & q_{11} & q_{21} & q_{31} \\ q_{02} & q_{12} & q_{22} & q_{32} \end{bmatrix} \begin{bmatrix} 1 & 1 & 1 & 1 \\ 1 & i & -1 & -i \\ 1 & -1 & 1 & -1 \\ 1 & -i & -1 & i \end{bmatrix} \tag{c}$$

The columns in the left-hand side of Eq. (c) are four displacement vectors $^{(k)}u, \ k \in I_4$, which can be expressed in terms of generalized displacement vectors $q_j, \ j \in I_4$. When $k = 1$, as an example, Eq. (c) gives

$$^{(1)}u = \begin{bmatrix} ^{(1)}u_1 \\ ^{(1)}u_2 \end{bmatrix} = \frac{1}{2} \left\{ \begin{bmatrix} q_{01} \\ q_{02} \end{bmatrix} + i \begin{bmatrix} q_{11} \\ q_{12} \end{bmatrix} - \begin{bmatrix} q_{21} \\ q_{22} \end{bmatrix} - i \begin{bmatrix} q_{31} \\ q_{32} \end{bmatrix} \right\} = \frac{1}{2} \sum_{j=0}^{3} \exp(ij\theta)q_j \tag{d}$$

Eq. (d) coincides with the expression in Eq. (6.2.12). From Eq. (6.2.15), it is easy to derive the displacement vector of sub-structure $^{(1)}S$ in terms of three independent generalized displacement vectors, namely,

$$^{(1)}\boldsymbol{u} = \begin{bmatrix} ^{(1)}u_1 \\ ^{(1)}u_2 \end{bmatrix} = \frac{1}{2} \sum_{j=0}^{2} a_j \mathrm{Re}\left[\exp(\mathrm{i}j\theta)\boldsymbol{q}_j\right] = \frac{1}{2}\mathrm{Re}\left\{\begin{bmatrix} q_{01} \\ q_{02} \end{bmatrix} + 2\mathrm{i}\begin{bmatrix} q_{11} \\ q_{12} \end{bmatrix} - \begin{bmatrix} q_{21} \\ q_{22} \end{bmatrix}\right\}$$

(e)

6.2.1.3 Transform of Interface Compatibility

In set J' of interfacial DoFs for sub-structure $^{(k)}S$, displacement vector $^{(k)}\boldsymbol{u}_{J'}$ for $^{(k)}S$ and displacement vector $^{(k+1)}\boldsymbol{u}_J$ for $^{(k+1)}S$ should be compatible. Noting that the two vectors come from different local frames of coordinates, we use orthogonal matrix $\boldsymbol{R}_J^{-1}$ to transform $^{(k+1)}\boldsymbol{u}_J$ to the local frame of coordinates for $^{(k)}S$, and derive the interfacial compatibility condition as follows

$$^{(k)}\boldsymbol{u}_{J'} - \boldsymbol{R}_J^{-1\,(k+1)}\boldsymbol{u}_J = \boldsymbol{0}, \quad k \in I_n \tag{6.2.16}$$

where $\boldsymbol{R}_J^{-1}$ is a diagonal block matrix of orthogonal transform with an order compatible to set J' of interfacial DoFs, that is,

$$\boldsymbol{R}_J^{-1} \equiv \mathrm{diag}\left[\begin{bmatrix} \cos(\theta) & -\sin(\theta) & 0 \\ \sin(\theta) & \cos(\theta) & 0 \\ 0 & 0 & 1 \end{bmatrix}\right] \in \mathbb{R}^{J \times J} \tag{6.2.17}$$

The sub-matrix in Eq. (6.2.17) is an orthogonal matrix for the rotation by an angle of $-\theta$.

Extending Eq. (6.2.16) to a matrix form of the displacement vectors for the entire sub-structure, we have

$$\boldsymbol{B}\begin{bmatrix} ^{(k)}\boldsymbol{u} \\ ^{(k+1)}\boldsymbol{u} \end{bmatrix} = \boldsymbol{0}, \quad \boldsymbol{B} \equiv \begin{bmatrix} 0 & 0 & I_J & 0 & -\boldsymbol{R}_J^{-1} & 0 \end{bmatrix} \in \mathbb{R}^{J \times 2m'}, \quad k \in I_n \tag{6.2.18}$$

where $\boldsymbol{B}$ is the interfacial compatibility matrix with $\mathrm{rank}(\boldsymbol{B}) = J$.

Now, we discuss how to use a set of independent generalized displacement vectors $\boldsymbol{q}_j$ to describe the interfacial compatibility condition in Eq. (6.2.18). Substituting Eq. (6.2.15) into Eq. (6.2.18), the linear independence of $\boldsymbol{q}_j$ leads to the interfacial compatibility conditions in each subspace S_j as follows

$$B \begin{bmatrix} \mathrm{Re}[\exp(ijk\theta)]I_{m'} \\ \mathrm{Re}\{\exp[ij(k+1)\theta]\}I_{m'} \end{bmatrix} q_j = 0, \quad j \in J_{[n/2]} \equiv \{0, 1, \ldots, [\frac{n}{2}]\}, \quad k \in I_n$$

$$(6.2.19)$$

By defining the interfacial compatibility matrix in subspace S_j as

$$B_j \equiv B \begin{bmatrix} I_{m'} \\ \exp(ij\theta)I_{m'} \end{bmatrix} = [0 \quad -\exp(ij\theta)R_J^{-1} \quad I_J] \in \mathbb{C}^{J \times m'}, \quad j \in J_{[n/2]}$$

$$(6.2.20)$$

Equation (6.2.19) is recast as

$$\mathrm{Re}\left[\exp(ijk\theta)B_j q_j\right] = 0, \quad j \in J_{[n/2]}, \quad k \in I_n \qquad (6.2.21)$$

Equation (6.2.21) can be discussed for different values of j and further simplified.
First, when $j = 0$ or $j = n/2 = [n/2]$ holds, B_j and q_j are a real matrix and a real vector such that the following relations hold

$$B_0 q_0 = \mathrm{Re}(B_0 q_0) = 0 \qquad (6.2.22)$$

$$\mathrm{Re}\left[\exp(ik\pi)B_{n/2}q_{n/2}\right] = (-1)^k \mathrm{Re}(B_{n/2}q_{n/2}) = 0 \qquad (6.2.23)$$

Equations (6.2.22) and (6.2.23) imply that $B_j q_j = 0$ holds for $j = 0$ and $n/2$.
Second, when $0 < j < n/2$ holds true, Eq. (6.2.21) can be recast as

$$\cos(jk\theta)\mathrm{Re}(B_j q_j) - \sin(jk\theta)\mathrm{Im}(B_j q_j) = 0, \quad k \in I_n \qquad (6.2.24)$$

As index k is arbitrary in Eq. (6.2.24), $k = 0$ and $k = 1$ lead to a set of linear algebraic equations as follows

$$\begin{bmatrix} I_J & 0 \\ \cos(j\theta)I_J & -\sin(j\theta)I_J \end{bmatrix} \begin{bmatrix} \mathrm{Re}(B_j q_j) \\ \mathrm{Im}(B_j q_j) \end{bmatrix} = 0 \qquad (6.2.25)$$

Given $0 < j < n/2, 0 < j\theta < \pi$ holds such that the determinant of the matrix in Eq. (6.2.25) is

$$\Delta = [-\sin(j\theta)]^J \neq 0 \qquad (6.2.26)$$

Thus, Eq. (6.2.25) has only trivial solutions $\mathrm{Re}(B_j q_j) = 0$ and $\mathrm{Im}(B_j q_j) = 0$ so that $B_j q_j = 0$ holds.

To this end, we have the interfacial compatibility equations of C_n structure in subspace S_j simplified as

$$B_j q_j = 0, \quad j \in J_{[n/2]} \tag{6.2.27}$$

Now, we focus on the independent entries in vector q_j. In view of the partition of vector $^{(k)}u$, the first m entries in vector $^{(k)}u$ are independent. From Eqs. (6.2.8) and (6.2.11), the first m entries in vector q_j are also independent. Using these independent entries to define a vector $q_{jd} \in \mathbb{C}^m$, and using the rest entries to define a vector $q_{jr} \in \mathbb{C}^J$, we recast Eq. (6.2.27) as a block form

$$B_j q_j = \begin{bmatrix} B_{jd} & B_{jr} \end{bmatrix} \begin{bmatrix} q_{jd} \\ q_{jr} \end{bmatrix} = 0, \quad j \in J_{[n/2]} \tag{6.2.28}$$

In view of a comparison between Eq. (6.2.20) and the block matrix B_j, we have

$$B_{jd} = \begin{bmatrix} 0 & -\exp(\mathrm{i}j\theta) R_J^{-1} \end{bmatrix} \in \mathbb{C}^{J \times m}, \quad B_{jr} = I_{JJ} \tag{6.2.29}$$

Hence, the interfacial compatibility relation in terms of independent vectors $q_{jd} \in \mathbb{C}^m$, $j \in J_{[n/2]}$ becomes available, namely,

$$q_j = \begin{bmatrix} q_{jd} \\ q_{jr} \end{bmatrix} = \begin{bmatrix} I_m \\ -B_{jd} \end{bmatrix} q_{jd} \equiv \tilde{B}_{jd} q_{jd} \quad j \in J_{[n/2]} \tag{6.2.30}$$

where $\tilde{B}_{jd}$ is the interfacial compatibility transform matrix in subspace S_j as follows

$$\tilde{B}_{jd} \equiv \begin{bmatrix} I_m \\ -B_{jd} \end{bmatrix} = \begin{bmatrix} I_I & 0 \\ 0 & I_J \\ 0 & \exp(\mathrm{i}j\theta) R_J^{-1} \end{bmatrix} \in \mathbb{C}^{m' \times m}, \quad j \in J_{[n/2]} \tag{6.2.31}$$

To this end, after the above two transforms in Eqs. (6.2.15) and (6.2.30), the displacement vectors $^{(k)}u \in \mathbb{R}^{m'}$, $k \in I_n$ with interfacial compatibility taken into account are expressed in terms of independent generalized displacement vectors $q_{jd} \in \mathbb{C}^m$, $j \in I_{[n/2]}$.

Example 6.2.3   Given the C_4 structure in Fig. 6.21, establish the interfacial compatibility condition in terms of the independent generalized displacement vectors.

To begin with, we write the interfacial compatibility condition for sub-structures $^{(k)}S$ and $^{(k+1)}S$ as follows

$$^{(k)}u_2 - {}^{(k+1)}u_1 = 0, \quad k \in I_4 \tag{a}$$

whereby the interfacial compatibility matrix, following Eq. (6.2.18), is defined as

$$\boldsymbol{B} = \begin{bmatrix} 0 & 1 & -1 & 0 \end{bmatrix} \tag{b}$$

Substitution of Eq. (b) into Eq. (6.2.20) gives

$$\boldsymbol{B}_j \equiv \begin{bmatrix} 0 & 1 & -1 & 0 \end{bmatrix} \begin{bmatrix} \boldsymbol{I}_2 \\ \exp(\mathrm{i}j\pi/2)\boldsymbol{I}_2 \end{bmatrix} = \begin{bmatrix} -\exp(\mathrm{i}j\pi/2) & 1 \end{bmatrix}, \quad j \in J_2 \tag{c}$$

Then, Eq. (6.2.27) leads to the interfacial compatibility relation as follows

$$\boldsymbol{B}_j \boldsymbol{q}_j = \begin{bmatrix} -\exp(\mathrm{i}j\pi/2) & 1 \end{bmatrix} \begin{bmatrix} q_{j1} \\ q_{j2} \end{bmatrix} = \boldsymbol{0}, \quad j \in J_2 \tag{d}$$

Selecting $q_{j1} \in \mathbb{C}^1$ in Eq. (d) as an independent generalized displacement, we derive the interfacial compatibility transform from q_{j1} to $\boldsymbol{q}_j$, namely,

$$\boldsymbol{q}_j = \begin{bmatrix} q_{j1} \\ q_{j2} \end{bmatrix} = \begin{bmatrix} 1 \\ \exp(\mathrm{i}j\pi/2) \end{bmatrix} q_{j1} \equiv \tilde{\boldsymbol{B}}_{jd} q_{j1}, \quad j \in J_2. \tag{e}$$

6.2.1.4 Dynamic Equations Decoupled

Now we turn to the dynamic problem of a C_n structure, and let all displacements and generalized displacements be functions of time t. Composing the two transforms in Eqs. (6.2.15) and (6.2.30) leads to

$$^{(k)}\boldsymbol{u}(t) = \frac{1}{\sqrt{n}} \sum_{j=0}^{[n/2]} a_j \mathrm{Re}\left[\exp(\mathrm{i}jk\theta)\tilde{\boldsymbol{B}}_{jd} \boldsymbol{q}_{jd}(t)\right], \quad k \in I_n \tag{6.2.32}$$

For convenience in subsequent deduction, we use conjugate relation $\boldsymbol{B}_{n-j} = \overline{\boldsymbol{B}}_j$ and recast Eq. (6.2.32) as a complex form, namely,

$$^{(k)}\boldsymbol{u}(t) = \frac{1}{\sqrt{n}} \sum_{j=0}^{n-1} \exp(\mathrm{i}jk\theta)\tilde{\boldsymbol{B}}_{jd} \boldsymbol{q}_{jd}(t), \quad k \in I_n \tag{6.2.33}$$

As all sub-structures have the same mass matrix $\boldsymbol{M} \in \mathbb{R}^{m' \times m'}$ and the same stiffness matrix $\boldsymbol{K} \in \mathbb{R}^{m' \times m'}$ in their local frame of coordinates, it is easy to write the kinetic and potential energies of the C_n structure according to Eq. (6.2.33). Using the conjugate transpose operation H and the following orthogonal relation

$$\sum_{k=0}^{n-1} \exp\left[i(j - j')k\theta\right] = n\delta_{jj'} \tag{6.2.34}$$

we derive the kinetic and potential energies as follows

$$
\begin{aligned}
T &= \frac{1}{2}\sum_{k=0}^{n-1} {}^{(k)}\dot{u}^{H}(t)M^{(k)}\dot{u}(t) \\
&= \frac{1}{2}\sum_{k=0}^{n-1}\left[\frac{1}{\sqrt{n}}\sum_{j=0}^{n-1}\exp(ijk\theta)\tilde{B}_{jd}\dot{q}_{jd}(t)\right]^{H} M \left[\frac{1}{\sqrt{n}}\sum_{j'=0}^{n-1}\exp(ij'k\theta)\tilde{B}_{j'd}\dot{q}_{j'd}(t)\right] \\
&= \frac{1}{2n}\sum_{j=0}^{n-1}\sum_{j'=0}^{n-1}\left\{\left[\dot{q}_{jd}^{H}(t)\left(\tilde{B}_{jd}^{H}M\tilde{B}_{j'd}\right)\dot{q}_{j'd}(t)\right]\sum_{k=0}^{n-1}\exp\left[i(j'-j)k\theta\right]\right\} \\
&= \frac{1}{2}\sum_{j=0}^{n-1}\dot{q}_{jd}^{H}(t)\left(\tilde{B}_{jd}^{H}M\tilde{B}_{jd}\right)\dot{q}_{jd}(t) = \frac{1}{2}\sum_{j=0}^{n-1}\dot{q}_{jd}^{H}(t)M_{j}\dot{q}_{jd}(t) \tag{6.2.35}
\end{aligned}
$$

$$
\begin{aligned}
V &= \frac{1}{2}\sum_{k=0}^{n-1} {}^{(k)}u^{H}(t)K^{(k)}u(t) = \frac{1}{2}\sum_{j=0}^{n-1}q_{jd}^{H}(t)\left(\tilde{B}_{jd}^{H}K\tilde{B}_{jd}\right)q_{jd}(t) \\
&= \frac{1}{2}\sum_{j=0}^{n-1}q_{jd}^{H}(t)K_{j}q_{jd}(t) \tag{6.2.36}
\end{aligned}
$$

where

$$M_{j} \equiv \tilde{B}_{jd}^{H}M\tilde{B}_{jd} \in \mathbb{C}^{m\times m}, \quad K_{j} \equiv \tilde{B}_{jd}^{H}K\tilde{B}_{jd} \in \mathbb{C}^{m\times m}, \quad j \in I_{n} \tag{6.2.37}$$

They are the mass matrix and stiffness matrix associated with the generalized displacement vector $q_{jd} \in \mathbb{C}^{m}$. It is easy to verify that the two matrices are Hermite matrices. In view of Eqs. (6.2.35) and (6.2.36), the dynamic problem of a C_{n} structure can be decoupled in subspace S_{j}.

To perform computations in real functions, define the real parts and imaginary parts of matrices in Eq. (6.2.37) as

$$M_{j}^{R} \equiv \text{Re}(M_{j}), \quad M_{j}^{I} \equiv \text{Im}(M_{j}), \quad K_{j}^{R} \equiv \text{Re}(K_{j}), \quad K_{j}^{I} \equiv \text{Im}(K_{j}) \quad j \in I_{n} \tag{6.2.38}$$

These sub-matricies satisfy the properties of a Hermite matrix as follows

$$\left(M_j^R\right)^{\mathrm{T}} = M_j^R, \quad \left(M_j^I\right)^{\mathrm{T}} = -M_j^I, \quad \left(K_j^R\right)^{\mathrm{T}} = K_j^R, \quad \left(K_j^I\right)^{\mathrm{T}} = -K_j^I, \quad j \in I_n \tag{6.2.39}$$

Let the real part and the imaginary part of vector $q_{jd}(t)$ be

$$q_{jd}^R(t) \equiv \mathrm{Re}\left[q_{jd}(t)\right], \quad q_{jd}^I(t) \equiv \mathrm{Im}\left[q_{jd}(t)\right], \quad j \in I_n \tag{6.2.40}$$

Using Eqs. (6.2.38) and (6.2.40) to recast Eqs. (6.2.35) and (6.2.36), we get the kinetic and potential energies in terms of real matrices and vectors, namely,

$$T = \frac{1}{2}\sum_{j=0}^{n-1} \dot{q}_{jd}^{\mathrm{H}}(t) M_j \dot{q}_{jd}(t) = \frac{1}{2}\sum_{j=0}^{[n/2]} a_j^2 \begin{bmatrix} \dot{q}_{jd}^R(t) \\ \dot{q}_{jd}^I(t) \end{bmatrix}^{\mathrm{T}} \begin{bmatrix} M_j^R & -M_j^I \\ M_j^I & M_j^R \end{bmatrix} \begin{bmatrix} \dot{q}_{jd}^R(t) \\ \dot{q}_{jd}^I(t) \end{bmatrix} \tag{6.2.41}$$

$$V = \frac{1}{2}\sum_{j=0}^{n-1} q_{jd}^{\mathrm{H}}(t) K_j q_{jd}(t) = \frac{1}{2}\sum_{j=0}^{[n/2]} a_j^2 \begin{bmatrix} q_{jd}^R(t) \\ q_{jd}^I(t) \end{bmatrix}^{\mathrm{T}} \begin{bmatrix} K_j^R & -K_j^I \\ K_j^I & K_j^R \end{bmatrix} \begin{bmatrix} q_{jd}^R(t) \\ q_{jd}^I(t) \end{bmatrix} \tag{6.2.42}$$

Substituting Eqs. (6.2.41) and (6.2.42) into Lagrange's equation of the second kind, we derive a set of dynamic equations of C_n structure decoupled in S_j, $j \in J_{[n/2]}$ as follows

$$\begin{bmatrix} M_j^R & -M_j^I \\ M_j^I & M_j^R \end{bmatrix} \begin{bmatrix} \ddot{q}_{jd}^R(t) \\ \ddot{q}_{jd}^I(t) \end{bmatrix} + \begin{bmatrix} K_j^R & -K_j^I \\ K_j^I & K_j^R \end{bmatrix} \begin{bmatrix} q_{jd}^R(t) \\ q_{jd}^I(t) \end{bmatrix} = 0, \quad j \in J_{[n/2]} \tag{6.2.43}$$

For a constrained C_n, the mass matrices in Eq. (6.2.43) are positive definite, while the stiffness matrices are symmetric. Let the natural vibration of the C_n structure be expressed as

$$^{(k)}u(t) = {}^{(k)}\varphi \sin(\omega t), \quad k \in I_n \tag{6.2.44}$$

From Eq. (6.2.32) and the decoupling property stated above, the natural vibration falls into a subspace S_j and yields the following form

$$^{(k)}u_j(t) = \frac{a_j}{\sqrt{n}}\mathrm{Re}[\exp(\mathrm{i}jk\theta)\tilde{B}_{jd}q_{jd}]\sin(\omega_j t), \quad k \in I_n, \quad j \in J_{[n/2]} \tag{6.2.45}$$

Here, vector q_{jd} is a complex constant vector satisfying the following eigenvalue problem

$$\left\{ \begin{bmatrix} K_j^R & -K_j^I \\ K_j^I & K_j^R \end{bmatrix} - \omega_j^2 \begin{bmatrix} M_j^R & -M_j^I \\ M_j^I & M_j^R \end{bmatrix} \right\} \begin{bmatrix} q_{jd}^R \\ q_{jd}^I \end{bmatrix} = \mathbf{0}, \quad j \in J_{[n/2]} \qquad (6.2.46)$$

Solving Eq. (6.2.46) gives natural frequency ω_j and the corresponding eigenvector $q_{jd} = q_{jd}^R + i q_{jd}^I$, which leads to, following Eq. (6.2.45), the mode shape as follows

$$^{(k)}\boldsymbol{\varphi}_j = \mathrm{Re}\left[\exp(ijk\theta)\tilde{\boldsymbol{B}}_{jd}q_{jd}\right] \in \mathbb{R}^{m'}, \quad k \in I_n, \quad j \in J_{[n/2]} \qquad (6.2.47)$$

Now, mode shape $^{(k)}\boldsymbol{\varphi}_j$ includes the interfacial displacement vector for J', which is redundant information. After canceling the last row in Eq. (6.2.31) to simplify Eq. (6.2.47) and still using $^{(k)}\boldsymbol{\varphi}_j$ for the mode shape without redundant information, we have

$$^{(k)}\boldsymbol{\varphi}_j = \mathrm{Re}\left[\exp(ijk\theta)q_{jd}\right] \in \mathbb{R}^{m}, \quad k \in I_n, \quad j \in J_{[n/2]} \qquad (6.2.48)$$

Finally, we describe the mode shape of the entire C_n structure in a global frame of coordinates. For this purpose, let the global frame be identical to the local frame $^{(0)}F$ of sub-structure $^{(0)}S$ and define the transform matrix from local frame $^{(k)}F$ to local frame $^{(0)}F$, that is, the orthogonal matrix of rotation through an angle $-k\theta$ about the central axis as follows

$$\boldsymbol{R}^{-k} = \begin{bmatrix} \cos(k\theta) & -\sin(k\theta) & 0 \\ \sin(k\theta) & \cos(k\theta) & 0 \\ 0 & 0 & 1 \end{bmatrix} \qquad (6.2.49)$$

From the direction of each DoF of $^{(0)}S$, we select $\boldsymbol{R}^{-k}$ or its sub-matrix and then construct the diagonal block matrix, the order of which coincides with the DoFs of $^{(0)}S$, as follows

$$\boldsymbol{R}_m^{-k} \equiv \mathrm{diag}\left[\boldsymbol{R}^{-k}\right] = \mathrm{diag}\left[\begin{bmatrix} \cos(k\theta) & -\sin(k\theta) & 0 \\ \sin(k\theta) & \cos(k\theta) & 0 \\ 0 & 0 & 1 \end{bmatrix}\right] \in \mathbb{R}^{m \times m} \qquad (6.2.50)$$

Hence, it is feasible to assemble the mode shapes of all sub-structures in Eq. (6.2.48) as a mode shape of the C_n structure in the global frame of coordinates, namely,

$$\boldsymbol{\varphi}_j \equiv \mathrm{col}_k\left[\boldsymbol{R}_m^{-k\,(k)}\boldsymbol{\varphi}_j\right] = \mathrm{col}_k\left[\boldsymbol{R}_m^{-k}\mathrm{Re}[\exp(ijk\theta)q_{jd}]\right] \in \mathbb{R}^{n \times m}, \quad j \in J_{[n/2]} \qquad (6.2.51)$$

Hereinafter, ω_j and $\boldsymbol{\varphi}_j$ are named a natural mode of the C_n structure in subspace S_j.

Remark 6.2.1 When $j = 0$ or $j = n/2 = [n/2]$, $\exp(0) = 1$ or $\exp(ik\pi) = (-1)^k$ holds such that B_j and $\tilde{B}_{jd}$ are real matrices, both M_j and K_j are real matrices, and both q_j and q_{jd} are real vectors. In this case, $M_j^I = 0$, $K_j^I = 0$ and $q_{jd}^I = 0$ hold in Eq. (6.2.46) such that Eq. (6.2.46) becomes an eigenvalue problem of a pair of real symmetric matrices as follows.

$$\left(K_j - \omega_j^2 M_j\right)q_{jd} = 0, \quad j = 0, \; n/2 = [n/2] \tag{6.2.52}$$

In this case, the order of eigenvalue problem decreases from $2m$ in Eq. (6.2.46) to m in Eq. (6.2.52).

Remark 6.2.2 Consider the case when n is an even number. After the above decoupling procedure, we need to solve two real symmetric eigenvalue problems of order m for $j = 0$ and $j = n/2$, and solve $n/2 - 1$ real symmetric eigenvalue problems of order $2m$ for $1 \le j \le n/2 - 1$. Thus, the summation of the orders of all eigenvalue problems to be solved is $\hat{N} = 2m + 2m \times (n/2 - 1) = m \times n$, which is equal to the total order N, given in Eq. (6.2.3), for the eigenvalue problem of the C_n structure.

As studied for the numerical methods of linear algebra, the computational cost of a real symmetric eigenvalue problem is proportional to the order of matrix pencil. As such, the computational cost of solving the eigenvalue problem for natural vibrations of a C_n structure approximately is

$$\text{cost}(N) \propto (n \times m)^3 = n^3 \times m^3 \tag{6.2.53}$$

whereas the computational cost after the decoupling procedure is about

$$\text{cost}(\hat{N}) \propto 2m^3 + (n/2 - 1) \times (2m)^3 = (4n - 6) \times m^3 \tag{6.2.54}$$

With an increase of n, $\text{cost}(\hat{N})$ becomes a very small fraction of $\text{cost}(N)$. Accordingly, the above estimations do not account for the number of matrix multiplications of generating Eq. (6.2.43) since it occupies a very small portion of the total computational cost.

Example 6.2.4 Study the natural vibrations of the C_4 structure in Fig. 6.21 by using the decoupling method of group theory.

The mass matrix and stiffness matrix of sub-structure $^{(k)}S$ take the following form

$$M = \frac{m}{2}\begin{bmatrix} 1 & 0 \\ 0 & 1 \end{bmatrix}, \quad K = \bar{k}\begin{bmatrix} 1 & \beta \\ \beta & 1 \end{bmatrix}, \quad \beta < 1 \tag{a}$$

Substitution of transform matrix $\tilde{B}_{jd}$ derived in Example 6.2.3 into Eq. (6.2.37) leads to the mass and stiffness matrices in subspace S_j as follows

$$\begin{cases} M_j = \tilde{B}_{jd}^{H} M \tilde{B}_{jd} = \dfrac{\overline{m}}{2} \begin{bmatrix} 1 & \exp(-\mathrm{i}j\pi/2) \end{bmatrix} \begin{bmatrix} 1 & 0 \\ 0 & 1 \end{bmatrix} \begin{bmatrix} 1 \\ \exp(\mathrm{i}j\pi/2) \end{bmatrix} = \overline{m} \\ K_j = \tilde{B}_{jd}^{H} K \tilde{B}_{jd} = 2\overline{k}\{1 + \beta \mathrm{Re}[\exp(\mathrm{i}j\pi/2)]\}, \quad j = 0,\ 1,\ 2 \end{cases} \quad (b)$$

Solving eigenvalue problems for $j = 0,\ 1,\ 2$ gives the natural frequencies and eigenvectors in subspace S_j, $j = 0,\ 1,\ 2$. Substitution of eigenvectors into Eq. (6.2.51) leads to the mode shapes of the C_4 structure in subspace S_j, $j = 0,\ 1,\ 2$. As $R_{m'}^{-k} = [1]$ holds, the mode shapes in Eq. (6.2.51) simply become

$$\tilde{\varphi}_j = \mathrm{col}_k \big[\mathrm{Re}[\exp(\mathrm{i}j k\pi/2) q_{j1}] \big], \quad j = 0, 1, 2 \quad (c)$$

According to Remark 6.2.1, we discuss three cases as follows.

When $j = 0$, we have a real eigenvalue problem of order one

$$\big(M_0 - \omega_0^2 K_0 \big) q_{01} = 0, \quad M_0 = \overline{m}, \quad K_0 = 2\overline{k}(1 + \beta) \quad (d)$$

The corresponding natural frequency and eigenvector are

$$\omega_{01} = \sqrt{\frac{2\overline{k}(1 + \beta)}{\overline{m}}}, \quad q_{01} = 1 \quad (e)$$

Substitution of $j = 0$ and $q_{01} = 1$ into Eq. (c) leads to the mode shape as follows

$$\varphi_{01} = \begin{bmatrix} 1 & 1 & 1 & 1 \end{bmatrix}^{\mathrm{T}} \quad (f)$$

As shown in Eq. (f), the four lumped masses in this case have identical vibration amplitudes in phase as shown in Fig. 6.22a, where symbol $\oplus$ represents the positive vibration amplitude.

When $j = 2$, we also have a real eigenvalue problem of order one as follows

$$\big(M_2 - \omega_2^2 K_2 \big) q_{21} = 0, \quad M_2 = \overline{m}, \quad K_2 = 2\overline{k}(1 - \beta) \quad (g)$$

There follow a natural frequency and an eigenvector as follows

$$\omega_{21} = \sqrt{\frac{2\overline{k}(1 - \beta)}{\overline{m}}}, \quad q_{21} = 1 \quad (h)$$

Substituting $j = 2$ and $q_{21} = 1$ into Eq. (c) gives the following mode shape

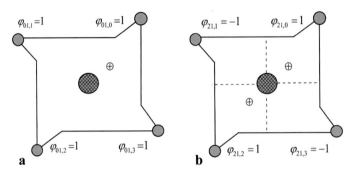

Fig. 6.22 Two natural modes with a single frequency of a C_4 structure in subspaces S_0 and S_2. **a** $\omega_{01} = \sqrt{2\bar{k}(1+\beta)/\bar{m}}$, $\boldsymbol{\varphi}_{01} = [1 \quad 1 \quad 1 \quad 1]^T$; **b** $\omega_{21} = \sqrt{2\bar{k}(1-\beta)/\bar{m}}$, $\boldsymbol{\varphi}_{21} = [1 \quad -1 \quad 1 \quad -1]^T$

$$\boldsymbol{\varphi}_{21} = \begin{bmatrix} 1 & -1 & 1 & -1 \end{bmatrix}^T \tag{i}$$

In this case, the adjacent lumped masses have the same vibration amplitude, but in opposite directions as shown in Fig. 6.22b, where two nodal lines appear.

When $j = 1$, we get a real eigenvalue problem of order two

$$\begin{bmatrix} K_1 - \omega_1^2 M_1 & 0 \\ 0 & K_1 - \omega_1^2 M_1 \end{bmatrix} \begin{bmatrix} \mathrm{Re}(q_{11}) \\ \mathrm{Im}(q_{11}) \end{bmatrix} = \begin{bmatrix} 0 \\ 0 \end{bmatrix}, \quad M_1 = \bar{m}, \quad K_1 = 2\bar{k} \tag{j}$$

Solving Eq. (j) leads to a pair of repeated natural frequencies and a pair of linearly independent eigenvectors as follows

$$\hat{\omega}_{11} = \check{\omega}_{11} = \sqrt{\frac{2\bar{k}}{\bar{m}}}, \quad \begin{bmatrix} \mathrm{Re}(\hat{q}_{11}) \\ \mathrm{Im}(\hat{q}_{11}) \end{bmatrix} = \begin{bmatrix} 1 \\ 0 \end{bmatrix}, \quad \begin{bmatrix} \mathrm{Re}(\check{q}_{11}) \\ \mathrm{Im}(\check{q}_{11}) \end{bmatrix} = \begin{bmatrix} 0 \\ 1 \end{bmatrix} \tag{k}$$

Substitution of $j = 1$ and the above eigenvectors into Eq. (c) leads to a pair of mode shapes with repeated frequencies as follows

$$\hat{\boldsymbol{\varphi}}_{11} = \begin{bmatrix} 1 & 0 & -1 & 0 \end{bmatrix}^T, \quad \check{\boldsymbol{\varphi}}_{11} = \begin{bmatrix} 0 & -1 & 0 & 1 \end{bmatrix}^T \tag{l}$$

As any linear combination of two mode shapes in Eq. (l) is also the mode shape with repeated frequencies, a new pair of mode shapes can be defined as

$$\hat{\boldsymbol{\psi}}_{11} = \hat{\boldsymbol{\varphi}}_{11} + \check{\boldsymbol{\varphi}}_{11} = \begin{bmatrix} 1 & -1 & -1 & 1 \end{bmatrix}^T, \quad \check{\boldsymbol{\psi}}_{11} = \hat{\boldsymbol{\varphi}}_{11} - \check{\boldsymbol{\varphi}}_{11} = \begin{bmatrix} 1 & 1 & -1 & -1 \end{bmatrix}^T \tag{m}$$

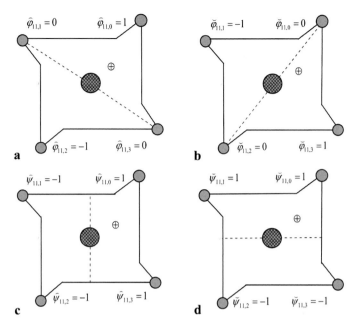

Fig. 6.23 Two pairs of natural modes with a pair of repeated frequencies of a C_4 structure in subspace S_1. **a** $\widehat{\omega}_{11} = \sqrt{2\overline{k}/m}$, $\widehat{\boldsymbol{\varphi}}_{11} = [1 \quad 0 \quad -1 \quad 0]^T$; **b** $\widecheck{\omega}_{11} = \sqrt{2\overline{k}/m}$, $\widecheck{\boldsymbol{\varphi}}_{11} = [0 \quad -1 \quad 0 \quad 1]^T$; **c** $\widehat{\omega}_{11} = \sqrt{2\overline{k}/m}$, $\widehat{\boldsymbol{\psi}}_{11} = [1 \quad -1 \quad -1 \quad 1]^T$; **d** $\widecheck{\omega}_{11} = \sqrt{2\overline{k}/m}$, $\widecheck{\boldsymbol{\psi}}_{11} = [1 \quad 1 \quad -1 \quad -1]^T$

It is worthy to notice that different symbols are used from Eqs. (k) to (m) to emphasize the natural modes with a pair of repeated frequencies.

Figures 6.23a and 6.23b demonstrate the nodal lines of a pair of mode shapes with repeated frequencies in Eq. (l), while Figs. 6.23c and 6.23d show the nodal lines of the other pair of mode shapes in Eq. (m). The number of such pairs of mode shapes is infinite. In the two pairs in Fig. 6.23, each mode shape remains unchanged when it rotates through $\pi/2$ counter-clockwise about the central axis. This property is similar to a pair of mode shapes with repeated frequencies for the square plate in Sect. 6.1.3.

Remark 6.2.3 For the above C_4 structure, the clamped condition at the central axis determines the sign of parameter β in the stiffness matrix of sub-structure $^{(k)}S$ and the ranking of natural frequencies. If one lumped mass in the structure is subject to an upward static load and the adjacent lumped mass moves downward, the cross flexibility between them is negative such that $\beta > 0$ holds. Hence, there is the following ranking of natural frequencies, i.e., $\omega_{21} < \widehat{\omega}_{11} = \widecheck{\omega}_{11} < \omega_{01}$. On the contrary, $\beta < 0$ holds such that the ranking becomes $\omega_{01} < \widehat{\omega}_{11} = \widecheck{\omega}_{11} < \omega_{21}$.

Fig. 6.24 The schematic of
a C_n structure deformable at
the central axis ($n = 4$)

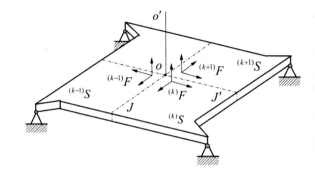

6.2.2   Decoupling a Cyclosymmetric Structure with a Central Axis

The subsection presents the detailed study on the natural vibrations of a C_n structure with a central axis, or called central DoFs, demonstrated in Fig. 6.24. In the 1970s, D. L. Thomas in the UK studied the standing waves of a cyclosymmetric structure in the circumferential direction. Many people believed, thus, that the cyclosymmetric structure with a central axis may not exhibit standing waves in the circumferential direction. As a consequence, a few studies dealt with the natural vibrations of a cyclosymmetric structure with a central axis.

The author studied the C_n structure with a central axis in 1986, and found that the central axis does not have the complicated influence on the standing waves of the structure, but only imposes the displacement compatibility conditions on all sub-structures at the central DoFs[11]. For example, all sub-structures $^{(k)}S,\ k \in I_n$ of the C_n structure in Fig. 6.24 share the interfacial DoFs at the central axis and require the compatible central displacements in subspace S_j. In this case, the other DoFs of $^{(k)}S,\ k \in I_n$ still satisfy the previous results in Sect. 6.2.1. This idea has seen the successful application of a recent study by B. Dong and R. G. Parker in the USA for computing the natural vibrations of the C_n structure with a central axis by using the method of cyclic matrix[12].

To begin with, denote the *central displacement vector* of sub-structure $^{(k)}S$ as $^{(k)}\boldsymbol{u}_o \in \mathbb{R}^o$, where o is the central DoF set and also denotes the number of DoFs in set o. Then, we assemble vector $^{(k)}\boldsymbol{u}_o$ and other displacement vector $^{(k)}\boldsymbol{u}$ of sub-structure $^{(k)}S$ as follows

$$^{(k)}\tilde{\boldsymbol{u}} \equiv \begin{bmatrix} ^{(k)}\boldsymbol{u}_o^{\mathrm{T}} & ^{(k)}\boldsymbol{u}^{\mathrm{T}} \end{bmatrix}^{\mathrm{T}} \in \mathbb{R}^{\tilde{m}'}, \quad \tilde{m}' \equiv o + I + J + J', \quad k \in I_N \qquad (6.2.55)$$

---

[11] Hu H Y, Cheng D L, Wang L (1986). Vibration analysis of a cyclosymmetric structure with central axis. Shanghai: The 2nd National Conference on Computational Mechanics, No. 2–236 (in Chinese).

[12] Dong B, Parker R G (2018). Modal properties of cyclically symmetric systems with central components vibrating as three-dimensional rigid bodies. Journal of Sound and Vibration, 435: 350–371.

In addition, let $M \in \mathbb{R}^{\tilde{m}' \times \tilde{m}'}$ be the mass matrix and $K \in \mathbb{R}^{\tilde{m}' \times \tilde{m}'}$ be the stiffness matrix for sub-structure $^{(k)}S$, respectively.

Because all sub-structures share central DoF set o and two adjacent sub-structures share interfacial DoF set J', the total DoFs of the C_n structure satisfy

$$\tilde{N} = o + n \times m = o + n \times (I + J) \tag{6.2.56}$$

6.2.2.1 Transform of Generalized Displacements

Following the ideas in Sect. 6.2.1, we use a set of generalized displacement vectors $\tilde{q}_j$, $j \in I_n$ to express displacement vectors $^{(k)}\tilde{u}$, $k \in I_n$ as follows

$$^{(k)}\tilde{u} = \frac{1}{\sqrt{n}} \sum_{j=0}^{n-1} \exp(\mathrm{i}jk\theta)\tilde{q}_j = \frac{1}{\sqrt{n}} \sum_{j=0}^{[n/2]} a_j \exp(\mathrm{i}jk\theta)\tilde{q}_j, \quad k \in I_n \tag{6.2.57}$$

where

$$\begin{cases} \tilde{q}_j \equiv \mathrm{col}_l\left[q_{jl}\right] \in \mathbb{C}^{\tilde{m}'} \\ q_{jl} \equiv \mathbf{e}_j^{\mathrm{H}} \mathbf{u}_l = \dfrac{1}{\sqrt{n}} \displaystyle\sum_{k=0}^{n-1} {}^{(k)}\tilde{u}_l \exp(-\mathrm{i}jk\theta) \quad j \in I_n, \quad l \in \tilde{I}_{\tilde{m}'} \end{cases} \tag{6.2.58}$$

It is worthy to notice that the first o entries in vectors $\tilde{q}_j$, $j \in I_n$ correspond to central displacement vector $^{(k)}\mathbf{u}_o$, $k \in I_n$, and require a special treatment for interfacial compatibility.

6.2.2.2 Transform of Interfacial Compatibility

Now, we study the interfacial compatibility of central displacements and other displacements in two steps.

As the first step, we analyze the displacement compatibility of the C_n structure at the central DoFs. The vectors $^{(k)}\mathbf{u}_o$, $k \in I_n$ describe the same displacements at central DoF set o, but look different from the local frame $^{(k)}F$ of sub-structure $^{(k)}S$. Taking $^{(0)}\mathbf{u}_o$ as the central displacement vector in subsequent computations, it easy to obtain $^{(k)}\mathbf{u}_o$ by rotating the local frame $^{(0)}F$ about the central axis to $^{(k)}F$ of sub-structure $^{(k)}S$.

In line with this idea, define the transform matrix from $^{(0)}F$ to $^{(k)}F$, that is, the orthogonal matrix through an angle $k\theta$ about the central axis as follows

$$\boldsymbol{R}^k = \begin{bmatrix} \cos(k\theta) & \sin(k\theta) & 0 \\ -\sin(k\theta) & \cos(k\theta) & 0 \\ 0 & 0 & 1 \end{bmatrix} \tag{6.2.59}$$

Here, Eq. (6.2.59) implies different rotational direction from the orthogonal matrix in Eq. (6.2.49) such that the superscript k has a different sign. From the direction of central DoFs, we select $\boldsymbol{R}^k$ or its sub-matrix to construct the diagonal block matrix of orthogonal transform from $^{(0)}\boldsymbol{u}_o \in \mathbb{R}^o$ to $^{(k)}\boldsymbol{u}_o \in \mathbb{R}^o$, namely,

$$\boldsymbol{R}_o^k \equiv \mathrm{diag}\big[\boldsymbol{R}^k\big] = \mathrm{diag}\left[\begin{bmatrix} \cos(k\theta) & \sin(k\theta) & 0 \\ -\sin(k\theta) & \cos(k\theta) & 0 \\ 0 & 0 & 1 \end{bmatrix}\right] \in \mathbb{R}^{o\times o} \tag{6.2.60}$$

There follows

$$^{(k)}\boldsymbol{u}_o = \boldsymbol{R}_o^{k\,(0)}\boldsymbol{u}_o, \quad k \in I_n \tag{6.2.61}$$

Letting R_{ls}^k be an entry in matrix $\boldsymbol{R}_o^k$, we use the displacements of sub-structures $^{(k)}S$, $k \in I_n$ at the l-th central DoF to form a vector as follows

$$\boldsymbol{u}_l = \mathrm{col}_k\big[^{(k)}u_l\big] = \mathrm{col}_k\big[\mathrm{row}_s[R_{ls}^k]^{(0)}\boldsymbol{u}_o\big], \quad l \in \tilde{I}_o \tag{6.2.62}$$

According to Eq. (6.2.58), we use central displacement vector $^{(0)}\boldsymbol{u}_o$ to describe the l-th entry q_{jl} in generalized displacement vectors $\tilde{\boldsymbol{q}}_j$, $j \in I_n$, namely,

$$q_{jl} \equiv \mathbf{e}_j^H \boldsymbol{u}_l = \frac{1}{\sqrt{n}} \sum_{k=0}^{n-1} \exp(-\mathrm{i}jk\theta)\,\mathrm{row}_s[R_{ls}^k]^{(0)}\boldsymbol{u}_o, \quad j \in I_n, \quad l \in \tilde{I}_o \tag{6.2.63}$$

Equation (6.2.63) enables one to cast the generalized displacement vector as follows

$$\begin{aligned} \boldsymbol{q}_{jo} &\equiv \mathrm{col}_l[q_{jl}] = \frac{1}{\sqrt{n}}\mathrm{col}_l\left[\sum_{k=0}^{n-1} \exp(-\mathrm{i}jk\theta)\,\mathrm{row}_s[R_{ls}^k]^{(0)}\boldsymbol{u}_o\right] \\ &= \frac{1}{\sqrt{n}}\left[\sum_{k=0}^{n-1} \exp(-\mathrm{i}jk\theta)\boldsymbol{R}_o^k\right]^{(0)}\boldsymbol{u}_o = \boldsymbol{D}_j^{(0)}\boldsymbol{u}_o, \quad j \in I_n \end{aligned} \tag{6.2.64}$$

Equation (6.2.64) presents the transform from central displacement vector $^{(0)}\boldsymbol{u}_o \in \mathbb{R}^o$ to generalized displacement vector $\boldsymbol{q}_{jo} \in \mathbb{C}^o$ in S_j, and $\boldsymbol{D}_j$ yields

$$D_j \equiv \frac{1}{\sqrt{n}} \left[\sum_{k=0}^{n-1} \exp(-ijk\theta)\, R_o^k \right] \in \mathbb{C}^{o \times o}, \quad j \in I_n \tag{6.2.65}$$

As the second step, we study the interfacial compatibility for other displacements, rather than the central displacements. If the first o entries in vector $\tilde{q}_j$ are canceled, the remaining sub-vector is identical to vector q_j in Sect. 6.2.1. Hence, we simply adopt the results in Sect. 6.2.1 and copy Eq. (6.2.30) as follows

$$q_j = \begin{bmatrix} q_{jd} \\ q_{jr} \end{bmatrix} = \begin{bmatrix} I_m \\ -B_{jd} \end{bmatrix} q_{jd} \quad j \in J_{[n/2]} \tag{6.2.66}$$

Assembling the results in the two steps leads to the interfacial compatibility transform described by independent generalized displacement vector q_{jD}, namely,

$$\tilde{q}_j \equiv \begin{bmatrix} q_{jo} \\ q_{jd} \\ q_{jr} \end{bmatrix} = \begin{bmatrix} D_j & 0 \\ 0 & I_m \\ 0 & -B_{jd} \end{bmatrix} \begin{bmatrix} {}^{(0)}u_o \\ q_{jd} \end{bmatrix} \equiv \tilde{B}_{jD} q_{jD} \quad j \in J_{[n/2]} \tag{6.2.67}$$

where $q_{jD} \in \mathbb{C}^{\tilde{m}}$, $D \equiv o \cup I \cup J$ and $\tilde{m} \equiv o + I + J$. The first o entries in generalized displacement vector $\tilde{q}_j$ satisfy Eq. (6.2.64).

6.2.2.3 Dynamic Equations Decoupled

To study the dynamic problem of the C_n structure, let all displacements and generalized displacements be functions of time t. Composing the above two transforms leads to

$${}^{(k)}u(t) = \frac{1}{\sqrt{n}} \sum_{j=0}^{[n/2]} a_j \mathrm{Re}\left[\exp(ijk\theta)\, \tilde{B}_{jD} q_{jD}(t) \right], \quad k \in I_n \tag{6.2.68}$$

According to the idea in Sect. 6.2.1, we derive the following real symmetric eigenvalue problem for the natural vibrations of the C_n structure

$$\left\{ \begin{bmatrix} K_j^R & -K_j^I \\ K_j^I & K_j^R \end{bmatrix} - \omega_j^2 \begin{bmatrix} M_j^R & -M_j^I \\ M_j^I & M_j^R \end{bmatrix} \right\} \begin{bmatrix} q_{jD}^R \\ q_{jD}^I \end{bmatrix} = 0, \quad j \in J_{[n/2]} \tag{6.2.69}$$

where q_{jD}^R and q_{jD}^I are constant vectors, and all sub-matrices are real square matrices of order $2\tilde{m}$ and satisfy the following property of Hermite matrices

$$\begin{cases} M_j^R \equiv \mathrm{Re}\left(\tilde{B}_{jD}^{\mathrm{H}} M \tilde{B}_{jD}\right) = \left(M_j^R\right)^{\mathrm{T}}, \quad M_j^I \equiv \mathrm{Im}\left(\tilde{B}_{jD}^{\mathrm{H}} M \tilde{B}_{jD}\right) = -\left(M_j^I\right)^{\mathrm{T}} \\ K_j^R \equiv \mathrm{Re}\left(\tilde{B}_{jD}^{\mathrm{H}} K \tilde{B}_{jD}\right) = \left(K_j^R\right)^{\mathrm{T}}, \quad K_j^I \equiv \mathrm{Im}\left(\tilde{B}_{jD}^{\mathrm{H}} K \tilde{B}_{jD}\right) = -\left(K_j^I\right)^{\mathrm{T}}, \quad j \in J_{[n/2]} \end{cases}$$

$$(6.2.70)$$

Remark 6.2.4 According to Remark 6.2.1, when $j = 0$ or $j = n/2 = [n/2]$, B_j and $\tilde{B}_{jd}$ are real matrices, and q_j and q_{jD} are real vectors. Hence, $M_j^I = 0$, $K_j^I = 0$ and $q_{jD}^I = 0$ such that Eq. (6.2.69) becomes the following real symmetric eigenvalue problem

$$\left(K_j - \omega_j^2 M_j\right)q_{jD} = 0, \quad j = 0, \ n/2 = [n/2] \qquad (6.2.71)$$

In this case, the order of eigenvalue problem decreases from $2\tilde{m} = 2(o + I + J)$ for Eq. (6.2.69) to $\tilde{m} = o + I + J$ for Eq. (6.2.71). Section 6.3.3 will further discuss the order of eigenvalue problems and provide more accurate order by eliminating redundant DoFs.

Solving Eq. (6.2.69) for eigenvalues and eigenvectors, and substituting the eigenvectors into Eq. (6.2.68), we get the natural frequencies ω_j and the corresponding mode shapes

$$^{(k)}\varphi_j = \mathrm{Re}\left[\exp(\mathrm{i}jk\theta)\tilde{B}_{jD}q_{jD}\right] \in \mathbb{R}^{\tilde{m}'}, \quad k \in I_n, \quad j \in J_{[n/2]} \qquad (6.2.72)$$

Similar to the analysis in Sect. 6.2.1, to remove the redundant information from vector $^{(k)}\varphi_j$ in interfacial DoF set J', we eliminate the last row of blocks in the transform of interfacial compatibility and define the following transform matrix

$$B_{jD} \equiv \begin{bmatrix} D_j & 0 \\ 0 & I_m \end{bmatrix} \in \mathbb{R}^{\tilde{m} \times \tilde{m}}, \quad j \in J_{[n/2]} \qquad (6.2.73)$$

Using the original symbol for mode shapes $^{(k)}\varphi_j$, we get the mode shapes of substructure $^{(k)}S$ without the redundant information as following

$$^{(k)}\varphi_j = \mathrm{Re}\left[\exp(\mathrm{i}jk\theta)B_{jD}q_{jD}\right] \in \mathbb{R}^{\tilde{m}}, \quad k \in I_n, \quad j \in J_{[n/2]} \qquad (6.2.74)$$

Following the final discussion in Sect. 6.2.1, we use the orthogonal matrix $R_{\tilde{m}}^{-k}$ to transform Eq. (6.2.74) to the global frame of coordinates and assemble the mode shapes of the entire C_n structures. Of course, all sub-structures share the central displacements and we ought to remove the redundant information again. In comparison with Eq. (6.2.67), we derive the mode shapes of entire structure C_n in the global frame of coordinates, namely,

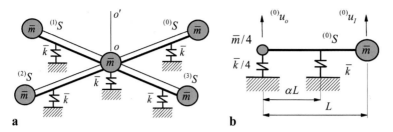

Fig. 6.25 The schematic of a C_4 structure with a central axis. **a** The division of sub-structures; **b** Sub-structure $^{(0)}S$ and its displacements

$$\boldsymbol{\varphi}_j = \begin{bmatrix} ^{(0)}\boldsymbol{u}_o \\ \mathrm{col}_k\big[\boldsymbol{R}_m^{-k}\mathrm{Re}[\exp(\mathrm{i}jk\theta)\boldsymbol{q}_{jd}]\big] \end{bmatrix}, \quad j \in J_{[n/2]} \tag{6.2.75}$$

The dimension of mode shape vector $\boldsymbol{\varphi}_j$ is $o + n \times m = o + n \times (I + J)$.

Example 6.2.5 A C_4 structure with a central axis is shown in Fig. 6.25a, where the five lumped masses with the same value $\overline{m}$ are mounted on four identical massless rigid bars supported by five identical springs with the same stiffness coefficient $\overline{k}$. Use the above decoupling method to study the natural vibrations of the C_4 structure.

To begin with, take $1/4$ part of the C_4 structure in Fig. 6.25a as the sub-structure $^{(0)}S$ shown in Fig. 6.25b, where $1/4$ part of the central mass and central spring fall into $^{(0)}S$, while set J and set J' for interfacial DoFs are empty. Then, we establish the local frame of coordinates $^{(0)}F$ for sub-structure $^{(0)}S$, and use the central displacement $^{(0)}u_o$ and the tip displacement $^{(0)}u_I$ to assemble the displacement vector for $^{(0)}S$ as follows

$$^{(0)}\tilde{\boldsymbol{u}} \equiv \begin{bmatrix} ^{(0)}u_o & ^{(0)}u_I \end{bmatrix}^{\mathrm{T}} \tag{a}$$

The kinetic energy and potential energy of sub-structure $^{(0)}S$ satisfy

$$T = \frac{1}{2}\left(\frac{\overline{m}}{4}\dot{u}_o^2 + \overline{m}\dot{u}_I^2\right), \quad V = \frac{1}{2}\left\{\frac{\overline{k}}{4}u_o^2 + \overline{k}[u_o + \alpha(u_I - u_o)]^2\right\} \tag{b}$$

whereby the mass and stiffness matrices of sub-structure $^{(0)}S$ are as follows

$$\boldsymbol{M} = \overline{m}\begin{bmatrix} 1/4 & 0 \\ 0 & 1 \end{bmatrix}, \quad \boldsymbol{K} = \overline{k}\begin{bmatrix} 1/4 + (1-\alpha)^2 & \alpha(1-\alpha) \\ \alpha(1-\alpha) & \alpha^2 \end{bmatrix} \tag{c}$$

Then, we study the natural vibrations of the C_4 structure in subspace S_j. Equation (6.2.67) gives the interfacial compatibility transform matrix in S_j, namely,

$$\tilde{\boldsymbol{B}}_{jD} = \begin{bmatrix} D_j & 0 \\ 0 & 1 \end{bmatrix}, \quad j \in J_2 \tag{d}$$

There follow the mass and stiffness matrices of the C_4 structure in subspace S_j

$$\begin{cases} \boldsymbol{M}_j = \tilde{\boldsymbol{B}}_{jD}^{\text{H}} \boldsymbol{M} \tilde{\boldsymbol{B}}_{jD} = \overline{m} \begin{bmatrix} D_j^2/4 & 0 \\ 0 & 1 \end{bmatrix} \\ \boldsymbol{K}_j = \tilde{\boldsymbol{B}}_{jD}^{\text{H}} \boldsymbol{K} \tilde{\boldsymbol{B}}_{jD} = \overline{k} \begin{bmatrix} [1/4 + (1-\alpha)^2] D_j^2 & \alpha(1-\alpha)\overline{D}_j \\ \alpha(1-\alpha)D_j & \alpha^2 \end{bmatrix} \end{cases} \tag{e}$$

where

$$D_j = \frac{1}{2}\left[1 + \exp\left(-\mathrm{i}\frac{j\pi}{2}\right) + \exp\left(-\mathrm{i}\frac{2j\pi}{2}\right) + \exp\left(-\mathrm{i}\frac{3j\pi}{2}\right)\right] = \begin{cases} 2, \ j = 0 \\ 0, \ j = 1 \\ 0, \ j = 2 \end{cases} \tag{f}$$

As an example, we study the natural vibrations of the C_4 structure for $\alpha = 3/4$.

When $j = 0$, substituting $D_0 = 2$, through Eq. (e), into Eq. (6.2.71) leads to the following eigenvalue problem

$$\left\{\overline{k}\begin{bmatrix} 1 + 4(1-\alpha)^2 & 2\alpha(1-\alpha) \\ 2\alpha(1-\alpha) & \alpha^2 \end{bmatrix} - \overline{m}\omega_0^2 \begin{bmatrix} 1 & 0 \\ 0 & 1 \end{bmatrix}\right\} \boldsymbol{q}_{0D} = \boldsymbol{0} \tag{g}$$

For $\alpha = 3/4$, we get two natural frequencies and their eigenvectors as follows

$$\begin{cases} \omega_{01} = 0.631\sqrt{\dfrac{\overline{k}}{\overline{m}}}, \quad \boldsymbol{q}_{0D1} = \begin{bmatrix} 0.440 & -1.000 \end{bmatrix}^{\text{T}} \\ \omega_{02} = 1.190\sqrt{\dfrac{\overline{k}}{\overline{m}}}, \quad \boldsymbol{q}_{0D2} = \begin{bmatrix} 1.000 & 0.440 \end{bmatrix}^{\text{T}} \end{cases} \tag{h}$$

Substituting the two eigenvectors into Eq. (6.2.74) leads to the following mode shapes of all sub-structures

$$\begin{cases} {}^{(k)}\boldsymbol{\varphi}_{01} = \boldsymbol{B}_{0D}\boldsymbol{q}_{0D1} = \begin{bmatrix} 0.880 & -1.000 \end{bmatrix}^{\text{T}} \\ {}^{(k)}\boldsymbol{\varphi}_{02} = \boldsymbol{B}_{0D}\boldsymbol{q}_{0D2} = \begin{bmatrix} 1.000 & 0.220 \end{bmatrix}^{\text{T}}, \quad k \in I_n \end{cases} \tag{i}$$

In view of Eq. (6.2.75), we assemble the above mode shapes, remove the redundant central displacements and obtain two mode shapes of the entire C_4 structure associated with subspace S_0 as follows

Fig. 6.26 Two natural modes of the C_4 structure with a central axis in subspace S_0.

a $\omega_{01} = 0.631\sqrt{k/m}$, φ_{01};

b $\omega_{02} = 1.190\sqrt{k/m}$, φ_{02}

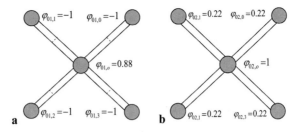

$$\begin{cases} \varphi_{01} = \begin{bmatrix} 0.880 & -1.000 & -1.000 & -1.000 & -1.000 \end{bmatrix}^T \\ \varphi_{02} = \begin{bmatrix} 1.000 & 0.220 & 0.220 & 0.220 & 0.220 \end{bmatrix}^T \end{cases} \tag{j}$$

Figure 6.26 depicts the nodal lines of the two mode shapes and shows that the natural vibrations of all sub-structures are identical. The mode shape in Fig. 6.26a exhibits a nodal line on each rigid bar, while the mode shape in Fig. 6.26b does not have any nodal lines. The above results indicate again that the first mode shape of a structure may have nodal lines, but the mode shape with higher natural frequency may not have any nodal lines.

When $j = 2$, substituting $D_2 = 0$ into Eq. (e) gives a singular mass matrix and a singular stiffness matrix as follows

$$M_2 = \bar{m} \begin{bmatrix} 0 & 0 \\ 0 & 1 \end{bmatrix}, \quad K_2 = \bar{k} \begin{bmatrix} 0 & 0 \\ 0 & \alpha^2 \end{bmatrix} \tag{k}$$

The first row and the first column in both matrices have null entries only such that the central DoF has no use in subspace S_2. Thus, we remove the corresponding row and column in Eq. (k) and discuss this treatment further in Remark 6.2.5. Substituting the remaining entries in Eq. (k) into Eq. (6.2.71), we get one natural frequency and the corresponding eigenvector as follows

$$\omega_{21} = \alpha\sqrt{\frac{k}{m}}, \quad q_{2D1} = \begin{bmatrix} 0 & 1 \end{bmatrix}^T \tag{l}$$

Substitution of the eigenvector into Eq. (6.2.74) gives the mode shapes of all substructures, namely,

$$^{(k)}\varphi_{21} = \begin{bmatrix} 0 & (-1)^k \end{bmatrix}^T, \quad k \in I_n \tag{m}$$

From Eq. (6.2.75), we eliminate redundant information and assemble the mode shape of the entire C_4 structure in subspace S_2 as follows

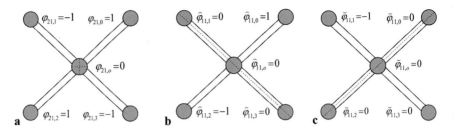

Fig. 6.27 Three natural modes of the C_4 structure with a central axis in subspace S_2 and subspace S_1. **a** $\omega_{21} = \alpha\sqrt{k/m}$, $\boldsymbol{\varphi}_{21}$, **b** $\hat{\omega}_{11} = \alpha\sqrt{k/m}$, $\hat{\boldsymbol{\varphi}}_{11}$, **c** $\breve{\omega}_{11} = \alpha\sqrt{k/m}$, $\breve{\boldsymbol{\varphi}}_{11}$

$$\boldsymbol{\varphi}_{21} \equiv \begin{bmatrix} 0 & 1 & -1 & 1 & -1 \end{bmatrix}^{\mathrm{T}} \tag{n}$$

As shown in Eq. (n), the natural vibrations of two adjacent sub-structures have the same amplitude but opposite directions. Figure 6.27a depicts the nodal lines of the mode shape. Of course, the central lumped mass is rigid and dose not have any deformation. The nodal lines here only represent the vibration direction of adjacent sub-structures.

When $j = 1$, substitution $D_1 = 0$ into Eq. (e) also leads to a singular mass matrix and a singular stiffness matrix as follows

$$\boldsymbol{M}_1 = \overline{m}\begin{bmatrix} 0 & 0 \\ 0 & 1 \end{bmatrix}, \quad \boldsymbol{K}_1 = \overline{k}\begin{bmatrix} 0 & 0 \\ 0 & \alpha^2 \end{bmatrix} \tag{o}$$

Different from case (2), the eigenvalue problem to be solved comes from Eq. (6.2.69) as follows

$$\left\{ \begin{bmatrix} \boldsymbol{K}_1 & \boldsymbol{0} \\ \boldsymbol{0} & \boldsymbol{K}_1 \end{bmatrix} - \omega_1^2 \begin{bmatrix} \boldsymbol{M}_1 & \boldsymbol{0} \\ \boldsymbol{0} & \boldsymbol{M}_1 \end{bmatrix} \right\} \begin{bmatrix} \mathrm{Re}(\boldsymbol{q}_{1D}) \\ \mathrm{Im}(\boldsymbol{q}_{1D}) \end{bmatrix} = \boldsymbol{0} \tag{p}$$

We eliminate the first row and the first column in Eq. (o), substitute the result into Eq. (p) and solve the eigenvalue problem. Then, we get a pair of repeated natural frequencies and two eigenvectors as follows

$$\hat{\omega}_{11} = \alpha\sqrt{\frac{k}{m}}, \quad \hat{\boldsymbol{q}}_{1D1} = \begin{bmatrix} 0 \\ 1 \end{bmatrix}, \quad \breve{\omega}_{11} = \alpha\sqrt{\frac{k}{m}}, \quad \breve{\boldsymbol{q}}_{1D1} = \begin{bmatrix} 0 \\ i \end{bmatrix} \tag{q}$$

Substitution of the eigenvectors in Eq. (q) into Eq. (6.2.74) gives a pair of mode shapes for all sub-structures, namely,

$$
\begin{cases}
{}^{(k)}\widehat{\boldsymbol{\varphi}}_{11} = \mathrm{Re}\left\{\exp(\mathrm{i}k\pi/2)\begin{bmatrix} 0 & 1 \end{bmatrix}^{\mathrm{T}}\right\} \\
{}^{(k)}\widecheck{\boldsymbol{\varphi}}_{11} = \mathrm{Re}\left\{\exp(\mathrm{i}k\pi/2)\begin{bmatrix} 0 & \mathrm{i} \end{bmatrix}^{\mathrm{T}}\right\}, \quad k \in I_n
\end{cases}
\tag{r}
$$

From Eq. (6.2.75), we remove the redundant information of the central displacements and assemble the two mode shapes with repeated frequencies of the entire C_4 structure in subspace S_1 as follows

$$
\begin{cases}
\widehat{\boldsymbol{\varphi}}_{11} = \begin{bmatrix} 0 & 1 & 0 & -1 & 0 \end{bmatrix}^{\mathrm{T}} \\
\widecheck{\boldsymbol{\varphi}}_{11} = \begin{bmatrix} 0 & 0 & -1 & 0 & 1 \end{bmatrix}^{\mathrm{T}}
\end{cases}
\tag{s}
$$

As shown in Figs. 6.27b and 6.27c, each mode shape has a nodal line such that two rigid bars undergo a rotation about the nodal line in such a natural vibration.

Remark 6.2.5 The C_4 structure in Example 6.2.5 has 5 DoFs, but the summation of orders of all eigenvalues problems is 8. That is, we solved an eigenvalue problem of order 2 in subspace S_0 and S_2 respectively, and solved an eigenvalue problem of order 4 in S_1. The reason is that the central displacement repeatedly appears in the dynamic equations decoupled so that both mass matrix and stiffness matrix in subspaces S_1 and S_2 are singular. Thus, we eliminated the row and column corresponding to the central displacement 3 times in total and the total order of all eigenvalue problems that we solved is 5, as same as five DoFs. This issue will be addressed in detail in Sect. 6.3.3.

6.2.3 High-Efficient Computation Based on Modal Reduction

The C_n structures in engineering are so complicated that their dynamic equations decoupled by using group theory may still have very high dimensions to be studied. For example, the bladed discs in an aero-engine have spatially twisting thin blades and the shell element model of a single sub-structure may have several thousands of DoFs. It is necessary, hence, to simplify the finite element models of those sub-structures before the decoupling procedure via group theory. Among the simplification methods, modal reductions are the most successful. They have two ways of dealing with the interface between two sub-structures. In the modal reduction methods of a sub-structure in C_n structure, the method of fixed-interface is simple[13] while the method of free-interface exhibits higher efficiency in reduction[14]. This subsection presents

[13] Hu H Y, Cheng D L (1986). Generalized mode synthesis of cyclosymmetrical structures. Vibration and Shock, 5(4): 1–7 (in Chinese).

[14] Wang W L, Zhu N S, Xu J H (1990). Double compatible mode synthesis of symmetric structures with C_n group. Journal of Aerospace Power, 5(4): 352–356 (in Chinese).

the generalized mode synthesis proposed by the author based on the method of fixed-interface and demonstrates the approach via the computation of natural vibrations of a bladed disc model.

6.2.3.1   Modal Reduction of Fixed-Interface

The basic idea of modal reduction is to use a few mode shapes of low order to approximately describe the dynamics of a sub-structure, or called a component for short. The most typical modal reduction method of fixed-interface is *the modal reduction of constrained component* proposed by R. R. Craig Jr. and M. M. C Bampton in the USA in 1968. They divided a complicated structure into a number of components, selected a few of modal coordinates of each component with interfaces fixed, together with the interfacial DoFs of the components, to approximately describe the dynamics of the component. As such, they established the dynamic equations of reduced order for the complicated structure after treatment of interfacial compatibility[15].

To explain the above idea in the vibration computation of a C_n structure without a central axis, consider sub-structure $^{(k)}S$ and denote two interfacial DoF sets J and J' as set Z. According to the study in Sect. 6.2.1, denote the displacement vector of sub-structure $^{(k)}S$ as

$$^{(k)}\boldsymbol{u} = \begin{bmatrix} ^{(k)}\boldsymbol{u}_I^{\mathrm{T}} & ^{(k)}\boldsymbol{u}_Z^{\mathrm{T}} \end{bmatrix}^{\mathrm{T}}, \quad k \in I_n \tag{6.2.76}$$

Now, we take sub-structure $^{(k)}S$ as "a super finite element" and use interfacial displacement vector $^{(k)}\boldsymbol{u}_Z \in \mathbb{R}^Z$ to approximately describe internal displacement vector $^{(k)}\boldsymbol{u}_I \in \mathbb{R}^I$ so as to reduce the computation cost. This is the *static reduction method* proposed by R. J. Guyan in the USA in 1965.

To be more specific, we partition the stiffness matrix $\boldsymbol{K}$ of sub-structure $^{(k)}S$ according to the block vector $^{(k)}\boldsymbol{u}$ in Eq. (6.2.76). If the internal DoFs in set I are not subject to any external loads, the static equations of sub-structure $^{(k)}S$ have a block matrix form as follows

$$\boldsymbol{K}^{(k)}\boldsymbol{u} = \begin{bmatrix} \boldsymbol{K}_{II} & \boldsymbol{K}_{IZ} \\ \boldsymbol{K}_{ZI} & \boldsymbol{K}_{ZZ} \end{bmatrix} \begin{bmatrix} ^{(k)}\boldsymbol{u}_I \\ ^{(k)}\boldsymbol{u}_Z \end{bmatrix} = \begin{bmatrix} \boldsymbol{0} \\ ^{(k)}\boldsymbol{f}_Z \end{bmatrix} \tag{6.2.77}$$

where $^{(k)}\boldsymbol{f}_Z$ is the interfacial force vector from adjacent sub-structures. The first row in Eq. (6.2.77) gives

$$^{(k)}\boldsymbol{u}_I = -\boldsymbol{K}_{II}^{-1}\boldsymbol{K}_{IZ}\,^{(k)}\boldsymbol{u}_Z \tag{6.2.78}$$

---

[15] Craig Jr R R, Bampton M M C (1968). Coupling of sub-structures for dynamic analysis. AIAA Journal, 6(7): 1313–1319.

That is, the interfacial displacement vector $^{(k)}u_Z$ is able to describe internal displacement vector $^{(k)}u_I$ in the static status. Thus, we have a linear transform from vector $^{(k)}u_Z \in \mathbb{R}^Z$ to vector $^{(k)}u \in \mathbb{R}^{I+Z}$ as follows

$$^{(k)}u = \begin{bmatrix} ^{(k)}u_I \\ ^{(k)}u_Z \end{bmatrix} = \begin{bmatrix} -K_{II}^{-1}K_{IZ} \\ I_Z \end{bmatrix} {}^{(k)}u_Z \tag{6.2.79}$$

It is easy to see that this transform enables one to greatly reduce the order of a finite element model of sub-structure $^{(k)}S$, but does not account for the dynamic effect of internal displacements and has poor accuracy. However, this idea of model reduction looks similar to the interpolation for the internal displacement of a finite element and stimulated the subsequent studies on the dynamic reduction, i.e., the modal reduction of constrained component.

The method proposed by Craig Jr. and Bampton includes the following three steps. As the first step, we compute the first D mode shapes of sub-structure $^{(k)}S$ with the interfacial DoFs fixed and assemble them as a matrix $\psi_{ID} \in \mathbb{R}^{I \times D}$, where $D \ll I$. If we denote $^{(k)}p_D$ as the modal displacement vector, vector $\psi_{ID}{}^{(k)}p_D$ approximately describes the internal displacement vector $^{(k)}u_I$ in the dynamic status. In the second step, we use the first row in Eq. (6.2.79) to describe the internal displacement vector $^{(k)}u_I$ in the static status as follows $-K_{II}^{-1}K_{IZ}{}^{(k)}u_Z$. The third step is to sum up the above two results. As such, we define the *transform of modal reduction* as following

$$^{(k)}u = \begin{bmatrix} ^{(k)}u_I \\ ^{(k)}u_Z \end{bmatrix} = \begin{bmatrix} \psi_{ID} & -K_{II}^{-1}K_{IZ} \\ 0 & I_Z \end{bmatrix} \begin{bmatrix} ^{(k)}p_D \\ ^{(k)}u_Z \end{bmatrix} \tag{6.2.80}$$

For the sake of unity, Craig Jr. and Bampton recast Eq. (6.2.80) as follows

$$\begin{bmatrix} ^{(k)}u_I \\ ^{(k)}u_Z \end{bmatrix} = \begin{bmatrix} \psi_{ID} & \psi_{IZ} \\ 0 & I_Z \end{bmatrix} \begin{bmatrix} ^{(k)}p_D \\ ^{(k)}p_Z \end{bmatrix} \equiv \Psi p \tag{6.2.81}$$

where they defined $^{(k)}p_Z \equiv {}^{(k)}u_Z \in \mathbb{R}^Z$ as the constrained modal displacement vector for sub-structure $^{(k)}S$, $\psi_{IZ} \equiv -K_{II}^{-1}K_{IZ} \in \mathbb{R}^{I \times Z}$ as the *constrained mode shape matrix* for $^{(k)}S$. Each column in ψ_{IZ} is an internal displacement vector of $^{(k)}S$ due to a unit static displacement at an interfacial DoF in set Z. The modal reduction in Eq. (6.2.81) enables one to approximate the dynamics of the sub-structure $^{(k)}S$, $k \in I_n$ and obtain the dynamic equation of the C_n structure after treatment of interfacial compatibility.

The study by Craig Jr. and Bampton, as well as many others, showed that a few of natural modes of constrained components are able to significantly improve the accuracy of the dynamic equation of the entire structure with greatly reduced order. That is, the modal reduction of constrained component works much better than the static reduction method.

6.2.3.2 Computational Scheme

Now, we study the high-efficient computation scheme of natural vibrations of a C_n structure without a central axis based on the above modal reduction of constrained components. Following the description about the sub-structure $^{(k)}S$ and the transform of modal reduction in Eq. (6.2.81), we describe the displacement vector $^{(k)}u \in \mathbb{R}^{m'}$ of $^{(k)}S$ approximately as follows

$$^{(k)}u = \begin{bmatrix} ^{(k)}u_I \\ ^{(k)}u_J \\ ^{(k)}u_{J'} \end{bmatrix} \approx \begin{bmatrix} \psi_{ID} & \psi_{IJ} & \psi_{IJ'} \\ 0 & I_J & 0 \\ 0 & 0 & I_{J'} \end{bmatrix} \begin{bmatrix} ^{(k)}p_I \\ ^{(k)}p_J \\ ^{(k)}p_{J'} \end{bmatrix} \equiv \psi\,^{(k)}p, \quad k \in I_n \qquad (6.2.82)$$

where $^{(k)}p \in \mathbb{R}^{\hat{m}}$ is the modal displacement vecctor and $\hat{m} \equiv D+2J$. In Eq. (6.2.82), two sets J and J' are resumed for the interfacial DoFs in set Z. From Eq. (6.2.81), the sub-matrices in Eq. (6.2.82) have clear meanings as follows.

(1) In the first column, $\psi_{ID} \in \mathbb{R}^{I \times D}$ consists of D vectors of normalized mode shapes of sub-structure $^{(k)}S$ with interfaces J and J' fixed such that the rest sub-matrices in this column are sub-matrix 0, and $D << I$ holds for the reduction purpose. Accordingly, $^{(k)}p_I \in \mathbb{R}^D$ is the modal displacement vector kept in subsequent computations.

(2) In the second column, $\psi_{IJ} \in \mathbb{R}^{IJ}$ is the internal displacement sub-matrix of $^{(k)}S$ when the interface $J' = {}^{(k)}S \cap {}^{(k+1)}S$ is fixed and each DoF in interface $J = {}^{(k)}S \cap {}^{(k-1)}S$ undergoes a unit displacement such that the other two sub-matrices are sub-matrix 0 and identity matrix I_J of order J, respectively. Accordingly, $^{(k)}p_J = {}^{(k)}u_J \in \mathbb{R}^J$ is the interfacial displacement vector for set J.

(3) In the third column, the three sub-matrices are similar to those in the second column, but exchange the subscripts between J and J'. As such, $^{(k)}p_{J'} = {}^{(k)}u_{J'} \in \mathbb{R}^J$ is the interfacial displacement vector for set J'.

In view of Eq. (6.2.82) and the meanings of the above sub-matrices, we get the mass matrix and stiffness matrix of sub-structure $^{(k)}S$ corresponding to modal displacement vector $^{(k)}p$, namely,

$$\begin{cases} m \equiv \psi^{\mathrm{T}}M\psi = \begin{bmatrix} I_D & m_{DJ} & m_{DJ'} \\ m_{JD} & m_{JJ} & m_{JJ'} \\ m_{J'D} & m_{J'J} & m_{J'J'} \end{bmatrix} \in \mathbb{R}^{\hat{m} \times \hat{m}} \\[18pt] k \equiv \psi^{\mathrm{T}}K\psi = \begin{bmatrix} k_{DD} & 0 & 0 \\ 0 & k_{JJ} & k_{JJ'} \\ 0 & k_{J'J} & k_{J'J'} \end{bmatrix} \in \mathbb{R}^{\hat{m} \times \hat{m}} \end{cases} \qquad (6.2.83)$$

where the sub-matrices in Eq. (6.2.83) are as follows

$$\begin{cases}
\boldsymbol{k}_{DD} = \boldsymbol{\psi}_{ID}^{\mathrm{T}} \boldsymbol{K}_{II} \boldsymbol{\psi}_{ID} = \underset{1 \le r \le D}{\mathrm{diag}} \left[\omega_r^2 \right] \\[4pt]
\begin{bmatrix} \boldsymbol{k}_{JJ} & \boldsymbol{k}_{JJ'} \\ \boldsymbol{k}_{J'J} & \boldsymbol{k}_{J'J'} \end{bmatrix} = \begin{bmatrix} \boldsymbol{K}_{JI} \\ \boldsymbol{K}_{J'I} \end{bmatrix} \boldsymbol{C}_{IJ} + \begin{bmatrix} \boldsymbol{K}_{JJ} & \boldsymbol{0} \\ \boldsymbol{0} & \boldsymbol{K}_{J'J'} \end{bmatrix} \\[8pt]
\left[\boldsymbol{m}_{DJ} \ \boldsymbol{m}_{DJ'} \right] = \boldsymbol{\psi}_{ID}^{\mathrm{T}} \left\{ \boldsymbol{M}_{II} \boldsymbol{C}_{IJ} + \left[\boldsymbol{M}_{IJ} \ \boldsymbol{M}_{IJ'} \right] \right\} \\[4pt]
\begin{bmatrix} \boldsymbol{m}_{JJ} & \boldsymbol{m}_{JJ'} \\ \boldsymbol{m}_{J'J} & \boldsymbol{m}_{J'J'} \end{bmatrix} = \boldsymbol{C}_{IJ}^{\mathrm{T}} \boldsymbol{M}_{II} \boldsymbol{C}_{IJ} + \begin{bmatrix} \boldsymbol{M}_{JI} \\ \boldsymbol{M}_{J'I} \end{bmatrix} \boldsymbol{C}_{IJ} + \boldsymbol{C}_{IJ}^{\mathrm{T}} \left[\boldsymbol{M}_{IJ} \ \boldsymbol{M}_{IJ'} \right] + \begin{bmatrix} \boldsymbol{M}_{JJ} & \boldsymbol{0} \\ \boldsymbol{0} & \boldsymbol{M}_{J'J'} \end{bmatrix} \\[8pt]
\boldsymbol{C}_{IJ} \equiv \left[\boldsymbol{\psi}_{IJ} \ \boldsymbol{\psi}_{IJ'} \right] = -\boldsymbol{K}_{II}^{-1} \left[\boldsymbol{K}_{IJ} \ \boldsymbol{K}_{IJ'} \right] \in \mathbb{R}^{I \times (2J)}
\end{cases}$$

$$(6.2.84)$$

In Eq. (6.2.84), ω_r, $r = 1, 2, .., D$ are the natural frequencies of sub-structure $^{(k)}S$ with interfaces J and J' fixed. Now, the internal displacements of $^{(k)}S$ are fully decoupled. In addition, the static internal displacements and static interfacial displacements are decoupled.

To derive the transform of interfacial compatibility, we take modal displacement vector $^{(k)}\boldsymbol{p} \in \mathbb{R}^{\hat{m}}$ of sub-structure $^{(k)}S$ as displacement vector $^{(k)}\boldsymbol{u} \in \mathbb{R}^{m'}$ of $^{(k)}S$, and conduct the decoupling procedure in subspace S_j. As the above modal reduction keeps all interfacial DoFs, the condition of interfacial compatibility is as same as that in Sect. 6.2.2. In view of Eq. (6.2.31), to reduce the internal DoFs from I to D, we simply replace m by $d \equiv D + J$, and then get the transform of interfacial compatibility in subspace S_j as follows

$$\tilde{\boldsymbol{B}}_{jd} \equiv \begin{bmatrix} \boldsymbol{I}_D & \boldsymbol{0} \\ \boldsymbol{0} & \boldsymbol{I}_J \\ \boldsymbol{0} & \exp(\mathrm{i}j\theta)\boldsymbol{R}_J^{-1} \end{bmatrix} \in \mathbb{C}^{\hat{m} \times d}, \quad j \in J_{[n/2]} \tag{6.2.85}$$

Recalling Eq. (6.2.82), we cast the displacement vector for $^{(k)}S$ as

$$^{(k)}\boldsymbol{u} = \boldsymbol{\Psi}^{(k)}\boldsymbol{p} = \boldsymbol{\Psi} \sum_{j=0}^{[n/2]} \mathrm{Re}\left[\exp(\mathrm{i}jk\theta)\tilde{\boldsymbol{B}}_{jd}\boldsymbol{q}_{jd} \right], \quad k \in I_n \tag{6.2.86}$$

Substitution of Eq. (6.2.85) into Eq. (6.2.37) leads to the mass matrix and stiffness matrix of the C_n structure in subspace S_j, namely,

$$\begin{cases}
\boldsymbol{m}_j = \begin{bmatrix} \boldsymbol{I}_D & \boldsymbol{m}_{DJ} + \exp(\mathrm{i}j\theta)\boldsymbol{m}_{DJ'}\boldsymbol{R}_J^{-1} \\ \mathrm{sym} & \boldsymbol{m}_{JJ} + \exp(\mathrm{i}j\theta)\boldsymbol{m}_{JJ'}\boldsymbol{R}_J^{-1} + \exp(-\mathrm{i}j\theta)\boldsymbol{R}_J^1\boldsymbol{m}_{J'J} + \boldsymbol{R}_J^1\boldsymbol{m}_{J'J'}\boldsymbol{R}_J^{-1} \end{bmatrix} \\[8pt]
\boldsymbol{k}_j = \begin{bmatrix} \mathrm{diag}[\omega_r^2] & \boldsymbol{0} \\ \boldsymbol{0} & \boldsymbol{k}_{JJ} + \exp(\mathrm{i}j\theta)\boldsymbol{k}_{JJ'}\boldsymbol{R}_J^{-1} + \exp(-\mathrm{i}j\theta)\boldsymbol{R}_J^1\boldsymbol{k}_{J'J} + \boldsymbol{R}_J^1\boldsymbol{k}_{J'J'}\boldsymbol{R}_J^{-1} \end{bmatrix}
\end{cases}$$

$$(6.2.87)$$

where sym stands for that the lower triangle entries are symmetric to the upper triangle entries. For computational convenience, we define the following real matrices corresponding to Eq. (6.2.87) as

$$
\left\{
\begin{aligned}
M_{jR} &=
\begin{bmatrix}
I_D & M_{j1} & 0 & M_{j3} \\
M_{j1}^{\mathrm{T}} & M_{j2} & -M_{j3}^{\mathrm{T}} & M_{j4} \\
0 & -M_{j3} & I_D & M_{j1} \\
M_{j3}^{\mathrm{T}} & M_{j4}^{\mathrm{T}} & M_{j1}^{\mathrm{T}} & M_{j2}
\end{bmatrix} \\
K_{jR} &=
\begin{bmatrix}
\operatorname*{diag}_{1\le r\le D}\left[\omega_r^2\right] & 0 & 0 & 0 \\
0 & K_{j1} & 0 & K_{j2} \\
0 & 0 & \operatorname*{diag}_{1\le r\le D}\left[\omega_r^2\right] & 0 \\
0 & K_{j2}^{\mathrm{T}} & 0 & K_{j1}
\end{bmatrix}, \quad j \in J_{[n/2]}
\end{aligned}
\right. \tag{6.2.88}
$$

where

$$
\left\{
\begin{aligned}
M_{j1} &\equiv m_{DJ} + \cos(j\theta)m_{DJ'}R_J^{-1} \\
M_{j2} &\equiv m_{JJ} + R_J^1 m_{J'J} R_J^{-1} + \cos(j\theta)(m_{JJ'}R_J^{-1} + R_J^1 m_{J'J}) \\
M_{j3} &\equiv -\sin(j\theta)m_{DJ}R_J^{-1} \\
M_{j4} &\equiv \sin(j\theta)(R_J^1 m_{J'J} - m_{JJ'}R_J^{-1}) \\
K_{j1} &\equiv k_{JJ} + R_J^1 k_{J'J} R_J^{-1} + \cos(j\theta)(k_{JJ'}R_J^{-1} + R_J^1 k_{J'J}) \\
K_{j2} &\equiv \sin(j\theta)(R_J^1 k_{J'J} - k_{JJ'}R_J^{-1}), \quad j \in J_{[n/2]}
\end{aligned}
\right. \tag{6.2.89}
$$

Solving the real symmetric eigenvalue problem of two matrices in Eq. (6.2.88) leads to the natural frequencies and eigenvectors $\omega_{jr}, q_{jr}, j \in J_{[n/2]}, r = 1, 2, 3, \ldots$. Substituting eigenvectors q_{jr} into Eq. (6.2.86), we obtain the mode shapes of sub-structure $^{(k)}S$, namely,

$$
^{(k)}\varphi_{jr} = \Psi \operatorname{Re}\left[\exp(ijk\theta)\tilde{B}_{jd}q_{jr}\right], \quad k \in I_n, \quad j \in J_{[n/2]}, \quad r = 1, 2, \ldots \tag{6.2.90}
$$

Removing the redundant interfacial displacements at set J' of $^{(k)}S$, we have the mode shapes of sub-structure $^{(k)}S$ in the local frame of coordinates as follows

$$
\left\{
\begin{aligned}
&^{(k)}\varphi_{jr} = \Psi\left[\cos(jk\theta)\operatorname{Re}(q_{jr}) - \sin(jk\theta)\operatorname{Im}(q_{jr})\right] \\
&k \in I_n, \quad j \in J_{[n/2]}, \quad r = 1, 2, ..
\end{aligned}
\right. \tag{6.2.91}
$$

6.2.3.3   Natural Vibrations of a Bladed Disc Model

Example 6.2.6 We recall the computation of natural vibrations of a bladed disc model with cyclosymmetry in group C_6 as shown in Fig. 6.28, where sub-structure

Fig. 6.28 A C_6 bladed disc model and the finite element model of sub-structure $^{(0)}S$

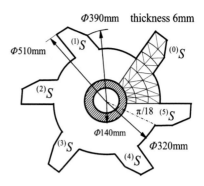

Table 6.1 The natural frequencies (Hz) of a C_6 bladed disk model

Frequency order	S_0	S_1	S_2	S_3
f_{j1}	245.6	245.5*	258.9*	277.9
f_{j2}	833.0	824.2*	802.5*	786.4
f_{j3}	1817.7	1594.8*	1289.1*	1156.9

$^{(0)}S$ was meshed by using thin triangle plate elements T-9. Thus, the discrete model had 23 internal nodes and 3 interfacial nodes at each interface, and each node had 3 DoFs. As such, one had $I = 69$ and $J = 9$. According to Eq. (6.2.3), the effective DoF of sub-structure $^{(0)}S$ was $m = I + J = 78$ and the total DoF of the bladed disc model was $N = m \times n = 468$.

As the first step, the decoupling method in Sect. 6.2.1 was used to compute the first three natural modes of the bladed disc model in subspace S_j, $j = 0, 1, 2, 3$ and 18 natural frequencies, including 6 pair of repeated natural frequencies, of the bladed disc model were obtained as shown in Table 6.1, where each frequency with an asterisk in two columns of S_1 and S_2 implies a pair of repeated natural frequencies. As shown in Table 6.1, the bladed disc model exhibits close natural modes in a lower frequency band. Once the bladed disc model is subject to any structural disturbance, each pair of repeated natural frequencies splits into a pair of close natural frequencies and enhances the modal density in frequency domain.

In order to validate the above results, the full finite element model was used to compute the first 18 natural frequencies of the bladed disc model. The natural frequencies computed by two methods got very good agreement for the first 6 decimal places.

To check the validity of modal reduction, Table 6.2 shows the relative errors of 18 natural frequencies, including 6 pairs of repeated ones with asterisk, of the bladed disc model based on the modal reduction with mode number D kept. Here, the natural frequencies are ranked according their values, while the double subscripts of a natural frequency f_{jr} stand for the order r in subspace S_j. The discussion about Table 6.2 leads to the following rules.

Table 6.2 The relative errors (%) of natural frequencies of a C_6 bladed disc model via different modal reductions

Natural frequency	$D = 0$	$D = 1$	$D = 2$	$D = 3$	$D = 5$
$f_{11}*$	6.93	0.0559	0.00385	0.00184	0.000737
f_{01}	6.77	0.0541	0.00508	0.00195	0.000650
$f_{21}*$	7.16	0.0439	0.00599	0.00168	0.000306
f_{31}	4.22	0.0269	0.00182	0.00043	0.000071
f_{32}	20.8	3.79	2.54	0.477	0.040
$f_{22}*$	31.4	6.28	1.73	0.377	0.049
$f_{12}*$	37.4	10.8	0.705	0.242	0.060
f_{02}	46.3	12.7	0.403	0.201	0.064
f_{33}	–	50.5	1.33	0.119	0.017
$f_{23}*$	–	52.5	4.77	0.351	0.032
$f_{13}*$	–	57.1	13.9	0.578	0.084
f_{03}	–	68.4	26.5	0.366	0.067

(1)   When $D = 0$, the dynamic contributions of all internal DoFs were neglected so that even the relative errors of the first 6 natural frequencies were not acceptable.

(2)   When $D = 1$, the relative errors of the first 6 natural frequencies, that is, the first natural frequency in subspaces S_j, $j = 0,\ 1,\ 2,\ 3$, were no more than 0.1%, indicating that the modal reduction began to work, but those of the second natural frequencies in S_j, $j = 0,\ 1,\ 2,\ 3$ were not small enough.

(3)   When $D = 2$, the relative errors of the first 12 natural frequencies, i.e., the first two natural frequencies in S_j, $j = 0,\ 1,\ 2,\ 3$, decreased and the maximal one was no more than 3%, exhibiting that the modal reduction worked better.

(4)   When $D = 3$, the relative errors of the first 18 natural frequencies, i.e., the first three natural frequencies in S_j, $j = 0,\ 1,\ 2,\ 3$, decreased to a level about 0.6%. In addition, when $D = 5$, the above errors reduced to a bound no more than 0.09%.

Table 6.3 lists the relative computation costs for different mode numbers D in the modal reduction compared with the case when no modal reduction was made. The table shows a promising result that the relative computation cost decreased to about 24 % when $D = 5$. This case study was a small problem that the sub-structure had no more than 10^2 DoFs. For a large problem that a sub-structure has $10^3 \sim 10^4$ DoFs, this relative computation cost is significantly important.

Table 6.3 The relative cost in the computation of natural vibrations of a bladed disc model via different modal reductions

Modes kept	$D = 0$	$D = 1$	$D = 2$	$D = 3$	$D = 5$
Relative cost (%)	9.44	14.45	15.19	17.39	23.53

6.2.4 Concluding Remarks

This section presented a systematic study on the computation of natural vibrations of a cyclosymmetric structure with n identical sectors, or called a C_n structure for short, by using the representative theory of cyclic group. The study has solved Problem 5B proposed in Chap. 1 and led to the following conclusions.

(1) Given a C_n structure without a central axis, one can select one of n sectors of the C_n structure, establish a finite element model of m-DoFs for the sector and use the generalized displacements in the irreducible representative subspace S_j to decouple the dynamic equations of the C_n structure with $(m \times n)$-DoFs. When n is an even number, the original eigenvalue problem of order $m \times n$ for the C_n structure is decoupled as two eigenvalue problems of order m, and $n/2-1$ eigenvalue problems of order $2m$. When n is an odd number, the original eigenvalue problem of order $m \times n$ is decoupled as one eigenvalue problem of order m, and $[n/2]$ eigenvalue problems of order $2m$. Solving those eigenvalue problems gives natural frequencies of the C_n structure and the eigenvectors in subspace S_j, and the eigenvectors can be transformed to the mode shapes of the C_n structure.

(2) The above decoupling procedure can be extended to the vibration problem of a C_n structure with a central axis, which all sub-structures share. In this case, one can select the central displacements of any sub-structure as unknowns to be solved and let the central displacements of other sub-structures be compatible with them. After this treatment, the decoupling procedure is similar to that for a C_n structure without a central axis. As the central displacements appear in some decoupled dynamic equations repeatedly, one ought to eliminate the central displacements repeated as redundant information.

(3) For a complicated C_n structure, it is feasible to use the modal reduction with fixed-interface to reduce the number of internal DoFs of a sub-structure before starting the decoupling procedure in order to achieve higher-efficient computations. The computation of natural vibrations of a bladed disc model showed that the mode syntheses based on 3 to 5 mode shapes of a sub-structure with fixed interfaces and the interfacial displacements provided a very good model reduction and accurate results, while the computation cost decreased by 76% further on the basis of the above decoupling procedure.

6.3 Modal Properties of Cyclosymmetric Structures

In the early studies on natural vibrations of C_n structures, the mode shapes were treated by analogy with those of a circular annular plate, exhibiting the combination of circular and radial nodal lines. In the 1970s, the modal tests of bladed discs of aero-engines indicated that their mode shapes were quite different from those of a circular annular plate, exhibiting very close natural frequencies[16]. Subsequent studies revealed some modal properties in theory[17]. For instance, the sufficient condition of the mode shape of a single natural frequency of the C_n structure is that either the mode shapes of all sub-structures are the same or the mode shapes are identical every two sub-structures when n is an even number.

Compared with the high-efficient computation of natural vibrations of cyclosymmetric structures, the study on modal properties of cyclosymmetric structures was far from sufficient, and some problems were not clear before the author paid attention to them in the 1980s. For example, what is the necessary condition of the mode shape of a single natural frequency? What about the properties of two mode shapes with repeated frequencies? What about the modal properties for a cyclosymmetric structure with a central axis?

Following the basic ideas of the author in the references[18,19], this section presents detailed studies on the modal properties and some criteria of an arbitrary C_n structure. The section begins with the vibration properties of generalized displacements of a C_n structure in subspace S_j, then discusses the properties of displacements of the C_n structure in physical space and the properties of central displacements because of their unique role. These studies will help readers solve Problem 5A proposed in Chap. 1, as well as the symmetry-breaking of C_n structures, such as a bladed disc for aero-engines[20], and the vibration of C_n structures with multiple stages[21].

6.3.1   Modal Properties in Representative Subspaces of Cyclic Group

In Sect. 6.2, we decoupled the natural vibrations of a C_n structure as those in the irreducible subspace S_j in terms of generalized displacement vectors, and derived the real symmetric eigenvalue problems governing the generalized displacement vectors in Eqs. (6.2.46) and (6.2.69), which are applicable to the C_n structure without a central axis and with a central axis, respectively.

---

[16] Ewins D J (1973). Vibration characteristics of bladed disk assemblies. Journal of Mechanical Engineering Science, 15(3): 165–186.

[17] Thomas D L (1979). Dynamics of rotationally periodic structures. International Journal of Numerical Methods in Engineering, 14(1): 81–102.

[18] Hu H Y, Cheng D L (1988). Investigation on modal characteristics of cyclosymmetric structures. Chinese Journal of Applied Mechanics, 5(3): 1–8 (in Chinese).

[19] Hu H Y, Cheng D L, Wang L (1986). Vibration analysis of a cyclosymmetric structure with central axis. Shanghai: The 2nd National Conference on Computational Mechanics, No. 2–236 (in Chinese).

[20] Yan Y J, Cui P L, Hao H N (2008). Vibration mechanism of a mistuned bladed disk. Journal of Sound and Vibration, 317(1–2): 294–307.

[21] Nyssen F, Epureanub B, Golinval J C (2017). Experimental modal identification of mistuning in an academic two-stage drum. Mechanical Systems and Signal Processing, 88: 428–444.

This subsection begins with Eq. (6.2.69) since it covers Eq. (6.2.46). Equation (6.2.70) enables one to verify that the complex form of Eq. (6.2.69) reads

$$\left(K_j - \omega_j^2 M_j\right) q_{jD} = 0, \quad j \in J_{[n/2]} \tag{6.3.1}$$

where M_j and K_j are Hermite matrices satisfying

$$M_j^H = M_j \equiv \tilde{B}_{jD}^H M \tilde{B}_{jD} \in \mathbb{C}^{\tilde{m} \times \tilde{m}}, \quad K_j^H = K_j \equiv \tilde{B}_{jD}^H K \tilde{B}_{jD} \in \mathbb{C}^{\tilde{m} \times \tilde{m}},$$
$$j \in J_{[n/2]} \tag{6.3.2}$$

Based on Eq. (6.3.1), the subsection discusses the properties of generalized displacement vector q_{jD} and does not account for the case when Eq. (6.3.1) becomes a real eigenvalue problem if $j = 0$ and $n/2 = [n/2]$. As following discussion has nothing to do with the second subscript of q_{jD}, we provisionally replace q_{jD} with q_j for brevity.

From linear algebra, if Hermite matrix M_j is positive definite, Eq. (6.3.1) has positive eigenvalues ranked as follows

$$0 < \omega_{j1}^2 \le \omega_{j2}^2 \le \cdots \le \omega_{j\tilde{m}}^2 \tag{6.3.3}$$

The corresponding eigenvectors q_{jr}, $r \in \tilde{I}_{\tilde{m}}$ are complex vectors in a unitary space and satisfy the following orthogonal relations

$$q_{jr}^H M_j q_{js} = 0, \quad q_{jr}^H K_j q_{js} = 0, \quad \omega_{jr} \neq \omega_{js}, \quad r, s = 1, 2, \ldots, \tilde{m}, \quad j \in J_{[n/2]}. \tag{6.3.4}$$

Given a complex eigenvector q_j with norm $\|q_j\| \equiv \sqrt{q_j^H q_j}$ in the unitary space, an arbitrary normalized eigenvector in the unitary space yields

$$\tilde{q}_j = \exp(i\phi) q_j, \quad \phi \in [0, 2\pi), \quad j \in J_{[n/2]} \tag{6.3.5}$$

Let the real vectors corresponding to the above two complex vectors be defined as

$$q_{jR} \equiv \begin{bmatrix} \text{Re}(q_j) \\ \text{Im}(q_j) \end{bmatrix}, \quad \tilde{q}_{jR} \equiv \begin{bmatrix} \text{Re}(\tilde{q}_j) \\ \text{Im}(\tilde{q}_j) \end{bmatrix}, \quad j \in J_{[n/2]} \tag{6.3.6}$$

It is easy to verify that the normalization of these two real vectors in Euclidian space is equivalent to the normalization of the two complex vectors in the unitary space. In addition, Eq. (6.3.5) can be recast as the following orthogonal transform

$$\tilde{\boldsymbol{q}}_{jR} = \begin{bmatrix} (\cos\phi)\boldsymbol{I}_{\tilde{m}} & -(\sin\phi)\boldsymbol{I}_{\tilde{m}} \\ (\sin\phi)\boldsymbol{I}_{\tilde{m}} & (\cos\phi)\boldsymbol{I}_{\tilde{m}} \end{bmatrix} \boldsymbol{q}_{jR}, \quad j \in J_{[n/2]} \tag{6.3.7}$$

Substituting $\phi = -\pi/2$ into Eq. (6.3.7) and defining the result as

$$\hat{\boldsymbol{q}}_{jR} \equiv \begin{bmatrix} \boldsymbol{0} & \boldsymbol{I}_{\tilde{m}} \\ -\boldsymbol{I}_{\tilde{m}} & \boldsymbol{0} \end{bmatrix} \begin{bmatrix} \mathrm{Re}(\boldsymbol{q}_j) \\ \mathrm{Im}(\boldsymbol{q}_j) \end{bmatrix} = \begin{bmatrix} \mathrm{Im}(\boldsymbol{q}_j) \\ -\mathrm{Re}(\boldsymbol{q}_j) \end{bmatrix}, \quad j \in J_{[n/2]} \tag{6.3.8}$$

we have the following orthogonal relation and linear expression

$$\hat{\boldsymbol{q}}_{jR}^{\mathrm{T}} \boldsymbol{q}_{jR} = 0, \quad j \in J_{[n/2]} \tag{6.3.9}$$

$$\tilde{\boldsymbol{q}}_{jR} = \left\{ \begin{bmatrix} (\cos\phi)\boldsymbol{I}_{\tilde{m}} & \boldsymbol{0} \\ \boldsymbol{0} & (\cos\phi)\boldsymbol{I}_{\tilde{m}} \end{bmatrix} - \begin{bmatrix} \boldsymbol{0} & (\sin\phi)\boldsymbol{I}_{\tilde{m}} \\ -(\sin\phi)\boldsymbol{I}_{\tilde{m}} & \boldsymbol{0} \end{bmatrix} \right\} \boldsymbol{q}_{jR}$$
$$= (\cos\phi)\boldsymbol{q}_{jR} - (\sin\phi)\hat{\boldsymbol{q}}_{jR}, \quad j \in J_{[n/2]} \tag{6.3.10}$$

The above analysis can be summarized as following.

Property 6.3.1 Each eigenvalue of Eq. (6.2.69) for a C_n structure corresponds to a two-dimensional real eigen-subspace V_j, $j \in J_{[n/2]}$. An arbitrary normalized eigenvector $\boldsymbol{q}_{jR}$ of Eq. (6.2.69) and the eigenvector $\hat{\boldsymbol{q}}_{jR}$ constructed from Eq. (6.3.8) serve as a pair of orthogonal base vectors for subspace V_j such that any normalized eigenvector $\tilde{\boldsymbol{q}}_{jR}$ in V_j can be expressed as the form in Eq. (6.3.10).

According to Eqs. (6.3.9) and (6.3.10), the following property can be proved.

Property 6.3.2 Given a C_n structure without any rigid-body motions, the orthogonality of two real eigenvectors in subspace V_j is equivalent to their orthogonality with respect to the real mass matrix and stiffness matrix in Eq. (6.2.69).

Proof As these orthogonal relations bear analogy, it is sufficient to prove the case of real stiffness matrix for an arbitrary $j \in J_{[n/2]}$.

Sufficiency: If two real vectors in two-dimensional subspace V_j are orthogonal, their orthogonal relation is as same as Eq. (6.3.9) except for a real scaling constant. Without loss of generality, we take vectors $\boldsymbol{q}_{jR}$ and $\hat{\boldsymbol{q}}_{jR}$. From Eq. (6.2.70), it is easy to verify

$$\begin{cases} \mathrm{Re}(\boldsymbol{q}_j)^{\mathrm{T}} \boldsymbol{K}_j^R \mathrm{Im}(\boldsymbol{q}_j) = \mathrm{Im}(\boldsymbol{q}_j)^{\mathrm{T}} \boldsymbol{K}_j^R \mathrm{Re}(\boldsymbol{q}_j) \\ \mathrm{Re}(\boldsymbol{q}_j)^{\mathrm{T}} \boldsymbol{K}_j^I \mathrm{Re}(\boldsymbol{q}_j) = \mathrm{Im}(\boldsymbol{q}_j)^{\mathrm{T}} \boldsymbol{K}_j^I \mathrm{Im}(\boldsymbol{q}_j) = 0 \end{cases} \tag{a}$$

From Eq. (6.3.8) and Eq. (a), we derive the orthogonal relation with respect to the real stiffness matrix as following

$$[\,\mathrm{Im}(q_j)^{\mathrm{T}} \ -\mathrm{Re}(q_j)^{\mathrm{T}}\,] \begin{bmatrix} K_j^R & -K_j^I \\ K_j^I & K_j^R \end{bmatrix} \begin{bmatrix} \mathrm{Re}(q_j) \\ \mathrm{Im}(q_j) \end{bmatrix}$$
$$= \mathrm{Im}(q_j)^{\mathrm{T}} K_j^R \mathrm{Re}(q_j) - \mathrm{Re}(q_j)^{\mathrm{T}} K_j^I \mathrm{Re}(q_j)$$
$$- \mathrm{Im}(q_j)^{\mathrm{T}} K_j^I \mathrm{Im}(q_j) - \mathrm{Re}(q_j)^{\mathrm{T}} K_j^R \mathrm{Im}(q_j) = 0$$

$$\text{(b)}$$

Necessity: Take a pair of normalized vectors q_{jR} and $\tilde{q}_{jR}$ in two-dimensional subspace V_j such that they are orthogonal with respect to the real stiffness matrix. Following Eq. (6.3.8), we construct vector $\hat{q}_{jR}$ as an orthogonal counterpart of vector q_{jR} and express vector $\tilde{q}_{jR}$ in the form of Eq. (6.3.10). There follows

$$0 = \tilde{q}_{jR}^{\mathrm{T}} \begin{bmatrix} K_j^R & -K_j^I \\ K_j^I & K_j^R \end{bmatrix} q_{jR} = \big[(\cos\phi)q_{jR}^{\mathrm{T}} - (\sin\phi)\hat{q}_{jR}^{\mathrm{T}}\big] \begin{bmatrix} K_j^R & -K_j^I \\ K_j^I & K_j^R \end{bmatrix} q_{jR}$$
$$= (\cos\phi)q_{jR}^{\mathrm{T}} \begin{bmatrix} K_j^R & -K_j^I \\ K_j^I & K_j^R \end{bmatrix} q_{jR} \propto (\cos\phi)\omega_j^2 \qquad \text{(c)}$$

If the C_n structure does not have any rigid-body motion, then $\omega_j^2 > 0$ and $\cos\phi = 0$ hold true. Then, we can complete the proof since

$$\tilde{q}_{jR}^{\mathrm{T}} q_{jR} = (\cos\phi)q_{jR}^{\mathrm{T}} q_{jR} - (\sin\phi)\hat{q}_{jR}^{\mathrm{T}} q_{jR} = (\cos\phi)q_{jR}^{\mathrm{T}} q_{jR} = 0 \qquad \text{(d)}$$

Remark 6.3.1 The commercial software usually provides users with a pair of eigenvectors with orthogonality with respect to mass matrix and stiffness matrix. Property 6.3.2 implies their orthogonality for a constrained C_n structure.

6.3.2 Modal Properties in Physical Space

This subsection turns to the detailed studies on the modal properties of a C_n structure in physical space. To begin with, we go back to the generalized displacement vector q_{jD}. Following the notations in Eq. (6.3.7), let $q_{jR} = [\,(q_{jD}^R)^{\mathrm{T}} \ (q_{jD}^I)^{\mathrm{T}}\,]^{\mathrm{T}}$ and $\hat{q}_{jR} = [\,(\hat{q}_{jD}^R)^{\mathrm{T}} \ (\hat{q}_{jD}^I)^{\mathrm{T}}\,]^{\mathrm{T}}$ be a pair of eigenvectors computed from Eq. (6.2.69) via commercial software, and their corresponding complex vectors be q_{jD} and $\hat{q}_{jD}$, respectively. Substituting q_{jD} and $\hat{q}_{jD}$ into Eq. (6.2.74), we get the mode shapes of sub-structure $^{(k)}S$ of the C_n structure in local frame $^{(k)}F$ as follows

$$\begin{cases} {}^{(k)}\varphi_j = \mathrm{Re}\big[\exp(\mathrm{i}jk\theta)B_{jD}q_{jD}\big] \in \mathbb{R}^{\tilde{m}} \\ {}^{(k)}\hat{\varphi}_j = \mathrm{Re}\big[\exp(\mathrm{i}jk\theta)B_{jD}\hat{q}_{jD}\big] \in \mathbb{R}^{\tilde{m}}, \quad k \in I_n, \quad j \in J_{\lfloor n/2 \rfloor} \end{cases} \qquad (6.3.11)$$

As real vectors q_{jR} and $\hat{q}_{jR}$ are orthogonal with respect to the real mass matrix and real stiffness matrix in subspace S_j, vectors $^{(k)}\varphi_j$ and $^{(k)}\hat{\varphi}_j$ are referred to as a pair of *base mode shapes* of sub-structure $^{(k)}S$ in subspace S_j. Equation (6.3.11) enables one to prove the following orthogonal transform between the base mode shapes of different sub-structures.

Property 6.3.3 The base mode shapes of sub-structure $^{(k)}S$ and those of sub-structure $^{(0)}S$ satisfy the following orthogonal transform.

$$\begin{bmatrix} ^{(k)}\boldsymbol{\varphi}_j \\ ^{(k)}\hat{\boldsymbol{\varphi}}_j \end{bmatrix} = \begin{bmatrix} \cos(jk\theta)\boldsymbol{I}_{\bar{m}} & -\sin(jk\theta)\boldsymbol{I}_{\bar{m}} \\ \sin(jk\theta)\boldsymbol{I}_{\bar{m}} & \cos(jk\theta)\boldsymbol{I}_{\bar{m}} \end{bmatrix} \begin{bmatrix} ^{(0)}\boldsymbol{\varphi}_j \\ ^{(0)}\hat{\boldsymbol{\varphi}}_j \end{bmatrix} \tag{6.3.12}$$

Proof Equation (6.3.11) gives.

$$^{(0)}\boldsymbol{\varphi}_j = \mathrm{Re}(\boldsymbol{B}_{jD}\boldsymbol{q}_{jD}), \quad ^{(0)}\hat{\boldsymbol{\varphi}}_j = \mathrm{Re}(\boldsymbol{B}_{jD}\hat{\boldsymbol{q}}_{jD}) \tag{a}$$

Multiplication of two complex matrices leads to

$$\begin{cases} \mathrm{Re}(\boldsymbol{B}_{jD}\boldsymbol{q}_{jD}) = \mathrm{Re}(\boldsymbol{B}_{jD})\mathrm{Re}(\boldsymbol{q}_{jD}) - \mathrm{Im}(\boldsymbol{B}_{jD})\mathrm{Im}(\boldsymbol{q}_{jD}) = -\mathrm{Im}(\boldsymbol{B}_{jD}\hat{\boldsymbol{q}}_{jD}) \\ \mathrm{Im}(\boldsymbol{B}_{jD}\boldsymbol{q}_{jD}) = \mathrm{Re}(\boldsymbol{B}_{jD})\mathrm{Im}(\boldsymbol{q}_{jD}) + \mathrm{Im}(\boldsymbol{B}_{jD})\mathrm{Re}(\boldsymbol{q}_{jD}) = \mathrm{Re}(\boldsymbol{B}_{jD}\hat{\boldsymbol{q}}_{jD}) \end{cases} \tag{b}$$

Applying Euler's formula to Eq. (6.3.11) and then using Eqs. (b) and (a), we derive

$$\begin{aligned} ^{(k)}\boldsymbol{\varphi}_j &= \cos(jk\theta)\mathrm{Re}(\boldsymbol{B}_{jD}\boldsymbol{q}_{jD}) - \sin(jk\theta)\mathrm{Im}(\boldsymbol{B}_{jD}\boldsymbol{q}_{jD}) \\ &= \cos(jk\theta)\mathrm{Re}(\boldsymbol{B}_{jD}\boldsymbol{q}_{jD}) - \sin(jk\theta)\mathrm{Re}(\boldsymbol{B}_{jD}\hat{\boldsymbol{q}}_{jD}) \\ &= \cos(jk\theta)^{(0)}\boldsymbol{\varphi}_j - \sin(jk\theta)^{(0)}\hat{\boldsymbol{\varphi}}_j \end{aligned} \tag{c}$$

$$\begin{aligned} ^{(k)}\hat{\boldsymbol{\varphi}}_j &= \cos(jk\theta)\mathrm{Re}(\boldsymbol{B}_{jD}\hat{\boldsymbol{q}}_{jD}) - \sin(jk\theta)\mathrm{Im}(\boldsymbol{B}_{jD}\hat{\boldsymbol{q}}_{jD}) \\ &= \cos(jk\theta)\mathrm{Re}(\boldsymbol{B}_{jD}\hat{\boldsymbol{q}}_{jD}) + \sin(jk\theta)\mathrm{Re}(\boldsymbol{B}_j\boldsymbol{q}_{jD}) \\ &= \sin(jk\theta)^{(0)}\boldsymbol{\varphi}_j + \cos(jk\theta)^{(0)}\hat{\boldsymbol{\varphi}}_j \end{aligned} \tag{d}$$

This completes the proof since Eqs. (c) and (d) are identical to Eq. (6.3.12).

In what follows, we study the natural vibrations with a single frequency or a pair of repeated frequencies for a C_n structure according to Property 6.3.3.

6.3.2.1 Natural Vibration with a Single Frequency

From the discussion in Sect. 6.3.1, the natural vibration problem of a C_n structure corresponds to an eigenvalue problem of a pair of Hermite matrices, the single real eigenvalue of which has a complex eigenvector such that the corresponding real eigenvalue problem has a pair of natural modes with repeated frequencies. As such, the natural vibration with a single frequency does not mean that the eigen-value problem of the Hermite matrices has other single real eigenvalues, but that the normalized mode shape of the C_n structure corresponds to a single frequency.

Property 6.3.4 If the sub-structures of a C_n have adjacent sub-structures coupled at their interfaces, the sufficient and necessary condition for the natural vibration with a single frequency can be classified as two cases.

(1)   The mode shapes of the C_n structure fall into subspace S_0 or the mode shapes of all sub-structures are identical.
(2)   When n is an even number, the mode shapes of the C_n structure fall into subspace $S_{n/2}$ or the mode shapes of two adjacent sub-structures have the same geometric shape but in opposite directions such that the mode shapes of sub-structures are identical for every two sub-structures.

Proof The proof includes two parts as follows.

Sufficiency: As stated in Remark 6.2.4, when $j = 0$ or $j = n/2$, $\exp(0) = 1$ or $\exp(ik\pi) = (-1)^k$ holds so that $\boldsymbol{B}_j$ and $\tilde{\boldsymbol{B}}_{jD}$ are real matrices. Thus, the imaginary parts of $\boldsymbol{M}_j$ and $\boldsymbol{K}_j$ are null matrices, while $\boldsymbol{q}_j$ and $\boldsymbol{q}_{jD}$ are real vectors. Consequently, Eq. (6.2.69) becomes an eigenvalue problem of two real symmetric matrices as follows

$$\left(\boldsymbol{K}_j - \omega_j^2 \boldsymbol{M}_j\right)\boldsymbol{q}_{jD} = \boldsymbol{0}, \quad j = 0, \ n/2 \tag{a}$$

This excludes the repeated natural frequencies due to the eigenvalue problem of two Hermite matrices.

For $j = 0$ and $j = n/2$, Eq. (6.3.12) leads to the assertion to be proved as follows

$$^{(k)}\boldsymbol{\varphi}_0 = {}^{(0)}\boldsymbol{\varphi}_0, \quad k \in I_n \tag{b}$$

$$^{(k)}\boldsymbol{\varphi}_{n/2} = (-1)^k \, {}^{(0)}\boldsymbol{\varphi}_{n/2}, \quad k \in I_n \tag{c}$$

As for the condition of adjacent sub-structures coupled at their interfaces, the physical explanations will be given in Remark 6.3.3 shortly later.

Necessity: If the C_n structure undergoes the natural vibration with a single frequency, the two normalized base mode shapes have two possible cases. One is $^{(k)}\hat{\boldsymbol{\varphi}}_j = {}^{(k)}\boldsymbol{\varphi}_j$, $k \in I_n$. The other is $^{(k)}\boldsymbol{\varphi}_j = \boldsymbol{0}$, $k \in I_n$ or $^{(k)}\hat{\boldsymbol{\varphi}}_j = \boldsymbol{0}$, $k \in I_n$.
In the first case, substituting $^{(0)}\hat{\boldsymbol{\varphi}}_j = {}^{(0)}\boldsymbol{\varphi}_j$ and $^{(1)}\hat{\boldsymbol{\varphi}}_j = {}^{(1)}\boldsymbol{\varphi}_j$ into Eq. (6.3.12) leads to

$$\boldsymbol{0} = {}^{(1)}\hat{\boldsymbol{\varphi}}_j - {}^{(1)}\boldsymbol{\varphi}_j$$
$$= \left[\sin(j\theta)^{(0)}\boldsymbol{\varphi}_j + \cos(j\theta)^{(0)}\hat{\boldsymbol{\varphi}}_j\right] - \left[\cos(j\theta)^{(0)}\boldsymbol{\varphi}_j - \sin(j\theta)^{(0)}\hat{\boldsymbol{\varphi}}_j\right] = 2\sin(j\theta)^{(0)}\boldsymbol{\varphi}_j \tag{d}$$

As $^{(0)}\varphi_j \neq 0$ holds, there follows $\sin(j\theta) = 0$, $j \in J_{[n/2]}$ such that $j = 0$ or $j = n/2$ holds.

In the second case, substituting the corresponding relations into Eq. (6.3.12) gives

$$\mathbf{0} = {}^{(1)}\hat{\varphi}_j = \sin(j\theta){}^{(0)}\varphi_j \quad \text{or} \quad \mathbf{0} = {}^{(1)}\varphi_j = \sin(j\theta){}^{(0)}\hat{\varphi}_j \tag{e}$$

As the mode shapes in Eq. (e) are not null vectors, we have $\sin(j\theta) = 0$ and get $j = 0$ or $j = n/2$.

In addition, if all sub-structures have identical mode shapes, there follow

$$^{(1)}\varphi_0 = {}^{(0)}\varphi_0, \quad {}^{(1)}\hat{\varphi}_0 = {}^{(0)}\hat{\varphi}_0 \tag{f}$$

From Eq. (6.3.12), the orthogonal transform matrix must be an identity matrix, namely,

$$\cos(j\theta) = 1, \quad \sin(j\theta) = 0, \quad j \in J_{[n/2]} \tag{g}$$

There follows $j = 0$. Similarly, if two adjacent sub-structures have the same geometric shape but opposite directions, $j = n/2$ holds and n should be an even number. This completes the proof.

Remark 6.3.2 A practical criterion for the natural vibration of a single frequency is that the mode shapes of any adjacent sub-structures have the same geometric shape and in the same or opposite directions.

Remark 6.3.3 As assumed in Sect. 6.2.1, we mainly study the C_n structure with two adjacent sub-structures coupled at their interface. If such a coupling does not exist, the C_n structure becomes n independent sub-structures and will undergo at least one natural vibration with n identical frequencies. When the coupling of two adjacent sub-structures is not at their interfaces, but at the central axis, a natural vibration of the C_n structure may have three or more identical frequencies. Recalling Example 6.2.5 in Sect. 6.2.2, the four sub-structures of the C_4 structure have the unique coupling at the central mass, rather than in their interfaces such that the natural vibrations in subspaces S_2 and S_1 have the same frequencies, exhibiting three identical frequencies.

6.3.2.2 Natural Vibration with a Pair of Repeated Frequencies

As proved above, the C_n structure exhibits the circumferentially periodic mode shapes with a single frequency if $j = 0$ or $j = n/2$ holds. That is, the mode shapes of sub-structures look identical for every sub-structure or every two sub-structures. Now, we study the circumferential periodicity of the mode shapes with a pair of repeated frequencies under some extra conditions.

Property 6.3.5 Given $0 < j < n/2$, if $p = n/j$ is an integer, then any mode shapes of the C_n structure in subspace S_j look identical for every p sub-structures, and vice versa.

Proof Without loss of generality, let an arbitrary mode shape of the C_n structure be expressed as ${}^{(k)}\boldsymbol{\varphi}_j$, $k \in I_n$ for all sub-structures and prove the above assertion in two parts.

Sufficiency: The condition $p = n/j$ leads to $jp\theta = n\theta = 2\pi$. From Eq. (6.3.12), it is easy to derive the sufficiency result as follows

$$
\begin{aligned}
{}^{(k+p)}\boldsymbol{\varphi}_j &= \cos[j(k+p)\theta]{}^{(0)}\boldsymbol{\varphi}_j - \sin[j(k+p)\theta]{}^{(0)}\hat{\boldsymbol{\varphi}}_j \\
&= \cos(jk\theta){}^{(0)}\boldsymbol{\varphi}_j - \sin(jk\theta){}^{(0)}\hat{\boldsymbol{\varphi}}_j = {}^{(k)}\boldsymbol{\varphi}_j, \quad k+p \in I_n
\end{aligned}
\tag{a}
$$

Necessity: Assume that there is an integer p such that the mode shapes ${}^{(k)}\boldsymbol{\varphi}_j$, $k \in I_n$ of sub-structures satisfy

$$
{}^{(k+p)}\boldsymbol{\varphi}_j = {}^{(k)}\boldsymbol{\varphi}_j, \quad k+p \in I_n, \quad 0 < j < n/2
\tag{b}
$$

In view of Eq. (6.3.12), write Eq. (b) as

$$
{}^{(k+p)}\boldsymbol{\varphi}_j = \cos(jp\theta){}^{(k)}\boldsymbol{\varphi}_j - \sin(jp\theta){}^{(k)}\hat{\boldsymbol{\varphi}}_j = {}^{(k)}\boldsymbol{\varphi}_j, \quad k+p \in I_n
\tag{c}
$$

From the triangle formula of a half angle, we recast Eq. (c) as

$$
2\sin\left(\frac{jp\theta}{2}\right)\left[\sin\left(\frac{jp\theta}{2}\right){}^{(k)}\boldsymbol{\varphi}_j + \cos\left(\frac{jp\theta}{2}\right){}^{(k)}\hat{\boldsymbol{\varphi}}_j\right] = \mathbf{0}, \quad k+p \in I_n
\tag{d}
$$

If $\sin(jp\theta/2) \neq 0$ holds, there follows

$$
\sin\left(\frac{jp\theta}{2}\right){}^{(k)}\boldsymbol{\varphi}_j + \cos\left(\frac{jp\theta}{2}\right){}^{(k)}\hat{\boldsymbol{\varphi}}_j = \mathbf{0}, \quad k+p \in I_n
\tag{e}
$$

Pre-multiplying both sides of Eq. (e) by ${}^{(k)}\boldsymbol{\varphi}_j^{\mathrm{T}}$ leads to the following scalar relation

$$
\sin\left(\frac{jp\theta}{2}\right){}^{(k)}\boldsymbol{\varphi}_j^{\mathrm{T}(k)}\boldsymbol{\varphi}_j + \cos\left(\frac{jp\theta}{2}\right){}^{(k)}\boldsymbol{\varphi}_j^{\mathrm{T}(k)}\hat{\boldsymbol{\varphi}}_j = 0, \quad k+p \in I_n
\tag{f}
$$

As ${}^{(k)}\boldsymbol{\varphi}_j^{\mathrm{T}(k)}\boldsymbol{\varphi}_j > 0$ always holds, there follows

$$
\tan\left(\frac{jp\theta}{2}\right) = -\frac{{}^{(k)}\boldsymbol{\varphi}_j^{\mathrm{T}(k)}\hat{\boldsymbol{\varphi}}_j}{{}^{(k)}\boldsymbol{\varphi}_j^{\mathrm{T}(k)}\boldsymbol{\varphi}_j}, \quad k+p \in I_n
\tag{g}
$$

Given an integer p, Eq. (g) does not hold for all $k + p \in I_n$ such that we have

$$\sin\left(\frac{jp\theta}{2}\right) = 0, \quad 0 < j < n/2 \tag{h}$$

As the final step, solving Eq. (h) leads to the result for the necessity, namely,

$$\frac{jp\theta}{2} = \frac{jp\pi}{n} = \pi \quad \Rightarrow \quad p = \frac{n}{j} \tag{i}$$

Property 6.3.6 If a non-negative integer r makes the positive integer s satisfy

$$s = \frac{n}{j}\left(r + \frac{1}{4}\right) \in \tilde{I}_n \tag{6.3.13}$$

the mode shapes $^{(k)}\hat{\boldsymbol{\varphi}}_j$, $k \in I_n$ of sub-structures of the C_n structure in subspace S_j are identical to the mode shapes $^{(k)}\boldsymbol{\varphi}_j$, $k \in I_n$ when $^{(k)}\hat{\boldsymbol{\varphi}}_j$, $k \in I_n$ rotate through an angle $s\theta$ about the central axis, and vice versa.

Proof The proof includes two parts as follows.

Sufficiency: If Eq. (6.3.13) holds, there follows

$$sj\theta = \frac{n}{j}\left(r + \frac{1}{4}\right) \cdot j \cdot \frac{2\pi}{n} = 2r\pi + \frac{\pi}{2} \tag{a}$$

It is easy, from Eq. (6.3.12), to derive the sufficiency result, namely,

$$\begin{aligned}
^{(k+s)}\hat{\boldsymbol{\varphi}}_j &= \sin(js\theta)^{(k)}\boldsymbol{\varphi}_j + \cos(js\theta)^{(k)}\hat{\boldsymbol{\varphi}}_j \\
&= \sin\left(2r\pi + \frac{\pi}{2}\right)^{(k)}\boldsymbol{\varphi}_j + \cos\left(2r\pi + \frac{\pi}{2}\right)^{(k)}\hat{\boldsymbol{\varphi}}_j = {}^{(k)}\boldsymbol{\varphi}_j, \\
k + s &\in I_n, \quad 0 < j < n/2
\end{aligned} \tag{b}$$

Necessity: Eq. (6.3.12) leads to

$$\begin{aligned}
^{(k+s)}\hat{\boldsymbol{\varphi}}_j &= \sin(js\theta)^{(k)}\boldsymbol{\varphi}_j + \cos(js\theta)^{(k)}\hat{\boldsymbol{\varphi}}_j = {}^{(k)}\boldsymbol{\varphi}_j, \\
k + s &\in I_n, \quad 0 < j < n/2
\end{aligned} \tag{c}$$

Utilizing the triangle formula of a half angle, Eq. (c) can be recast as

$$\cos\left(js\theta - \frac{\pi}{2}\right)^{(k)}\boldsymbol{\varphi}_j - \sin\left(js\theta - \frac{\pi}{2}\right)^{(k)}\hat{\boldsymbol{\varphi}}_j = {}^{(k)}\boldsymbol{\varphi}_j, \quad k + s \in I_n, \quad 0 < j < n/2 \tag{d}$$

Eq. (d) looks like Eq. (c) in the proof of Property 6.3.5, then similar deduction gives

$$\sin\left(\frac{js\theta}{2} - \frac{\pi}{4}\right) = 0, \quad 0 < j < n/2 \tag{e}$$

Solving Eq. (e) gives a non-negative integer r such that the following relation holds

$$\frac{js\theta}{2} - \frac{\pi}{4} = r\pi \quad \Rightarrow \quad s = \frac{n}{j}\left(r + \frac{1}{4}\right) \tag{f}$$

Given $0 < s < n$, Eq. (6.3.13) leads to the interval for non-negative integer r as follows

$$0 \le r < j - \frac{1}{4} < j \tag{6.3.14}$$

This completes the proof.

Example 6.3.1 Discuss the computation results of the C_4 structure in Example 6.2.4 according to the above properties proved.

The mode shapes of the C_4 structure computed in Example 6.2.4 can be classified as three kinds. The first kind is the mode shape of a single frequency in subspace S_0 and the four lumped masses have the identical natural vibrations. The second kind is the mode shape of a single frequency in subspace S_2 and the natural vibrations of two adjacent lumped masses have the same amplitude but opposite directions. The above two natural vibrations fully agree with Property 6.3.4. The third kind is the mode shapes with a pair of repeated frequencies and their rotation period is $p = n/j = 4$, satisfying Property 6.3.5.

Remark 6.3.4 The proofs of Properties 6.3.4, 6.3.5 and 6.3.6 have nothing to do with central DoFs. Thus, these properties are applicable to the C_n structures with a central axis.

6.3.3 Central Displacements in Mode Shapes

6.3.3.1 Structure of Compatibility Transform Matrix at Central Axis

We start with the compatibility transform matrix D_j for the central displacements of different sub-structures and discuss the internal structure of the matrix in subspace S_j.

Using Kronecker symbol, we can write the following triangle formulae

$$
\begin{cases}
\displaystyle\sum_{k=0}^{n-1}\cos(jk\theta) = n\delta_{j0}, \quad \sum_{k=0}^{n-1}\sin(jk\theta) = 0 \\[3mm]
\displaystyle\sum_{k=0}^{n-1}\cos(jk\theta)\cos(k\theta) = \frac{n}{2}\big(\delta_{j1} + \delta_{jn-1}\big) \\[3mm]
\displaystyle\sum_{k=0}^{n-1}\sin(jk\theta)\sin(k\theta) = \frac{n}{2}\big(\delta_{j1} - \delta_{jn-1}\big) \\[3mm]
\displaystyle\sum_{k=0}^{n-1}\cos(jk\theta)\sin(k\theta) = 0 \\[3mm]
\displaystyle\sum_{k=0}^{n-1}\sin(jk\theta)\cos(k\theta) = 0, \quad j \in I_n
\end{cases}
\tag{6.3.15}
$$

Substitution of Eq. (6.3.15) into matrix D_j defined in Eq. (6.2.65) leads to

$$
\begin{aligned}
D_j &= \frac{1}{\sqrt{n}}\left[\sum_{k=0}^{n-1}\exp(-\mathrm{i}jk\theta)R_o^k\right] \\[2mm]
&= \frac{1}{\sqrt{n}}\mathrm{diag}\left[\sum_{k=0}^{n-1}[\cos(jk\theta) - \mathrm{i}\sin(jk\theta)]\begin{bmatrix}\cos(k\theta) & \sin(k\theta) & 0 \\ -\sin(k\theta) & \cos(k\theta) & 0 \\ 0 & 0 & 1\end{bmatrix}\right] \\[2mm]
&= \frac{\sqrt{n}}{2}\mathrm{diag}\left[\begin{bmatrix}\delta_{j1}+\delta_{jn-1} & -\mathrm{i}(\delta_{j1}-\delta_{jn-1}) & 0 \\ \mathrm{i}(\delta_{j1}-\delta_{jn-1}) & \delta_{j1}+\delta_{jn-1} & 0 \\ 0 & 0 & 2\delta_{j0}\end{bmatrix}\right] \in \mathbb{C}^{o\times o}, \quad j \in I_n
\end{aligned}
\tag{6.3.16}
$$

There follows the expression of D_j as below

$$
\begin{cases}
D_0 = \dfrac{\sqrt{n}}{2}\mathrm{diag}\left[\begin{bmatrix}0&0&0\\0&0&0\\0&0&2\end{bmatrix}\right], \quad D_1 = \overline{D}_{n-1} = \dfrac{\sqrt{n}}{2}\mathrm{diag}\left[\begin{bmatrix}1&-\mathrm{i}&0\\ \mathrm{i}&1&0\\0&0&0\end{bmatrix}\right] \\[4mm]
D_j = 0, \quad 1 < j \le [n/2]
\end{cases}
\tag{6.3.17}
$$

From the analysis in Sect. 6.2, the independent generalized displacements occur in subspaces S_j, $j \in J_{[n/2]}$. Hence, the subsequent analysis does not take $\overline{D}_{n-1}$ into consideration. In what follows, we discuss the central displacements in each subspace S_j for $j \in J_{[n/2]}$.

When $j = 0$, substituting D_0 in Eq. (6.3.17) into Eq. (6.2.67) and using Eq. (6.2.68), we derive the central displacement vectors of all sub-structures as follows

$$^{(k)}\boldsymbol{u}_{o0} = \frac{1}{\sqrt{n}}a_0\boldsymbol{D}_0{}^{(0)}\boldsymbol{u}_o = \mathrm{diag}\left[\begin{bmatrix} 0 & 0 & 0 \\ 0 & 0 & 0 \\ 0 & 0 & 1 \end{bmatrix}\right]{}^{(0)}\boldsymbol{u}_o, \quad k \in I_n \qquad (6.3.18)$$

As defined in Sect. 6.2, the third axis z of local frame $^{(k)}F$ is parallel to the central axis. Equation (6.3.18) indicates that after filtering by matrix $\boldsymbol{D}_0$, the non-zero central displacement in vector $^{(0)}\boldsymbol{u}_{o0}$ in a natural vibration of the C_n structure is either a rectilinear displacement along the central axis or an angular displacement about the central axis. The null entries in vector $^{(0)}\boldsymbol{u}_{o0}$ can be eliminated when solving the natural vibration.

Remark 6.3.5 In many C_n structures, a local neighborhood of the central axis is a part of thin shell, which can be considered as the combination of a part of thin plate and a part of membrane. In view of the deformation relation of a thin plate, vector $^{(0)}\boldsymbol{u}_{o0}$ includes rectilinear displacements, rather than rotation angles. It is not hard to show that the above rectilinear displacements get extreme values if and only if no nodal line of a mode shape passes through the central axis, and that the central axis becomes a saddle node if and only if rn nodal lines pass through the central axis. Here, r is a positive integer. The second case will be discussed in Example 6.3.2.

When $j = 1$, substituting $\boldsymbol{D}_1$ in Eq. (6.3.17) into Eq. (6.2.67) and using Eq. (6.2.68), we have the central displacement vectors of all sub-structures as follows

$$^{(k)}\boldsymbol{u}_{o1} = \frac{1}{\sqrt{n}}a_1\mathrm{Re}\big[\exp(\mathrm{i}k\theta)\boldsymbol{D}_1{}^{(0)}\boldsymbol{u}_o\big]$$

$$= \mathrm{diag}\left[\begin{bmatrix} \cos(k\theta) & \sin(k\theta) & 0 \\ -\sin(k\theta) & \cos(k\theta) & 0 \\ 0 & 0 & 0 \end{bmatrix}\right]{}^{(0)}\boldsymbol{u}_o, \quad k \in I_n \qquad (6.3.19)$$

As the third axis z of local frame $^{(k)}F$ is parallel to the central axis, Eq. (6.3.19) indicates that after filtering by matrix $\boldsymbol{D}_1$, the non-zero central displacement in vector $^{(0)}\boldsymbol{u}_{o1}$ in a natural vibration of the C_n structure is either a rectilinear displacement in a plane normal to the central axis or an angular displacement about an axis in the above plane. The null entries in vector $^{(0)}\boldsymbol{u}_{o1}$ can be removed when solving the natural vibration.

Remark 6.3.6 When a local neighborhood of the central axis is a part of thin shell, the displacement continuity of C_n structure ensures that there are nodal lines passing through the central axis. In addition, the mode shapes with one nodal line passing through the central axis must correspond to Eq. (6.3.19), but not vice versa[22].

---

[22] Irie T, Yawada G (1978). Free vibration of circular plate elastically supported at some points. Bulletin of Japanese Society of Mechanical Engineers, 21(61): 1602.

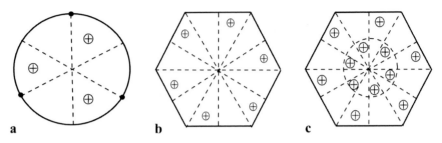

Fig. 6.29 Three mode shapes of high order of a C_{3v} structure and a C_{6v} structure in subspace S_0. **a** The circular plate; **b** The hexagon membrane; **c** The hexagon membrane

When $1 < j \leq [n/2]$, Eq. (6.3.17) gives $\boldsymbol{D}_j = 0$ such that $^{(0)}\boldsymbol{u}_{oj} = 0$ holds. That is, the central DoFs are at rest due to the filtering by $\boldsymbol{D}_j = 0$. In this case, $^{(0)}\boldsymbol{u}_{oj}$ ought to be removed when solving the natural vibration.

Remark 6.3.7 When a neighborhood of the central axis of the C_n structure is a part of thin shell, there are at least two nodal lines passing through the central axis. Otherwise, there is one nodal line passing through the central axis since $^{(k)}\boldsymbol{u}_{oj} = {}^{(0)}\boldsymbol{u}_{oj} = \boldsymbol{0}$, $k \in I_n$ hold. If this is the case, a central displacement becomes the rotation about the tangential direction of the nodal line. This contradicts to the assertion that the central DoFs are at rest.

Example 6.3.2 Figure 6.29 shows a mode shape of a circular plate with three simple support points and two mode shapes of an equilateral hexagon membrane with all edges fixed[23,24]. The two structures exhibit symmetry in group C_3 and group C_6, respectively. Furthermore, each sub-structure of them has mirror-symmetry. As stated in Sect. 6.2, they have symmetry in groups C_{3v} and C_{6v}, respectively. Discuss the modal properties of the two structures as a C_3 structure and a C_6 structure, respectively.

As shown in Fig. 6.29a, the mode shapes of all sub-structures of the circular plate are identical and fall into subspace S_0. Meanwhile, three nodal lines in Fig. 6.29a pass through the central axis of the circular plate. This behavior agrees with the prediction of a saddle node in Remark 6.3.5. The two mode shapes of the equilateral hexagon membrane in Figs. 6.29b and 6.29c have the similar property, but six nodal lines pass through the central axis. It is interesting that the nodal lines coincide with the mirror-symmetry plane of each sub-structure in Fig. 6.29. In fact, the C_{nv} structure has $2n$ identical sub-structures, and the mode shape with a single frequency may have the same geometric shape but opposite directions for two adjacent sub-structures. The three mode shapes here well agree with this property.

[23] Evensen D A (1976). Vibration analysis of multi-symmetric structures. AIAA Journal, 14(4): 446–453.

[24] Chi C (1972). Modes of vibration in a circular plate with three simple support points. AIAA Journal, 10(2): 142–147.

6.3.3.2 Order of Eigenvalue Problem with Central Displacements

Based on the above analysis, we discuss the order of eigenvalue problems in Sect. 6.2.2. Equation (6.3.17) shows that transform matrix D_j is always singular such that both mass matrix and stiffness matrix in Eq. (6.2.70) are singular, too. Thus, it is necessary to remove the rows and columns, which cause singularity, in mass and stiffness matrices. The treatment includes three cases.

(1) When $j = 0$, the central displacements to be computed are the rectilinear displacements along the central axis and the rotation angles about the central axis. Let the total number of those central displacements be $\bar{o}$, remove the rows and columns associated with other central displacements being null in matrices M_0^R and K_0^R. Then, both orders of matrix M_0^R and matrix K_0^R in Eq. (6.2.70) are $N_0 = \bar{o} + I + J$.

(2) When $j = 1$, the central displacements to be computed are the rectilinear displacements in a plate normal to the central axis and the rotation angles about an axis in the plane. Let the total number of those central displacements be $\tilde{o}$, where $\tilde{o} = o - \bar{o}$. To this end, two central displacements seem to be computed for each matrix D_1 in Eq. (6.3.17), but the rank of D_1 is one. That is, the multiplication of the first row in D_1 by i becomes the second row in D_1. Thus, the central displacement to be computed is only one complex unknown and the total number of central displacements is $\tilde{o}/2$. After removing the rows and columns associated with other central displacements being null in matrices in Eq. (6.2.70), the order of the eigenvalue problem in Eq. (6.2.69) is $N_1 = \tilde{o} + 2(I + J)$.

(3) When $1 < j \leq [n/2]$, no central displacement needs to be computed and the C_n structure with a central axis is identical to the C_n structure without a central axis. If $j = n/2 = [n/2]$ holds, the order of eigenvalue problem in Eq. (6.2.71) is $N_2 = I + J$. If $1 < j < [n/2]$ holds, one should solve $n/2 - 2$ eigenvalue problems in Eq. (6.2.69) and the total order of them is $N_3 = (n/2 - 2) \times 2(I + J) = (n - 4) \times (I + J)$.

In summary, consider, as an example, that n is an even number, the total order of eigenvalue problems to be solved satisfies

$$
\begin{aligned}
\hat{N} &= N_0 + N_1 + N_2 + N_3 \\
&= (\bar{o} + I + J) + [\tilde{o} + 2(I + J)] + (I + J) + (n - 4) \times (I + J) \\
&= o + n \times (I + J)
\end{aligned}
\tag{6.3.20}
$$

Equation (6.3.20) exactly coincides with Eq. (6.2.56).

6.3.3.3 Physical Meaning of Central Displacement Vector

In Sect. 6.2.2, the central displacement vector $^{(0)}u_o$ of sub-structure $^{(0)}S$ was taken as a sub-vector in vector q_{jD} to be computed. Here are the discussions about the physical meanings of central displacement vector $^{(0)}u_o$.

When $j = 0$, the natural vibration of the C_n corresponds to an eigenvalue problem of a real symmetric matrix pencil in Eq. (6.2.71) and the eigenvector q_{0D} solved is a real vector, including the sub-vector proportional to central displacement vector $^{(0)}u_{o0}$.

When $j = 1$, the natural vibration of the C_n structure corresponds to an eigenvalue problem of two complex Hermite matrices in Eq. (6.2.69) and eigenvector q_{1D} to be solved is a complex vector. The discussion in Sect. 6.3.1 and Eq. (6.3.5) have shown that the difference between two normalized eigenvectors in unitary space is $\exp(\mathrm{i}\phi)$. According to Eq. (6.2.68), the central displacement vector takes the following form

$$
^{(0)}\tilde{u}_{o1} = \frac{1}{\sqrt{n}} a_1 \mathrm{Re}\big[D_1 \exp(\mathrm{i}\phi)\,^{(0)}u_o \big] = \frac{1}{\sqrt{n}} a_1 \mathrm{Re}\big[\exp(\mathrm{i}\phi) D_1\,^{(0)}u_o \big]
$$

$$
= \mathrm{diag}\left[\begin{bmatrix} \cos\phi & \sin\phi & 0 \\ -\sin\phi & \cos\phi & 0 \\ 0 & 0 & 0 \end{bmatrix} \right]\,^{(0)}u_o = \mathrm{diag}\left[\begin{bmatrix} \cos\phi & \sin\phi & 0 \\ -\sin\phi & \cos\phi & 0 \\ 0 & 0 & 0 \end{bmatrix} \right]\,^{(0)}u_{o1}
$$

$$(6.3.21)$$

where the last step comes from the fact that vector $^{(0)}u_{o1}$ is equal to vector $^{(0)}u_o$ when the third entry of each DoF is zero. As such, the central displacement vector $^{(0)}\tilde{u}_{o1}$ computed may be different from vector $^{(0)}u_{o1}$ by a rotation angle ϕ. When $\phi = \pm\pi/2$ holds, there follows

$$
^{(0)}\hat{u}_{o1} = \mathrm{diag}\left[\begin{bmatrix} 0 & \pm 1 & 0 \\ \mp 1 & 0 & 0 \\ 0 & 0 & 0 \end{bmatrix} \right]\,^{(0)}u_{o1}
$$

$$(6.3.22)$$

In view of Eqs. (6.3.19) and (6.3.22), we derive

$$
^{(k)}\hat{u}_{o1}^{\mathrm{T}}\,^{(k)}u_{o1} = ^{(0)}\hat{u}_{o1}^{\mathrm{T}}\,^{(0)}u_{o1} = ^{(0)}u_{o1}^{\mathrm{T}}\mathrm{diag}\left[\begin{bmatrix} 0 & \mp 1 & 0 \\ \pm 1 & 0 & 0 \\ 0 & 0 & 0 \end{bmatrix} \right]\,^{(0)}u_{o1} = 0, \quad k \in I_n
$$

$$(6.3.23)$$

That is, two central displacement vectors in a pair of mode shapes with repeated frequencies of a C_n structure are orthogonal if the rotation angle in complex plane is $\phi = \pm\pi/2$.

Remark 6.3.8 According to Remark 6.3.6, each mode shape of a C_n structure in subspace S_1 has a nodal line passing through the central axis. Thus, the nodal lines

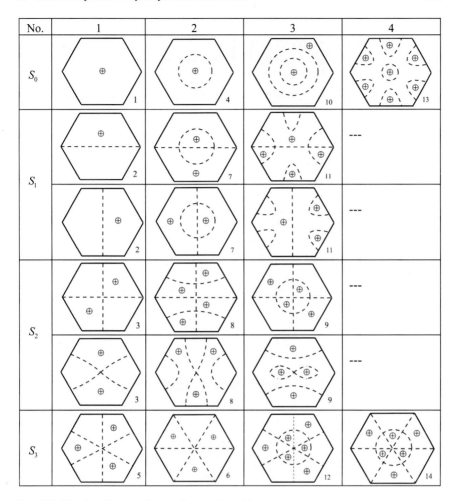

Fig. 6.30 The first 20 mode shapes of an equilateral hexagon plate

of a pair of mode shapes with repeated frequencies in S_1 are perpendicular to each other as a pair of mode shapes computed has orthogonality.

Example 6.3.3 L. Bauer and E. L. Reiss in Germany computed the first 20 natural modes of an equilateral hexagon plate with all edges hinged and ranked them in ascending frequencies as shown in the bottom-right number of each subfigure in Fig. 6.30[25]. Discuss the classification of the 20 mode shapes based on the above study.

As stated in Sect. 6.2, the equilateral hexagon plate with all edges hinged is a C_{6v} structure and has also all modal properties of a C_6 structure. According to

---

[25] Bauer L, Reiss E L (1978). Cutoff wave-numbers and modes of hexagonal waveguides. SIAM Journal of Applied Mathematics, 35(3): 508–514.

Properties 6.3.4 and 6.3.5, as well as associated remarks, we classify the 20 mode shapes according to their properties in subspace S_j, $j = 0, 1, 2, 3$, respectively. Figure 6.30 shows the classification of the 20 mode shapes of the hexagon plate, and the number in each subfigure is the order of the corresponding natural frequency.

(1)   In subspace S_0, the hexagon plate exhibits four mode shapes with a single frequency. These mode shapes are identical in all sub-structures and show an extreme value at the central axis, as predicted in Property 6.3.4 and Remark 6.3.5.

(2)   In subspace S_1, the hexagon plate has three pairs of mode shapes with repeated frequencies and each mode shape has one nodal line passing through the central axis, as stated in Property 6.3.5 and Remark 6.3.6.

(3)   In subspace S_2, the hexagon plate has also three pairs of node shapes with repeated frequencies. Each mode shape exhibits the circumferential period of three ($p = 6/2$) and has two nodal lines passing through the central axis such that the hexagon plate at the central axis is immovable, as predicted in Property 6.3.5 and Remark 6.3.7.

(4)   In subspace S_3, the hexagon plate has four mode shapes with a single frequency. Each mode shape exhibits the circumferential period of two ($p = 6/3$) and has three nodal lines intersecting the central axis so that the plate at the central axis is fixed, as stated in Property 6.3.4 and Remark 6.3.7.

6.3.4   Orthogonality of Mode Shapes with Repeated Frequencies

6.3.4.1   The C_n Structure Without a Central Axis

Section 6.2.1 presented how to assemble the mode shapes of the sub-structures of a C_n structure without a central axis as the mode shape of the entire structure in Eq. (6.2.51). Given the base mode shapes of a sub-structure in Eq. (6.3.11) for $0 < j < n/2$, we remove the central displacements and assemble two base mode shapes of the C_n structure without a central axis as done in Eq. (6.2.51), namely,

$$\boldsymbol{\varphi}_j = \mathrm{col}_k\left[\boldsymbol{R}_m^{-k\,(k)}\boldsymbol{\varphi}_j\right], \quad \hat{\boldsymbol{\varphi}}_j = \mathrm{col}_k\left[\boldsymbol{R}_m^{-k\,(k)}\hat{\boldsymbol{\varphi}}_j\right], \quad 0 < j < n/2 \qquad (6.3.24)$$

where $\boldsymbol{R}_m^{-k}$ is the orthogonal transform matrix through an angle $-k\theta$ about the central axis, while the order and entries of $\boldsymbol{R}_m^{-k}$ are consistent with vector $^{(k)}\hat{\boldsymbol{\varphi}}_j$. Now, we study the orthogonality of a pair of base mode shapes with repeated frequencies.

Property 6.3.7 When $0 < j < n/2$, the base mode shapes $\boldsymbol{\varphi}_j$ and $\hat{\boldsymbol{\varphi}}_j$ of a C_n structure without a central axis are orthogonal to each other.

Proof According to Eqs. (6.3.24) and (6.3.12), as well as the property of orthogonal matrix $\boldsymbol{R}_m^{-k}$, we derive

$$\hat{\boldsymbol{\varphi}}_j^T \boldsymbol{\varphi}_j = \sum_{k=0}^{n-1} {}^{(k)}\hat{\boldsymbol{\varphi}}_j^T (R_m^{-k})^T R_m^{-k}\, {}^{(k)}\boldsymbol{\varphi}_j = \sum_{k=0}^{n-1} {}^{(k)}\hat{\boldsymbol{\varphi}}_j^T\, {}^{(k)}\boldsymbol{\varphi}_j$$

$$= \sum_{k=0}^{n-1} \left[\sin(jk\theta)^{(0)}\boldsymbol{\varphi}_j + \cos(jk\theta)^{(0)}\hat{\boldsymbol{\varphi}}_j\right]^T \left[\cos(jk\theta)^{(0)}\boldsymbol{\varphi}_j - \sin(jk\theta)^{(0)}\hat{\boldsymbol{\varphi}}_j\right]$$

$$= \sum_{k=0}^{n-1} \sin(jk\theta)\cos(jk\theta)\left[{}^{(0)}\boldsymbol{\varphi}_j^T\, {}^{(0)}\boldsymbol{\varphi}_j - {}^{(0)}\hat{\boldsymbol{\varphi}}_j^T\, {}^{(0)}\hat{\boldsymbol{\varphi}}_j\right]$$

$$+ \sum_{k=0}^{n-1} \cos^2(jk\theta)^{(0)}\hat{\boldsymbol{\varphi}}_j^T\, {}^{(0)}\boldsymbol{\varphi}_j - \sum_{k=0}^{n-1} \sin^2(jk\theta)^{(0)}\boldsymbol{\varphi}_j^T\, {}^{(0)}\hat{\boldsymbol{\varphi}}_j \qquad (a)$$

Noting the inner product $^{(0)}\hat{\boldsymbol{\varphi}}_j^T\, {}^{(0)}\boldsymbol{\varphi}_j = {}^{(0)}\boldsymbol{\varphi}_j^T\, {}^{(0)}\hat{\boldsymbol{\varphi}}_j$ and using the following summation results of triangle functions

$$\sum_{k=0}^{n-1} \sin(jk\theta)\cos(jk\theta) = 0, \quad \sum_{k=0}^{n-1}\left[\cos^2(jk\theta) - \sin^2(jk\theta)\right] = 0, \quad 0 < j < n/2$$

$$(b)$$

we get the orthogonality relation to be proved as follows

$$\hat{\boldsymbol{\varphi}}_j^T \boldsymbol{\varphi}_j = 0, \quad 0 < j < n/2 \qquad (c)$$

6.3.4.2 The C_n Structure with a Central Axis

From Eq. (6.2.75), the base mode shapes of a C_n structure with a central axis take the following form

$$\begin{cases} \boldsymbol{\varphi}_j = \begin{bmatrix} {}^{(0)}\boldsymbol{u}_{oj} \\ \text{col}_k\left[R_m^{-k}\text{Re}[\exp(\mathrm{i}jk\theta)\boldsymbol{q}_{jd}]\right] \end{bmatrix} \equiv \begin{bmatrix} {}^{(0)}\boldsymbol{u}_{oj} \\ \boldsymbol{\psi}_j \end{bmatrix} \\ \hat{\boldsymbol{\varphi}}_j = \begin{bmatrix} {}^{(0)}\hat{\boldsymbol{u}}_{oj} \\ \text{col}_k\left[R_m^{-k}\text{Re}[\exp(\mathrm{i}jk\theta)\hat{\boldsymbol{q}}_{jd}]\right] \end{bmatrix} \equiv \begin{bmatrix} {}^{(0)}\hat{\boldsymbol{u}}_{oj} \\ \hat{\boldsymbol{\psi}}_j \end{bmatrix} \end{cases} \qquad (6.3.25)$$

where sub-vectors $\boldsymbol{\psi}_j$ and $\hat{\boldsymbol{\psi}}_j$ are the mode shapes of a C_n structure with the central axis fixed. It is easy to prove the following property.

Property 6.3.8 For $0 < j < n/2$, the base mode shapes of a C_n structure with a central axis are orthogonal to each other.

Proof In view of the analysis in Sect. 6.3.3, the central displacements of the C_n structure are zeros for $2 \leq j < n/2$. Thus, it is only necessary to study the mode shapes with repeated frequencies for $j = 1$. In this case, as shown in Eq. (6.3.23), the central displacement vectors in Eq. (6.3.25) satisfy

$$^{(0)}\hat{\boldsymbol{u}}_{o1}^{\mathrm{T}}\,^{(0)}\boldsymbol{u}_{o1} = 0 \tag{a}$$

From the definition of vectors $\boldsymbol{\psi}_j$ and $\hat{\boldsymbol{\psi}}_j$ in Eq. (6.3.25), by analogy with the proof of Property 6.3.7, we get the following orthogonality relation

$$\hat{\boldsymbol{\psi}}_1^{\mathrm{T}}\boldsymbol{\psi}_1 = 0 \tag{b}$$

There follows the final orthogonality to be proved, namely,

$$\hat{\boldsymbol{\phi}}_1^{\mathrm{T}}\boldsymbol{\varphi}_1 = {}^{(0)}\hat{\boldsymbol{u}}_{o1}^{\mathrm{T}}\,^{(0)}\boldsymbol{u}_{o1} + \hat{\boldsymbol{\psi}}_1^{\mathrm{T}}\boldsymbol{\psi}_1 = 0 \tag{c}$$

6.3.4.3   Relation of Arbitrary Mode Shapes with Repeated Frequencies

According to Properties 6.3.7 and 6.3.8, the base mode shapes with repeated frequencies are orthogonal and can describe any other mode shapes corresponding to the repeated frequency via a linear combination.

Property 6.3.9 The normalized mode shape with repeated frequencies of a C_n structure can be expressed as

$$\tilde{\boldsymbol{\varphi}}_j = (\cos\phi)\boldsymbol{\varphi}_j - (\sin\phi)\hat{\boldsymbol{\varphi}}_j, \quad \phi \in [0,\ 2\pi), \quad 0 < j < n/2 \tag{6.3.26}$$

Proof As shown in Eq. (6.3.5), an arbitrary normalized eigenvector of the C_n structure in subspace S_j yields

$$\tilde{\boldsymbol{q}}_{jD} = \exp(\mathrm{i}\phi)\boldsymbol{q}_{jD}, \quad \phi \in [0,\ 2\pi), \quad 0 < j < n/2 \tag{a}$$

Equation (6.2.74) and Eq. (a) lead to the mode shape of sub-structure $^{(k)}S$ in S_j as follows

$$
\begin{aligned}
^{(k)}\tilde{\boldsymbol{\varphi}}_j &= \mathrm{Re}\big[\exp(\mathrm{i}jk\theta)\boldsymbol{B}_{jD}\tilde{\boldsymbol{q}}_{jD}\big] = \mathrm{Re}\big\{\exp[\mathrm{i}(jk\theta + \phi)]\boldsymbol{B}_{jD}\boldsymbol{q}_{jD}\big\} \\
&= \cos(jk\theta + \phi)^{(0)}\boldsymbol{\varphi}_j - \sin(jk\theta + \phi)^{(0)}\hat{\boldsymbol{\varphi}}_j \\
&= (\cos\phi)^{(k)}\boldsymbol{\varphi}_j - (\sin\phi)^{(k)}\hat{\boldsymbol{\varphi}}_j, \quad k \in I_n, \quad 0 < j < n/2
\end{aligned} \tag{b}
$$

Using Eq. (6.2.75) to assemble the mode shape of the entire C_n structure, we arrive at Eq. (6.3.26) and complete the proof.

Finally, we use Property 6.3.9 to study the possible circumferential period of any two mode shapes with repeated frequencies as follows.

Property 6.3.10 Given two mode shapes $^{(k)}\boldsymbol{\varphi}_j$, $k \in I_n$ and $^{(k)}\tilde{\boldsymbol{\varphi}}_j$, $k \in I_n$ with repeated frequencies of a C_n structure for $0 < j < n/2$, if an integer $s + k \in I_n$ makes $^{(k)}\boldsymbol{\varphi}_j$, $k \in I_n$, after rotated by an angle $s\theta$, are identical to $^{(k)}\tilde{\boldsymbol{\varphi}}_j$, $k \in I_n$, then there must be an integer $r > 0$ and an angle $\phi > 0$ such that the integer s yields

$$s = \frac{n}{j}\left(r - \frac{\phi}{2\pi}\right), \quad 0 < j < \frac{n}{2} \tag{6.3.27}$$

Proof Given the mode shape $^{(k)}\boldsymbol{\varphi}_j$, $k \in I_n$ of a C_n structure for $0 < j < n/2$, select the mode shape $^{(k)}\hat{\boldsymbol{\varphi}}_j$, $k \in I_n$ such that $^{(k)}\boldsymbol{\varphi}_j$, $k \in I_n$ and $^{(k)}\hat{\boldsymbol{\varphi}}_j$, $k \in I_n$ are a pair of base mode shapes of the C_n structure. In view of Eq. (b) in the proof of Property 6.3.9, the mode shape $^{(k)}\tilde{\boldsymbol{\varphi}}_j$, $k \in I_n$ can be expressed as

$$^{(k)}\tilde{\boldsymbol{\varphi}}_j = (\cos\phi)^{(k)}\boldsymbol{\varphi}_j - (\sin\phi)^{(k)}\hat{\boldsymbol{\varphi}}_j, \quad k \in I_n, \quad 0 < j < n/2 \tag{a}$$

where ϕ is a real number. If mode shape $^{(k)}\boldsymbol{\varphi}_j$, after rotated through an angle $s\theta$ about the central axis, is identical to mode shape $^{(k+s)}\tilde{\boldsymbol{\varphi}}_j$, Eq. (6.3.12) enables one to derive the following result

$$
\begin{aligned}
^{(k+s)}\tilde{\boldsymbol{\varphi}}_j &= (\cos\phi)^{(k+s)}\boldsymbol{\varphi}_j - (\sin\phi)^{(k+s)}\hat{\boldsymbol{\varphi}}_j \\
&= [(\cos\phi)\boldsymbol{I}_{\tilde{m}} \quad -(\sin\phi)\boldsymbol{I}_{\tilde{m}}] \begin{bmatrix} \cos(js\theta)\boldsymbol{I}_{\tilde{m}} & -\sin(js\theta)\boldsymbol{I}_{\tilde{m}} \\ \sin(js\theta)\boldsymbol{I}_{\tilde{m}} & \cos(js\theta)\boldsymbol{I}_{\tilde{m}} \end{bmatrix} \begin{bmatrix} ^{(k)}\boldsymbol{\varphi}_j \\ ^{(k)}\hat{\boldsymbol{\varphi}}_j \end{bmatrix} \\
&= \cos(js\theta + \phi)^{(k)}\boldsymbol{\varphi}_j - \sin(js\theta + \phi)^{(k)}\hat{\boldsymbol{\varphi}}_j = {}^{(k)}\boldsymbol{\varphi}_j \tag{b}
\end{aligned}
$$

Eq. (b) looks similar to Eq. (c) in the proof of Property 6.3.5, the analogous deduction leads to

$$\sin\left(\frac{js\theta + \phi}{2}\right) = 0 \tag{c}$$

Solving Eq. (c) for s gives the final result in Eq. (6.3.27) as follows

$$js\theta + \phi = 2r\pi \quad \Rightarrow \quad s = \frac{n}{j}\left(r - \frac{\phi}{2\pi}\right) \tag{d}$$

This completes the proof.

Remark 6.3.9 A comparison between Eqs. (6.3.27) and (6.3.13) indicates that they are the same when $\phi = -\pi/2$. As ϕ can take a continuous value, Eq. (6.3.27) seems to hold easily. Yet, the mode shapes with repeated frequencies computed for a C_n structure are a pair of base mode shapes when $\phi = -\pi/2$, while an arbitrary mode vector $^{(k)}\tilde{\boldsymbol{\varphi}}_j$, $k \in I_n$ usually comes from a modal test so that Eq. (6.3.27) may hardly hold.

6.3.5   Modal Test of a Bladed Disc Model

This subsection presents how to use the theoretical results in Sects. 6.3.2 and 6.3.4 to reduce blindness in the modal test of a bladed disc model with six short blades and the central hollow clamped in Fig. 6.31.

6.3.5.1   Predictions Before the First Modal Test

In the early modal tests, the bladed disc like the model in Fig. 6.31 was usually treated by analogy with a circular annular plate such that the natural modes measured were classified in view of the number of nodal lines along the circumferential direction and the radial direction, respectively. When the nodal lines had complicated patterns, the above classification failed to work. As the bladed disc model here is a C_6 structure, the theoretical studies in Sect. 6.3.2 enabled one to predict some modal properties of the bladed disc model in order to reduce blindness before the modal test as much as possible.

Before the first modal test, the natural modes of the bladed disc model to be measured were classified into four types according to the subspace S_j, $j = 0, 1, 2, 3$ of group C_6, and ranked ascendingly. From Properties 6.3.4 and 6.3.5, the following predictions were made before the modal test.

(1)   According to Property 6.3.4, the mode shapes would be classified into S_0, corresponding to those with a single frequency, if they exhibited the same geometric shape for every sub-structure.

Fig. 6.31 A bladed disc model with six short blades and central hollow clamped

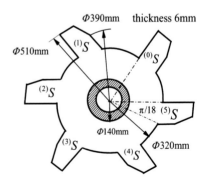

(2) In view of Property 6.3.5, the mode shapes would be put into S_1, corresponding to those with a pair of repeated frequencies, if they did not exhibit any circumferential period from sub-structures $^{(0)}S$ to $^{(5)}S$ since $j = n/p = 6/6 = 1$.

(3) In view of Property 6.3.5, once the mode shape exhibited the circumferential period three, that is, $j = n/p = 6/3 = 2$, they had repeated frequencies and would be put into subspace S_2.

(4) From Property 6.3.4, the mode shapes would be put into subspace S_3, corresponding to those with a single frequency, if they had the same geometric shape in two adjacent sub-structures, but opposite directions.

(5) From Sect. 6.2.3, the major difference among the natural frequencies of the same order in different subspace S_j comes from the terms associated with $\cos(j\theta)$ and $\sin(j\theta)$ in mass and stiffness matrices in Eq. (6.2.89). These terms rely on the interface between two adjacent sub-structures. If the coupling between two adjacent sub-structures is weak, the natural frequencies of the same order in different subspace S_j get close in frequency domain. As such, a bladed disc model with short blades here might not have extremely close natural frequencies since the interfacial coupling of the bladed disc model is not weak enough.

6.3.5.2 Discussions After the First Modal Test

According to the above prediction in (5), the vibration test was made through a single sweeping harmonic excitation and the nodal lines were recorded via sand patterns. In the first modal test, nine natural vibrations were observed and their natural frequencies were classified according to the predictions before the modal test and listed in Table 6.4, where an asterisk represents a pair of possible mode shapes with repeated frequencies.

In view of Table 6.4, the discussions for further tests were made as follows.

(1) Nine natural vibrations were missed in the first modal test if the maximal natural frequency, i.e., 1813 Hz, was taken as the upper bound and a possible pair of repeated natural frequencies was taken into account.

(2) According to the discussion before the first modal test, the first natural frequency in subspace S_1 would be about 200 Hz and had a nodal line since the mode shape missed would not exhibit any circumferential period from sub-structures $^{(0)}S$ to $^{(5)}S$. In addition, the second natural frequency in S_0 missed

Table. 6.4 The nature frequencies (Hz) of a bladed disc model in the first vibration test

Frequency order	S_0	S_1	S_2	S_3
1	213	–	257	292
2	–	–	–	766
3	1813	1620*	1283	1099

would fall into a frequency band of 700 Hz ~ 1000 Hz, but would not be higher than the minimal natural frequency 1099 Hz among the third-order natural frequencies in all S_j, and the corresponding mode shape would have a nodal line.

(3) The mode shape corresponding to 1620 Hz with the circumferential period after it was rotated through an angle π was observed by changing excitation location. This evidence verified that the mode shape would fall into S_1 and initiated a guess that adjusting excitation location might enable one to observe more natural vibrations with repeated frequencies.

Following the above discussions, the second modal test was made for the bladed disc model. Finally, 18 natural modes were measured by adjusting the excitation locations and identifying the measured mode shapes from their nodal lines. Figure 6.32 presents the nodal lines of those mode shapes classified in subspace S_j, $j = 0, 1, 2, 3$.

6.3.5.3   Discussions After the Second Modal Tests

According to the measured results in Fig. 6.32, further discussions about the modal tests of the bladed disc model are as follows.

(1) As shown in the modal tests, the bladed disc model exhibited close natural frequencies in frequency domain. For instance, six natural modes of the bladed disc model appeared in a frequency band from 712 to 766 Hz, and the natural frequencies 763 and 766 Hz were very close, but their mode shapes were quite different. If the number of blades and the length of blades were increased further, more natural frequencies would fall into an even narrow frequency

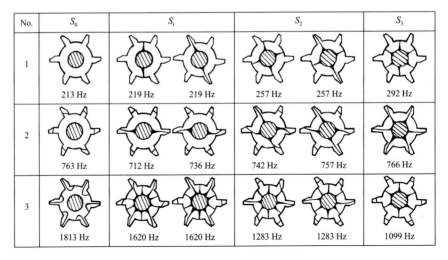

Fig. 6.32 The natural modes of a bladed disc model measured in modal tests

band. Without any theoretical studies on the modal properties of a C_n structure, it would be impossible to get completed experimental results in modal tests.

(2)   Different mode shapes with a pair of repeated frequencies in subspaces S_1 and S_2 were observed by adjusting the excitation locations. The mode shapes in Fig. 6.32 are a pair of mode shapes with approximately orthogonality. Similar to the statement of Property 6.3.10, the measured mode shapes would rotate about the central axis when the excitation location was moved along the circumferential direction.

(3)   The manufacturing errors of this bladed disc model were significant such that some pairs of natural modes with repeated frequencies did split into those with a pair of close frequencies. For instance, the mode shapes with 742 and 757 Hz exhibited asymmetry in Fig. 6.32 due to manufacturing errors.

(4)   As shown in Fig. 6.32, the ranking of the same order of natural frequencies in subspace S_j, $j = 1, 2, 3$ is not unique. For example, the natural frequencies in S_0 may be the lowest one (213 Hz), or the middle one (763 Hz), or the highest one (1813 Hz). This evidence contradictes the previous assertion that the mode shapes had the lowest natural frequencies when all blades underwent the same vibration shapes[26].

6.3.6 Concluding Remarks

This section presented a systematic study on the modal properties of a C_n structure according to the representative theory of group C_n. An important finding is that the modal properties of the C_n structure without or with a central axis are almost the same. This completed the study on Problem 5B proposed in Chap. 1. The major conclusions of the section are as follows.

(1)   The natural mode of a C_n structure has a single frequency in two cases. In the first case, the natural vibration falls into subspace S_0 and the mode shapes in all sub-structures are identical. In the second case when n is an even number, the vibration falls into subspace $S_{n/2}$ and the mode shapes in two adjacent sub-structures have the same geometric shape but opposite directions.

(2)   A C_n structure has abundant natural vibrations with a pair of repeated frequencies. A pair of mode shapes has the repeated frequencies if and only if the natural vibration occurs in a subspace S_j, $0 < j < n/2$ and the mode shape exhibits a circumferential period of every p sub-structures with $p = n/j$.

(3)   In a subspace S_j, $0 < j < n/2$, the mass and stiffness matrices of a C_n structure are Hermite matrices, and each eigenvalue of corresponding real matrices is a repeated one. The commercial software provides a pair of orthogonal eigenvectors with repeated natural frequencies and a pair of base modal shapes. The

---

[26] Ewins D J (1980). Bladed disc vibration - A review of techniques and characteristics. Southampton: International Conference on Recent Advances in Structural Dynamics.

base mode shapes can describe any mode shape of the same frequency, and the rotation between any two mode shapes with repeated frequencies.

(4) The natural vibrations of a C_n sub-structure with a central axis include the following three cases. The first case is the natural vibration with a single frequency in subspace S_0. In this case, the central displacements are either the rectilinear displacements along the central axis or the rotation angles about the central axis, and the C_n structure at the central axis exhibits either an extreme amplitude or a saddle node of rn nodal lines, where r is a positive integer. The second case is a pair of natural vibrations with repeated frequencies in subspace S_1. In this case, the central displacements occur at a plane normal to the central axis and at least one nodal line passes through the central axis. The third case is the natural vibrations with a single frequency in subspace $S_{n/2}$ or a pair of repeated frequencies in subspace S_j, $1 < j < n/2$. In this case, at least two nodal lines intersect the central axis such that the C_n structure at the central axis is fixed.

(5) The above theoretical results get good agreement with numerical and experimental studies on C_n structures, including C_{nv} structures, and have successfully guided the modal tests of a bladed disc model.

6.4   Further Reading and Thinking

(1) Given a thin rectangular plate with a pair of opposite edges hinged and the other two edges free, prove that any two mode shapes with different frequencies satisfy an orthogonal relation. In addition, discuss the natural vibrations with repeated frequencies when the above plate becomes a square plate.

(2) Among the computational methods for the natural vibrations of a C_n structure in the references[27,28,29], select one method and compare it with the method in this chapter in consideration of simplicity and regularity from the beauty of science.

(3) Read the reference[30] for a thin circular plate with clamped boundary, make a comparison of mode shapes with repeated frequencies for such a circular plate and the cyclosymmetric structure with a central axis, and sort out the difference between the two structures. Let R be the radius of the above plate and a lumped mass m be attached at a radial location $R/2$ from the central axis, discuss the influence of the lumped mass on the natural vibrations with repeated frequencies.

[27] Thomas D L (1979). Dynamics of rotationally periodic structures. International Journal of Numerical Methods in Engineering, 14(1): 81–102.

[28] Olson B J, Shaw S W, Shi C Z, et al. (2014). Circulant matrices and their application to vibration analysis. Applied Mechanics Review, 66(4): 040803.

[29] Wang D J, Wang Q S, He B C (2019). Qualitative Theory in Structural Mechanics. Singapore: Springer Nature, 271–326.

[30] Rao S S (2007). Vibration of Continuous Systems. Hoboken: John Wiley & Sons Inc., 485–495.

(4) Read the reference[31] or reference[32], and discuss the modal properties of an equilateral triangle plate with hinged boundaries or a square plate with hinged boundaries according to the representative theory of group C_{3v} or group C_{4v}.

(5) Read the references[33,34], and discuss the flowchart of a high-efficient computation procedure for the natural vibrations of a complicated C_n structure.

---

[31] Evensen D A (1976). Vibration analysis of multi-symmetric structures. AIAA Journal, 14(4): 446–453.

[32] Joshi A W (1977). Elements of Group Theory for Physicists. New Delhi: Wiley Eastern Inc., 57–102.

[33] Tran D M (2014). Reduced models of multi-stage cyclic structures using cyclic symmetry reduction and component mode synthesis. Journal of Sound and Vibration, 333(21): 5443–5463.

[34] Rong B, Lu K, Ni X J (2020). Hybrid finite element transfer matrix method and its parallel solution for fast calculation of large-scale structural eigenproblem. Applied Mathematical Modeling, 77(1): 169–181.

Chapter 7
Waves and Vibrations
of One-Dimensional Structures

Both studies on vibrations and waves fall into the scope of dynamics in engineering, but take different points of view. Taking a slender structure as an example, the study of the vibrations of the structure deals with the oscillatory motions around the static equilibrium of the structure and usually begins with the analysis of natural vibrations, which are the synchronous motions of the structure and relatively simple. The research of the waves of the structure, however, deals with the temporal-spatial evolution of deformations of the structure and starts with the analysis of traveling waves, which are the asynchronous motions of the structure and relatively complicated. As a consequence, vibration analysis is easier than wave analysis in general, and uses simpler mathematical tools.

In engineering education, the elementary courses associated with vibration mechanics mainly serve for undergraduate programs of aerospace engineering, mechanical engineering and civil engineering, focusing on the dynamic problems in finite sizes. Meanwhile, the elementary courses of elastic waves are for advanced programs of geophysics, earthquake engineering and non-destructive detections with an emphasis on the dynamic problems with infinite or semi-infinite size. The integrated courses of vibrations and waves, as well as their textbooks, are not very popular.

Recent years have seen integrated developments of vibration mechanics and wave mechanics. For example, ultrasonic motors for precise actuation[1] and metamaterials for acoustic wave reduction[2] have well demonstrated the integrated developments. The dynamic analysis of extremely large space structures in Sect. 1.2.6 also requires a good understanding of both vibrations and waves. To meet the requirements of future engineers, this chapter will deal with both waves and vibrations.

---

[1] Zhao C S (2011). Ultrasonic Motor Technologies and Applications. Heidelberg: Springer, 118–299.

[2] Ma G C, Sheng P (2016). Acoustic metamaterials: From local resonances to broad horizons. Science Advances, 2: e1501595.

© Science Press 2022   325
H. Hu, *Vibration Mechanics*,
https://doi.org/10.1007/978-981-16-5457-2_7

Different from the books of elasto-dynamics of three-dimensional waves, the chapter focuses on the dynamics of one-dimensional structures, such as rods and beams, with an emphasis on the relation between waves and vibrations associated with Problems 6A and 6B proposed in Chap. 1. The appendix of the book will provide a detailed introduction to the elasto-dynamics of three-dimensional waves.

7.1 Non-dispersive Waves of Rods

This section presents the dynamic analysis of longitudinal response of a uniform rod made of a linear homogeneous elastic material. The focus of the section is the relation between waves and vibrations. As the first step of wave analysis, the section deals with the *non-dispersive wave*, the speed of which does no depend on the wave frequency. Then, the next section will present the dispersive wave with frequency-dependent wave speed. The section begins with the analysis of harmonic waves in an infinitely long rod, and then study the stead-state vibrations of a finitely long rod under an axial harmonic force. The section features a comparison between different methods of wave analysis and vibration analysis so as to address Problem 6A in part.

7.1.1 Wave Analysis of an Infinitely Long Rod

This subsection deals with the dynamics of an infinitely long uniform rod under an external force. As studied in Sect. 3.3.1, the longitudinal displacement $u(x, t)$ of a uniform rod to an axial force $f(x, t)$ yields an inhomogeneous partial differential equation as follows

$$\rho A \frac{\partial^2 u(x, t)}{\partial t^2} - EA \frac{\partial^2 u(x, t)}{\partial x^2} = f(x, t), \quad x \in (-\infty, +\infty), \quad t \in [0, +\infty)$$

$$(7.1.1)$$

where A is the cross-sectional area, ρ is the material density, and E is Young's modulus. The study begins with the free waves of the rod, and then deals with the forced waves of the rod.

7.1.1.1 Free Traveling Waves of a Rod

If $f(x, t) = 0$, Eq. (7.1.1) becomes a homogeneous wave equation recast as

$$\frac{\partial^2 u(x, t)}{\partial t^2} - c_0^2 \frac{\partial^2 u(x, t)}{\partial x^2} = 0, \quad c_0 \equiv \sqrt{\frac{E}{\rho}}, \quad x \in (-\infty, +\infty), \quad t \in [0, +\infty)$$

$$(7.1.2)$$

where c_0 is the *longitudinal wave speed* of the uniform rod. Using the following transform of independent variables

$$\xi = c_0 t - x, \quad \eta = c_0 t + x \tag{7.1.3}$$

we derive

$$\frac{\partial^2 u}{\partial x^2} = \frac{\partial^2 u}{\partial \xi^2} - 2\frac{\partial^2 u}{\partial \xi \partial \eta} + \frac{\partial^2 u}{\partial \eta^2}, \quad \frac{\partial^2 u}{\partial t^2} = c_0^2\left(\frac{\partial^2 u}{\partial \xi^2} + 2\frac{\partial^2 u}{\partial \xi \partial \eta} + \frac{\partial^2 u}{\partial \eta^2}\right) \tag{7.1.4}$$

Equation (7.1.4) enables one to recast the partial differential equation in Eq. (7.1.2) as

$$\frac{\partial^2 u(\xi, \eta)}{\partial \xi \partial \eta} = 0 \tag{7.1.5}$$

Integrating Eq. (7.1.5) with respect to ξ and η respectively, we get the solution, which was first derived by J. L. R. d'Alembert, a French scientist, in 1746, as follows

$$u(\xi, \eta) = u_R(\xi) + u_L(\eta) \tag{7.1.6}$$

where $u_R(\xi)$ and $u_L(\eta)$ are arbitrary functions with the second-order smoothness. Further study on partial differential equations has shown that the smoothness of the two functions can be weakened to piecewise differentiable. In this case, the solution is a *weak solution* or a *generalized solution*[3], which will be discussed in Sect. 7.3 for an impacting rod.

Substituting Eq. (7.1.3) into Eq. (7.1.6) yields

$$u(x, t) = u_R(c_0 t - x) + u_L(c_0 t + x), \quad x \in (-\infty, +\infty), \quad t \in [0, +\infty) \tag{7.1.7}$$

In physics, $u_R(c_0 t - x)$ is a free *traveling wave* with speed c_0 towards right direction, while $u_L(c_0 t + x)$ is a free traveling wave with speed c_0 in left direction. As such, Eq. (7.1.7) is called *d'Alembert's solution* or the *traveling wave solution*.

Figure 7.1 demonstrates the evolution of two traveling waves in the frame of coordinates (x, t, u) in a temporal-spatial domain. As shown in the figure, $u_R(\xi) = [1 + \cos(\xi\pi)]/2$ and $u_L(\eta) = [1 + \cos(\eta\pi)]/2$ keep constant along two straight lines $\xi = c_0 t - x = $ const and $\eta = c_0 t + x = $ const, respectively, in plane (x, t). When t goes on, the wave $u_L(c_0 t + x)$ travels along straight line $c_0 t + x = $ const in the left direction shown in Fig. 7.1a, while the wave $u_R(c_0 t - x)$ travels along straight line $c_0 t - x = $ const in the right direction in Fig. 7.1b. The two straight lines are named the *characteristic lines* of Eq. (7.1.2). The two traveling waves form two cylindrical surfaces in the temporal-spatial domain in Fig. 7.1, where the traveling

---

[3] Courant R, Hilbert D (1989). Methods of Mathematical Physics. Volume II. New York: John Wiley & Sons Inc., 486–488.

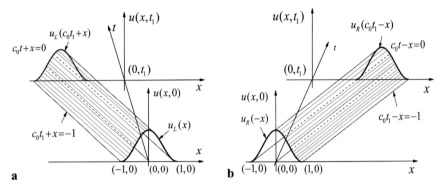

Fig. 7.1 The temporal-spatial surfaces of two waves of a uniform rod. **a** the left traveling wave $u_L(c_0t + x) = \{1 + \cos[\pi(c_0t + x)]\}/2$; **b** the right traveling wave $u_R(c_0t - x) = \{1 + \cos[\pi(c_0t - x)]\}/2$

waves only affect the zones beneath the two cylindrical surfaces. These zones are called the *domain of influence*.

The free traveling waves of a uniform rod have the following non-dispersive properties. That is, the wave speed c_0 is a constant and independent of wave frequency and wave number. Hence, if the Fourier spectrum of an initial disturbances $u_R(x)$ and $u_L(x)$ have several frequency components, the harmonic waves of these frequency components propagate with the same speed c_0 such that the wave shapes of $u_R(c_0t - x)$ and $u_L(c_0t + x)$ do not change with time t. As of Sect. 7.2, the chapter will present the *dispersive wave*, the speed of which depends on both wave frequency and wave number such that the wave shape of a dispersive wave varies with t during the wave propagation.

In the vibration analysis of linear systems, harmonic vibrations are not only the simpliest ones, but also offer the linear superposition to approximate any other vibrations. In the wave analysis of linear continuous systems, harmonic waves play the same role. As a consequence, we shall analyze the forced harmonic waves of a semi-infinitely long rod and an infinitely long rod, respectively, by using d'Alembert's solution as follows.

7.1.1.2 Traveling Waves of a Rod to a Harmonic Axial Force at One End

Figure 7.2 demonstrates a semi-finitely long uniform rod, which is subject to an axial harmonic force $f(t) = F\cos(\omega_0 t)$ at the left end of the rod, while the right end of the rod extends to infinity. We establish a frame of coordinate ox measured from the left end of the rod and denote $u(x, t)$ as the longitudinal displacement of cross-section x in the right direction. Now, the wave under force $f(t)$ travels towards the right direction and does not have any reflection.

According to Eq. (7.1.1), the dynamic equation of the rod reads

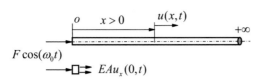

Fig. 7.2 A semi-infinitely long uniform rod to an axial harmonic force at the left end and the force analysis of an infinitesimal segment of the rod

$$\rho A \frac{\partial^2 u(x,t)}{\partial t^2} - EA \frac{\partial^2 u(x,t)}{\partial x^2} = \delta(x)F\cos(\omega_0 t), \quad x \in [0,\ +\infty), \quad t \in [0,\ +\infty)$$

$$(7.1.8)$$

where $\delta(x)$ is Dirac delta function. Given $x > 0$, Eq. (7.1.8) becomes a homogeneous wave equation and has the general form of solution in Eq. (7.1.7). As the harmonic force to the left end of the rod excites a right traveling wave, which undergoes no reflection, the solution in Eq. (7.1.7) reads

$$u(x,t) = u_R(c_0 t - x), \quad x \in [0,\ c_0 t], \quad t \in [0,\ +\infty) \qquad (7.1.9)$$

Now, we use the excitation force to determine the function form $u_R(\xi) = u_R(c_0 t - x)$ in Eq. (7.1.9). In Fig. 7.2, the forces applied to an infinitesimal segment of the rod at $x = 0$ satisfy the equilibrium equation as follows

$$EAu_x(0,t) + F\cos(\omega_0 t) = -EA \frac{du_R(\xi)}{d\xi}\bigg|_{\xi=c_0 t} + F\cos(\omega_0 t) = 0 \qquad (7.1.10)$$

Replacing the independent variable t in Eq. (7.1.10) with $t = \xi/c_0$ leads to

$$EA \frac{du_R(\xi)}{d\xi} = F\cos(\kappa\xi), \quad \kappa \equiv \frac{\omega_0}{c_0}, \quad \xi = c_0 t \geq 0 \qquad (7.1.11)$$

where κ is the *wave number*. Integrating both sides of Eq. (7.1.11) gives the following solution

$$u_R(\xi) = \frac{F}{\kappa EA} \sin(\kappa\xi), \quad \xi = c_0 t \geq 0 \qquad (7.1.12)$$

Replacing augment ξ in Eq. (7.1.12) with $\xi = c_0 t - x$, we derive the right traveling wave of the rod to excitation $F\cos(\omega_0 t)$ at the left end as follows

$$u(x,t) = u_R(c_0 t - x) = \frac{F}{\kappa EA} \sin(\omega_0 t - \kappa x)$$

$$= \frac{F}{\kappa EA} \sin[\kappa(c_0 t - x)], \quad x \in [0,\ c_0 t], \quad t \in [0,\ +\infty) \qquad (7.1.13)$$

Fig. 7.3 A semi-infinitely long uniform rod to an axial harmonic force at the right end and the force analysis of an infinitesimal segment of the rod

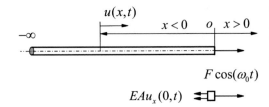

Compared with the harmonic force $F \cos(\omega_0 t)$, the longitudinal displacement $u(0, t) = F \sin(\omega_0 t)/(\kappa EA)$ at the left end of the rod has a phase shift $\pi/2$. This results from the equilibrium of elastic restoring force and excitation force at the left end of the rod, and also implies the stead-state harmonic wave, instead of a transient wave, of the rod to an axial harmonic force.

To understand how the steady-state harmonic wave evolutes, consider the case when the excitation frequency ω_0 near zero. When $\omega_0 \to 0^+$, the excitation force approaches a static compressional force. As the wave number $\kappa = O(\omega_0) \to 0$, the limit process of Eq. (7.1.13) gives the quasi-static displacement and strain as follows

$$\lim_{\omega_0 \to 0^+} u(x, t) \approx \frac{F}{EA}(c_0 t - x) \geq 0, \quad \lim_{\omega_0 \to 0^+} \varepsilon(x, t) \approx -\frac{F}{EA} < 0, \quad x \in [0, \, c_0 t]$$

$$(7.1.14)$$

This implies that in the beginning of a static excitation, the tip displacement of the rod yields $u(0, t) \approx F c_0 t/(EA)$ and the rod displacement becomes the harmonic wave given in Eq. (7.1.13) with an increase of excitation frequency ω_0.

Next, we study a semi-infinitely long rod in Fig. 7.3, where the axial harmonic force is applied to the right end of the rod. For brevity, we first consider the case when the positive excitation direction is assumed towards the left. This case is the mirror of Fig. 7.2. If the frame of coordinate ox in the left direction is defined, it is easy to use Eq. (7.1.13) to describe the left traveling wave.

Now, we establish the frame of coordinate ox in the right direction in Fig. 7.3 and study the case when the positive excitation direction is towards the right. Replacing x in Eq. (7.1.13) with $-x$, we derive the left traveling wave to a right harmonic force as follows

$$u(x, t) = u_L(c_0 t + x) = \frac{F}{\kappa EA} \sin[\kappa(c_0 t + x)], \quad x \in [-c_0 t, \, 0], \quad t \in [0, \, +\infty)$$

$$(7.1.15)$$

Different from the excitation at the left end, the harmonic force in this case approaches a static tension force when $\omega_0 \to 0^+$. As the wave number $\kappa = O(\omega_0) \to 0$, the limit process of Eq. (7.1.15) leads to the quasi-static displacement and static strain as following

$$\lim_{\omega_0 \to 0^+} u(x,t) \approx \frac{F}{EA}(c_0 t + x) \geq 0, \quad \lim_{\omega_0 \to 0^+} \varepsilon(x,t) = \frac{F}{EA} > 0, \quad x \in [-c_0 t, 0]$$

(7.1.16)

7.1.1.3 Traveling Waves of an Infinitely Long Rod to a Harmonic Axial Force

Figure 7.4 shows an infinitely long uniform rod subject to an axial harmonic force $F\cos(\omega_0 t)$ with its positive direction towards the right. We establish the frame of coordinate ox measured from the excitation position and study the longitudinal displacement $u(x,t)$ to the excitation force.

If the rod is divided into two parts at the excitation position, the problem of concern can be decomposed into the two problems studied above. That is, the right half rod is subject to a right force $F\cos(\omega_0 t)/2$ at the left end and undergoes a right traveling wave, while the left half rod is subject to a right force $F\cos(\omega_0 t)/2$ at the right end and undergoes a left travel wave. As a consequence, the longitudinal displacement of the rod satisfies

$$\begin{aligned} u(x,t) &= \frac{F}{2\kappa EA} \sin(\omega_0 t - \kappa|x|) \\ &= \frac{F}{2\kappa EA} \sin[\kappa(c_0 t - |x|)], \quad x \in [-c_0 t, c_0 t], \quad t \in [0, +\infty) \end{aligned}$$

(7.1.17)

When the harmonic force is applied to position x_0, the longitudinal displacement reads

$$\begin{aligned} u(x,t) &= \frac{F}{2\kappa EA} \sin(\omega_0 t - \kappa|x - x_0|) \\ &= \frac{F}{2\kappa EA} \sin[\kappa(c_0 t - |x - x_0|)], \quad x \in [x_0 - c_0 t, x_0 + c_0 t], \quad t \in [0, +\infty) \end{aligned}$$

(7.1.18)

Similarly, if $\omega_0 \to 0^+$, then $\kappa = O(\omega_0) \to 0$ holds. The harmonic force at the left part of position x_0 approaches a static tension and leads to a tension strain, while the harmoinc force at the right part of position x_0 becomes a static compression and gives a compressional strain. The displacement of the whole rod keeps positive. In

Fig. 7.4 An infinitely long uniform rod to an axial harmonic force and the force analysis of an infinitesimal segment of the rod

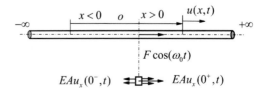

the beginning of loading, the quasi-static displacement and strain are

$$
\begin{cases}
\lim_{\omega_0 \to 0^+} u(x, t) \approx \dfrac{F}{2EA}(c_0 t - |x - x_0|) \geq 0, \\[4mm]
\lim_{\omega_0 \to 0^+} \varepsilon(x, t) = -\dfrac{F}{2EA}\,\mathrm{sgn}(x - x_0), \quad x \in [x_0 - c_0 t,\ x_0 + c_0 t], \quad t \in [0, +\infty)
\end{cases}
$$

$$(7.1.19)$$

7.1.2 Complex Function Analysis of Waves of an Infinitely Long Rod

The wave analysis in Sect. 7.1.1 focused on the temporal-spatial evolution of the dynamic response of a semi-infinitely or infinitely long rod to an axial harmonic force, and looks a little bit complicated. As the traveling waves of the above rods keep harmonic in time domain, it will be beneficial to use a complex function to describe the harmonic traveling wave so as to simplify the analysis and focus on the variation of wave amplitude in space.

For brevity, the subsection deals with the wave of an infinitely long rod to an axial harmonic force at given position shown in Fig. 7.4. Now, the dynamic equation of the rod is

$$
\rho A \frac{\partial^2 u(x, t)}{\partial t^2} - EA \frac{\partial^2 u(x, t)}{\partial x^2} = \delta(x) F \cos(\omega_0 t), \quad x \in (-\infty, +\infty), \quad t \in [0, +\infty)
$$

$$(7.1.20)$$

We use $\mathrm{Re}[\exp(i\omega_0 t)]$ to express the cosine function in Eq. (7.1.20) so that Eq. (7.1.20) can be recast as a partial differential equation of a complex function, namely,

$$
\rho A \frac{\partial^2 u_c(x, t)}{\partial t^2} - EA \frac{\partial^2 u_c(x, t)}{\partial x^2} = \delta(x) F \exp(i\omega_0 t), \quad x \in (-\infty, +\infty), \quad t \in [0, +\infty)
$$

$$(7.1.21)$$

At present, Eq. (7.1.20) is the real part of Eq. (7.1.21) and has a solution $u(x, t) = \mathrm{Re}[u_c(x, t)]$, while solving Eq. (7.1.21) for $u_c(x, t)$ may be easier than solving Eq. (7.1.20) for $u(x, t)$. The above procedure of converting a real problem to a complex problem is called the *complex function method* throughout the chapter. As a counterpart, the solution procedure in Sect. 7.1.1 is the *real function method*.

Let the complex solution of Eq. (7.1.21) be

$$
u_c(x, t) = \tilde{u}_c(x) \exp(i\omega_0 t), \quad x \in (-\infty, +\infty), \quad t \in [0, +\infty) \qquad (7.1.22)
$$

where $\tilde{u}_c(x)$ is the complex amplitude to be determined. Substitution of Eq. (7.1.22) into Eq. (7.1.21) leads to an ordinary differential equation of a complex function $\tilde{u}_c(x)$ as follows

$$\frac{\partial^2 \tilde{u}_c(x)}{\partial x^2} + \kappa^2 \tilde{u}_c(x) = -\frac{F}{EA}\delta(x), \quad x \in (-\infty, +\infty) \tag{7.1.23}$$

where $\kappa = \omega_0/c_0$ is the wave number defined in Eq. (7.1.11). Several methods have been available for solving Eq. (7.1.23). The section will present two of them. One is about the force analysis at the excitation position from the viewpoint of mechanics, while the other comes from Fourier transform in mathematics.

7.1.2.1 Force Analysis at Excitation Position

When $x \neq 0$, Eq. (7.1.23) becomes a linear homogeneous ordinary differential equation, which has the following general solution

$$\tilde{u}_c(x) = c_1 \exp(-i\kappa x) + c_2 \exp(i\kappa x), \quad x \in (-\infty, +\infty) \tag{7.1.24}$$

Substitution of Eq. (7.1.24) into Eq. (7.1.22) leads to a complex solution of Eq. (7.1.21)

$$\begin{cases} u_c(x, t) = \left[c_1 \exp(-i\kappa x) + c_2 \exp(i\kappa x)\right] \exp(i\omega_0 t) \\ \qquad = c_1 \exp[i\kappa(c_0 t - x)] + c_2 \exp[i\kappa(c_0 t + x)], \\ x \in [-c_0 t, c_0 t], \quad t \in [0, +\infty) \end{cases} \tag{7.1.25}$$

In the right part of the rod with $x > 0$ measured from the excitation position, the rod undergoes only a right traveling wave $c_1 \text{Re}\{\exp[i\kappa(c_0 t - x)]\}$ such that $c_2 = 0$ holds. In the left part of the rod with $x < 0$, the rod undergoes a left traveling wave $c_2 \text{Re}\{\exp[i\kappa(c_0 t + x)]\}$ and $c_1 = 0$ holds. Therefore, Eq. (7.1.24) becomes

$$\tilde{u}_c(x) = \begin{cases} c_1 \exp(-i\kappa x), & x \in [0, c_0 t] \\ c_2 \exp(i\kappa x), & x \in [-c_0 t, 0] \end{cases} \tag{7.1.26}$$

In view of the continuity of displacement $u_c(x, t)$ at $x = 0$, $\tilde{u}_c(x)$ is continuous at $x = 0$ so that $c_1 = c_2$ holds. Noting that the strain is not continuous at $x = 0$, we have

$$\frac{\partial u_c(x, t)}{\partial x}\bigg|_{x=0} = \exp(i\omega_0 t)\frac{d\tilde{u}_c(x)}{dx}\bigg|_{x=0} = \begin{cases} -i\kappa c_1 \exp(i\omega_0 t), & x = 0^+ \\ i\kappa c_1 \exp(i\omega_0 t), & x = 0^- \end{cases} \tag{7.1.27}$$

As shown in Fig. 7.4, when the length dx of an infinitesimal segment at $x = 0$ approaches zero, the inertial force vanishes and the equilibrium equation of the infinitesimal segment becomes

$$\left\{EA\left[\frac{d\tilde{u}_c(x)}{dx}\bigg|_{x=0^+} - \frac{d\tilde{u}_c(x)}{dx}\bigg|_{x=0^-}\right] + F\right\}\exp(i\omega_0 t) = 0 \qquad (7.1.28)$$

Substitution of Eq. (7.1.27) into Eq. (7.1.28) leads to

$$c_1 = c_2 = -\frac{iF}{2\kappa EA} \qquad (7.1.29)$$

Substituting Eq. (7.1.29) into Eq. (7.1.26), we get the following complex solution

$$\tilde{u}_c(x) = -\frac{iF}{2\kappa EA}\exp(-i\kappa|x|), \quad x \in [-c_0 t, c_0 t] \qquad (7.1.30)$$

The corresponding real solution of the harmonic wave reads

$$\begin{aligned}u(x, t) &= \mathrm{Re}\left[-\frac{iF\exp(-i\kappa|x|)}{2\kappa EA}\exp(i\omega_0 t)\right] \\ &= -\frac{F}{2\kappa EA}\mathrm{Re}\{i\exp[i(\omega_0 t - \kappa|x|)]\} \\ &= \frac{F}{2\kappa EA}\sin(\omega_0 t - \kappa|x|), \quad x \in [-c_0 t, c_0 t], \quad t \in [0, +\infty) \qquad (7.1.31)\end{aligned}$$

If the harmonic force is applied to position x_0, Eq. (7.1.30) can be recast as

$$\tilde{u}_c(x) = -\frac{iF}{2\kappa EA}\exp(-i\kappa|x - x_0|), \quad x \in [x_0 - c_0 t, x_0 + c_0 t] \qquad (7.1.32)$$

There follows the real solution of the harmonic wave

$$u(x, t) = \frac{F}{2\kappa EA}\sin(\omega_0 t - \kappa|x - x_0|), \quad x \in [x_0 - c_0 t, x_0 + c_0 t], \quad t \in [0, +\infty) \qquad (7.1.33)$$

These results are the same as those derived via the real function method in Sect. 7.1.1.

Remark 7.1.1 Now both sides of an infinitesimal segment at the excitation position are subject to internal forces. This differs from the infinitesimal segment at one end of the rod. Thus, if the problem of an infinitely long rod to an axial harmonic force is divided into two problems of semi-infinitely long rods, the excitation force should be allocated to both ends of the semi-infinitely long rods.

Remark 7.1.2 To deal with a semi-infinitely long rod to an axial harmonic force at the end of the rod, one can use Eq. (7.1.30) or Eq. (7.1.31) to describe the traveling wave. Following Remark 7.1.1 and the equilibrium equation at the end of the rod, however, one ought to replace the complex amplitude in Eq. (7.1.30) with $-iF/(\kappa EA)$ and replace the wave amplitude in Eq. (7.1.31) by $F/(\kappa EA)$.

7.1.2.2 Method Based on Fourier Transforms

Applying Fourier transform to both sides of Eq. (7.1.23) gives

$$\tilde{U}_c(\omega) = \frac{F}{EA(\omega^2 - \kappa^2)} = \frac{iF}{2\kappa EA}\left[\frac{1}{i(\omega - \kappa)} - \frac{1}{i(\omega + \kappa)}\right], \quad \omega \in (-\infty, +\infty).$$

$$(7.1.34)$$

By applying the inverse Fourier transforms to both sides of Eq. (7.1.34) and using Heaviside function $s(x)$, we have

$$\tilde{u}_c(x) = \mathscr{F}^{-1}[\tilde{U}_c(\omega)] = \frac{iF}{2\kappa EA}\mathscr{F}^{-1}\left[\frac{1}{i(\omega - \kappa)} - \frac{1}{i(\omega + \kappa)}\right]$$

$$= \frac{iF}{2\kappa EA}[\exp(i\kappa x) - \exp(-i\kappa x)]s(x) \qquad (7.1.35)$$

In Eq. (7.1.35), the exponential functions $\exp(i\kappa x)$ and $\exp(-i\kappa x)$ describe the complex amplitude of the left traveling wave for $x \in (-\infty, 0]$ and the complex amplitude of the right traveling wave for $x \in [0, +\infty)$, respectively. Yet, the Heaviside function $s(x)$ makes Eq. (7.1.35) describe the right traveling wave only. Recalling the propagation time of a right traveling wave, we recast Eq. (7.1.35) as

$$\tilde{u}_c(x) = -\frac{iF}{2\kappa EA}\exp(-i\kappa x), \quad x \in [0, \ c_0 t] \qquad (7.1.36)$$

The symmetry of the problem gives the complex amplitude of the left traveling wave as follows

$$\tilde{u}_c(x) = -\frac{iF}{2\kappa EA}\exp(i\kappa x), \quad x \in [-c_0 t, \ 0] \qquad (7.1.37)$$

The combination of Eqs. (7.1.36) and (7.1.37) leads to the same result as Eq. (7.1.32), namely,

$$\tilde{u}_c(x) = -\frac{iF}{2\kappa EA}\exp(-i\kappa|x|), \quad x \in [-c_0 t, \ c_0 t] \qquad (7.1.38)$$

Remark 7.1.3 A comparison of the two methods shows that the method based on Fourier transforms avoids the force analysis which may lead to careless mistakes in force directions, but has less intuitive understanding of mechanics.

Fig. 7.5 A uniform rod
fixed at one end and subject
to an axial harmonic force at
the other end

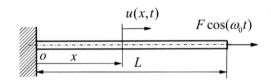

7.1.3 Harmonic Wave Analysis of a Finitely Long Rod

The traveling waves of an infinitely long or semi-infinitely long rod studied in the preceding subsections do not undergo any reflection at infinitely far ends and look simple. This subsection deals with the wave of a finitely long rod to an axial harmonic force. In this case, the waves undergo reflections at two ends of the rod and become complicated. The subsection focuses on the steady-state wave of a finitely long rod to an axial harmonic force since the transient wave is usually damped out in practice.

Figure 7.5 shows a uniform rod of length L, which is fixed at the left end and subject to an axial harmonic force at the right end. The problem is to determine the steady-state response $u(x, t)$ of the rod in the frame of coordinates in Fig. 7.5.

To study the steady-state response of the rod, it is not necessary to consider any initial conditions of the rod and deal with the dynamic boundary value problem of a partial differential equation as follows

$$\begin{cases} \rho A \dfrac{\partial^2 u(x, t)}{\partial t^2} - EA \dfrac{\partial^2 u(x, t)}{\partial x^2} = 0, \quad x \in [0, L] \\[3mm] u(0, t) = 0, \quad EA \dfrac{\partial u(x, t)}{\partial x}\bigg|_{x=L} = F \cos(\omega_0 t) \end{cases} \tag{7.1.39}$$

Many elementary textbooks[4] presented the steady-state solution of Eq. (7.1.39) by using the vibration analysis, which will be further addressed in Sect. 7.1.4. This subsection presents how to solve Eq. (7.1.39) via the wave analysis in order to reveal how several traveling waves form a standing wave, that is, the steady-state response to an axial harmonic force. To assist readers to understand the wave analysis well, the subsection uses the real function method and the complex function method, respectively.

7.1.3.1 Real Function Method

In view of the linear superposition principle, the solution of Eq. (7.1.39) is the linear superposition of the following two kinds of traveling waves. One kind is the left traveling wave excited by the axial harmonic force at the right end of the rod. The other is the right traveling wave comes from the reflection of the forced traveling

---

[4] Timoshenko S, Young D H, Weaver Jr W (1974). Vibration Problems in Engineering. 4th Edition. New York: John Wiley and Sons Inc., 379.

wave at the left end of the rod, i.e., d'Alembert's wave solution, and undergoes further reflection at the right end of the rod. Accordingly, their superposition should satisfy the boundary conditions in Eq. (7.1.39).

To describe the left traveling $u_{1L}(x, t)$ excited by the axial harmonic force, we recall the left traveling wave of a semi-infinitely long rod to an axial harmonic force at the right end, i.e., Eq. (7.1.15). Assuming a temporary origin of coordinate at the right end of the rod and replacing x with $x - L$, we have the forced left traveling wave $u_{1L}(x, t)$ from Eq. (7.1.15) as follows

$$u_{1L}(x, t) = \frac{F}{\kappa EA} \sin[\omega_0 t + \kappa(x - L)], \quad x \in [0, L] \tag{7.1.40}$$

Now, we turn to the study of two free traveling waves. The right traveling wave $u_{2R}(x, t)$ comes from the reflection of the forced left traveling wave $u_{1L}(x, t)$ at the left end, and the left traveling wave $u_{2L}(x, t)$ comes from the reflection of the right traveling wave $u_{2R}(x, t)$ at the right end. As studied in Sect. 7.1.1, both free traveling waves of the uniform rod satisfy d'Alembert's solution. As such, they lead to a free harmonic wave with frequency ω_0 and wave number κ as follows

$$u_2(x, t) = u_{2R}(x, t) + u_{2L}(x, t) = c_1 \cos(\omega_0 t - \kappa x) + c_2 \cos(\omega_0 t + \kappa x) \tag{7.1.41}$$

Of course, the above harmonic functions can also be sinusoidal functions.

From the above analysis, the solution of Eq. (7.1.39) takes the following form

$$u(x, t) = c_1 \cos(\omega_0 t - \kappa x) + c_2 \cos(\omega_0 t + \kappa x) + \frac{F}{\kappa EA} \sin[\omega_0 t + \kappa(x - L)] \tag{7.1.42}$$

Substitution of Eq. (7.1.42) into the two boundaries in Eq. (7.1.39) leads to

$$u(0, t) = c_1 \cos(\omega_0 t) + c_2 \cos(\omega_0 t) + \frac{F}{\kappa EA} \sin(\omega_0 t - \kappa L) = 0 \tag{7.1.43}$$

$$EA \left. \frac{\partial u(x, t)}{\partial x} \right|_{x=L} = c_1 EA\kappa \sin(\omega_0 t - \kappa L) - c_2 EA\kappa \sin(\omega_0 t + \kappa L) + F \cos(\omega_0 t)$$

$$= F \cos(\omega_0 t) \tag{7.1.44}$$

Solving Eq. (7.1.44) gives

$$c_2 = c_1 \frac{\sin(\omega_0 t - \kappa L)}{\sin(\omega_0 t + \kappa L)} \tag{7.1.45}$$

Substitution of Eq. (7.1.45) into Eq. (7.1.43) leads to

$$c_1 = \frac{F}{\kappa EA} \frac{\sin(\omega_0 t + \kappa L) \sin(\kappa L - \omega_0 t)}{[\sin(\omega_0 t + \kappa L) + \sin(\omega_0 t - \kappa L)] \cos(\omega_0 t)}$$

$$= \frac{F \sin(\omega_0 t + \kappa L) \sin(\kappa L - \omega_0 t)}{2\kappa EA \cos(\kappa L) \sin(\omega_0 t) \cos(\omega_0 t)} \tag{7.1.46}$$

Substituting Eqs. (7.1.45) and (7.1.46) into Eq. (7.1.42), via simplification in software Maple, we finally derive the standing wave to an axial harmonic force as follows

$$u(x, t) = \frac{F \sin(\kappa x)}{\kappa EA \cos(\kappa L)} \cos(\omega_0 t), \quad x \in [0, L] \tag{7.1.47}$$

Remark 7.1.4 As shown in Eq. (7.1.43), the three displacement waves counteract each other to satisfy the fixed boundary at the left end of the rod, while Eq. (7.1.44) implies that a part of left strain wave and the right strain wave counteract each other so that the other part of the left stain wave generates an internal force to balance the axial harmonic force at the right end of the rod.

Example 7.1.1 Given the following parameters

$$\omega_0 = \pi, \quad \kappa = \frac{\omega_0}{c_0} = \pi, \quad \frac{F}{\kappa EA} = 1 \tag{a}$$

study the forced traveling wave in Eq. (7.1.40) and two free traveling waves

$$u_{2R}(x, t) = c_1 \cos(\omega_0 t - \kappa x), \quad u_{2L}(x, t) = c_2 \cos(\omega_0 t + \kappa x) \tag{b}$$

where c_1 and c_2 satisfy Eqs. (7.1.46) and (7.1.45), repectively.

Figure 7.6 shows the variation of three displacement waves in Eq. (7.1.40) and Eq. (b) with time. When $t = 0$, $u_{2R}(x, 0) = u_{2L}(x, 0) = 0$ holds such that the thick solid line coincides with the abscissa in Fig. 7.6b and Fig. 7.6c.

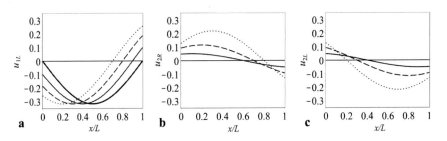

Fig. 7.6 The evolution of three displacement waves of the rod (thick solid line: $t = 0$, thin solid line: $t = 0.1$, dashed line: $t = 0.15$, dot line: $t = 0.2$). **a** $u_{1L}(x, t)$; **b** $u_{2R}(x, t)$; **c** $u_{2L}(x, t)$

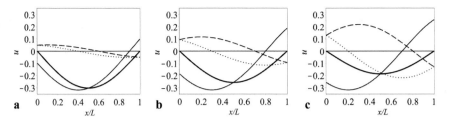

Fig. 7.7 The evolution of three displacement waves and their superposition of the rod (thick solid line: $u(t)$, thin solid line: $u_{1L}(x, t)$, dashed line: $u_{2R}(x, t)$, dot line: $u_{2L}(x, t)$). **a** $t = 0.1$; **b** $t = 0.2$; **c** $t = 0.3$

Figure 7.7 shows the above three displacement waves at three moments and their superposition. Here, no matter how the three displacement waves vary, their superposition is always a standing wave in Eq. (7.1.47). At moments $t = 0.1$, 0.2 and 0.3, the amplitude of the standing wave decrease. Meanwhile, the slops of $u_{2R}(x, t)$ denoted by dashed line and $u_{2L}(x, t)$ denoted by dot line are opposite at the right end of the rod such that they counteract each other. Consequently, the internal force caused by the left strain wave of $u_{1L}(x, t)$ balances the axial harmonic force at the right end of the rod.

7.1.3.2 Determination of Solution

The above wave analysis stems from d'Alembert's solution. It is reasonable to question whether d'Alembert's solution can directly be used to determine the above solution since all solutions of the one-dimensional wave equation take the form of d'Alembert's solution.

To answer this question, we recall d'Alembert's solution in Eq. (7.1.7) as follows

$$u(x, t) = u_R(c_0 t - x) + u_L(c_0 t + x) = u_R(\xi) + u_L(\eta) \tag{7.1.48}$$

Substituting Eq. (7.1.48) into the left boundary $u(0, t) = 0$ of the rod leads to the reflection condition of left and right traveling waves as follows

$$u_L(c_0 t) = -u_R(c_0 t) \tag{7.1.49}$$

As Eq. (7.1.49) holds for all $t \geq 0$, we replace $c_0 t$ in Eq. (7.1.49) with η and get

$$u_L(\eta) = -u_R(\eta), \quad \eta \geq 0 \tag{7.1.50}$$

Consequently, Eq. (7.1.48) becomes

$$u(x, t) = u_R(\xi) - u_R(\eta) = u_R(c_0 t - x) - u_R(c_0 t + x) \tag{7.1.51}$$

To make a comparison, we write the right traveling wave and the left traveling wave in Eq. (7.1.42) as follows

$$
\begin{cases}
u_R(c_0 t - x) = c_1 \cos(\omega_0 t - \kappa x), \\
u_L(c_0 t + x) = c_2 \cos(\omega_0 t + \kappa x) + \dfrac{F}{\kappa EA} \sin[\omega_0 t + \kappa(x - L)] \\
\quad = \left[c_2 - \dfrac{F \sin(\kappa L)}{\kappa EA} \right] \cos(\omega_0 t + \kappa x) + \dfrac{F \cos(\kappa L)}{\kappa EA} \sin(\omega_0 t + \kappa x)
\end{cases}
\tag{7.1.52}
$$

Compared with Eq. (7.1.51), the right traveling wave in Eq. (7.1.52) is in terms of $\cos(\omega_0 t - \kappa x)$, while the left traveling wave is not only in terms of $\cos(\omega_0 t + \kappa x)$, but also in terms of $\sin(\omega_0 t + \kappa x)$ if $\cos(\kappa L) \neq 0$. The condition $\cos(\kappa L) = 0$ only holds for the discrete wave numbers κ_r of natural vibrations of the fixed-free rod, and is also the resonance condition in Eq. (7.1.47) when the resonant peaks approach infinity. That is, it is impossible, from d'Alembert's solution and homogeneous boundary conditions, to derive Eq. (7.1.42), which describes a resultant wave superposed by one forced traveling wave and two free traveling waves.

Remark 7.1.5 In mathematics, d'Alembert's solution serves as a general form of all solutions for a one-dimensional homogeneous wave equation, but not a sufficient condition to determine a solution. Most books of equations of mathematical physics only dealt with the initial value problems of such an equation based on d'Alembert's solution[5] because the initial value problem satisfies causality and provides sufficient conditions for determining the solution. It is easy to see that both space coordinate x and time t play the same role in a one-dimensional wave equation. Thus, Eq. (7.1.39) can be understood as a two-point boundary value problem with respect to time t, which has been exchanged with space coordinate x. This problem usually fails to satisfy causality and can not provide sufficient conditions for determining a solution. As such, it is necessary to determine the left traveling wave to the harmonic force at the right end of the rod first, and then to solve the above boundary value problem.

7.1.3.3 Complex Function Method

The complex function method and real function method share the same idea in solution procedure, but have different formulations. Let the wave of the rod to an axial harmonic force be expressed in the following complex form

$$
u_c(x, t) = \tilde{u}_c(x) \exp(i\omega_0 t), \quad x \in [0, L]
\tag{7.1.53}
$$

where the complex function $\tilde{u}_c(x)$ includes two parts, that is, the complex amplitude of the wave to a harmonic force, and the complex amplitude of free waves. The first

---

[5] Sauvigny F (2006). Partial Differential Equations. Volume I. Berlin: Springer-Verlag, 395–397.

part can be determined from Eq. (7.1.32) and Remark 7.1.2 according to $0 \le x \le L = x_0$, while the second part is simply the complex form of d'Alembert's solution. Therefore, we have

$$\tilde{u}_c(x) = c_1 \exp(i\kappa x) + c_2 \exp(-i\kappa x) - \frac{iF}{EA\kappa} \exp[i\kappa(x - L)], \quad x \in [0, L]$$

$$(7.1.54)$$

where constants c_1 and c_2 satisfy the boundary conditions.

From Eq. (7.1.39), Eq. (7.1.54) yields the following boundary conditions

$$\tilde{u}_c(0) = 0, \quad EA \left. \frac{d\tilde{u}_c(x)}{dx} \right|_{x=L} = F \qquad (7.1.55)$$

Substitution of Eq. (7.1.54) into (7.1.55) leads to

$$\begin{cases} c_1 + c_2 - \dfrac{iF \exp(-i\kappa L)}{\kappa EA} = 0 \\ i\kappa EA\big[c_1 \exp(i\kappa L) - c_2 \exp(-i\kappa L)\big] + F = F \end{cases} \qquad (7.1.56)$$

Solving Eq. (7.1.56) gives

$$c_1 = \frac{iF \exp(-2i\kappa L)}{2\kappa EA \cos(\kappa L)}, \quad c_2 = \frac{iF}{2\kappa EA \cos(\kappa L)} \qquad (7.1.57)$$

By substituting Eq. (7.1.57) into Eq. (7.1.54) and simplifying the result, we derive

$$\tilde{u}_c(x) = \frac{F \sin(\kappa x)}{\kappa EA \cos(\kappa L)}, \quad x \in [0, \ L] \qquad (7.1.58)$$

Finally, substitution of Eq. (7.1.58) into Eq. (7.1.53) leads to the same result in Eq. (7.1.47), namely,

$$u(x, t) = \text{Re}\big[\tilde{u}_c(x) \exp(i\omega_0 t)\big] = \frac{F \sin(\kappa x)}{\kappa EA \cos(\kappa L)} \cos(\omega_0 t), \quad x \in [0, L] \quad (7.1.59)$$

Remark 7.1.6 As shown in the first equation in Eq. (7.1.56), the forced left traveling wave to an axial harmonic force at the right end of the rod counteracts two free traveling waves at the left end to satisfy the fixed boundary, while two free traveling waves counteract each other at the right end, as shown in the second equation in Eq. (7.1.56), so that the forced left traveling wave generates an internal force to balance the axial harmonic force. Thus, the complex function method also reveals the superposition result of three traveling waves, but deals with simpler formulations of exponential functions, instead of the tedious deduction of triangle functions.

7.1.4 Harmonic Vibration Analysis of a Finitely Long Rod

The vibration analysis of the steady-state response of a rod to an axial harmonic force in elementary textbooks followed the idea of separation of variables, and included two methods. One is the *semi-inverse method*, where a harmonic standing wave with unknown amplitude is assumed and substituted into the boundary value problem to determine the amplitude. The other is the *modal solution method* or called the *modal superposition method*, which leads to the modal series solution. In general, the first one is simple, while the second one is universal.

7.1.4.1 Semi-inverse Method

Let the steady-state solution of Eq. (7.1.39) be

$$u(x, t) = \tilde{u}(x)\cos(\omega_0 t), \quad x \in [0, L] \tag{7.1.60}$$

Substituting Eq. (7.1.60) into Eq. (7.1.39), we have the following boundary value problem of a homogeneous ordinary differential equation of amplitude $\tilde{u}(x)$

$$\begin{cases} \dfrac{d^2\tilde{u}(x)}{dx^2} + \kappa^2\tilde{u}(x) = 0, & x \in [0, L] \\ \tilde{u}(0) = 0, \quad EA\dfrac{d\tilde{u}(x)}{dx}\bigg|_{x=L} = F \end{cases} \tag{7.1.61}$$

The general solution of the differential equation in Eq. (7.1.61) reads

$$\tilde{u}(x) = c_1\cos(\kappa x) + c_2\sin(\kappa x), \quad x \in [0, L] \tag{7.1.62}$$

Substitution of Eq. (7.1.62) into the boundary conditions in Eq. (7.1.61) leads to

$$c_1 = 0, \quad c_2 = \frac{F}{\kappa EA\cos(\kappa L)} \tag{7.1.63}$$

By substituting Eq. (7.1.63), through Eq. (7.1.62), into Eq. (7.1.60), we get the same steady-state response in Eq. (7.1.47), namely,

$$u(x, t) = \frac{F\sin(\kappa x)}{\kappa EA\cos(\kappa L)}\cos(\omega_0 t), \quad x \in [0, L] \tag{7.1.64}$$

Remark 7.1.7 The solution candidate in Eq. (7.1.60) for Eq. (7.1.39) comes from the fundamental property of linear systems. That is, a linear system to a harmonic input of given frequency gives rise to the steady-state output of the same frequency. The semi-inverse method has been not only accepted in engineering circle, but also collected into the world-renowned book of mathematical physics[6]. Yet, the semi-inverse method is not able to reveal how several traveling waves form a standing wave.

7.1.4.2 Modal Solution Method

To solve the homogeneous wave equation under inhomogeneous boundary conditions in Eq. (7.1.39), the standard procedure in mathematics is to introduce an auxiliary function $w(x, t)$ to transform the above problem to an inhomogeneous wave equation under homogeneous boundary conditions[7], then solve the new problem for $w(x, t)$ via the method of separation of variables, and finally get $u(x, t)$ from $w(x, t)$.

To be specific, define the following auxiliary function

$$w(x, t) \equiv EAu(x, t) - x F \cos(\omega_0 t), \quad x \in [0, L] \tag{7.1.65}$$

and derive the second-order derivatives to be used as follows

$$\frac{\partial^2 u(x, t)}{\partial t^2} = \frac{1}{EA} \frac{\partial^2 w(x, t)}{\partial t^2} - \frac{\omega_0^2 Fx \cos(\omega_0 t)}{EA}, \quad \frac{\partial^2 u(x, t)}{\partial x^2} = \frac{1}{EA} \frac{\partial^2 w(x, t)}{\partial x^2} \tag{7.1.66}$$

Substitution of Eq. (7.1.66) into Eq. (7.1.39) leads to an inhomogeneous wave equation under homogeneous boundary condition, namely,

$$\begin{cases} \dfrac{\partial^2 w(x, t)}{\partial t^2} - c_0^2 \dfrac{\partial^2 w(x, t)}{\partial x^2} = F\omega_0^2 x \cos(\omega_0 t), \quad x \in [0, L] \\ w(0, t) = EAu(0, t) = 0, \quad \dfrac{\partial w(x, t)}{\partial x}\bigg|_{x=L} = EA \dfrac{\partial u(x, t)}{\partial x}\bigg|_{x=L} - F \cos(\omega_0 t) = 0 \end{cases} \tag{7.1.67}$$

Substituting the solution $w(x, t)$ of Eq. (7.1.67) into Eq. (7.1.65) gives $u(x, t)$.

From Table 2.1, we get the wave numbers, natural frequencies and mode shapes of a fixed-free uniform rod as following

$$\kappa_r = \frac{(2r - 1)\pi}{2L}, \quad \omega_r = \kappa_r c_0, \quad \overline{\varphi}_r(x) = \sqrt{\frac{2}{L}} \sin(\kappa_r x), \quad r = 1, 2, \ldots \tag{7.1.68}$$

where the mode shapes $\overline{\varphi}_r(x)$ have been normalized with respect to the modal mass coefficient. Thus, we define a modal transform for Eq. (7.1.67) as follows

---

[6] Courant R, Hilbert D (1989). Methods of Mathematical Physics. Volume II. New York: John Wiley & Sons Inc., 517–528.

[7] Courant R, Hilbert D (1989). Methods of Mathematical Physics. Volume I. New York: John Wiley & Sons Inc., 277.

$$w(x, t) = \sum_{r=1}^{+\infty} \overline{\varphi}_r(x) q_r(t) = \sqrt{\frac{2}{L}} \sum_{r=1}^{+\infty} \sin(\kappa_r x) q_r(t) \qquad (7.1.69)$$

Substitution of Eq. (7.1.69) into the inhomogeneous partial differential equation in Eq. (7.1.67) leads to

$$\sum_{r=1}^{+\infty} \overline{\varphi}_r(x) \frac{\partial^2 q_r(t)}{\partial t^2} - c_0^2 \sum_{r=1}^{+\infty} \frac{\partial^2 \overline{\varphi}_r(x)}{\partial x^2} q_r(t) = F \omega_0^2 x \cos(\omega_0 t) \qquad (7.1.70)$$

Let Eq. (7.1.70) be multiplied by $\overline{\varphi}_r(x)$ and then integrated over $[0, L]$, we derive a set of decoupled ordinary differential equations in view of the orthogonality of normalized mode shapes, namely,

$$\frac{d^2 q_r(t)}{dt^2} + \omega_r^2 q_r(t) = f_r \cos(\omega_0 t), \quad r = 1, 2, \ldots \qquad (7.1.71)$$

where the inhomogeneous term in Eq. (7.1.71) is the *modal excitation* as follows

$$f_r \equiv F \omega_0^2 \sqrt{\frac{2}{L}} \int_0^L x \sin(\kappa_r x)\, dx = \frac{4FL^2 \omega_0^2 (-1)^r}{(2r-1)^2 \pi^2} \sqrt{\frac{2}{L}}, \quad r = 1, 2, \ldots \qquad (7.1.72)$$

The steady-state solution of Eq. (7.1.71) reads

$$q_r(t) = \frac{f_r \cos(\omega_0 t)}{\omega_r^2 - \omega_0^2}, \quad r = 1, 2, \ldots \qquad (7.1.73)$$

Substituting Eq. (7.1.73), through Eq. (7.1.69), into Eq. (7.1.65), we get the modal series solution, or called a *formal solution* at this stage, for the steady-state harmonic response as follows

$$u(x, t) = \frac{1}{EA}[w(x, t) + xF \cos(\omega_0 t)]$$

$$= \frac{F}{EA}\left[\frac{8L}{\pi^2} \sum_{r=1}^{+\infty} \frac{(-1)^r \omega_0^2}{(2r-1)^2 (\omega_r^2 - \omega_0^2)} \sin(\kappa_r x) + x \right] \cos(\omega_0 t), \quad x \in [0, L]$$

$$(7.1.74)$$

Obviously, when the excitation ω_0 coincides with any natural frequency ω_r of the fixed-free uniform rod, the steady-state response $u(x, t)$ undergoes a resonance.

Remark 7.1.8 Given $\omega_0 \neq \omega_r$, an arbitrary term in Eq. (7.1.74), for a sufficiently large r, satisfies

$$\left| \frac{(-1)^r \omega_0^2}{(2r-1)^2(\omega_r^2 - \omega_0^2)} \sin(\kappa_r x) \right| \le \frac{1}{(2r-1)^2} \quad (7.1.75)$$

From the M-criterion of a function series in mathematical analysis, the formal solution in Eq. (7.1.74) is uniformly convergent. That is, the modal series solution is the right solution for the steady-state response.

Remark 7.1.9 If the auxiliary function is not introduced, but the modal transform similar to Eq. (7.1.69) is constructed for $u(x, t)$ as follows

$$u(x, t) = \sum_{r=1}^{+\infty} \bar{\varphi}_r(x) q_r(t) = \sqrt{\frac{2}{L}} \sum_{r=1}^{+\infty} \sin(\kappa_r x) q_r(t) \quad (7.1.76)$$

After decoupling inhomogeneous wave equation in Eq. (7.1.39), the dynamic boundary at the right end of the rod becomes

$$0 = EA \left. \frac{\partial u(x, t)}{\partial x} \right|_{x=L} - F \cos(\omega_0 t) = \sqrt{\frac{2}{L}} \sum_{r=1}^{+\infty} \kappa_r \cos(\kappa_r L) q_r(t) - F \cos(\omega_0 t)$$

$$(7.1.77)$$

It is not feasible to use modal analysis to deal with Eq. (7.1.77).

7.1.4.3 Equivalence of Two Methods

We first demonstrate that the method of separation of variables also gives the solution in Eq. (7.1.64) derived via the semi-inverse method. Let $u(x, t) = \tilde{u}(x)q(t)$ be a solution candidate with separated variables for Eq. (7.1.39). Substituting $u(x, t) = \tilde{u}(x)q(t)$ into Eq. (7.1.39) and using the procedure of separation of variables, we convert Eq. (7.1.39) to the following boundary value problem

$$\begin{cases} \dfrac{d^2\tilde{u}(x)}{dx^2} + \kappa^2\tilde{u}(x) = 0, & \dfrac{d^2q(t)}{dt^2} + \omega^2 q(t) = 0 \\[2mm] u(0, t) = \tilde{u}(0)q(t) = 0, & EA\dfrac{d\tilde{u}(x)}{dx}q(t)\bigg|_{x=L} = F\cos(\omega_0 t) \end{cases} \quad (7.1.78)$$

Solving the second differential equation under the second boundary condition in Eq. (7.1.78) leads to $q(t) = q_0 \cos(\omega t)$, where q_0 is a non-zero constant. Substitution of $q(t) = q_0 \cos(\omega_0 t)$ into Eq. (7.1.78) gives

$$\begin{cases} \dfrac{d^2\tilde{u}(x)}{dx^2} + \kappa^2\tilde{u}(x) = 0 \\[2mm] \tilde{u}(0) = 0, \quad EAq_0\dfrac{d\tilde{u}(x)}{dx}\bigg|_{x=L} = F \end{cases} \qquad (7.1.79)$$

Equation (7.1.79) is identical to Eq. (7.1.61) if $q_0 = 1$. Then, following the procedure of solving Eq. (7.1.61) arrives at the same solution in Eq. (7.1.64). Accordingly, this procedure is to solve the unique solution of a set of inhomogeneous linear algebraic equations, rather than the independent solutions of a set of homogeneous linear algebraic equations for natural vibrations of a rod.

Next, we verify that the solution of semi-inverse method is identical to the modal series solution. We return to Eq. (7.1.74) and expand x as a Fourier series

$$x = \sum_{r=1}^{+\infty} a_r \sin(\kappa_r x), \quad x \in [0,\ L] \qquad (7.1.80)$$

where the Fourier coefficients are

$$a_r = \frac{2}{L}\int_0^L x \sin(\kappa_r x)\,dx = \frac{8L(-1)^r}{\pi^2(2r-1)^2} \quad r = 1, 2, \dots \qquad (7.1.81)$$

Substitution of Eq. (7.1.81) into Eq. (7.1.74) leads to

$$\begin{aligned} u(x,t) &= \frac{F}{EA}\left\{\frac{8L}{\pi^2}\sum_{r=1}^{+\infty}\sin(\kappa_r x)\left[\frac{(-1)^r\omega_0^2}{(2r-1)^2(\omega_r^2-\omega_0^2)} + \frac{(-1)^r}{(2r-1)^2}\right]\right\}\cos(\omega_0 t) \\[2mm] &= \frac{F}{EA}\left\{\frac{8L}{\pi^2}\sum_{r=1}^{+\infty}\sin(\kappa_r x)\left[\frac{(-1)^r\omega_r^2}{(2r-1)^2(\omega_r^2-\omega_0^2)}\right]\right\}\cos(\omega_0 t) \\[2mm] &= \frac{2Fc_0^2}{EAL}\sum_{r=1}^{+\infty}\frac{(-1)^r}{\omega_r^2-\omega_0^2}\sin(\kappa_r x)\cos(\omega_0 t), \quad x \in [0,\ L] \qquad (7.1.82) \end{aligned}$$

Meanwhile, we expand Eq. (7.1.64) derived by using the semi-inverse method as a Fourier series, namely,

$$u(x,t) = \frac{F\sin(\kappa x)}{\kappa EA\cos(\kappa L)}\cos(\omega_0 t) = \sum_{r=1}^{+\infty}b_r\sin(\kappa_r x)\cos(\omega_0 t), \quad x \in [0,\ L] \qquad (7.1.83)$$

Calculating the Fourier coefficients in Eq. (7.1.83) and using the expression of natural frequencies in Eq. (7.1.68), we derive

$$b_r = \frac{2}{L} \int_0^L \frac{F \sin(\kappa x)}{\kappa EA \cos(\kappa L)} \sin(\kappa_r x) \, dx$$

$$= \frac{2Fc_0^2}{EAL} \frac{(-1)^r}{\frac{\pi^2(2r-1)^2 c_0^2}{4L^2} - c_0^2 \kappa^2} = \frac{2Fc_0^2}{EAL} \frac{(-1)^r}{\omega_r^2 - \omega_0^2} \tag{7.1.84}$$

Obviously, the Fourier coefficients in Eqs. (7.1.84) and (7.1.82) are exactly identical.

Recalling the above four methods enables one to draw a conclusion for the dynamic analysis of a harmonically forced uniform rod. The real function method and complex function method in wave analysis, as well as the semi-inverse method, lead to the same form of exact solution, while the modal superposition method in vibration analysis gives a modal series solution, which is uniformly convergent to the exact solution.

Example 7.1.2 To make a comparison between the exact solution and approximate solution truncated from the modal series for the steady-state response of the uniform rod to a harmonic force, the amplitude of steady-state vibration of the rod at the right end is scaled by the static deformation $FL/(EA)$ of the rod at the right end to a static force F, and defined as the dimensionless amplitude of the steady-state vibration of the rod at the right end. Discuss the dimensionless amplitude of direct frequency response of the rod at the right end.

We get the dimensionless amplitude of direct frequency response of the rod at the right end from the exact solution in Eq. (7.1.64) as follows

$$|H(\omega_0)| \equiv \left| \frac{\sin(\kappa L)}{\kappa L \cos(\kappa L)} \right| = \left| \frac{c_0}{\omega_0 L} \tan\left(\frac{\omega_0 L}{c_0}\right) \right| \tag{a}$$

Let Eq. (7.1.74) be scaled by $FL/(EA)$ and substituted with $x = L$, we have the approximate solution of Eq. (a) with the first n modes as following

$$|H_n(\omega_0)| \equiv \left| \frac{8}{\pi^2} \sum_{r=1}^{n} \frac{(-1)^r \omega_0^2}{(2r-1)^2(\omega_r^2 - \omega_0^2)} \sin(\kappa_r L) + 1 \right| \tag{b}$$

Figure 7.8 shows a comparison of Eqs. (a) and (b) for $n = 4$ and indicates that the approximate modal solution truncated for $n = 4$ well agrees with the exact solution in a frequency band of the first three resonant peaks. To tell the small difference between the two solutions in anti-resonance regions, the ordinate in Fig. 7.8 uses the logarithmic scale.

Fig. 7.8 The dimensionless amplitude of direct frequency response of the rod at the right end (solid line: exact solution $\lg|H|$, dashed line: modal series solution $\lg|H_4|$ of the first 4 modes)

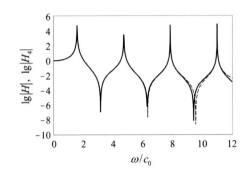

7.1.5 Concluding Remarks

This section presented how to study the non-dispersive waves and steady-state harmonic vibrations of uniform rods with infinite, semi-infinite and finite lengths, respectively. The section focused on the relation between wave analysis and vibration analysis for the steady-state response of a rod to an axial harmonic force, and demonstrated their equivalence associated with Problem 6A proposed in Chap. 1. The major conclusions are as follows.

(1) In the methods of wave analysis, the real function method based on d'Alembert's solution shows a clear picture of traveling waves and is suitable for the wave propagation with discontinuous wavefront. The complex function method has simple formulations, but is only applicable to harmonic waves. To compute arbitrary waves efficiently, it is possible to develop an integrated approach based on both complex function method and fast Fourier transform.

(2) In the methods of vibration analysis, the semi-inverse method is simple but only valid for the steady-state harmonic vibration, while the modal superposition method is superior to the semi-inverse method, but involves tedious formulations. The success of modal superposition method, accordingly, depends on the convergence of modal series solution, which will be further studied in Sect. 7.3.

7.2 Dispersive Waves of Rods

As studied in Sect. 7.1, the non-dispersive waves based on Eq. (7.1.1) are a kind of one-dimensional waves of a simplified uniform rod, which have to satisfy several assumptions. For example, the cross-section of the rod is assumed to remain plane during the dynamic deformation of the rod, and the transverse deformations and shear deformations are assumed to be negligible. The study on the one-dimensional waves of such a simplified uniform rod, thus, is referred to as the *elementary theory*.

As a matter of fact, many wave propagations exhibit *dispersion*. That is, the speed of a wave depends on wave frequency. If an initial wave includes several frequency components, the different frequency components in a dispersive wave propagate with their own speeds such that the wavefront undergoes distortion in the wave propagation.

For a dispersive wave, the wave speeds of two frequency components with very close frequencies usually have a tiny difference so that the two frequency components modulate each other and exhibit a *wave packet* as demonstrated in the following example.

Example 7.2.1 Study two harmonic traveling waves, with close frequencies and wave numbers, towards the right as follows

$$u_{R1}(x, t) = a \cos(\omega_1 t - \kappa_1 x), \quad u_{R2}(x, t) = a \cos(\omega_2 t - \kappa_2 x) \tag{a}$$

In view of the triangle formula between summation and multiplication, we recast the superposition of the two waves as

$$u_R(x, t) = u_{R1}(x, t) + u_{R2}(x, t) = a \cos(\omega_1 t - \kappa_1 x) + a \cos(\omega_2 t - \kappa_2 x)$$

$$= 2a \cos\left(\frac{\omega_1 - \omega_2}{2} t - \frac{\kappa_1 - \kappa_2}{2} x\right) \cos\left(\frac{\omega_1 + \omega_2}{2} t - \frac{\kappa_1 + \kappa_2}{2} x\right)$$

$$= 2a \cos(\Delta\omega t - \Delta\kappa x) \cos(\overline{\omega} t - \overline{\kappa} x) \tag{b}$$

where

$$\Delta\omega \equiv \frac{\omega_1 - \omega_2}{2}, \quad \Delta\kappa \equiv \frac{\kappa_1 - \kappa_2}{2}, \quad \overline{\omega} \equiv \frac{\omega_1 + \omega_2}{2}, \quad \overline{\kappa} \equiv \frac{\kappa_1 + \kappa_2}{2} \tag{c}$$

Figure 7.9 shows the two harmonic waves and their wave packet from the top to the bottom. A comparison between Eq. (b) and the wave packet in Fig. 7.9 shows that the term $2a \cos(\Delta\omega t - \Delta\kappa x)$ is a *modulated wave*, exhibiting a slow amplitude

Fig. 7.9 Two harmonic traveling waves and the wave packet composed

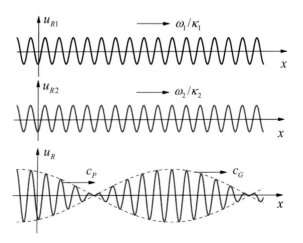

variation of the wave packet, while the term $\cos(\bar{\omega} t - \bar{\kappa} x)$ is a *carrier wave*, showing a phase variation of the wave packet. Eq. (c) indicates that the frequency and wave number of the carrier wave are $\bar{\omega} \approx \omega_1 \approx \omega_2$ and $\bar{\kappa} \approx \kappa_1 \approx \kappa_2$, while those of the modulated wave are small quantities $\Delta\omega$ and $\Delta\kappa$. Thus, the carrier wave varies much faster than the modulated wave. As such, two velocities are introduced for the wave packet

$$c_P = \frac{\bar{\omega}}{\bar{\kappa}} = \frac{\omega_1 + \omega_2}{\kappa_1 + \kappa_2} \approx \frac{\omega_1}{\kappa_1} \approx \frac{\omega_2}{\kappa_2}, \quad c_G = \frac{\Delta\omega}{\Delta\kappa} = \frac{\omega_1 - \omega_2}{\kappa_1 - \kappa_2} \tag{d}$$

where c_P represents the ratio of $\bar{\omega}$ (phase derivative in time) to $\bar{\kappa}$ (phase derivative in space) of the carrier wave, and c_G represents the ratio of the same counterparts for the slow amplitude variation of the modulated wave.

Now, we generalize the above concepts in Example 7.2.1. When the frequencies of two traveling waves are getting infinitely close and so are the two wave numbers of them, we define the rate of phase derivative in time to phase derivative in space for the carrier wave as the *phase velocity*, and the rate of amplitude variation in time to amplitude variation in space for the wave packet as the *group velocity*. In reference to Eq. (d) in Example 7.2.1, the two velocities are expressed as

$$c_P \equiv \frac{\omega}{\kappa}, \quad c_G \equiv \frac{d\omega}{d\kappa} = \frac{d(\kappa c_P)}{d\kappa} = c_P + \kappa \frac{dc_P}{d\kappa} \tag{7.2.1}$$

The dispersion is named the *normal dispersion* if $c_G < c_P$ holds, and is called the *abnormal dispersion* if $c_P < c_G$ holds. For non-dispersive waves, $c_P = c_G$ holds. As studied in Sect. 7.1.1, the longitudinal wave of a uniform rod is non-dispersive such that both phase velocity c_P and group velocity c_G are identical to wave speed c_0 though they carry different meanings. Physically speaking, the wave energy is proportional to the squared amplitude of the wave packet. Hence, the group velocity of a dispersive wave embodies the propagation velocity of the wave energy.

Based on the above description of wave velocities of a dispersive wave, this section presents the wave dispersions caused by the non-uniform cross-section, effect of transverse inertia and elastic boundary of a rod, respectively so as to help readers lay a theoretical foundation for studying the non-harmonic wave in a wide frequency band associated with Problem 6A proposed in Chap. 1.

7.2.1 Dispersive Waves in a Rod Due to Non-uniform Cross-section

The wave of a slender non-uniform rod under a longitudinal disturbance or excitation is mainly one-dimensional longitudinal wave. Under this precondition, this subsection deals with the longitudinal wave of a slender rod with the cross-sectional area being an exponential function. Such a rod is named the rod with an *exp-cross-section* for short. The subsection begins with understanding the dispersive wave of a semi-infinitely long rod with an exp-cross-section, and then deals with the longitudinal wave of an infinitely long rod with an exp-cross-section to a harmonic force, and finally presents the steady-state response of a finitely long rod with an exp-cross-section to a harmonic force.

7.2.1.1 Dispersive Waves of a Semi-infinitely Long Rod

Consider the dynamic equation of a semi-infinitely long non-uniform rod with the left end fixed as follows

$$
\begin{cases}
\rho A(x) \dfrac{\partial^2 u(x,t)}{\partial t^2} - \dfrac{\partial}{\partial x}\left[EA(x) \dfrac{\partial u(x,t)}{\partial x} \right] = 0 \\
u(0,t) = 0, \quad x \in [0, +\infty)
\end{cases}
\tag{7.2.2}
$$

where ρ is the material density, E is Young's modulus, and $A(x) > 0$ is the cross-sectional area.

Assume that the solution of Eq. (7.2.2) is in a form of separated variables

$$
u(x,t) = \tilde{u}(x) q(t)
\tag{7.2.3}
$$

Substitution of Eq. (7.2.3) into Eq. (7.2.2) leads to

$$
\frac{1}{q(t)} \frac{d^2 q(t)}{dt^2} = \frac{1}{\rho A(x) \tilde{u}(x)} \frac{d}{dx}\left[EA(x) \frac{d\tilde{u}(x)}{dx} \right] = \gamma
\tag{7.2.4}
$$

It is easy to verify that solution $q(t)$ is divergent with time t when $\gamma \geq 0$. Hence, we take $\gamma = -\omega^2 < 0$ and get

$$
\begin{cases}
\dfrac{d}{dx}\left[A(x) \dfrac{d\tilde{u}(x)}{dx} \right] + \dfrac{\omega^2}{c_0^2} A(x)\tilde{u}(x) = 0, \quad \tilde{u}(0) = 0 \\
\dfrac{d^2 q(t)}{dt^2} + \omega^2 q(t) = 0
\end{cases}
\tag{7.2.5}
$$

In Eq. (7.2.5), the second differential equation has a vibration solution $q(t) = a \sin(\omega t + \theta)$ in time domain, but the wave amplitude governed by the solution of the first differential equation in space domain needs to be discussed. As such, parameter $c_0 \equiv \sqrt{E/\rho}$ looks like the longitudinal wave speed, but is only used as a scaling factor. In general, c_0 is not able to describe the longitudinal wave speed of a non-uniform rod.

Consider the rod with an exp-cross-section studied in Sect. 5.2 and take the following cross-sectional area

$$A(x) = A_0 \exp(\alpha x), \quad A_0 > 0, \quad x \in [0, +\infty) \tag{7.2.6}$$

where the case $\alpha = 0$ stands for a uniform rod, which is a trivial case hereinafter. Substitution of Eq. (7.2.6) into the first differential equation in Eq. (7.2.5) and its boundary condition yields

$$\begin{cases} \dfrac{d^2\tilde{u}(x)}{dx^2} + \alpha\dfrac{d\tilde{u}(x)}{dx} + \dfrac{\omega^2}{c_0^2}\tilde{u}(x) = 0 \\[2mm] \tilde{u}(0) = 0 \end{cases} \tag{7.2.7}$$

We discuss the three cases for the eigenvalue problem of Eq. (7.2.7) as follows.

The case when $\alpha^2 - 4(\omega/c_0)^2 = 0$ leads to a pair of repeated eigenvalues $\lambda_{1,2} = -\alpha/2$ and the solution of Eq. (7.2.7) as following

$$\tilde{u}(x) = c_1 x \exp\left(-\frac{\alpha x}{2}\right) \tag{7.2.8}$$

where c_1 is an integration constant not claimed later. There are two sub-cases for $\alpha \neq 0$ as follows. First, for $\alpha > 0$, $\tilde{u}(x) \to 0$ when $x \to +\infty$ such that the wave can not propagate far away. This wave has the name of *near-field wave* or *evanescent wave*. Second, for $\alpha < 0$, $|\tilde{u}(x)| \to +\infty$ when $x \to +\infty$. The wave does not make sense for the semi-infinitely long rod.

The case when $\alpha^2 - 4(\omega/c_0)^2 > 0$ results in a pair of real eigenvalues

$$\lambda_{1,2} = -\frac{\alpha}{2} \pm \beta, \quad \beta \equiv \sqrt{\left(\frac{\alpha}{2}\right)^2 - \left(\frac{\omega}{c_0}\right)^2} > 0 \tag{7.2.9}$$

whereby we have the solution of Eq. (7.2.7) as follows

$$\tilde{u}(x) = c_1\left[\exp(\lambda_1 x) - \exp(\lambda_2 x)\right] \tag{7.2.10}$$

Given $\alpha > 0$, $\lambda_1 < 0$ and $\lambda_2 < 0$ hold such that $\tilde{u}(x) \to 0$ when $x \to +\infty$. That is, the wave is also an evanescent wave. Given $\alpha < 0$, $\lambda_1 > 0$ and $\lambda_2 > 0$ hold so that $|\tilde{u}(x)| \to +\infty$ when $x \to +\infty$. The wave is not meaningful for the semi-infinitely long rod.

The case when $\alpha^2 - 4(\omega/c_0)^2 < 0$ leads to a pair of conjugate complex eigenvalues

$$\lambda_{1,2} = -\frac{\alpha}{2} \pm i\kappa, \quad \kappa \equiv \sqrt{\left(\frac{\omega}{c_0}\right)^2 - \left(\frac{\alpha}{2}\right)^2} > 0 \tag{7.2.11}$$

where κ is the wave number. The solution of Eq. (7.2.7) reads

$$\tilde{u}(x) = c_1 \exp\left(-\frac{\alpha x}{2}\right) \sin(\kappa x) \tag{7.2.12}$$

As shown in Eq. (7.2.12), the wave amplitude exhibits the harmonic oscillation with an exponential envelope along the rod.

It is worthy to notice that the case when $\alpha^2 - 4(\omega/c_0)^2 < 0$ imposes a condition on wave frequency ω as follows

$$\omega > \omega_c, \quad \omega_c \equiv \frac{|\alpha|}{2} c_0 \tag{7.2.13}$$

That is, the longitudinal wave propagates far way only when the wave frequency yields $\omega > \omega_c$, where ω_c is *the cut-off circular frequency*, while the *cut-off frequency* is $f_c \equiv \omega_c/(2\pi)$.

When $\omega > \omega_c$, we define the *phase velocity* of the longitudinal wave of a rod with an exp-cross-section, from Eq. (7.2.1), as

$$c_{LP} \equiv \frac{\omega}{\kappa} = \frac{\omega}{\sqrt{\left(\dfrac{\omega}{c_0}\right)^2 - \left(\dfrac{\alpha}{2}\right)^2}} = \frac{c_0}{\sqrt{1 - \left(\dfrac{\omega_c}{\omega}\right)^2}} > c_0 \tag{7.2.14}$$

Equation (7.2.14) has an equivalent relation in terms of wave number, namely,

$$c_{LP} \equiv \frac{\omega}{\kappa} = \frac{c_0\sqrt{\kappa^2 + (\alpha/2)^2}}{\kappa} = c_0\sqrt{1 + \left(\frac{\alpha}{2\kappa}\right)^2} > c_0 \tag{7.2.15}$$

The above two equations show that the phase velocity c_{LP} depends on frequency ω or wave number κ and receives notations $c_{LP}(\omega)$ or $c_{LP}(\kappa)$. In addition, we define the *group velocity* of the longitudinal wave of a rod with an exp-cross-section, from Eq. (7.2.1), as

$$c_{LG} = \frac{d\omega}{d\kappa} = \frac{\kappa c_0}{\sqrt{\kappa^2 + (\alpha/2)^2}} \le c_0 \tag{7.2.16}$$

Consequently, the longitudinal wave of a rod with an exp-cross-section exhibits normal dispersion, and the phase velocity and group velocity satisfy the following relation

$$c_{LP}c_{LG} = c_0\sqrt{1 + \left(\frac{\alpha}{2\kappa}\right)^2} \cdot \frac{\kappa c_0}{\sqrt{\kappa^2 + (\alpha/2)^2}} = c_0^2 \tag{7.2.17}$$

Remark 7.2.1 In Eq. (7.2.14), $\omega \to \omega_c^+$ leads to $c_{LP} \to +\infty$ and $\kappa \to 0$ such that $c_{LG} \to 0$ from Eq. (7.2.16). Noting that $c_{LP} \to +\infty$ implies an infinitely fast change of wave phase and that $c_{LG} \to 0$ means a null propagation velocity of wave energy, we conclude that the longitudinal wave can not travel in this case.

Example 7.2.2 Given a fixed-free aluminum rod with length $L = 1$ m and an exp-cross-section, where the ratio of cross-sectional areas at two ends is $A(L)/A_0 = 2$, study the relation between the cut-off frequency and natural frequencies.

From Eq. (7.2.6), the rate in exponential function is

$$\alpha = \frac{\ln[A(L)/A_0]}{L} = \ln(2) = 0.6931 \text{ /m} \tag{a}$$

Let $\rho = 2700$ kg/m^3 be matierial density and $E = 72.7$ GPa be Young's modulus for the aluminum, then we have

$$c_0 = \sqrt{\frac{E}{\rho}} = 5189.02 \text{ m/s} \tag{b}$$

Substituting Eq. (a) and Eq. (b) into Eq. (7.2.13) leads to the cut-off frequency, which does not rely on the rod boundaries, as follows

$$f_c = \frac{\omega_c}{2\pi} = \frac{\alpha c_0}{4\pi} \approx 286.2 \text{ Hz} \tag{c}$$

The right boundary condition of the rod reads

$$EA(x)\frac{\partial u(x,t)}{\partial x}\bigg|_{x=L} = 0 \tag{d}$$

Substitution of Eq. (7.2.12) into Eq. (d) gives

$$c_1 EA(L)\tilde{u}_x(L) = c_1 EA_0 \exp\left(\frac{\alpha L}{2}\right)\left[\kappa \cos(\kappa L) - \frac{\alpha}{2}\sin(\kappa L)\right] = 0 \tag{e}$$

Substituting Eq. (a) into Eq. (e), we derive the characteristic equation of the rod as

$$0.6931L \tan(\kappa L) - 2\kappa L = 0 \tag{f}$$

Solving Eq. (f) for eigenvalues leads to

$$\kappa_1 = 1.3126/L, \quad \kappa_2 = 4.6378/L, \quad \kappa_3 = 7.8096/L, \quad \cdots \tag{g}$$

whereby we have the natural frequencies of the rod with an exp-cross-section as following

$$f_r = \frac{c_0}{2\pi}\sqrt{\left(\frac{\alpha}{2}\right)^2 + \kappa_r^2}, \quad r = 1, 2, \ldots \tag{h}$$

Table 7.1 shows a comparison of the first three natural frequencies and the dimen-

Table 7.1 The natural frequencies and dimensionless group velocities of a uniform rod and a rod with an exp-cross-section

Natural frequency order	1st	2nd	3rd
Uniform rod (Hz)	1297.25	3891.75	6486.25
Rod with an exp-cross-section (Hz)	1121.23	3840.83	6455.97
Dimensionless group velocity (c_{LG}/c_0)	0.9668	0.9972	0.9990

sionless group velocities associated for the uniform rod and non-uniform rod. The difference between the two rods decreases with an increase of frequency order.

Remark 7.2.2 It is similar to study the wave dispersion of other non-uniform rods. For instance, S. Q. Guo and S. P. Yang in China studied the rods with cross-sectional areas of a quadratic function and a hyperbolic triangle function, respectively[8]. The cut-off frequency of a rod with quadratic cross-sectional area is zero and the longitudinal wave of the rod is not dispersive, while the longitudinal wave of a rod with hyperbolic triangle cross-sectional area is dispersive since the rod has a positive cut-off frequency.

Remark 7.2.3 The above analysis on the dispersive wave of a rod with an exp-cross-section is based on the elementary theory and does not account for three-dimensional effects, such as the transverse inertia and shear deformation of a rod. The study based on an improved model of Mindlin-Herrmann predicted that the cut-off frequency of the rod in Example 7.2.2 was 340 Hz, while the cut-off frequency measured in a test was about 360 Hz. They both were much higher than 286.2 Hz in Eq. (c) based on the elementary theory[9].

7.2.1.2 Harmonic Waves of an Infinitely Long Rod

Given a rod with an exp-cross-section, study the steady-state wave of the rod to an axial harmonic force at given position. The dynamic equation of the rod described by the complex displacement in Sect. 7.1.2 reads

$$\rho A_0 \exp(\alpha x) \frac{\partial^2 u_c(x,t)}{\partial t^2} - \frac{\partial}{\partial x}\left[EA_0 \exp(\alpha x) \frac{\partial u_c(x,t)}{\partial x}\right] = \delta(x) F \exp(i\omega_0 t) \tag{7.2.18}$$

Assume that the complex solution of Eq. (7.2.18) yields

$$u_c(x,t) = \tilde{u}_c(x) \exp(i\omega_0 t) \tag{7.2.19}$$

---

[8] Guo S Q, Yang S P (2012). Wave motions in non-uniform one-dimensional waveguides. Journal of Vibration and Control, 18(1): 92–100.

[9] Wei Y M, Yang S X, Gan C B (2016). Experimental study of longitudinal wave propagation in rod with variable cross-section. Journal of Vibration, Measurement and Diagnosis, 36(3): 498–504 (in Chinese).

where $\tilde{u}_c(x)$ is the complex amplitude to be determined. Substitution of Eq. (7.2.19) into Eq. (7.2.18) leads to an ordinary differential equation as follows

$$\frac{d^2\tilde{u}_c(x)}{dx^2} + \alpha\frac{d\tilde{u}_c(x)}{dx} + \left(\frac{\omega_0}{c_0}\right)^2\tilde{u}_c(x) = -\frac{F}{EA_0\exp(\alpha x)}\delta(x) \qquad (7.2.20)$$

When $x \neq 0$, Eq. (7.2.20) is as same as the homogeneous ordinary differential equation in Eq. (7.2.7). Thus, when the wave frequency is higher than the cut-off circular frequency, i.e., $\omega_0 > \omega_c = |\alpha|c_0/2$, the general solution of Eq. (7.2.20) reads

$$\tilde{u}_c(x) = \exp\left(-\frac{\alpha x}{2}\right)[c_1\exp(i\kappa x) + c_2\exp(-i\kappa x)], \quad x \in (-\infty, +\infty) \qquad (7.2.21)$$

where κ is the wave number defined in Eq. (7.2.11). Substituting Eq. (7.2.21) into Eq. (7.2.19) gives the complex solution

$$\begin{cases} u_c(x, t) = \exp\left(-\frac{\alpha x}{2}\right)\{c_1\exp[i\kappa(c_{LP}t - x)] + c_2\exp[i\kappa(c_{LP}t + x)]\} \\ x \in [-c_{LP}t, \; c_{LP}t], \quad t \in [0, +\infty) \end{cases} \qquad (7.2.22)$$

where c_{LP} is the phase velocity of longitudinal wave defined in Eq. (7.2.14).

In the right part of the rod, i.e., $x > 0$, the rod undergoes the right traveling wave only such that $c_2 = 0$ holds. In the left part of rod, i.e., $x < 0$, $c_1 = 0$ holds since no right traveling wave appears. Hence, Eq. (7.2.21) becomes

$$\tilde{u}_c(x) = \begin{cases} c_1\exp\left(-\frac{\alpha x}{2}\right)\exp(-i\kappa x), & x \in [0, \; c_{LP}t] \\ c_2\exp\left(-\frac{\alpha x}{2}\right)\exp(i\kappa x), & x \in [-c_{LP}t, \; 0] \end{cases} \qquad (7.2.23)$$

The longitudinal displacement $u_c(x, t)$ is continuous at $x = 0$ such that $c_1 = c_2$ holds. In addition, the longitudinal strain is not continuous at $x = 0$ and leads to

$$\left.\frac{\partial u_c(x, t)}{\partial x}\right|_{x=0} = \exp(i\omega_0 t)\left.\frac{d\tilde{u}_c(x)}{dx}\right|_{x=0} = \begin{cases} \left(-\frac{\alpha}{2} - i\kappa\right)c_1\exp(i\omega_0 t), & x = 0^+ \\ \left(-\frac{\alpha}{2} + i\kappa\right)c_1\exp i(\omega_0 t), & x = 0^- \end{cases}$$
$$(7.2.24)$$

Similar to the force analysis of an infinitesimal segment of a uniform rod at $x = 0$, we have

$$EA_0\left[\left(-\frac{\alpha}{2} - i\kappa\right)c_1 - \left(-\frac{\alpha}{2} + i\kappa\right)c_1\right] + F = 0 \qquad (7.2.25)$$

Substituting $c_1 = c_2 = -\mathrm{i}F/(2\kappa EA_0)$, obtained from Eq. (7.2.25), into Eq. (7.2.23) leads to

$$\tilde{u}_c(x) = -\frac{\mathrm{i}F}{2\kappa EA_0} \exp\left(-\frac{\alpha x}{2}\right) \exp(-\mathrm{i}\kappa|x|), \quad x \in [-c_{LP}t, \ c_{LP}t] \qquad (7.2.26)$$

Taking the real part of Eq. (7.2.19) with Eq. (7.2.26) substituted, we have the real solution as follows

$$u(x, t) = \frac{F}{2\kappa EA_0} \exp\left(-\frac{\alpha x}{2}\right) \sin(\omega_0 t - \kappa|x|), \quad x \in [-c_{LP}t, \ c_{LP}t], \quad t \in [0, \ +\infty) \qquad (7.2.27)$$

Obviously, Eq. (7.2.27) becomes the harmonic wave of a uniform rod when $\alpha = 0$. It is worthy to notice that the above discussion is only valid when $\omega_0 > \omega_c$, and the case when $\omega_0 \leq \omega_c$ requires further discussion shortly later.

7.2.1.3 Steady-State Response of a Finitely Long Rod

Consider the dynamic equation of a rod with an exp-cross-section, which is fixed at the left end and subject to an axial harmonic force at the right end, as follows

$$\begin{cases} \rho A \dfrac{\partial^2 u(x, t)}{\partial t^2} - \dfrac{\partial}{\partial x}\left[EA_0 \exp(\alpha x) \dfrac{\partial u(x, t)}{\partial x}\right] = 0, \quad x \in [0, \ L] \\ u(0, t) = 0, \quad EA(L)u_x(L, t) = F\cos(\omega_0 t) \end{cases} \qquad (7.2.28)$$

In what follows, the excitation frequency does not need to satisfy $\omega_0 > \omega_c$.

As the steady-state output of a linear system to the harmonic input must be harmonic, we use the semi-inverse method in Sect. 7.1.4 to assume the following steady-state response of Eq. (7.2.2.28), namely,

$$u(x, t) = \tilde{u}(x)\cos(\omega_0 t), \quad x \in [0, \ L] \qquad (7.2.29)$$

Substitution of Eq. (7.2.29) into Eq. (7.2.28) leads to the boundary value problem of differential equation of amplitude $\tilde{u}(x)$ as follows

$$\begin{cases} \dfrac{\mathrm{d}^2\tilde{u}(x)}{\mathrm{d}x^2} + \alpha \dfrac{\mathrm{d}\tilde{u}(x)}{\mathrm{d}x} + \left(\dfrac{\omega_0}{c_0}\right)^2 \tilde{u}(x) = 0, \quad x \in [0, L] \\ \tilde{u}(0) = 0, \quad EA(x)\dfrac{\mathrm{d}\tilde{u}(x)}{\mathrm{d}x}\bigg|_{x=L} = F \end{cases} \qquad (7.2.30)$$

The general solution of the homogeneous differential equation in Eq. (7.2.30) reads

$$\tilde{u}(x) = \exp\left(-\frac{\alpha x}{2}\right)\left[c_1 \exp(\tilde{\kappa}x) + c_2 \exp(-\tilde{\kappa}x)\right] \tag{7.2.31}$$

where $\tilde{\kappa}$ is the complex wave number defined as

$$\tilde{\kappa} \equiv \sqrt{\left(\frac{\alpha}{2}\right)^2 - \left(\frac{\omega_0}{c_0}\right)^2} \tag{7.2.32}$$

Substituting Eq. (7.2.31) into the left boundary condition in Eq. (7.2.30) gives $c_2 = -c_1$, there follows

$$\tilde{u}(x) = c_1 \exp\left(-\frac{\alpha x}{2}\right)\left[\exp(\tilde{\kappa}x) - \exp(-\tilde{\kappa}x)\right] \tag{7.2.33}$$

Substitution of Eq. (7.3.33) into the right boundary condition in Eq. (7.2.30) gives

$$EA(x)\frac{d\tilde{u}(x)}{dx}\bigg|_{x=L}$$
$$= c_1 EA_0 \exp\left(\frac{\alpha L}{2}\right)\left[\left(\tilde{\kappa} - \frac{\alpha}{2}\right)\exp(\tilde{\kappa}L) + \left(\tilde{\kappa} + \frac{\alpha}{2}\right)\exp(-\tilde{\kappa}L)\right] = F \tag{7.2.34}$$

There follows

$$c_2 = \frac{F}{EA_0 \exp\left(\frac{\alpha L}{2}\right)\left[\left(\tilde{\kappa} - \frac{\alpha}{2}\right)\exp(\tilde{\kappa}L) + \left(\tilde{\kappa} + \frac{\alpha}{2}\right)\exp(-\tilde{\kappa}L)\right]} \tag{7.2.35}$$

Substituting Eq. (7.2.35), through Eq. (7.2.33), into Eq. (7.2.29), we get the steady-state harmonic response of the rod as follows

$$u(x,t) = \frac{F \exp\left(-\frac{\alpha L}{2}\right)\exp\left(-\frac{\alpha x}{2}\right)\left[\exp(\tilde{\kappa}x) - \exp(-\tilde{\kappa}x)\right]}{EA_0\left[\left(\tilde{\kappa} - \frac{\alpha}{2}\right)\exp(\tilde{\kappa}L) + \left(\tilde{\kappa} + \frac{\alpha}{2}\right)\exp(-\tilde{\kappa}L)\right]}\cos(\omega_0 t) \tag{7.2.36}$$

According to the cut-off circular frequency $\omega_c \equiv c_0|\alpha|/2$ defined in (7.2.13), we discuss the following three cases for different relations between excitation frequency ω_0 and cut-off circular frequency ω_c

Case (1): When $\omega_0 > \omega_c$ holds, Eq. (7.2.32) leads to $\tilde{\kappa} = i\kappa$, where κ is the wave number defined in Eq. (7.2.11), such that Eq. (7.2.36) becomes

$$u(x,t) = \frac{F \exp\left(-\frac{\alpha L}{2}\right)\exp\left(-\frac{\alpha x}{2}\right)\sin(\kappa x)}{EA_0\left[\kappa\cos(\kappa L) - \frac{\alpha}{2}\sin(\kappa L)\right]}\cos(\omega_0 t) \tag{7.2.37}$$

The term $\exp(-\alpha x/2)\sin(\kappa x)$ in the numerator of Eq. (7.2.37) represents the amplitude variation with x. When $\alpha > 0$, the amplitude at the right end is smaller than that at the left end, while the result is on the contrary when $\alpha < 0$. From Eq. (e) and Eq. (g) in Example 7.2.2, when κ is an eigenvalue of a fix-free rod with an exp-cross-section, the denominator in Eq. (7.2.37) vanishes and the rod undergoes a resonance. When $\alpha = 0$, Eq. (7.2.37) degenerates to Eq. (7.1.64) for the steady-state response of a uniform rod. As such, the exp-cross-section of a rod only affects the amplitude variation of a steady-state harmonic response along the rod when $\omega_0 > \omega_c$. Of course, the dispersion of a non-harmonic wave in such a rod may be more complicated.

Case (2): When $\omega_0 < \omega_c$ holds, Eq. (7.2.32) gives $\tilde{\kappa} > 0$. The term involving x in the numerator of Eq. (7.2.36) can be expressed as

$$
\exp\left(-\frac{\alpha x}{2}\right)\left[\exp(\tilde{\kappa}x) - \exp(-\tilde{\kappa}x)\right]
$$
$$
= \exp\left(\tilde{\kappa}L - \frac{\alpha x}{2}\right)\left\{\exp[\tilde{\kappa}(x - L)] - \exp[-\tilde{\kappa}(x + L)]\right\}
$$
(7.2.38)

The second term $\exp[-\tilde{\kappa}(x + L)]$ in braces in Eq. (7.2.38) monotonically decreases with an increase of x from $x = 0$ and represents a right evanescent wave due to the left end boundary, while the first term $\exp[\tilde{\kappa}(x - L)]$ in the same braces monotonically decreases with a decrease of x from $x = L$ and is the left evanescent wave due to the right end boundary. Even though they are near-field waves, their difference multiplied by $\exp(\tilde{\kappa}L - \alpha x/2)$ describes the amplitude variation of the harmonic response along the rod. Now, we recast Eq. (7.2.36) as

$$
u(x, t) = \frac{F \exp\left(-\dfrac{\alpha L}{2}\right)\exp\left(-\dfrac{\alpha x}{2}\right)\sinh(\tilde{\kappa}x)}{EA_0\left[\tilde{\kappa}\cosh(\tilde{\kappa}L) - \dfrac{\alpha}{2}\sinh(\tilde{\kappa}L)\right]}\cos(\omega_0 t)
$$
(7.2.39)

The amplitude of the harmonic response monotonically varies along the rod since both terms of $\exp(-\alpha x/2)$ and $\sinh(\tilde{\kappa}x)$ monotonically vary with an increase of x. In view of Eq. (7.2.32), $\tilde{\kappa} \to |\alpha|/2$ holds when $\omega_0 \to 0$. Therefore, the limit of Eq. (7.2.39) when $\tilde{\kappa} \to |\alpha|/2$ gives the deformation of the rod to a static force F, namely,

$$
\lim_{\omega_0 \to 0}\tilde{u}_s(x) = \lim_{\tilde{\kappa} \to |\alpha|/2}\frac{F \exp\left(-\dfrac{\alpha L}{2}\right)\exp\left(-\dfrac{\alpha x}{2}\right)\sinh(\tilde{\kappa}x)}{EA_0\left[\tilde{\kappa}\cosh(\tilde{\kappa}L) - \dfrac{\alpha}{2}\sinh(\tilde{\kappa}L)\right]} = \frac{F[1 - \exp(-\alpha x)]}{\alpha EA_0}
$$
(7.2.40)

Letting Eq. (7.2.28) be substituted with $\rho = 0$ and $\omega_0 = 0$ and then integrated twice, one can verify the correctness of Eq. (7.2.40) at ease.

Case (3): When $\omega_0 = \omega_c$ holds, Eq. (7.2.32) leads to $\tilde{\kappa} = \kappa = 0$ and Eq. (7.2.36) exhibits a discontinuity at $\omega_0 = \omega_c$. However, no matter whether the left limit

$\omega_0 \to \omega_c^-$ in Eq. (7.2.39) or the right limit $\omega_0 \to \omega_c^+$ in Eq. (7.2.37) is taken, the rule of L'Hôpital leads to the longitudinal displacement of the rod as follows

$$\lim_{\omega_0 \to \omega_c^-} u(x, t) = \lim_{\omega_0 \to \omega_c^+} u(x, t) = \frac{2F \exp\left(-\dfrac{\alpha L}{2}\right) x \exp\left(-\dfrac{\alpha x}{2}\right)}{EA_0(2 - \alpha L)} \cos(\omega_c t)$$

$$(7.2.41)$$

Remark 7.2.4 Although the wave numbers defined in Eqs. (7.2.11) and (7.2.32) are zero, the steady-state harmonic response does not vanish. The term $x \exp(-\alpha x/2)$ in the numerator in Eq. (7.2.41) describes the amplitude variation along the rod and takes the same form of Eq. (7.2.8), which corresponds to a pair of repeated eigenvalues.

Example 7.2.3 For the aluminum rod with an exp-cross-section in Example 7.2.2, study the steady-state response of the rod to an axial harmonic force at the right end.

According to the rod parameters given in Example 7.2.2 as follows

$$L = 1 \text{ m}, \quad \alpha = 0.6931 \text{ /m}, \quad E = 72.7 \text{ GPa}, \quad \rho = 2700 \text{ kg/m}^3 \qquad \text{(a)}$$

the cut-off frequency of the rod yields

$$f_c = \frac{\omega_c}{2\pi} = \frac{\alpha c_0}{4\pi} = \frac{\alpha}{4\pi}\sqrt{\frac{E}{\rho}} \approx 286.2 \text{ Hz} \qquad \text{(b)}$$

From Eqs. (7.2.37) and (7.2.39), we get the dimensionless amplitude of the direct frequency response $|\tilde{u}(L)/\tilde{u}_s(L)|$ of the rod at the right end, while $\tilde{u}_s(L)$ is the rod deformation at the right end to a static force from Eq. (7.2.40). In Fig. 7.10, the solid line represents the above frequency response and the dashed line is the

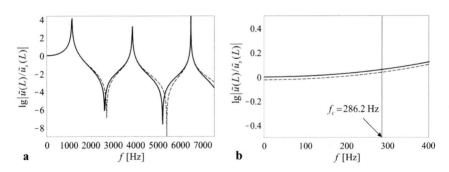

Fig. 7.10 The dimensionless amplitude of direct frequency response of an exp-cross-section rod with left end fixed and a harmonic force at the right end (solid line: exact solution, dashed line: approximate modal solution). **a** the global view for a frequency band of the first three resonance; **b** the zoomed view in a frequency band of cut-off frequency

approximate result based on the modal superposition of the first six modes. Compared with the approximate modal solution of a uniform rod in Fig. 7.8, the exp-cross-section of the rod reduces the accuracy of modal superposition. The zoomed view in Fig. 7.10b shows that the amplitude of frequency response is smooth around the cut-off frequency $f_c = \omega_c/(2\pi)$. That is, the cut-off frequency is a normal singularity.

7.2.2 Dispersive Waves of a Rod Due to Transverse Inertia

The early studies in elasto-dynamics revealed that the wave of a slender rod under an axial disturbance or excitation is not an idealized one-dimensional wave, but complicated three-dimensional waves instead. The three-dimensional elasto-dynamic effects, such as the transverse inertia and shear deformation, play important roles of the rod dynamics with an increase of wave frequency. The design of precise products, such as a waveguide associated with higher frequencies, has to take those effects into consideration.

The studies of three-dimensional elastic waves of a rod can be traced to the seminal work of L. Pochhammer, a German mathematician, who first found an infinite number of exact solution branches of the steady-state waves of an infinitely long circular rod with free surface in 1876[10]. At present, these exact solutions are mainly applicable to checking various approximate solutions since they are very complicated and only valid for a circular rod. In the studies of approximate three-dimensional waves of a rod, important achievements include, but not limited to, the rod theory established by A. E. H. Love, a British scientist, in 1944 to account for the transverse inertia, and the rod theory proposed by R. D. Mindlin and G. Herrmann in the USA in 1950 to take the shear deformations of a rod into account[11].

Among these achievements, the rod theory established by Love looks simple, but significantly improves the elementary theory of one-dimensional waves of a rod in a middle frequency band or even a higher frequency band. This subsection, thus, presents the theory established by Love and Sect. 7.3 will show how to use the theory to deal with an impacting rod associated with waves of higher frequencies.

7.2.2.1 Love's Rod and Dispersive Waves

Section 3.1.1 presented how to establish the dynamic equation for Love's rod under fixed-free boundaries so as to account for the transverse inertia of the rod. The corresponding dynamic boundary value problem reads

[10] Pochhammer L (1876). Über die fortpflanzungsgeschwindigkeiten kleiner schwingungen in einem unbegrenzten isotropen kreiscylinder. Journal für die Reine und Angewandte Mathematik, 81(3): 324–336.

[11] Graff K F (1975). Wave Motion in Elastic Solids. Columbus: Ohio State University Press, 431–575.

$$\begin{cases} \dfrac{\partial^2 u(x,t)}{\partial t^2} = c_0^2 \dfrac{\partial^2 u(x,t)}{\partial x^2} + v^2 r_p^2 \dfrac{\partial^4 u(x,t)}{\partial x^2 \partial t^2}, \quad x \in [0, L], \quad c_0 \equiv \sqrt{\dfrac{E}{\rho}} \\[2mm] u(0,t) = 0, \quad c_0^2 u_x(L,t) + v^2 r_p^2 u_{xtt}(L,t) = 0 \end{cases} \quad (7.2.42)$$

where v is Poisson's ratio, r_p is the polar radius of gyration of cross-sectional area of the rod. It is worthy to notice that the boundary condition in Eq. (7.2.42) implies the effect of Poisson's ratio of three-dimensional elasticity such that the longitudinal strain has to counteract the transverse inertia at the right end of rod.

Assume that the complex traveling wave solution for Eq. (7.2.42) reads

$$u(x,t) = \hat{u} \exp[i(\omega t - \kappa x)] \qquad (7.2.43)$$

where $\hat{u}$ is a constant. Substitution of Eq. (7.2.43) into the partial differential equation in Eq. (7.2.42) leads to a relation between wave frequency and wave number as follows

$$\omega^2 + v^2 r_p^2 \kappa^2 \omega^2 - c_0^2 \kappa^2 = 0 \qquad (7.2.44)$$

Then, we discuss Eq. (7.2.44) in following two aspects.

At first, solving Eq. (7.2.44) for the wave number leads to

$$\kappa = \frac{\omega}{\sqrt{c_0^2 - v^2 r_p^2 \omega^2}} \qquad (7.2.45)$$

Equation (7.2.45) implies that $c_0^2 - v^2 r_p^2 \omega^2 > 0$ holds such that the wave frequency ω should be lower than the cut-off circular frequency ω_c, namely,

$$\omega < \omega_c \equiv \frac{c_0}{v r_p} \qquad (7.2.46)$$

If $\omega > \omega_c$, then the wave number in Eq. (7.2.45) is an imaginary number, implying an evanescent longitudinal wave. That is, when the rod is subject to a harmonic force above this critical frequency, the longitudinal wave can not travel far away. In this case, Love's rod may exhibit a longitudinal wave of small amplitude, like a rod with an exp-cross-section, or convert the axial input energy to transverse waves.

Remark 7.2.5 From the viewpoint of three-dimensional elasto-dynamics, the axial input energy can not fully become the energy of normal strains in transverse direction, but a part of energy of shear strains instead. This is a shortcoming of Love's rod model. It is lucky that the cut-off frequency of Love's rod is usually very high. Given the steel rod with a circular cross-section, for example, let the cross-sectional diameter of the rod be $d = 0.1$ m and Poisson's ratio be $v = 0.28$. We can get the polar radius of gyration $r_p = d/\sqrt{8} \approx 3.54 \times 10^{-2}$ m, the cut-off circular frequency $\omega_c =$

$c_0/\nu r_p \approx 5.23 \times 10^6$ rad/s and the cut-off frequency $f_c = \omega_c/(2\pi) \approx 83.3$ kHz. Hence, the available frequency band of Love's rod is quite wide.

Second, solving Eq. (7.2.44) for the wave frequency gives

$$\omega = \frac{c_0 \kappa}{\sqrt{1 + \nu^2 r_p^2 \kappa^2}} \tag{7.2.47}$$

whereby we have the phase velocity and group velocity of the longitudinal wave of the rod

$$c_{LP} = \frac{\omega}{\kappa} = \frac{c_0}{\sqrt{1 + \nu^2 r_p^2 \kappa^2}}, \quad c_{LG} = \frac{d\omega}{d\kappa} = \frac{c_0}{(1 + \nu^2 r_p^2 \kappa^2)^{3/2}} < c_{LP} \tag{7.2.48}$$

As $c_{LG} < c_{LP}$ holds, Love's rod exhibits the normal dispersion of longitudinal waves due to the transverse inertia.

To understand the above theoretical prediction of Love's rod, we refer to it as *Love's theory* for short, and make a comparison among the elementary theory, Love's theory and the exact theory of three-dimensional elasto-dynamics for the dimensionless phase velocity c_{LP}/c_0 of the longitudinal wave of a rod with an increase of dimensionless wave number $\nu r_p \kappa$ in Fig. 7.11.

As shown in the figure, the difference between the elementary theory and the exact theory significantly increases with increasing of $\nu r_p \kappa$. In the range of $\nu r_p \kappa \in [0, 1]$, the dimensionless phase velocities predicted by Love's theory and the exact theory are close, and Love's theory works not well when $\nu r_p \kappa > 1$. In particular, the dimensionless phase velocity predicted by Love's theory approaches zero when $\kappa \to +\infty$. As shown in Eq. (7.2.47), $\omega \to \omega_c$ holds when $\kappa \to +\infty$ such that the longitudinal wave fails to propagate. Yet, the exact theory of three-dimensional elasto-dynamics does not have any cut-off frequency and can predict the longitudinal traveling waves.

In practice, Love's theory can be used as an improvement for the elementary theory when studying the dispersive waves of a rod. If $\nu r_p \kappa = 1$ is taken as an upper bound for Love's theory, the corresponding bound of wave number is

Fig. 7.11 The phase velocity of the longitudinal wave of a rod with an increase of wave number

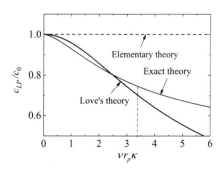

$$\kappa_{max} = \frac{1}{vr_p} \qquad (7.2.49)$$

Substituting Eq. (7.2.49) into Eq. (7.2.47) and using Eq. (7.2.45), we get the upper bound of wave frequency, namely,

$$\omega_{max} = \frac{c_0 \kappa_{max}}{\sqrt{2}} = \frac{c_0}{\sqrt{2}vr_p} = \frac{\omega_c}{\sqrt{2}} \qquad (7.2.50)$$

Consider, as an example agian, the circular rod with cross-sectional diameter $d = 0.1$ m and Poisson's ratio $v = 0.28$. From Remark 7.2.5 and Eq. (7.2.50), we get $f_{max} = \omega_{max}/(2\pi) = f_c/\sqrt{2} \approx 58.96$ kHz. This result shows again that the valid frequency band of Love's theory is wide enough for most wave problems of rods in engineering.

7.2.2.2 Free Vibration of Love's Rod

Consider the natural vibrations of Love's rod under fixed-free boundaries and assume that the solution of Eq. (7.2.42) satisfies a form of separated variables

$$u(x, t) = \varphi(x)q(t) \qquad (7.2.51)$$

Substitution of Eq. (7.2.51) into Eq. (7.2.42) leads to

$$\left[c_0^2 q(t) + v^2 r_p^2 \frac{d^2 q(t)}{dt^2} \right]^{-1} \frac{d^2 q(t)}{dt^2} = [\varphi(x)]^{-1} \frac{d^2 \varphi(x)}{dx^2} \qquad (7.2.52)$$

Thus, both sides of Eq. (7.2.52) should be a constant. From the boundary conditions in Eq. (7.2.42), the constant must be negative and can be taken as $-\kappa^2$. Consequently, Eq. (7.2.52) becomes two differential equations as follows

$$\frac{d^2 \varphi(x)}{dx^2} + \kappa^2 \varphi(x) = 0, \quad (1 + v^2 r_p^2 \kappa^2) \frac{d^2 q(t)}{dt^2} + c_0^2 \kappa^2 q(t) = 0 \qquad (7.2.53)$$

Substituting Eq. (7.2.51) into the boundary conditions in Eq. (7.2.42) gives

$$\begin{cases} u(0, t) = \varphi(0)q(t) = 0 \\ c_0^2 u_x(L, t) + v^2 r_p^2 u_{xtt}(L, t) = \varphi_x(L)\left[c_0^2 q(t) + v^2 r_p^2 q_{tt}(t) \right] = 0 \end{cases} \qquad (7.2.54)$$

There follow two boundary conditions simplified as

$$\varphi(0) = 0, \quad \varphi_x(L) = 0 \qquad (7.2.55)$$

We solve the first differential equation in Eq. (7.2.53) and get the following general solution

$$\varphi(x) = a_1 \cos(\kappa x) + a_2 \sin(\kappa x) \tag{7.2.56}$$

Substitution of Eq. (7.2.56) into Eq. (7.2.55) leads to

$$a_1 = 0, \quad a_2 \kappa \cos(\kappa L) = 0 \tag{7.2.57}$$

Solving the triangle equation in Eq. (7.2.57) for wave numbers gives

$$\kappa_r = \frac{(2r-1)\pi}{2}, \quad r = 1, 2, \dots \tag{7.2.58}$$

By substituting Eq. (7.2.58) into Eq. (7.2.56), we get the normalized mode shapes

$$\varphi_r(x) = \sin(\kappa_r x) = \sin\left[\frac{(2r-1)\pi x}{2}\right], \quad r = 1, 2, \dots \tag{7.2.59}$$

Next, we solve the second differential equation in Eq. (7.2.53) under given wave number κ_r and get the free vibration as follows

$$q_r(t) = b_{1r} \cos(\omega_r t) + b_{2r} \sin(\omega_r t), \quad \omega_r \equiv \frac{c_0 \kappa_r}{\sqrt{1 + v^2 r_p^2 \kappa_r^2}}, \quad r = 1, 2, \dots \tag{7.2.60}$$

Therefore, we arrive at the general solution of free vibration of Love's rod under fixed-free boundaries, namely,

$$u(x,t) = \sum_{r=1}^{+\infty} \sin(\kappa_r x) \left[b_{1r} \cos\left(\frac{c_0 \kappa_r t}{\sqrt{1 + v^2 r_p^2 \kappa_r^2}}\right) + b_{2r} \sin\left(\frac{c_0 \kappa_r t}{\sqrt{1 + v^2 r_p^2 \kappa_r^2}}\right) \right] \tag{7.2.61}$$

It is worthy to notice that the natural frequencies in Eq. (7.2.61) satisfy

$$\lim_{r \to +\infty} \omega_r \equiv \frac{c_0 \kappa_r}{\sqrt{1 + v^2 r_p^2 \kappa_r^2}} = \frac{c_0}{v r_p} = \omega_c \tag{7.2.62}$$

As shown in Eq. (7.2.61), the free vibration shape of a fixed-free rod with transverse inertia is still the superposition of the mode shapes of a rod without transverse inertia, and the wave numbers in Eq. (7.2.58) form an arithmetic sequence. Equation (7.2.62) shows that the natural frequencies ω_r, $r = 1, 2, \dots$ approach the cut-off frequency ω_c with an increase of wave numbers κ_r, $r = 1, 2, \dots$.

Remark 7.2.6 The natural frequencies ω_r, $\quad r = 1, 2, \ldots$ approach the cut-off frequency ω_c with an increase of frequency order r such that Eq. (7.2.61) fails to describe a free vibration of high frequency. Section 7.3.4 will discuss this shortcoming of Love's rod model when dealing with an impacting rod associated with waves of higher frequencies.

7.2.3 Dispersive Waves of a Rod Due to an Elastic Boundary

Figure 7.12 shows a semi-infinitely long uniform rod, the right end of which is connected via a spring of stiffness coefficient k to a rigid wall. The right end boundary of the rod, thus, is an elastic boundary and can also describe either a free boundary when $k = 0$ or a fixed boundary when $k \to +\infty$. This subsection deals with the reflection problem of a traveling wave at the right end of the rod so as to give a simple description about the dispersion in reflected waves.

At first, let a harmonic wave incident upon the rod in the right direction. The incident wave propagates and undergoes a reflection at the right end of the rod, and then produces a harmonic traveling wave towards to the left direction such that the two traveling waves are superposed. In the frame of coordinates shown in Fig. 7.12, the resultant wave of the two waves reads

$$u(x, t) = a \exp[i(\omega t - \kappa x)] + b \exp[i(\omega t + \kappa x)] \tag{7.2.63}$$

which satisfies the right boundary of the rod as follows

$$E A u_x(0, t) = -k u(0, t) \tag{7.2.64}$$

Substitution of Eq. (7.2.63) into Eq. (7.2.64) leads to

$$i E A \kappa (-a + b) = -k(a + b) \tag{7.2.65}$$

Solving Eq. (7.2.65) for b/a gives a ratio, associated with stiffness coefficient k, of the reflected wave amplitude to the incident wave amplitude as follows

$$\gamma_T(k) \equiv \frac{b}{a} = \frac{i E A \kappa - k}{i E A \kappa + k} \tag{7.2.66}$$

The ratio is also named the *reflection coefficient* for two traveling waves.

Fig. 7.12 A semi-infinitely long uniform rod with an elastic boundary

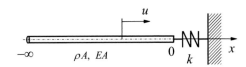

To understand the reflection coefficient, we discuss Eq. (7.2.66) in three cases as follows.

Case (1): The right end of the rod becomes a free boundary when $k = 0$ such that the reflection coefficient reads

$$\gamma_T(0) = 1 \tag{7.2.67}$$

Now, the reflected wave and incident wave have the same amplitude and the same phase such that the amplitude of resultant wave is doubled compared with the incident wave amplitude. It is easy to derive that the reflected stain wave and incident strain wave have the same amplitude but an opposite phase so that the resultant strain amplitude is zero, exhibiting the free boundary. In addition, the above waves are non-dispersive as discussed for a uniform rod with a free end in Sect. 7.1.

Case (2): The right end of the rod is a fixed boundary when $k \rightarrow +\infty$, and the reflection coefficient is

$$\gamma_T(+\infty) = -1 \tag{7.2.68}$$

That is, the reflected wave and incident wave have the same amplitude but an opposite phase such that the resultant wave amplitude is zero, showing the fixed boundary. The reflected stain wave and incident strain wave have the same amplitude and the same phase so that the resultant strain amplitude is doubled. All those waves are also non-dispersive.

Case (3): The right end of the rod is an elastic boundary when $k \in (0, +\infty)$. Now, Eq. (7.2.66) can be recast as

$$\gamma_T(k) = \frac{i\beta\kappa - 1}{i\beta\kappa + 1}, \quad \beta \equiv \frac{EA}{k} \tag{7.2.69}$$

where $\beta\kappa$ is an dimensionless wave number.

Figure 7.13 presents the real part and imaginary part of reflection coefficient γ_T with respect to $\beta\kappa$. It is easy to see that case (1) and case (2) correspond to the limits $\beta \rightarrow +\infty$ and $\beta \rightarrow 0$ in the figure. When $\beta\kappa = 1$, $\mathrm{Re}(\gamma_T) = 0$ and

Fig. 7.13 The reflection coefficient with an increase of dimensionless wave number (solid line: real part, dashed line: imaginary part)

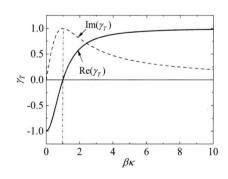

$\text{Im}(\gamma_T) = \max\{\text{Im}(\gamma_T)\} = 1$ hold such that the imaginary part of the reflection coefficient reaches an extreme value, exhibiting the most serious dispersion, while the phase of reflection coefficient yields

$$\arg(\gamma_T) = \tan\left[\frac{\text{Im}(\gamma_T)}{\text{Re}(\gamma_T)}\right] = \frac{\pi}{2} \tag{7.2.70}$$

Example 7.2.4 Let the incident right traveling wave be a triangle wave in terms of $\xi > 0$ as follows

$$u_R(\xi) = \begin{cases} \xi, & \xi \in [0.0,\ 0.5] \\ 1 - \xi, & \xi \in [0.5,\ 1.0] \\ 0 & \xi \notin [0.0,\ 1.0] \end{cases} \tag{a}$$

discuss the reflected waves for different stiffness coefficient k of the spring.

In view of $\kappa = \omega/c_0$, we recast the reflection coefficient in Eq. (7.2.66) as the following frequency response function, namely,

$$\gamma_T(\omega) = \frac{i\bar{\beta}\omega - 1}{i\bar{\beta}\omega + 1}, \quad \bar{\beta} \equiv \frac{EA}{kc_0} > 0 \tag{b}$$

Applying Fourier transform to both sides of Eq. (a) leads to the Fourier spectrum of $u_R(\xi)$ as follows

$$U_R(\omega) = \frac{4\sin^2(\omega/4)\exp(-i\omega/2)}{\omega^2} \tag{c}$$

There follows the Fourier spectrum of the reflected wave

$$U_L(\omega) = \gamma_T(\omega)U_R(\omega) = \frac{4(i\bar{\beta}\omega - 1)\sin^2(\omega/4)\exp(-i\omega/2)}{\omega^2(i\bar{\beta}\omega + 1)} \tag{d}$$

Applying the inverse Fourier transform to both sides of Eq. (d), we can obtain the reflected wave in terms of $\eta < 0$. To make an easy comparison with the incident triangle wave in terms of $\xi > 0$, we use both posstive argument $|\eta|$ and Heaviside function $s(\cdot)$, and write the reflected wave in terms of $|\eta|$ as follows

$$\begin{aligned} u_L(|\eta|) &= \left[2\bar{\beta}\exp\left(-\frac{|\eta|}{\bar{\beta}}\right) + |\eta| - 2\bar{\beta}\right]s(-|\eta|) \\ &+ \left[2\bar{\beta}\exp\left(\frac{1 - |\eta|}{\bar{\beta}}\right) + |\eta| - 1 - 2\bar{\beta}\right]s(1 - |\eta|) \\ &+ \left[-4\bar{\beta}\exp\left(\frac{1 - 2|\eta|}{2\bar{\beta}}\right) - 2|\eta| + 4\bar{\beta} + 1\right]s\left(\frac{1}{2} - |\eta|\right) \end{aligned}$$

Fig. 7.14 The reflected waves for different stiffness coefficients of the spring (thick solid line: $\overline{\beta} = 50$ s, dashed line: $\overline{\beta} = 5$ s, dot line: $\overline{\beta} = 0.5$ s, thin solid line: $\overline{\beta} = 0.05$ s)

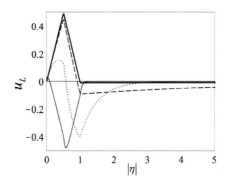

$$- 4\overline{\beta} \sinh^2\left(\frac{\overline{\beta}}{4}\right) \exp\left(\frac{1 - 2|\eta|}{2\overline{\beta}}\right) \qquad \text{(e)}$$

To check the influence of parameter $\overline{\beta}$ on the reflection wave, select a steel rod with following parameters

$$\rho = 7800 \text{ kg/m}^3, \quad E = 210 \text{ GPa}, \quad A = 10^{-4} \text{ m}^2, \quad L = 1 \text{ m} \qquad \text{(f)}$$

There follows the stiffness coefficient $k_R = EA/L = 2.1 \times 10^7$ N/m for the rod. Here are the discussions about Eq. (e) and Fig. 7.14, where the abscissa is $|\eta|$ for an easy comparison with the incident triangle wave.

(1) $\overline{\beta} = 50$ s: The stiffness coefficient of spring is $k = EA/(c_0\overline{\beta}) = \sqrt{\rho E}A/\overline{\beta} \approx 80.94$ N/m and leads to $k/k_R = 3.84 \times 10^{-6}$. That is, the spring is very soft such that the right end of the rod looks like a free end and does not induce wave dispersion as discussed for Eq. (7.2.67). In addition, the reflected wave is almost identical to the incident triangle wave in both amplitude and phase as shown by the thick solid line in Fig. 7.14.

(2) $\overline{\beta} = 0.05$ s: The stiffness coefficient of spring is $k = 80.94$ kN/m and the stiffness ratio is $k/k_R = 3.84 \times 10^{-3}$. That is, the spring is so hard that the right end of the rod seems a fixed and does not produce wave dispersion as discussed for Eq. (7.2.68). Compared with the incident triangle wave, the reflected wave, as shown by the thin solid line in Fig. 7.14, has almost the same amplitude and an opposite phase in this case.

(3) $\overline{\beta} = 5$ s and $\overline{\beta} = 0.5$ s: The stiffness coefficients of spring are $k = 809.4$ N/m and $k = 8094$ N/m, respectively, and imply two middle values of the stiffness coefficient. As shown by the dashed line in Fig. 7.14, the reflected wave shape deviates from the incident triangle wave for $|\eta| > 1$. The dot line in Fig. 7.14 indicates that the reflected wave greatly deviates from the incident triangle wave for $0 < \eta < 3$ because of dispersion. They are the dispersion effects due to the middle stiffness of the spring.

7.2.4 Concluding Remarks

This section presented the studies on the longitudinal dispersive waves of a rod based on the elementary theory of one-dimensional waves. The study demonstrated the wave dispersions of a rod due to the non-uniform cross-section in a lower frequency band, to the transverse inertia in a higher frequency band, and to an elastic boundary in a middle frequency band, respectively. The study laid a theoretical foundation for further studies of longitudinal waves of a rod in a wide frequency band in Sect. 7.3 although it did not touch upon other causes of wave dispersions of a rod, such as the shear deformation, the free surface and the inhomogeneous material. The major conclusions of the section are summarized as follows.

(1)   The non-uniform cross-section of a rod gives rise to the dispersion of longitudinal waves of the rod. For example, the rod with an exp-cross-section has a cut-off frequency such that the wave with a frequency lower than the cut-off frequency is an evanescent wave and the wave with a frequency higher than the cut-off frequency is dispersive. The phase velocity and group velocity of the longitudinal waves of a rod with an exp-cross-section decreasingly and increasingly approach the longitudinal wave speed of a uniform rod respectively with an increase of wave number. Given a finitely long rod with the left end fixed and the right end to an axial harmonic force, when the excitation frequency approaches the cut-off frequency, the evanescent waves form the steady-state harmonic vibration although both wave number and group velocity are zeros.

(2)   The transverse inertia of a uniform rod induces the dispersion of longitudinal waves. Love's rod model well predicts the wave dispersion of the rod when the dimensionless wave number yields $\nu r_p \kappa < 1$. Yet, Love's rod model has a cut-off frequency and does not work in a higher frequency band.

(3)   An elastic spring boundary of a uniform rod induces the dispersion of reflected longitudinal waves at the boundary. When the stiffness coefficient k of the spring approaches zero or positive infinity, the boundary is equivalent to a free boundary or a fixed boundary so that no dispersion occurs. Otherwise, the spring induces the above dispersion, and the most serious dispersion occurs when the wave number is $\kappa = k/(EA)$, where EA represents the axial stiffness of the rod.

7.3 Waves in a Rod Impacting a Rigid Wall

The dynamics of an elastic rod impacting a rigid wall with low velocity is a classical problem in mechanics and of significance in engineering. The studies on this problem stemmed from two branches of mechanics and have led to two kinds of approaches.

The first kind of approaches came from the rigid-body dynamics under the assumption that the contact duration of the rod and the wall is so short that the rebounding velocity of the rod can be determined according to the momentum conservation and a coefficient of restitution measured in tests. These approaches are applicable to engineering problems at ease, but depend on the coefficients of restitution measured for various rods and can hardly offer any information about the dynamic deformation of the rod during a collision.

The second kind of approaches stemmed from the elasto-dynamics or simplified wave theories, dealing with the traveling waves back and forth in the rod. The first study on the wave problem of an elastic rod impacting a rigid wall was due to B. D. St. Venant and M. Flamant, two French mechanicists, in 1883[12]. They found that the contact duration is the time when the compressional wave travels to the free end and returns to the contact end. The book by W. Goldsmith in the UK well surveyed the early studies on this topic[13]. Recent years have seen many advanced studies in this field, including, but not limited to, the problems of a non-uniform rod[14], three-dimensional effects of a thick cylindrical rod[15], and the transverse vibration of a rod to an axial impact[16]. These approaches based on theoretical models and computations have revealed the wave properties and dynamic deformations of the rod under influence of rod parameters and material parameters, but are relatively complicated.

This section deals with the wave dynamics of an elastic rod impacting a rigid wall so as to address a part of Problem 6B proposed in Chap. 1. The section begins with the wave analysis of the impacting rod and gives an exact solution of the impact response of the rod, then turns to the feasibility of modal superposition method and discusses the influence of non-uniform cross-section and transverse inertia on both wave dispersion and convergence of modal series solutions.

7.3.1 Wave Analysis of an Impacting Uniform Rod

As shown in Fig. 7.15, an elastic uniform rod is horizontally moving and impacting a rigid wall with low velocity v_0, and then is rebounding from the wall after a short contact duration. This subsection presents how to analyze the wave propagation of the rod when the rod contacts with the rigid wall.

[12] St. Venant B D, Flamant M (1883). Résistance vive ou dynamique des solids. Représentation graphique des lois du choc longitudinal. Comptes rendus hebdomadaires des Séances de l'Académie des Sciences, 97: 127–353.

[13] Goldsmith W (1960). Impact: The Theory and Physical Behavior of Colliding Solids. London: Edward Arnold Ltd., 1–379.

[14] Gan C B, Wei Y M, Yang S X (2016). Longitudinal wave propagation in a multi-step rod with variable cross-section. Journal of Vibration and Control, 22(3): 837–852.

[15] Cerv J, Adamek V, Vales F, et al. (2016). Wave motion in a thick cylindrical rod undergoing longitudinal impact. Wave Motion, 66(1): 88–105.

[16] Morozov N F, Tovstik P E (2013). Transverse rod vibration under a short-term longitudinal impact. Doklady Physics, 58(9): 387–391.

Fig. 7.15 The schematic of
a uniform rod impacting a
rigid wall

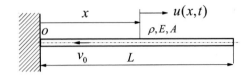

7.3.1.1   Problem to Be Studied

Let L be the rod length, A be the cross-sectional area, ρ be the material density and E be Young's modulus. We define the coordinate x measured from the left end of the rod and time t measured from the beginning of contact between the rod and the wall, and denote the dynamic deformation of the rod at cross-section x as $u(x, t)$. As such, $u(x, t)$ yields the dynamic equation as follows

$$\frac{\partial^2 u(x, t)}{\partial t^2} = c_0^2 \frac{\partial^2 u(x, t)}{\partial x^2}, \quad c_0 \equiv \sqrt{\frac{E}{\rho}}, \quad x \in [0, L], \quad t \in [0, +\infty) \quad (7.3.1)$$

It is easy to derive the following conditions for Eq. (7.3.1). That is, the initial conditions of the rod starting contact with the wall are

$$\text{C1:} \quad u(x, 0) = 0, \quad \text{C2:} \quad u_t(x, 0) = -v_0, \quad x \in [0, L] \quad (7.3.2a)$$

The boundary condition of the free end of the rod is

$$\text{C3:} \quad u_x(L, t) = 0, \quad t \in [0, +\infty) \quad (7.3.2b)$$

The boundary condition of the left end of the rod during contact is

$$\text{C4:} \quad u(0, t) = 0, \quad t \in [0, t_c] \quad (7.3.2c)$$

where t_c is the contact duration to be determined and satisfies the switching condition between the contact and the rebound as follows

$$\text{C5:} \quad u_x(0, t_c^-) \leq 0, \quad u_x(0, t_c^+) = 0, \quad u_t(0, t_c^+) > 0 \quad (7.3.2d)$$

7.3.1.2   Wave Analysis

In what follows, we first solve Eq. (7.3.2) for the traveling waves in the rod impacting the rigid wall and then discuss how an impact wave propagates.

(1) Solution of impact waves

In the solution procedure, we assume that condition C4 holds at any time, and then determine the switching moment t_c in view of condition C5 between the contact and the rebound in order to define the contact duration. Hence, $t \in [0, +\infty)$ becomes a temporary release for $t \in [0, t_c]$ in Eq. (7.3.2c).

As analyzed in Sect. 7.1.1, d'Alembert's solution of Eq. (7.3.1) reads

$$u(x, t) = u_R(c_0 t - x) + u_L(c_0 t + x)$$
$$\equiv u_R(\xi) + u_L(\eta), \quad \xi \in [-L, +\infty), \quad \eta \in [0, +\infty) \tag{7.3.3}$$

where $u_R(c_0 t - x)$ is the right traveling wave and $u_L(c_0 t + x)$ is the left traveling wave. They keep constant in the characteristic lines $\xi = c_0 t - x$ and $\eta = c_0 t + x$, respectively.

As the first step, substituting Eq. (7.3.3) into condition C4 of the left boundary of rod leads to

$$u(0, t) = u_R(c_0 t) + u_L(c_0 t) = 0, \quad t \in [0, +\infty) \tag{7.3.4}$$

Replacing the independent variable $c_0 t \in [0, +\infty)$ in Eq. (7.3.4) with argument $\eta \in [0, +\infty)$, we have

$$u_L(\eta) = -u_R(\eta), \quad \eta \in [0, +\infty) \tag{7.3.5}$$

Substitution of Eq. (7.3.5) into Eq. (7.3.3) gives

$$\begin{cases} u(x, t) = u_R(c_0 t - x) - u_R(c_0 t + x) \\ u_t(x, t) = c_0 u_{R\xi}(c_0 t - x) - c_0 u_{R\xi}(c_0 t + x) \\ u_x(x, t) = -u_{R\xi}(c_0 t - x) - u_{R\xi}(c_0 t + x), \quad x \in [0, L], \quad t \in [0, +\infty] \end{cases} \tag{7.3.6}$$

where $u_{R\xi}$ is the derivative of function $u_R(\xi)$ with respect to argument ξ.

To start the second step, substituting the first two equations in Eq. (7.3.6) into conditions C1 and C2 respectively leads to

$$\begin{cases} u_R(-x) - u_R(x) = 0 \\ c_0 u_{R\xi}(-x) - c_0 u_{R\xi}(x) = -v_0, \quad x \in [0, L] \end{cases} \tag{7.3.7}$$

Integrating the second equation in Eq. (7.3.7) gives

$$u_R(-x) + u_R(x) = \frac{v_0 x}{c_0} + a_1, \quad x \in [0, L] \tag{7.3.8}$$

where a_1 is an integration constant. Solving Eq. (7.3.8) and the first equation in Eq. (7.3.7) simultaneously, we have

$$u_R(x) = \frac{v_0 x}{2c_0} + \frac{a_1}{2}, \quad u_R(-x) = \frac{v_0 x}{2c_0} + \frac{a_1}{2}, \quad x \in [0,\ L] \tag{7.3.9}$$

Replacing independent variables $x \in [0,\ L]$ in $u_R(x)$ and $-x \in [-L,\ 0]$ in $u_R(-x)$ above by argument $\xi \in [-L,\ L]$, we find, from Eq. (7.3.9), that $u_R(\xi)$ is an even function for $\xi \in [-L,\ L]$ such that $u_{R\xi}(\xi)$ is an odd function in the same interval. That is, Eq. (7.3.9) becomes

$$u_R(\xi) = \frac{v_0}{2c_0}|\xi| + \frac{a_1}{2}, \quad u_{R\xi}(\xi) = \frac{v_0}{2c_0}\mathrm{sgn}(\xi), \quad \xi \in [-L,\ L] \tag{7.3.10}$$

Substitution of the first equation in Eq. (7.3.10) into Eq. (7.3.6) shows that the integration constant has no influence on $u(x, t)$. Thus, let $a_1 = 0$ for simplicity.

In the final step, substituting the third equation in Eq. (7.3.6) into condition C3 of right boundary, we have

$$u_{R\xi}(c_0 t - L) + u_{R\xi}(c_0 t + L) = 0, \quad t \in [0,\ +\infty) \tag{7.3.11}$$

Let $\zeta = c_0 t - L$, then $\zeta + 2L = c_0 t + L$ holds so that Eq. (7.3.11) is equivalent to

$$u_{R\xi}(\zeta + 2L) = -u_{R\xi}(\zeta), \quad \zeta \in [-L,\ +\infty) \tag{7.3.12}$$

Equation (7.3.12) implies that function $u_{R\xi}(\zeta)$ exhibits periodicity of $4L$ and can be periodically extended to infinity. Integrating Eq. (7.3.12) gives

$$u_R(\zeta + 2L) = -u_R(\zeta) + a_i, \quad \zeta \in [-L,\ +\infty) \tag{7.3.13}$$

where a_i, $i = 2, 3, \ldots$ are integration constants.

In view of Eq. (7.3.10) and Eq. (7.3.13), we have the periodic extension of $u_R(\xi)$ from interval $[-L,\ L]$ to $[L,\ 3L), [3L,\ 5L), \ldots +\infty$, and derive the expressions of $u_R(\eta)$ in intervals $[0,\ 2L)$ and $[2L,\ 4L)$ as follows

$$u_R(\xi) = \begin{cases} -v_0\xi/2c_0, & \xi \in [-L, 0) \\ v_0\xi/2c_0, & \xi \in [0, 2L) \\ -v_0\xi/2c_0 + a_2, & \xi \in [2L, 4L), \quad \ldots \end{cases} \tag{7.3.14a}$$

$$u_R(\eta) = \begin{cases} v_0\eta/2c_0, & \eta \in [0,\ 2L) \\ -v_0\eta/2c_0 + a_3, & \eta \in [2L, 4L), \quad \cdots \end{cases} \tag{7.3.14b}$$

where a_2 and a_3 are the integration constants to be determined from the displacement continuity.

The traveling displacement waves described by Eq. (7.3.14) are two piecewise continuous functions, while the velocity waves and strain waves derived from Eq. (7.3.6) and Eq. (7.3.14) exhibit the first kind of discontinuities in the characteristic lines $\xi = 0,\ 2L,\ 4L, \ldots$ and $\eta = 0,\ 2L,\ 4L, \ldots$. Hence, they are non-smooth weak solutions as defined before.

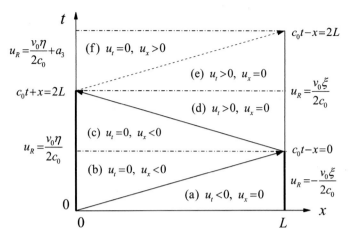

Fig. 7.16 The regions divided by characteristic lines on plane (x, t), where arrows are characteristic lines and expressions come from Eqs. (7.3.14a) and (7.3.14b)

To study the traveling waves in Eq. (7.3.14), consider the regions divided by characteristic lines shown in plane (x, t) in Fig. 7.16, where the waves in the first four regions in time sequence are discussed as follows.

(1) In region (a) for $0 \le c_0 t < x \le L$: As $c_0 t - x < 0$ holds true on one hand, the right traveling wave arrives, along the characteristic line $\xi \equiv c_0 t - x = \text{const}$, at the right end of the rod denoted by a thick solid segment in Fig. 7.16. Thus, Eq. (7.3.14a) gives $u_R(c_0 t - x) = -v_0(c_0 t - x)/(2c_0)$. On the other hand, as $0 < x + c_0 t < 2L$ holds, the left traveling wave reaches, along characteristic line $\eta \equiv x + c_0 t = \text{const}$, the left end of the rod denoted by a thick solid segment in Fig. 7.16. Hence, Eq. (7.3.14b) gives $u_R(c_0 t + x) = v_0(c_0 t + x)/(2c_0)$. Substituting the two solutions into Eq. (7.3.6) leads to displacement $u(x, t) = -v_0 t$, velocity $u_t(x, t) = -v_0$ and strain $u_x(x, t) = 0$ of the rod, respectively.

(2) In region (b) for $0 \le x \le c_0 t < L$: As $c_0 t - x > 0$ and $c_0 t - x < 2L$ hold true on one hand, the right wave arrives at the right end of the rod denoted by a thin solid segment along the characteristic line, thus, $u_R(c_0 t - x) = v_0(c_0 t - x)/(2c_0)$ holds. On the other hand, the left traveling wave reaches the left end of the rod denoted by a thick solid segment as $0 < c_0 t + x < 2L$ holds. Thus, $u_R(c_0 t + x) = v_0(c_0 t + x)/(2c_0)$ holds. Substituting the two solutions into Eq. (7.3.6) leads to displacement $u(x, t) = -v_0 x/c_0$, velocity $u_t(x, t) = 0$ and strain $u_x(x, t) = -v_0/c_0$ of the rod. Noting that the characteristic line $c_0 t - x = 0$ serves as an interface between region (a) and region (b), we can verify the displacement continuity in the characteristic line by substituting $x = c_0 t$ into $u(x, t) = -v_0 x/c_0$.

(3) In region (c) for $L \le c_0 t < 2L - x$: There is no characteristic line between region (b) and region (c) so that the solutions keep unchanged in the two regions.

(4)   In region (d) for $2L - x \leq c_0 t < 2L$: As $c_0 t - x \geq c_0 t - L > 0$ and $c_0 t - x < 2L$ hold, the right traveling wave arrives, along the characteristic line, at the right end of the rod denoted by a thin solid segment, thus, $u_R(c_0 t - x) = v_0(c_0 t - x)/(2c_0)$ holds, while the left traveling wave arrives at the left end of the rod also denoted by a thin solid segment since $2L < c_0 t + x < 3L$ holds true. Therefore, $u_R(c_0 t + x) = -v_0(c_0 t + x)/(2c_0) + a_3$ holds. Substitution of two solutions into Eq. (7.3.6) gives displacement $u(x, t) = v_0 t - a_3$. In view of the displacement continuity in the characteristic line $c_0 t + x = 2L$, substituting the displacement in region (c) into the about solution, we have $v_0 t - a_3 = -v_0 x/c_0 = v_0(c_0 t - 2L)/c_0$ such that $a_3 = 2L/c_0$ and $u(x, t) = v_0(t - 2L/c_0)$ hold. The velocity and strain are $u_t(x, t) = v_0$ and $u_x(x, t) = 0$, respectively.

Remark 7.3.1 In this problem, it is sufficient to discuss the above four regions for $t \leq t_c = 2L/c_0$. The regions above region (d) in Fig. 7.16 can be used to study the wave propagation of the rod when it keeps contact with the rigid wall when $t > t_c^+$. For example, if the rod vertically impacts a rigid wall under gravity, the contact duration may be longer than $2L/c_0$ such that the waves travel into region (e) and region (f).

In summary, the displacement, velocity and strain in the above four regions are expressed as follows

$$u(x, t) = \begin{cases} -v_0 t, & 0 \leq c_0 t < x \leq L \\ -v_0 x/c_0, & 0 \leq x \leq c_0 t < 2L - x \\ v_0(t - 2L/c_0), & 2L - x \leq c_0 t \leq 2L \end{cases} \qquad (7.3.15)$$

$$u_t(x, t) = \begin{cases} -v_0, & 0 \leq c_0 t < x \leq L \\ 0, & 0 \leq x \leq c_0 t < 2L - x \\ v_0, & 2L - x \leq c_0 t \leq 2L \end{cases} \qquad (7.3.16)$$

$$u_x(x, t) = \begin{cases} 0, & 0 \leq c_0 t < x \leq L \\ -v_0/c_0, & 0 \leq x \leq c_0 t < 2L - x \\ 0, & 2L - x \leq c_0 t \leq 2L \end{cases} \qquad (7.3.17)$$

(2)   Discussions about the wave propagation

To understand the displacement, velocity and strain waves of the rod, we recall the above results in time sequence as following.

(1)   At moment $t = 0^+$, the left end of the rod impacts the rigid wall and undergoes a velocity jump from $u_t(0, 0^-) = -v_0$ to $u_t(0, 0^+) = 0$. Meanwhile, the left end of the rod also undergoes a strain jump from $u_x(0, 0^-) = 0$ to compressional strain $u_x(0, 0^+) = -v_0/c_0$.

(2)   When $c_0 t \in (0, L)$, the traveling velocity wave of amplitude v_0 counteracts the rod velocity $-v_0$ such that the left part of the rod for $x \in [0, c_0 t)$ is at rest,

while the right part of the rod for $x \in [c_0t, L]$ keeps moving with velocity $-v_0$ such that the left part of the rod is compressed. The compressed strain wave travels with speed c_0 in the right direction so that the left part of the rod for $x \in [0, c_0t)$ has compressional strain of $-v_0/c_0$ while the right part of the rod for $x \in [c_0t, L]$ has no strain at all.

(3) At moment $c_0t = L$, the whole rod is at rest simultaneously, and the strain wave arrives at the right end of the rod and undergoes a reflection.

(4) When $c_0t \in (L, 2L)$, the right part of the rod for $x \in [2L - c_0t, L]$ moves with velocity v_0 in the right direction due to potential energy, while the left part of the rod for $x \in [0, 2L - c_0t)$ is at rest. The right traveling strain wave counteracts the left traveling strain wave such that the right part of the rod for $x \in [2L - c_0t, L]$ has no strain, but the left part of the rod for $x \in [0, 2L - c_0t)$ is still subject to compressional strain $-v_0/c_0$.

(5) At moment $t = 2L/c_0$, the strain in the whole rod disappears and the left end of the rod satisfies both non-negative strain condition $u_x(0, 2L/c_0) \geq 0$ and positive velocity condition $u_t(x, 2L/c_0) = v_0 > 0$ such that the rod begins rebounding in the right direction.

Finally, we collect the above results in Fig. 7.17, where the arrows present the propagation directions of the strain wave. The above analysis indicates that the duration of contact between the rod and the rigid wall yields

$$t_c \equiv \frac{2L}{c_0} \tag{7.3.18}$$

In the contact duration when $0 \leq t < t_c^-$, the strain of the left end of the rod keeps

$$u_x(0, t) = -\frac{v_0}{c_0} < 0, \quad t \in [0, t_c) \tag{7.3.19}$$

At moment $t = t_c^+$, the strain jumps to zero. Then, the rod rebounds with velocity v_0 in the right direction when $t > t_c$ and behaves like a rigid-body without any strain.

As shown in Eq. (7.3.18), the contact duration t_c has nothing to do with impact velocity v_0. Yet, if v_0 is large enough, the elastic strain in the rod may exceed the yielding limit of materials and becomes plastic, and the mathematical model in

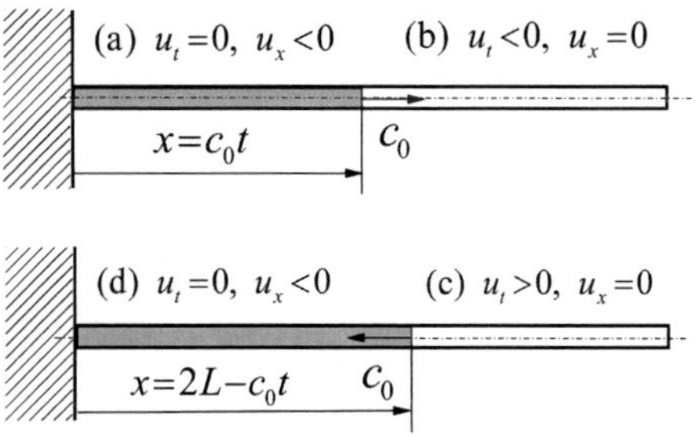

Fig. 7.17 The velocity and strain waves of the rod during the contact with a rigid wall (gray: compressional strain, white: no strain)

Eq. (7.3.1) fails to work. Consider, as an example, carbon steel Q235 with the lower bound of yielding stress $\sigma_s = 216$ MPa, Young's modulus $E = 210$ GPa and material density $\rho = 7800$ kg/m^3. Then, the upper bound of impact velocity of the rod for elastic deformation is only $v_c = c_0\sigma_s/E = \sigma_s/\sqrt{\rho E} \approx 5.337$ m/s.

7.3.2 Modal Analysis of an Impacting Uniform Rod

The analysis in Sect. 7.3.1 gave an exact wave solution of the rod impacting a rigid wall, but is only applicable to a few wave problems satisfying d'Alembert's solution, such as the axial impact of a uniform rod and the torsional impact of a uniform shaft, and involves tedious mathematics. It is reasonable to wonder whether the modal superposition method works for an impacting problem as questioned in Problem 6B in Chap. 1. This subsection, hence, presents how to use the modal superposition method to analyze the wave problem, such as a rod horizontally impacting a rigid wall.

7.3.2.1   Modal Superposition Method

Assume that the solution of Eq. (7.3.1) has a form of separated variables

$$u(x, t) = \varphi(x)q(t) \tag{7.3.20}$$

Substitution of Eq. (7.3.20) into Eq. (7.3.1) and two boundary conditions C3 and C4 in Eq. (7.3.2) leads to two differential equations as follows

$$\frac{d^2\varphi(x)}{dx^2} + \kappa^2\varphi(x) = 0, \quad \varphi(0) = 0, \quad \varphi_x(L) = 0 \tag{7.3.21}$$

$$\frac{dq(t)}{dt^2} + c_0^2\kappa^2q(t) = 0 \tag{7.3.22}$$

Equation (7.3.21) describes the vibration shape of a fixed-free rod in space domain and κ is the wave number satisfying two boundary conditions in Eq. (7.3.21), while Eq. (7.3.22) describes the vibration history in time domain and the corresponding solutions superposed should satisfy two initial conditions C1 and C2 in Eq. (7.3.2).

Let the general solution of Eq. (7.3.21) be expressed as

$$\varphi(x) = \alpha\cos(\kappa x) + \beta\sin(\kappa x) \tag{7.3.23}$$

Substituting Eq. (7.3.23) into the boundary conditions in Eq. (7.3.21) leads to the characteristic equation as follows

$$\varphi(0) = \alpha = 0, \quad \varphi_x(L) = \beta\kappa\cos(\kappa L) = 0 \tag{7.3.24}$$

There follow the wave numbers and the normalized mode shapes, namely,

$$\kappa_r = \frac{(2r-1)\pi}{2L}, \quad \varphi_r(x) = \sin(\kappa_r x), \quad r = 1, 2, \ldots \tag{7.3.25}$$

Given a wave number κ_r, solving Eq. (7.3.22) leads to the corresponding time history

$$q_r(t) = a_r\cos(c_0\kappa_r t) + b_r\sin(c_0\kappa_r t), \quad r = 1, 2, \ldots \tag{7.3.26}$$

In view of the solution structure of a linear partial differential equation, the solution of Eq. (7.3.1) yields

$$u(x, t) = \sum_{r=1}^{+\infty}\varphi_r(x)q_r(t) = \sum_{r=1}^{+\infty}\sin(\kappa_r x)[a_r\cos(c_0\kappa_r t) + b_r\sin(c_0\kappa_r t)] \tag{7.3.27}$$

where coefficients a_r, b_r, $r = 1, 2, \ldots$ are constants to be determined from initial conditions C1 and C2.

To calculate those coefficients, assume that the function series in Eq. (7.3.27) is time differentiable for each term such that we have

$$u_t(x, t) = \sum_{r=1}^{+\infty}\sin(\kappa_r x)[b_r c_0\kappa_r\cos(c_0\kappa_r t) - a_r c_0\kappa_r\sin(c_0\kappa_r t)] \tag{7.3.28}$$

The rationality from Eqs. (7.3.27) to (7.3.28) will be discussed in Sect. 7.3.2.2 shortly later. Substitution of Eqs. (7.3.27) and (7.3.28) into the initial conditions C1 and C2 in Eq. (7.3.2) leads to

$$u(x, 0) = \sum_{r=1}^{+\infty}a_r\sin(\kappa_r x) = 0, \quad u_t(x, 0) = \sum_{r=1}^{+\infty}c_0 b_r\kappa_r\sin(\kappa_r x) = -v_0 \tag{7.3.29}$$

Let subscript r in Eq. (7.3.29) be replaced with s, Eq. (7.3.29) be multiplied by mode shape $\sin(\kappa_r x)$ and then integrated over $[0, L]$, we derive, in view of the orthogonality of mode shapes, the following two relations

$$\begin{cases} \dfrac{a_r L}{2} = \sum_{s=1}^{+\infty}\left[a_s\int_0^L\sin(\kappa_s x)\sin(\kappa_r x)dx\right] = 0 \\[2mm] \dfrac{c_0 b_r\kappa_r L}{2} = \sum_{s=1}^{+\infty}\left[\int_0^L c_0 b_s\kappa_s\sin(\kappa_s x)\sin(\kappa_r x)dx\right] = -\int_0^L v_0\sin(\kappa_r x)dx, \\[2mm] r = 1, 2, \ldots \end{cases}$$

$$\tag{7.3.30}$$

Thereby, we get

$$
\begin{cases}
a_r = 0, \\[2mm]
b_r = -\dfrac{2v_0}{c_0 \kappa_r L} \displaystyle\int_0^L \sin(\kappa_r x)\,dx = -\dfrac{2v_0}{c_0 L \kappa_r^2} = -\dfrac{8v_0 L}{c_0 \pi^2 (2r-1)^2}, \quad r = 1, 2, ..
\end{cases}
\tag{7.3.31}
$$

Substitution of Eq. (7.3.31) into Eqs. (7.3.27) and (7.3.28) leads to the modal series solutions, i.e., the formal solutions, of displacement and velocity of the rod as follows

$$
u(x,t) = -\frac{8v_0 L}{\pi^2 c_0} \sum_{r=1}^{+\infty} \frac{1}{(2r-1)^2} \sin(\kappa_r x)\sin(c_0 \kappa_r t)
\tag{7.3.32}
$$

$$
u_t(x,t) = -\frac{4v_0}{\pi} \sum_{r=1}^{+\infty} \frac{1}{(2r-1)} \sin(\kappa_r x)\cos(c_0 \kappa_r t)
\tag{7.3.33}
$$

Of course, it is necessary to discuss the relation between the modal series solutions and the exact solutions in Sect. 7.3.1, as well as the rationality from Eq. (7.3.27) to Eq. (7.3.28) via the time differentiation of Eq. (7.3.27) for each term in the right-hand side.

7.3.2.2   Relation of Modal Series Solution and Exact Solution

We first convert the product of triangle functions in each term in Eq. (7.3.32) to a sum of two triangle functions as follows

$$
\sin(\kappa_r x)\sin(c_0 \kappa_r t) = \frac{1}{2}\{\cos[\kappa_r(c_0 t - x)] - \cos[\kappa_r(c_0 t + x)]\}
\tag{7.3.34}
$$

Substituting Eq. (7.3.34) into Eq. (7.3.32), hence, the displacement solution in Eq. (7.3.32) is the superposition of an infinite number of left traveling waves and right traveling waves, namely,

$$
u(x,t) = \frac{4v_0 L}{\pi^2 c_0} \sum_{r=1}^{+\infty} \frac{1}{(2r-1)^2}\{\cos[\kappa_r(c_0 t + x)] - \cos[\kappa_r(c_0 t - x)]\}
\tag{7.3.35}
$$

That is, the modal series solution includes the traveling waves of the rod impacting a rigid wall and exhibits a reasonable picture of physics.

Next, we study the convergence of Eq. (7.3.32), where the general term satisfies

$$
\left| \frac{1}{(2r-1)^2} \sin(\kappa_r x)\sin(c_0 \kappa_r t) \right| \le \frac{1}{(2r-1)^2}
\tag{7.3.36}
$$

As the series $\sum_{r=1}^{+\infty}[1/(2r-1)^2] = \pi/8$ is convergent, Eq. (7.3.32) is uniformly convergent according to the M-criterion of function series in mathematics. The solution uniqueness of the mixed initial-boundary value problem of the wave equation, which was proved via energy inequality[17] for partial differential equations of hybolic type, enables one to conclude that Eq. (7.3.32) is uniformly convergent to the exact solution. As the coefficients in Eq. (7.3.32) are convergent to zero at decreasing rate $O[1/(2r-1)^2]$ with mode order r, the terms of large wave number in Eq. (7.3.32) make tiny contributions such that the modal series solution has very fast convergence.

Now, we study the convergence of velocity series solution in two steps. The first step is to prove the rationality from Eq. (7.3.27) to Eq. (7.3.28) by differentiating each term in Eq. (7.3.27) and the second step is to prove the convergence of Eq. (7.3.33).

Recalling the wave analysis in Sect. 7.3.1, Eq. (7.3.10) indicates that the right traveling wave $u_{R\xi}(\xi)$ is absolutely integrable for $\xi \in [-L, \ L]$, but has a finite discontinuity at $\xi = 0$, while Eq. (7.3.12) indicates that the right traveling wave $u_{R\xi}(\xi)$ is also integrable, and has periodicity of $4L$ and finite discontinuities at $\xi = 0, \ 2L, \ 4L, \ 6L, \cdots$. According to Eq. (7.3.6), therefore, $u_t(x, t) = c_0[u_{R\xi}(\xi) - u_{R\xi}(\eta)]$ is an absolutely integrable periodic function with discontinuities at $\xi = 0, \ 2L, \ 4L, \ 6L, \cdots$ and $\eta = 0, \ 2L, \ 4L, \ 6L, \cdots$. Given any x, Eq. (7.3.27) can be regarded as a Fourier series of $u(x, t)$ for $t \in [0, \ t_c]$. From the theory of Fourier series[18], it is reasonable to perform the time differentiation for each term in Eq. (7.3.27) so as to derive the Fourier series of $u_t(x, t)$ in Eq. (7.3.28) since $u(x, t)$ is a continuous function and $u_t(x, t)$ is an integrable functions with finite discontinuities.

To start the second step, we return to Eq. (7.3.10) and its periodic extension, and reach the right traveling wave of period $4L$ as follows

$$u_{R\xi}(\xi) = \frac{v_0}{2c_0}\mathrm{sgn}(\xi), \quad \xi \in [-2L, \ 2L] \tag{7.3.37}$$

In view of the theory of Fourier series, $u_{R\xi}(\xi)$ is an odd function with finite discontinuities and can be expanded as a following convergent series

$$u_{R\xi}(\xi) = \frac{v_0}{2c_0}\sum_{r=1}^{+\infty} a_r \sin(\kappa_r\xi) = \frac{2v_0}{\pi c_0}\sum_{r=1}^{+\infty}\frac{1}{2r-1}\sin(\kappa_r\xi) \tag{7.3.38}$$

where the Fourier coefficients satisfy

$$a_r = \frac{1}{2L}\int_{-2L}^{2L}\mathrm{sgn}(\xi)\sin(\kappa_r\xi)\mathrm{d}\xi = \frac{2}{L}\int_0^L \sin(\kappa_r\xi)\mathrm{d}\xi = \frac{4}{\pi(2r-1)}, \quad r = 1, 2, \dots \tag{7.3.39}$$

---

[17] Courant R, Hilbert D (1989). Methods of Mathematical Physics. Volume II. New York: John Wiley & Sons Inc., 656–661.

[18] Yan Z D (1989). Fourier Series Solutions in Structure Mechanics. Tianjin: Press of Tianjin University, 35–66 (in Chinese).

From Eq. (7.3.6), we have the sinusoidal series convergent to the exact velocity solution as follows

$$u_t(x, t) = c_0 \big[u_{R\xi}(\xi) - u_{R\xi}(\eta) \big] = \frac{2v_0}{\pi} \sum_{r=1}^{+\infty} \frac{1}{2r-1} [\sin(\kappa_r \xi) - \sin(\kappa_r \eta)]$$

$$= \frac{2v_0}{\pi} \sum_{r=1}^{+\infty} \frac{1}{2r-1} \{\sin[\kappa_r(c_0 t - x)] + \sin[\kappa_r(c_0 t + x)]\} \qquad (7.3.40)$$

Obviously, Eq. (7.3.40) is exactly identical to Eq. (7.3.33) where the term $\sin(\kappa_r x)\cos(c_0\kappa_r t)$ can be replaced with term $\{\sin[\kappa_r(c_0 t - x)] + \sin[\kappa_r(c_0 t + x)]\}/2$ in Eq. (7.3.40). As a consequence, the velocity series solution in Eq. (7.3.33) is convergent to the exact velocity solution in Eq. (7.3.16).

Remark 7.3.2 From Eq. (7.3.37), $u_t(x, 0) = c_0[u_{R\xi}(-x) - u_{R\xi}(x)] = -v_0$ holds for $x \in [0, 2L]$ such that $u_t(x, 0)$ is symmetric about $x = L$ and its Fourier series contains only the odd super-harmonics of a sinusoidal function[19]. From Eq. (7.3.29), the Fourier series is a sinusoidal series of $u_t(x, 0)$ of period $4L$ for $x \in [0, L]$ and the fundamental wave length is $2\pi/\kappa_1 = 2\pi/(\pi/2L) = 4L$.

In the study of impact problems of a structure, it is essential to analyze the propagation of stress and strain waves, which determine the dynamic strength of the structure. For this purpose, one should differentiate the modal series solution of displacement in Eq. (7.3.32) with respect to x for each term and derive the modal series solution for strain, i.e., the formal strain solution, as following

$$u_x(x, t) = -\frac{4v_0}{\pi c_0} \sum_{r=1}^{+\infty} \frac{1}{2r-1} \cos(\kappa_r x) \sin(c_0 \kappa_r t)$$

$$= \frac{2v_0}{\pi c_0} \sum_{r=1}^{+\infty} \frac{1}{2r-1} \{\sin[\kappa_r(c_0 t - x)] - \sin[\kappa_r(c_0 t + x)]\} \qquad (7.3.41)$$

By analogy with the analysis of the velocity series solution, one can prove the rationality of partial differentiation for each term in Eq. (7.3.32) with respect to x and the convergence of the formal strain solution in Eq. (7.3.41) to the exact strain solution in Eq. (7.3.17).

Remark 7.3.3 Equation (7.3.6) indicates that both strain wave $u_x(x, t)$ and velocity wave $u_t(x, t)$ take an equivalent position in Eq. (7.3.1). That is, they both are piecewise continuous and have finite discontinuities in characteristic lines such that their modal series solutions are convergent at decreasing rate $O[1/(2r-1)]$ with mode order r.

---

[19] Yan Z D (1989). Fourier Series Solutions in Structure Mechanics. Tianjin: Press of Tianjin University, 9–13 (in Chinese).

7.3.2.3 Case Studies and Convergence Rates

Example 7.3.1 Given a steel rod with following parameters

$$E = 210 \text{ GPa}, \quad \rho = 7800 \text{ kg/m}^3, \quad L = 5.189 \text{ m}, \quad v_0 = 5.189 \text{ m/s} \qquad \text{(a)}$$

validate the above modal series solutions.

Substitution of Eq. (a) into Eqs. (7.3.1) and (7.3.18) leads to the longitudinal wave speed and contact duration as follows

$$c_0 = \sqrt{\frac{E}{\rho}} \approx 5189 \text{ m/s}, \quad t_c = \frac{2L}{c_0} \approx 0.002 \text{ s} \qquad \text{(b)}$$

Let n be the truncated mode order in computing the displacement, velocity and strain waves, we study the convergence of Eqs. (7.3.32), (7.3.33) and (7.3.41).

At first, consider the convergence of displacement series solution in Eq. (7.3.32). Figure 7.18 shows the displacement distributions of the rod at two typical moments. Here, the approximate solution has fast convergence. When $n = 3$ increases to $n = 6$, for instance, the coefficient of highest super-harmonics in Eq. (7.3.32) decreases from $1/(2n-1)^2 = 1/25$ to $1/(2n-1)^2 = 1/121$ such that the approximate solution of $n = 6$ is much better than that of $n = 3$, exhibiting only a slight difference at the corner of displacement in Fig. 7.18b.

Next, we check the approximate velocity solution. From the convergence rate in Eq. (7.3.33), we take $n = (25 + 1)/2 = 13$ and $n = (121 + 1)/2 = 61$ to make the accuracy of velocity solution in Eq. (7.3.33) compatible to that of displacement solution in Eq. (7.3.32). Figure 7.19 shows the velocity histories at two typical cross-sections of the rod. One is at the middle of the rod and the other is near the right end of the rod. Because the right traveling wave jumps at the right end of the rod when $t = t_c/2$ such that the neighboring velocity history exhibits strong oscillations. As

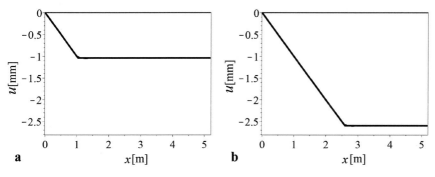

Fig. 7.18 The displacement distributions of a uniform rod at two moments (thick solid line: the exact solution, thin solid line: an approximate solution of $n = 6$, dashed line: an approximate solution of $n = 3$). **a** $t = 0.1\, t_c$; **b** $t = 0.25\, t_c$

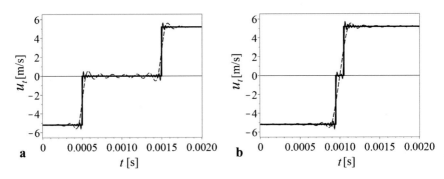

Fig. 7.19 The velocity histories of a uniform rod at two cross-sections (thick solid line: the exact solution, thin solid line: an approximate solution of $n = 61$, dashed line: an approximate solution of $n = 13$). **a** $x = L/2$; **b** $x = 0.95L$

shown in Fig. 7.19b, thus, the convergence of velocity series solution is not so good as that of displacement series solution. The approximate velocity solution of $n = 13$ can hardly describe the sudden change of velocity near the right end of the rod, and the approximate velocity solution of $n = 61$, with 48 mode orders increased, gives an improved result as shown in Fig. 7.19b.

Remark 7.3.4 Engineers prefer the finite element method and numerical integration to compute the dynamic response of an impacting structure. For the rod impacting a rigid wall here, the finite element model with very fine meshes does not work well to describe the sudden jump of the velocity wave near the right end of the rod. For instance, the numerical results of a rod model meshed by 1000 finite elements exhibit irregular fluctuations near the sudden jump. Hence, it is necessary to design special algorithms to compute discontinuous waves.

Finally, we check the convergence of approximate strain solution. From Remark 7.3.3 and the truncated mode orders for approximate velocity solution, we take $n = (25 + 1)/2 = 13$ and $n = (121 + 1)/2 = 61$ again. Figure 7.20 presents the strain distributions of the rod at two typical moments, displaying the equivalent computation accuracy in comparison with the approximate velocity solution.

Remark 7.3.5 The reasons why the velocity series solution and the strain series solution exhibit slow convergence are due to their discontinuities in characteristic lines. As proved in mathematics[20], if the right traveling wave $u_R(\xi)$ and the left traveling wave $u_L(\eta)$ have the p-th order piecewise continuous derivatives and the $(p-1)$-th order of continuous derivatives, the decreasing rate of Fourier coefficients is no more than $O(1/\kappa^p)$, where κ is the wave number in Fourier series. In the present case, $u_R(\xi)$ is a piecewise smooth function with $p = 1$ and $u_{R\xi}(\xi)$ is a piecewise continuous function with $p = 0$. Hence, the convergence rates of both velocity and

[20] Courant R, Hilbert D (1989). Methods of Mathematical Physics. Volume I. New York: John Wiley & Sons Inc., 74.

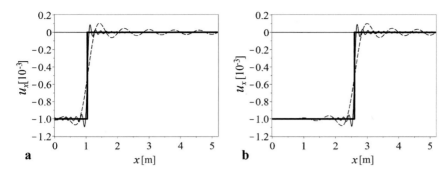

Fig. 7.20 The strain distributions of a uniform rod at two moments (thick solid line: the exact solution, thin solid line: an approximate solution of $n = 61$, dashed line: an approximate solution of $n = 13$). **a** $t = 0.1\, t_c$; **b** $t = 0.25\, t_c$

strain series solutions are slower than that of displacement series solution by an order. Hence, it is possible, but not efficient, to improve the accuracy of an approximate solution by increasing the truncated mode order n. In particular, the improvement is poor near the discontinuity of the exact solution of displacement.

Remark 7.3.6 The waves in this problem are non-dispersive, thus, it is possible to improve the accuracy of an approximate solution by increasing the truncated mode order n. For other problems, one ought to check whether the dynamic model established is able to describe the wave dispersion of large wave numbers. Otherwise, an improvement of approximate solution by increasing mode order in mathematics may deteriorate the physical fidelity of approximate solution. Section 7.3.4 will address this issue further.

Remark 7.3.7 In Figs. 7.19 and 7.20, the approximate solutions exhibit remarkable fluctuations at the vicinity of a discontinuity of exact solution. If one increases the truncated mode order n, the maximal peak of fluctuation moves towards to the discontinuity position of exact solution, but keeps about 9% of the jump amplitude of the exact solution at the discontinuity and does not decrease further. This is the well-known *Gibbs phenomenon* or *Gibbs effect*[21] in the truncated Fourier series of a discontinuous function.

7.3.3 Influence of Non-uniform Cross-section

This subsection is about the wave problem of a non-uniform rod horizontally impacting a rigid wall with a low velocity v_0 and the influence of wave dispersion on impact dynamics.

---

[21] Courant R, Hilbert D (1989). Methods of Mathematical Physics. Volume I. New York: John Wiley & Sons Inc., 105–107.

As studied in Sect. 7.2.1, the dynamic equation of a non-uniform rod reads

$$\rho A(x)\frac{\partial^2 u(x,t)}{\partial t^2} - \frac{\partial}{\partial x}\left[EA(x)\frac{\partial u(x,t)}{\partial x}\right] = 0, \quad x \in [0, L], \quad t \in [0, +\infty)$$

$$(7.3.42)$$

where ρ is material density, E is Young's modulus of material, $A(x) > 0$ is the cross-sectional area. Similar to the study in Sect. 7.3.1, the initial conditions of contact dynamics of the rod are

$$C1: \; u(x, 0) = 0, \quad C2: \; u_t(x, 0) = -v_0, \quad x \in [0, L] \qquad (7.3.43a)$$

The boundary conditions of the rod in contact stage are

$$C3: \; u_x(L, t) = 0, \quad C4: \; u(0, t) = 0, \quad t \in [0, t_c] \qquad (7.3.43b)$$

where t_c is the contact duration to be determined.

7.3.3.1   Modal Series Solution

Let the solution of Eq. (7.3.42) be in the following form of separated variables

$$u(x, t) = \varphi(x)q(t) \qquad (7.3.44)$$

Substituting Eq. (7.3.44) into Eq. (7.3.42), and following Eq. (7.2.5), we have two differential equations as follows

$$\begin{cases} \dfrac{d}{dx}\left[A(x)\dfrac{d\varphi(x)}{dx}\right] + \dfrac{\omega^2}{c_0^2}A(x)\varphi(x) = 0, \quad c_0 \equiv \sqrt{\dfrac{E}{\rho}} \\[4mm] \dfrac{d^2 q(t)}{dt^2} + \omega^2 q(t) = 0 \end{cases} \qquad (7.3.45)$$

Substitution of Eq. (7.3.44) into boundary conditions C3 and C4 in Eq. (7.4.43b) results in the boundary conditions of the first equation in Eq. (7.3.45), namely,

$$\varphi(0) = 0, \quad \varphi_x(L) = 0 \qquad (7.3.46)$$

This subsection focuses on the impact problem of a rod with an exp-cross-section studied in Sect. 7.2.1, where the cross-sectional area was defined as

$$A(x) = A_0 \exp(\alpha x), \quad A_0 > 0, \quad \alpha \geq 0. \qquad (7.3.47)$$

The following analysis deals with the case when $\alpha > 0$, while the numerical case studies will touch upon both cases when $\alpha > 0$ and $\alpha < 0$.

The first step now is to solve the first differential equation in Eq. (7.3.45). As studied in Sect. 7.2.1, the characteristic equation associated with the differential equation has a pair of conjugate complex eigenvalues when $\omega > \omega_c \equiv \alpha c_0/2$, namely,

$$\lambda_{1,2} = -\frac{\alpha}{2} \pm i\kappa, \quad \kappa \equiv \sqrt{\left(\frac{\omega}{c_0}\right)^2 - \left(\frac{\alpha}{2}\right)^2} > 0 \qquad (7.3.48)$$

As such, the general solution of the differential equation reads

$$\varphi(x) = \exp\left(-\frac{\alpha x}{2}\right)[a_1 \cos(\kappa x) + a_2 \sin(\kappa x)] \qquad (7.3.49)$$

Substitution of Eq. (7.3.49) into Eq. (7.3.46) leads to

$$a_1 = 0, \quad \frac{a_2}{2} \exp\left(-\frac{\alpha L}{2}\right)[2\kappa \cos(\kappa L) - \alpha \sin(\kappa L)] = 0 \qquad (7.3.50)$$

There follows the characteristic equation governing wave number κ, namely,

$$\tan(\kappa L) = \frac{2}{\alpha L}(\kappa L) \qquad (7.3.51)$$

For example, if the uniform rod of length $L = 5.189$ m in Example 7.3.1 becomes the rod with an exp-cross-section, then $\alpha = 0.1$ m^{-1} and $\alpha = 0.2$ m^{-1} result in two sets of eigenvalues of (7.3.51) respectively as follows

$$\begin{cases} \alpha L = 0.5189 : & \kappa_1 = 1.386/L, \quad \kappa_2 = 4.657/L, \quad \kappa_3 = 7.821/L \\ \alpha L = 1.0378 : & \kappa_1 = 1.145/L, \quad \kappa_2 = 4.600/L, \quad \kappa_3 = 7.787/L \end{cases} \qquad (7.3.52a)$$

$$\kappa_r \approx \frac{(2r-1)\pi}{2L}, \quad r = 4, 5, 6, ... \qquad (7.3.52b)$$

That is, the exp-cross-section has a negligible influence on the large wave numbers. This property will be discussed later. From Eq. (7.3.52), we have the natural frequencies and mode shapes of the rod with an exp-cross-section as follows

$$\omega_r = c_0\sqrt{\left(\frac{\alpha}{2}\right)^2 + \kappa_r^2}, \quad \varphi_r(x) = \exp\left(-\frac{\alpha x}{2}\right)\sin(\kappa_r x), \quad r = 1, 2, ... \qquad (7.3.53)$$

For the subsequent modal analysis, we derive the orthogonality relations of mode shapes now. Let the first differential equation in Eq. (7.3.45) be substituted with $\varphi_r(x)$ and multiplied by $\varphi_s(x)$, and then integrated over the rod length, we derive, via the integration by parts and the boundary condition in Eq. (7.3.46), the following relation

$$\int_0^L \frac{d}{dx}\left[A(x)\frac{d\varphi_r(x)}{dx}\right]\varphi_s(x)dx + \left(\frac{\omega_r}{c_0}\right)^2\int_0^L A(x)\varphi_r(x)\varphi_s(x)dx$$

$$= \left(\frac{\omega_r}{c_0}\right)^2\int_0^L A(x)\varphi_r(x)\varphi_s(x)dx - \int_0^L A(x)\frac{d\varphi_r(x)}{dx}\frac{d\varphi_s(x)}{dx}dx = 0$$

(7.3.54)

Exchanging the subscripts in Eq. (7.3.54) leads to

$$\left(\frac{\omega_s}{c_0}\right)^2\int_0^L A(x)\varphi_r(x)\varphi_s(x)dx - \int_0^L A(x)\frac{d\varphi_r(x)}{dx}\frac{d\varphi_s(x)}{dx}dx = 0 \qquad (7.3.55)$$

Let Eq. (7.3.54) be subtracted by Eq. (7.3.55), we have two orthogonal relations as follows

$$\begin{cases} \int_0^L A(x)\varphi_r(x)\varphi_s(x)dx = 0, \\ \int_0^L A(x)\frac{d\varphi_r(x)}{dx}\frac{d\varphi_s(x)}{dx}dx = 0, \quad r \neq s, \quad r,s = 1,2,\dots \end{cases} \qquad (7.3.56)$$

In addition, according to Eqs. (7.3.48) and (7.3.53), we get

$$\int_0^L A(x)\,\varphi_r^2(x)\,dx = A_0\int_0^L \sin^2(\kappa_r x)\,dx = \frac{A_0[2L\kappa_r - \sin(2\kappa_r L)]}{4\kappa_r}, \quad r = 1,2,\dots$$

(7.3.57)

To start the second step, we solve the second differential equation in Eq. (7.3.45) for a given natural frequency ω_r and get the corresponding time history, namely,

$$q_r(t) = b_{1r}\cos(\omega_r t) + b_{2r}\sin(\omega_r t), \quad r = 1,2,\dots \qquad (7.3.58)$$

where the coefficients b_{1r}, b_{2r}, $r = 1,2,\dots$ are the integration constants to be determined. Hence, the general solution of Eq. (7.3.42) under the boundary conditions C3 and C4 in Eq. (7.3.43b) is

$$u(x,t) = \sum_{r=1}^{+\infty}\varphi_r(x)q_r(t) = \sum_{r=1}^{+\infty}\exp\left(-\frac{\alpha x}{2}\right)\sin(\kappa_r x)[b_{1r}\cos(\omega_r t) + b_{2r}\sin(\omega_r t)]$$

(7.3.59)

Now, we determine the coefficients in Eq. (7.3.59) from initial conditions C1 and C2 in Eq. (7.3.43a). Hence, we take Eq. (7.3.59) as a Fourier series of coordinate

t and differentiate each term with respect to t. It is hard to prove the rationality of this procedure in mathematics. In the study of equations of mathematical physics[22], however, it is a usual practice to derive the formal solutions first and then verify the correctness.

Substitution of Eq. (7.3.59) and its partial derivative with respect to time t into initial conditions C2 and C2 in Eq. (7.3.43a) leads to

$$
\begin{cases}
u(x,0) = \displaystyle\sum_{r=1}^{+\infty} b_{1r} \exp\left(-\frac{\alpha x}{2}\right) \sin(\kappa_r x) = 0 \\
u_t(x,0) = \displaystyle\sum_{r=0}^{+\infty} b_{2r}\omega_r \exp\left(-\frac{\alpha x}{2}\right) \sin(\kappa_r x) = -v_0
\end{cases}
\tag{7.3.60}
$$

Let Eq. (7.3.60) have a new subscript s replaced, be multiplied by $A(x)\varphi_r(x)$ and integrated over the rod length, we derive, via the first equation in Eq. (7.3.56) and Eq. (7.3.57), the following relations

$$
\begin{cases}
b_{1r} = 0, \\
\dfrac{A_0[2L\kappa_r - \sin(2\kappa_r L)]\omega_r b_{2r}}{4\kappa_r} = -v_0 A_0 \displaystyle\int_0^L \exp\left(\frac{\alpha x}{2}\right) \sin(\kappa_r x)\mathrm{d}x, \quad r = 1, 2, \ldots
\end{cases}
\tag{7.3.61}
$$

Performing the integration in Eq. (7.3.61) and using Eq. (7.3.50) and Eq. (7.3.57) for simplification, we have

$$
\begin{aligned}
b_{2r} &= \frac{8v_0\kappa_r\{\exp(\alpha L/2)[2\kappa_r \cos(\kappa_r L) - \alpha \sin(\kappa_r L)] - 2\kappa_r\}}{\omega_r[2L\kappa_r - \sin(2\kappa_r L)](\alpha^2 + 4\kappa_r^2)} \\
&= -\frac{16v_0\kappa_r^2}{\omega_r(\alpha^2 + 4\kappa_r^2)[2L\kappa_r - \sin(2\kappa_r L)]}, \quad r = 1, 2, \ldots
\end{aligned}
\tag{7.3.62}
$$

Substitution of the first equation in Eq. (7.3.61) and Eq. (7.3.62) into Eq. (7.3.60) leads to the modal series solution of the rod displacement as follows

$$
u(x,t) = -32v_0 \sum_{r=1}^{+\infty} \frac{\kappa_r^2 \exp(-\alpha x/2) \sin(\kappa_r x) \sin(\omega_r t)}{\omega_r(\alpha^2 + 4\kappa_r^2)[2L\kappa_r - \sin(2\kappa_r L)]}
\tag{7.3.63}
$$

The corresponding modal series solutions of velocity and strain are as following

$$
u_t(x,t) = -32v_0 \sum_{r=1}^{+\infty} \frac{\kappa_r^2 \exp(-\alpha x/2) \sin(\kappa_r x) \cos(\omega_r t)}{(\alpha^2 + 4\kappa_r^2)[2L\kappa_r - \sin(2\kappa_r L)]}
\tag{7.3.64}
$$

[22] Tveito A, Winther R (2005). Introduction to Partial Differential Equations. Berlin: Springer-Verlag, 90–106.

$$u_x(x, t) = -16v_0 \sum_{r=1}^{+\infty} \frac{\kappa_r^2 \exp(-\alpha x/2)[2\kappa_r \cos(\kappa_r x) - \alpha \sin(\kappa_r x)] \sin(\omega_r t)}{\omega_r(\alpha^2 + 4\kappa_r^2)[2L\kappa_r - \sin(2\kappa_r L)]}$$

$$(7.3.65)$$

As stated above, it is necessary to verify the above formal solutions.

7.3.3.2   Convergence of Modal Series Solutions

To check the convergence of displacement series solution, let Eq. (7.3.63) be multiplied by $\exp(\alpha x/2)$ such that the right-hand side of Eq. (7.3.63) becomes a Fourier series. From Eq. (7.3.53), the general term of this Fourier series, for a sufficiently large r, satisfies

$$\left| \frac{\kappa_r^2 \sin(\kappa_r x) \sin(\omega_r t)}{\omega_r(\alpha^2 + 4\kappa_r^2)[2L\kappa_r - \sin(2\kappa_r L)]} \right| \approx \left| \frac{\kappa_r^2 \sin(\kappa_r x) \sin(\omega_r t)}{2L\kappa_r \omega_r(\alpha^2 + 4\kappa_r^2)} \right| < \left| \frac{1}{16 c_0 L \kappa_r^2} \right|$$

$$(7.3.66)$$

In view of Eq. (7.3.52b), the series $\sum_{r=1}^{+\infty} 1/\kappa_r^2$ is convergent. Thus, the Fourier series is uniformly convergent to $\exp(\alpha x/2)u(x, t)$ according to the M-criterion of function series in mathematical analysis. Hence, Eq. (7.3.63) is uniformly convergent to $u(x, t)$. In addition, the convergence rate of Eq. (7.3.63) is $O(1/\kappa_r^2 L^2) = O[1/(2r - 1)^2]$ for a sufficiently large r.

It is not difficult to see that the coefficients of velocity series solution and the coefficients of strain series solution, like the impact problem of a uniform rod in Sect. 7.3.2, have a decreasing rate $O(1/\kappa_r L) = O[1/(2r - 1)]$ with mode order r. Yet, it is hard to prove the uniform convergence of the two series solutions in mathematics. The following case study indicates that their convergence is similar to those in the case of a uniform rod.

7.3.3.3   Case Studies and Wave Dispersions

Example 7.3.2 Take the uniform steel rod in Example 7.3.1 as a reference and study the wave problem of the impacting rod with an exp-cross-section.

For this purpose, copy the parameters in Example 7.3.1 as follows

$$E = 210 \text{ GPa}, \quad \rho = 7800 \text{ kg/m}^3, \quad L = 5.189 \text{ m}, \quad v_0 = 5.189 \text{ m/s} \quad \text{(a)}$$

Substitution of Eq. (a) into Eqs. (7.3.1) and (7.3.18) gives the longitudinal wave speed $c_0 \approx 5189$ m/s and the contact duration $t_c \approx 0.002$ s for the uniform rod.

Now, select two rods with an exp-cross-section for $\alpha = 0.1$ m^{-1} and $\alpha = 0.2$ m^{-1}, and name them a thin rod and a thick rod, respectively. In view of $L = 5.189$ m, the ratio of the cross-sectional areas of a rod at two ends yields

$$\frac{A(L)}{A_0} = \exp(\alpha L) = \begin{cases} 1.680, & \alpha = 0.1 \text{ m}^{-1} \\ 2.823, & \alpha = 0.2 \text{ m}^{-1} \end{cases} \tag{b}$$

Hence, the rod with an exp-cross-section in Example 7.2.2 is between the two rods.

From Eq. (7.2.15), we have the phase velocity of the r-th natural vibration as follows

$$c_{LP}(r) = c_0\sqrt{1 + \left(\frac{\alpha}{2\kappa_r}\right)^2} = c_0\sqrt{1 + \frac{(\alpha L)^2}{\pi^2(2r-1)^2}} \tag{c}$$

Take the thick rod as an example, $\alpha L = 1.0378$ leads to $c_{LP}(1) = 1.053c_0$. In addition, if the difference between the phase velocity of the thick rod with an exp-cross-section and that of a uniform rod is expected to satisfy $|c_{LP}(r)/c_0 - 1| < 0.001$, solving Eq. (c) for r gives $r = 5$. This implies that the wave dispersion of the thick rod with an exp-cross-section mainly occurs for a few natural vibrations with small wave numbers. According to the discussion about the convergence, the dispersion in a lower frequency band and the results in Example 7.3.1, we take the truncated mode order as $n = 60$.

Figure 7.21 presents the displacement distributions along the rod when $t = 0.1t_c$ and $t = 0.9t_c$ for a uniform rod with $\alpha = 0.0$ m^{-1}, the thin rod with $\alpha = 0.1$ m^{-1} and the thick rod with $\alpha = 0.2$ m^{-1}, respectively. At the fist moment, the waves travel a short moment and the wave dispersions of both non-uniform rods are not obvious. That is, the displacements of three rods are almost identical. At the second moment, the waves return to a vicinity of the left end after a reflection at the right end, and the influence of an exp-cross-section becomes obvious. In particular, the displacement of the thick rod with $\alpha = 0.2$ m^{-1} severely deviates from that of a uniform rod and exhibits strong wave dispersions.

Figure 7.22 displays the velocity histories of three rods at two cross-sections

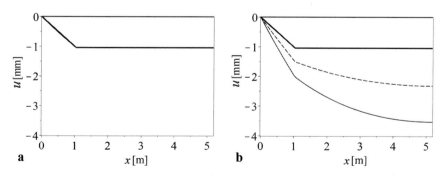

Fig. 7.21 A comparison of displacement distributions among rods at two moments (thick solid line: $\alpha = 0$ m^{-1}, dashed line: $\alpha = 0.1$ m^{-1}, thin solid line: $\alpha = 0.2$ m^{-1}). **a** $t = 0.1t_c$; **b** $t = 0.9t_c$

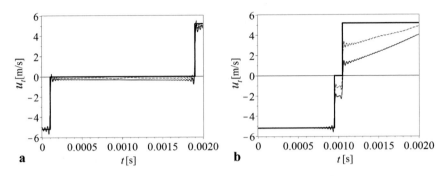

Fig. 7.22 A comparison of velocity histories of three rods at two positions (thick solid line: $\alpha = 0\,\mathrm{m}^{-1}$, dashed line: $\alpha = 0.1\,\mathrm{m}^{-1}$, thin solid line: $\alpha = 0.2\,\mathrm{m}^{-1}$). **a** $x = 0.1L$; **b** $x = 0.95L$

$x = 0.1L$ and $x = 0.95L$, respectively. In the left cross-section, the velocity waves of both non-uniform rods undergo upward jumps from $-v_0$ in the characteristic line but do not reach zero like a uniform rod, and their second upward jumps do not reach v_0 like a uniform rod either. In the right cross-section, the velocity waves of two non-uniform rods exhibit significant distortions and their upward jumps are much lower than the velocity jumps of a uniform rod. As shown in Fig. 7.22b, the thick rod exhibits a stronger wave dispersion.

Figure 7.23 shows the strain distributions of three rods at two moments $t = 0.1\,t_c$ and $t = 0.9t_c$, respectively. Similar to Fig. 7.21, the wave dispersions of two uniform rods are not obvious at the first moment, the fluctuations of two approximate solutions come from the modal truncation of Fourier series. At the second moment, however, the strain waves return to a vicinity of the left end after a reflection at the right end and wave dispersions of two non-uniform rods, especially the thick rod, are significant.

As shown in Fig. 7.23b, the compressional strains of two non-uniform rods at the left end are much larger than that of the uniform rod at moment $t = 0.9t_c$. In particular, the left ends of the non-uniform rods are under a compression when $t = t_c$ such that they can not rebound from the rigid wall.

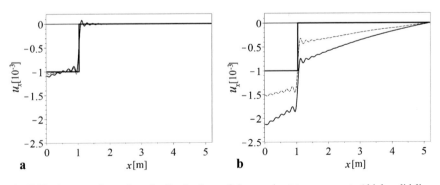

Fig. 7.23 A comparison of strain distributions of three rods at two moments (thick solid line: $\alpha = 0\,\mathrm{m}^{-1}$, dashed line: $\alpha = 0.1\,\mathrm{m}^{-1}$, thin solid line: $\alpha = 0.2\,\mathrm{m}^{-1}$). **a** $t = 0.1t_c$; **b** $t = 0.9t_c$

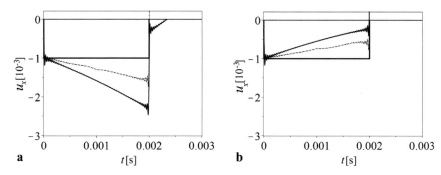

Fig. 7.24 A comparison of stain histories of three rods at the left end (thick solid line: $\alpha = 0\,\text{m}^{-1}$, dashed line: $|\alpha| = 0.1\,\text{m}^{-1}$, thin solid line: $|\alpha| = 0.2\,\text{m}^{-1}$). **a** $\alpha > 0$; **b** $\alpha < 0$

Remark 7.3.8 To gain an insight into this case, we check the strain histories of three rods at their left ends as shown in Fig. 7.24. As shown in Fig. 7.24a, the compressional strain of the thin rod at the left end disappears at moment $t = t_c = 0.002$ s, but that moment is postponed to $t \approx 1.17t_c = 0.00234$ s for the thick rod. From the intuition of mechanics, the cross-sectional area of the thick rod at the left end is much smaller than that at the right end. Hence, the compressional strain of the rod at the left end is larger and the rod can hardly rebound at moment $t = t_c = 0.002$ s. Figure 7.24b presents the strain histories of the rod with exp-cross-sections of $\alpha = -0.1\,\text{m}^{-1}$ and $\alpha = -0.2\,\text{m}^{-1}$. Now, the cross-sectional area of the rod at the left end is larger than that at the right end. As such, the compressional strain of the non-uniform rod at the left end is smaller than that of the uniform rod. Hence, the contact duration of the non-uniform rod becomes a little bit shorter.

7.3.4 Influence of Transverse Inertia

In the dynamic analysis of an impacting rod, it is essential to understand the validity of the elementary theory of one-dimensional waves because the waves associated with an impacting problem fall into a wide frequency band, while the elementary theory does not account for wave dispersions. This subsection presents how to study the wave dispersion due to the transverse inertia of Love's rod, which is horizontally impacting a rigid wall.

7.3.4.1 Modal Series Solution

As studied in Sect. 7.2.2, the dynamic equation of Love's rod with transverse inertia taken into account yields

$$\frac{\partial^2 u(x,t)}{\partial t^2} = c_0^2 \frac{\partial^2 u(x,t)}{\partial x^2} + v^2 r_p^2 \frac{\partial^4 u(x,t)}{\partial x^2 \partial t^2}, \quad x \in [0, L], \quad t \in [0, +\infty)$$

(7.3.67)

where $c_0 \equiv \sqrt{E/\rho}$, v is Poisson's ratio of material, r_p is the polar radius of gyration. Following the description about a uniform rod horizontally impacting a rigid wall, the initial conditions of Love's rod at the beginning of the contact with the wall satisfy

$$C1: \ u(x,0) = 0, \quad C2: \ u_t(x,0) = -v_0, \quad x \in [0, L] \qquad (7.3.68a)$$

while Love's rod during the contact with the wall yields the boundary conditions as follows

$$C3: \ c_0^2 u_x(L,t) + v^2 r_p^2 u_{xtt}(L,t) = 0, \quad C4: \ u(0,t) = 0, \quad t \in [0, t_c] \ (7.3.68b)$$

where t_c is the contact duration.

As a matter of fact, Eqs. (7.3.67) and (7.3.68) describe the free vibration of Love's rod under fixed-free boundaries studied in Sect. 7.2.2, where Eq. (7.2.61) gave the general solution of the free vibration as follows

$$u(x,t) = \sum_{r=1}^{+\infty} \sin(\kappa_r x)[b_{1r} \cos(\omega_r t) + b_{2r} \sin(\omega_r t)] \qquad (7.3.69)$$

where

$$\kappa_r = \frac{(2r-1)\pi}{2L}, \quad \omega_r = c_0 \kappa_r \sqrt{\frac{1}{1 + v^2 r_p^2 \kappa_r^2}}, \quad r = 1, 2, \dots \qquad (7.3.70)$$

They are the r-th wave number and the r-th natural frequency, respectively.

Substituting Eq. (7.3.69) and the time differivative of Eq. (7.3.69) into the two initial conditions C1 and C2 in Eq. (7.3.68a), we have

$$\begin{cases} u(x,0) = \sum_{r=1}^{+\infty} b_{1r} \sin(\kappa_r x) = 0 \\[2mm] u_t(x,0) = \sum_{r=0}^{+\infty} b_{2r}\omega_r \sin(\kappa_r x) = -v_0 \end{cases} \qquad (7.3.71)$$

Similar to the deduction procedure from Eqs. (7.3.29) to (7.3.31), we derive the Fourier coefficients in Eq. (7.3.71) as follows

$$b_{1r} = 0, \quad b_{2r} = -\frac{2v_0}{\omega_r L} \int_0^L \sin(\kappa_r x) dx = -\frac{2v_0}{L \omega_r \kappa_r}, \quad r = 1, 2, \dots \qquad (7.3.72)$$

Substitution of Eq. (7.3.72) into Eq. (7.3.69) leads to the modal series solution of displacement for the rod impacting a rigid wall, namely,

$$
\begin{aligned}
u(x, t) &= -\frac{2v_0}{L} \sum_{r=1}^{+\infty} \frac{1}{\omega_r \kappa_r} \sin(\kappa_r x) \sin(\omega_r t) \\
&= -\frac{2v_0}{c_0 L} \sum_{r=1}^{+\infty} \frac{\sqrt{1 + v^2 r_p^2 \kappa_r^2}}{\kappa_r^2} \sin(\kappa_r x) \sin\left(\frac{c_0 \kappa_r t}{\sqrt{1 + v^2 r_p^2 \kappa_r^2}}\right)
\end{aligned}
\tag{7.3.73}
$$

The modal series solutions of velocity and strain take the following forms

$$
u_t(x, t) = -\frac{2v_0}{L} \sum_{r=1}^{+\infty} \frac{1}{\kappa_r} \sin(\kappa_r x) \cos\left(\frac{c_0 \kappa_r t}{\sqrt{1 + v^2 r_p^2 \kappa_r^2}}\right)
\tag{7.3.74}
$$

$$
u_x(x, t) \sim -\frac{2v_0}{c_0 L} \sum_{r=1}^{+\infty} \frac{\sqrt{1 + v^2 r_p^2 \kappa_r^2}}{\kappa_r} \cos(\kappa_r x) \sin\left(\frac{c_0 \kappa_r t}{\sqrt{1 + v^2 r_p^2 \kappa_r^2}}\right)
\tag{7.3.75}
$$

The subsequent discussion will show that the modal series sulotion in Eq. (7.3.75) is not convergent. Thus, symbol $\sim$ is used here to stand for "may be equal to".

7.3.4.2 Convergence of Modal Series Solutions

We begin with the convergence of displacement series solution. As shown in Eq. (7.3.73), the dimensionless coefficients in the series approach zero at a decreasing rate $O(v r_p/\kappa_r L) = O[1/(2r - 1)]$ with mode order r such that the M-criterion used in previous subsections fails to justify the uniform convergence of Eq. (7.3.73). However, taking Eq. (7.3.73) as a Fourier series and utilizing the theory of Fourier series, we can confirm that the displacement series solution is uniformly convergent to $u(x, t)$.

Next, Eq. (7.3.74) indicates that the dimensionless coefficients approach zero at rate $O(1/\kappa_r L) = O[1/(2r - 1)]$. As the velocity solution is not continuous, the series solution is convergent at most, but not uniformly convergent, to $u_t(x, t)$.

Afterwards, as shown in Eq. (7.3.75), the dimensionless coefficients are convergent to a nonzero value such that the strain series solution is not convergent. This issue will be further discussed in Sect. 7.3.5.

In the above modal superposition method, the general solution of free vibrations of the rod in Eq. (7.3.69) is a Fourier series. Based on proper selection of the Fourier coefficients, the Fourier series is convergent to the initial displacement and velocity of the rod as shown in Eq. (7.3.71), but rather than the initial strain of the rod.

That is, the convergence of strain series solution depends on the convergence rate of displacement series solution.

Furthermore, Eq. (7.3.75) is the superposition of an infinite number of traveling waves, but wave dispersion makes the phase velocity of those traveling waves be different. In this case, the space coordinate x and time t are not in equivalent positions. That is, the convergence of strain series solution and that of velocity series solution are independent.

7.3.4.3 Case Studies and Wave Dispersions

Example 7.3.3 Consider the uniform rod impacting a rigid wall in Example 7.3.1 and study the wave propagations of Love's rod model with transverse inertia taken into account.

To check the effect of transverse inertia of Love's rod model, take two uniform rods, which have a circular cross-section in radius $d = 0.1$ m and $d = 0.2$ m respectively, and call them a thin Love's rod and a thick Love's rod for short. The polar radii of gyration for two Love's rods are as follows

$$r_p = \frac{d}{\sqrt{8}} = 0.0354 \text{ m}, \ 0.0708 \text{ m} \tag{a}$$

The other parameters of two rods are the same as those in Example 7.3.1, namely,

$$\rho = 7800 \text{ kg/m}^3, \quad E = 210 \text{ GPa}, \quad v = 0.28, \quad L = 5.189 \text{ m}, \quad v_0 = 5.189 \text{ m/s} \tag{b}$$

Recalling Example 7.3.1, we have the longitudinal wave speed $c_0 \approx 5189$ m/s and the contact duration $t_c \approx 0.002$ s for the impact problem of a uniform rod without transverse inertia taken into consideration.

As discussed in Sect. 7.2.2, Love's rod model is good enough if the wave number yields $\kappa_r \leq 1/(vr_p)$. Substitution of $\kappa_r \leq 1/(vr_p)$ into the first equation in Eq. (7.3.70) leads to an inequality for the maximal mode number r, which can serve as the maximal truncated mode order denoted as n_{max}. In the case of the thick Love's rod, for example, there follows

$$n_{max} = \left(\frac{L}{\pi v r_p} + \frac{1}{2} \right) \approx 84 \tag{c}$$

Regarding the limit of Eq. (c) and the convergence rates of displacement and velocity series solutions, we take $n = 13$ and $n = 26$ for Eqs. (7.3.73), (7.3.74) and (7.3.75). Figure 7.25 shows the displacement distributions of the thick Love's rod at moment $t = 0.25t_c$. Now, the thick solid line represents the solution of elementary theory in Eq. (7.3.15), the thin solid line and dashed line are the approximate modal solutions

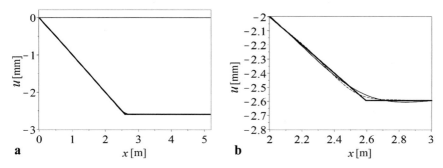

Fig. 7.25 The displacement distributions of a thick Love's rod when $t = 0.25t_c$ (thick solid line: the elementary theory, thin solid line: an approximate solution of $n = 13$, dashed line: an approximate solution of $n = 26$). **a** the global view; **b** the zoomed view aroud the conner

of Love's rod when $n = 13$ and $n = 26$, respectively. In Fig. 7.25a, the approximate modal solution is very close to the solution of elementary theory. In Fig. 7.25b, the difference between the approximate modal solution and the solution of elementary theory decreases with increasing of n, showing the convergence of displacement series solution, as well as a small influence of transverse inertia on the displacement waves.

Figure 7.26 presents the velocity and strain distributions of a thick Love's rod at moment $t = 0.25t_c$. Here, the solutions of elementray theory come from Eqs. (7.1.16) and (7.1.17) and do not take the transverse inertial into account. The approximate velocity and strain solutions exhibit similar phenomena when n increases. That is, the fluctuations of them decrease in a flat region of the solution of elementray theory, while the fluctuations do not decrease, but exhibit Gibbs effect and even increase near the discontinuity of the solution of elementray theory. In this case, the convergence of

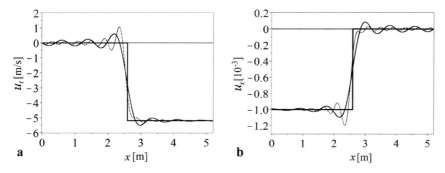

Fig. 7.26 The velocity and strain distributions of a thick Love's rod at moment $t = 0.25t_c$ (thick solid line: the elementary theory, thin solid line: an approximate solution of $n = 13$, dashed line: an approximate solution of $n = 26$). **a** the velocity distribution; **b** the strain distribution

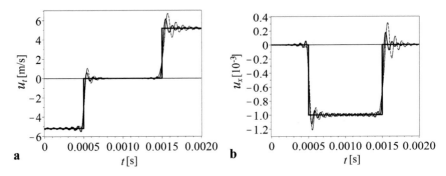

Fig. 7.27 A comparison of wave dispersions of a thin Love's rod and a thick Love's rod at the middle-span (thick solid line: the elementary theory, thin solid line: the thin Love's rod, dashed line: the thick Love's rod). **a** the velocity history; **b** the strain history

approximate velocity solution is slow when n goes up to n_{max}. The approximate strain solution, albeit divergent, behaves like the approximate velocity solution. Sect. 7.3.5 will address this issue further.

As the influence of transverse inertia on the displacement wave is not very significant, Fig. 7.27 gives the comparison of velocity and strain histories at the middle-span $x = L/2$ for the elementary theory, both thin and thick Love's rods. Now, both velocity and strain waves undergo distortions, and the wave distortion of the thick Love's rod looks significant, demonstrating the wave dispersion of higher frequencies caused by the transverse inertia.

7.3.5   Criteria of Modal Truncation

To address Problem 6B proposed in Chap. 1, this subsection reveals the rules behind the computational results of the impacting wave of a uniform rod without dispersion and the dissipersive impacting waves of a rod with an exp-cross-section and Love's rod, and addresses the problem of modal truncation.

In the impact problem of the above three rods with a rigid wall, both velocity and strain waves exhibit discontinuity, associated with Problem 6B proposed in Chap. 1. Alhough the following studies only focus on the longitudinal dynamics of three rod models, they deal with the weak solutions with discontinuity and play significant roles in the dynamic computations of structural impacts. Furthermore, the longitudinal wave of a uniform rod has an exact solution of elementary theory, and can serve as the reference for the modal truncations of an impacting rod with an exp-cross-section or with transverse inertia taken into account.

7.3.5.1 Convergence of Displacement Series Solutions

As studied in previous subsections, the displacement series solutions of the above three rods based on modal superposition are uniformly convergent, but have two different cases.

(1) Fast uniform convergence

In Sects. 7.3.2 and 7.3.3, it has been proved via the M-criterion of function series that the displacement series solutoins of a uniform rod and a rod with an exp-cross-section are uniformly convergent at a convergence rate $O[1/(2n-1)^2]$, where n is the truncated mode order.

In practice, the displacement series solution can be said to have the *fast uniform convergence* if the convergence rate is $O(1/n_c^\alpha)$, where $\alpha > 3/2$ and n_c is the *truncated number* of series solution. If the truncated number satisfies $1/n_c^\alpha = 1/100$, then $n_c = (100)^{1/\alpha}$ holds. For example, the displacement series solution of a uniform rod in Sect. 7.3.2 had a convergence rate $O(1/n_c^\alpha) = O[1/(2n-1)^2]$. When $2n - 1 = n_c = 100^{1/2} = 10$ is required, $n = 6$ ensures the convergence rate. This is just the truncated mode order taken in Example 7.3.1. If $\alpha = 3/2$ holds, then $2n - 1 = n_c = 100^{2/3} \approx 22$ leads to $n = 12$.

(2) Slow uniform convergence

As studied in Sect. 7.3.4, the convergence rate of the displacement series solution of Love's rod decreases to $O[1/(2n-1)]$ such that the M-criterion of function series fails to confirm the uniform convergence. From the theory of Fourier series, however, the displacement continuity ensures the *slow uniform convergence* of series solution.

In this case, if the truncated number of displacement series solution yields $1/n_c^\alpha = 1/100$, then $2n - 1 = n_c = 100$ leads to $n = 51$. In Example 7.3.3, $n = 26$ was selected so as to give consideration of the strain series solution. In fact, further improvement of solution accuracy requires much higher cost of computations.

The discussions above can be generalized to the impact computations of other structures. For example, the mode shapes of a hinged-hinged beam and a thin rectangular plate with four edges hinged yield either a sinusoidal function or a product of two sinusoidal functions, the theory of Fourier series guarantees the uniform convergence of their displacement series solutions in impacting problems. The displacements of other beams and plates at a cross-section far from their boundaries have the similar properties.

7.3.5.2 Convergence of Velocity and Strain Series Solutions

As the convergence rate of Fourier series of a periodic function mainly relies on the function smoothness such that the convergence rate of a discontinuous velocity series solution is slower than that of a displacement series solution by an order, so is that of a discontinuous strain series solution. The convergence rate of discontinuous solutions of impacting response can also be divided into two cases.

(1) Slow convergence

Consider a uniform rod described and a rod with an exp-cross-section, their velocity and strain series solutions are convergent, but the convergence rates decrease to $O[1/(2n - 1)]$, compared with the convergence rate $O[1/(2n - 1)^2]$ of their displacement series solutions. In particular, the velocity and strain series solutions exhibit fluctuations near their discontinuities. To increase the accuracy of a series solution, previous studies usually increased the truncated mode order n, and some studies even selected $n = 1000$.

In fact, the attempts in those studies have given or may give rise to the following problems. First, the Fourier series solution of a discontinuous solution exhibits Gibbs phenomenon so that increasing the truncated mode order can not reduce the fluctuation peaks. Second, the computation cost increases by an order if one tries to improve the accuracy of results by one decimal place. Third, most dynamic models are not able to provide reliable mode shapes with a large wave number. Thus, increasing mathematical accuracy does not enhance the physical fidelity of the results.

In Examples 7.3.1 and 7.3.2, hence, we did not select very high truncated mode orders and took, say, $n = 60$ only, but verified the model efficacy of the rod with an exp-cross-section.

(2) Quasi-convergence

Section 7.3.4 pointed out that the strain series solution of Love's rod may not be convergent, but still serves as an approximate solution to discuss the influence of transverse inertia on the wave dispersion. One may question whether the treatment makes sense. Now, we introduce the concept of quasi-convergence and discuss the above strain series solution.

The series solution is said to be *quasi-convergent* if it is not convergent, but the absolute values of dimensionless coefficients of the series have a decreasing rate $O(1/n)$ for a wide range of n. From a practical point of view, the quasi-convergent series solution can offer important information to engineers when the absolute values of truncated terms are small enough.

We recall the modal series solutions in Eqs. (7.3.73), (7.3.74) and (7.3.75), and denote the dimensionless coefficient of general term in displacement, velocity and strain series solutions as follows

$$
\begin{cases}
c_d(r) \equiv \dfrac{\sqrt{1 + v^2 r_p^2 \kappa_r^2}}{\kappa_r^2 L^2} = \sqrt{\dfrac{16}{\pi^4 (2r - 1)^4} + \dfrac{4}{\pi^2 (2r - 1)^2}\left(\dfrac{v r_p}{L}\right)^2} \\[4mm]
c_v(r) \equiv \dfrac{1}{\kappa_r L} = \dfrac{2}{\pi(2r - 1)} \\[4mm]
c_s(r) \equiv \dfrac{\sqrt{1 + v^2 r_p^2 \kappa_r^2}}{\kappa_r L} = \sqrt{\dfrac{4}{\pi^2 (2r - 1)^2} + \left(\dfrac{v r_p}{L}\right)^2}
\end{cases}
\tag{7.3.76}
$$

Consider, as an example, the thick Love's rod with $d = 0.2$ m, the dimensionless parameter in Eq. (7.3.76) reads

Fig. 7.28 The decreasing rates of dimensionless coefficients of series solutions of a thick Love's rod with an increase of mode order r (thick solid line: c_d, thin solid line: c_v, dashed line c_s)

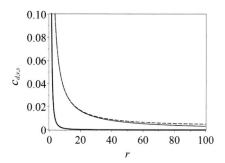

$$\frac{vr_p}{L} = \frac{vd}{\sqrt{8}L} \approx 0.0038 \tag{7.3.77}$$

As this is a very small parameter, the variations of three dimensionless coefficients in a range of $1 \le r \le 100$ are different from those when $r \to +\infty$. For instance, both convergence rates of displacement and velocity series solutions are $O(v\,r_p/\kappa_r L)$ $= O[1/(2r-1)]$ when $r \to +\infty$. In Fig. 7.28, however, the dimensionless coefficient c_d of displacement series solution decreases much faster than the dimensionless coefficients c_v and c_s for velocity and strain series solutions in a range of $1 \le r \le 100$. Figure 7.28 also shows that c_v and c_s have similar decreasing rates for $1 \le r \le 100$. That is, the quasi-convergent strain series solution has an equivalent accuracy of the convergent velocity series solution for $1 \le r \le 100$.

Recalling Eq. (c) in Example 7.3.3, the maximal truncated mode order is $n_{max} = 84$ for the thick Love's model. In this case, it is meaningless to increase the truncated mode order further. As a consequence, the quasi-convergent approximate solution offers practical information to engineers.

7.3.6 Concluding Remarks

This section presented a systematic study on the wave analysis of a rod horizontally impacting a rigid wall, focusing on the feasibility of modal superposition method so as to address Problem 6B proposed in Chap. 1. The major conclusions are as follows.

(1) Based on d'Alembert's solution of one-dimensional wave equation, it is feasible to derive the exact wave solution for the uniform rod impacting the rigid wall with low velocity. The exact displacement solution is a weak solution such that the corresponding velocity and strain solutions exhibit discontinuity in characteristic lines of the wave equation. As shown by the exact wave solution, when the compressional wave travels a round in the rod, the contact between the rod and the rigid wall is over, and the rod is rebounding.

(2) The modal superposition method enables one to derive the modal series solutions for the traveling waves of the uniform rod impacting the rigid wall. As

proved in the section, the displacement series solution is uniformly convergent to the exact solution at a convergence rate $O[1/(2n-1)^2]$ with truncated mode order n, while both velocity and strain series solutions have a convergence rate $O[1/(2n-1)]$ to the corresponding exact solutions.

(3)   The modal superposition method also gives the modal series solutions for the traveling waves of a rod with an exp-cross-section impacting the rigid wall. The modal series solutions enjoy the same convergence rate as those for a uniform rod, but the exp-cross-section of the rod brings the wave dispersion in a lower frequency band. If the small end of the rod impacts the rigid wall, the compressional strain in the rod is larger than that in a uniform rod so that the contact duration of the rod with the rigid wall is slightly longer. The contrary results come to be true if the large end of the rod impacts the rigid wall.

(4)   To check the influence of transverse inertia of a uniform rod impacting a rigid wall, the section presented the modal series solutions of Love's rod impacting the rigid wall, and the discussion about their convergence. The displacement series solution is uniformly convergent and the velocity series solution is convergent, both at a convergence rate $O[1/(2n-1)]$. Because of the cut-off frequency of Love's rod, the strain series solution is not convergent, but exhibits the quasi-convergence in a practical sense, defined in the section. The transverse inertia of Love's rod induces significant dispersions for both velocity and strain waves in a higher frequency band. As such, the truncated mode order has an upper bound.

(5)   In summary, the modal superposition method is able to provide reliable approximate solutions for a rod impacting the rigid wall though the velocity and strain waves exhibit discontinuity. To increase the accuracy of velocity and strain, some studies increased the truncated mode order to several hundreds or one thousand. It is worthy to notice that these efforts did not remove Gibbs effect of a modal series solution near the discontinuity. A more important issue is that the dynamic models used might not offer proper mode shapes of large wave numbers.

7.4   Free Waves and Vibrations of Beams

The bending waves of a beam are more complicated than the axial waves of a rod. For example, the bending waves of an Euler–Bernoulli beam, called an E-B beam or a beam for short, are dispersive. In particular, the model of an E-B beam fails to work if the ratio of the cross-sectional height of the beam to the wave length of a bending wave is not small enough such that both shear deformation and rotary inertia of cross-section of the beam have significant influences on the bending wave. If this is the case, it is necessary to use the model of a Timoshenko-Ehrenfest beam, or called a T-E beam for short.

This section presents the systematic study on the bending dynamics of uniform beams, including their free waves and vibrations, and their forced waves and vibrations. To address Problem 6A proposed in Chap. 1, the study focuses on a comparison between an E-B beam and a T-E beam, and the relations between waves and vibrations.

7.4.1 Free Waves of an Euler–Bernoulli Beam

Consider a uniform E-B beam with length L, cross-sectional area A, the second moment of cross-sectional area $I \equiv r_g^2 A$, material density ρ, and Young's modulus of material E. Establish a frame of coordinates oxv, where axis x, measured from the left end of the beam and coincides with the neutral axis of the beam. The dynamic deflection $v(x, t)$ of the beam at cross-section x amd moment t yields

$$\rho A \frac{\partial^2 v(x,t)}{\partial t^2} + EI \frac{\partial^4 v(x,t)}{\partial x^4} = 0, \quad x \in [0, \ L], \quad t \in [0, \ +\infty) \tag{7.4.1}$$

7.4.1.1 Wave Solution of Complex Functions

Let the dynamic deflection $v(x, t)$ of the beam be the real part of complex function $v_c(x, t)$ in a form of separated variables as follows

$$v(x, t) = \mathrm{Re}[v_c(x, t)], \quad v_c(x, t) = \tilde{v}_c(x) \exp(i\omega t) \tag{7.4.2}$$

where $\omega > 0$ is the wave frequency and $\tilde{v}_c(x)$ is the complex amplitude. Substitution of $v_c(x, t)$ into Eq. (7.4.1) leads to a homogeneous differential equation, valid for an arbitrary moment t, of complex amplitude $\tilde{v}_c(x)$, namely,

$$\frac{\mathrm{d}^4 \tilde{v}_c(x)}{\mathrm{d}x^4} - \kappa^4 \tilde{v}_c(x) = 0, \quad \kappa \equiv \sqrt[4]{\frac{\rho A \omega^2}{EI}} > 0 \tag{7.4.3}$$

where κ is the *wave number* associated with frequency ω.

As studied in the theory of linear ordinary differential equations, the general solution of Eq. (7.4.3) yields

$$\tilde{v}_c(x) = a_1 \exp(-i\kappa x) + a_2 \exp(i\kappa x) + a_3 \exp(-\kappa x) + a_4 \exp(\kappa x) \tag{7.4.4}$$

where a_k, $k = 1, 2, 3, 4$ are the integration constants to be determined from the boundary conditions of the beam. Substituting Eq. (7.4.4) into Eq. (7.4.2) leads to the completed solution of free waves as follows

$$v_c(x,t) = \left[a_1 \exp(-i\kappa x) + a_2 \exp(i\kappa x) + a_3 \exp(-\kappa x) + a_4 \exp(\kappa x)\right] \exp(i\omega t)$$
$$= a_1 \exp[i(\omega t - \kappa x)] + a_2 \exp[i(\omega t + \kappa x)]$$
$$+ a_3 \exp(-\kappa x) \exp(i\omega t) + a_4 \exp(\kappa x) \exp(i\omega t)$$
$$= a_1 \exp[i\kappa(c_{BP}t - x)] + a_2 \exp[i\kappa(c_{BP}t + x)]$$
$$+ a_3 \exp(-\kappa x) \exp(i\omega t) + a_4 \exp(\kappa L) \exp[-\kappa(L - x)] \exp(i\omega t) \quad (7.4.5)$$

where c_{BP} is the *phase velocity* of a bending wave defined as

$$c_{BP} \equiv \frac{\omega}{\kappa} \qquad\qquad (7.4.6)$$

The first two terms in the right-hand side in Eq. (7.4.5) are the *traveling waves* with an oscillatory property, while the rest two terms are the *evanescent waves*, which decay away from the two ends of the beam. These two types of waves are further discussed below.

(1) Evanescent waves

The third term in Eq. (7.4.5) gets the maximum at the left end $x = 0$ of the beam and decays with an increase of x, while the fourth term has the maximum at the right end $x = L$ of the beam and decays with increasing of $L - x$. The waves corresponding to the two terms are evanescent waves, which are similar to the evanescent waves in a rod with an exp-cross-section in a frequency band lower than the cur-off frequency. As such, the two waves are named *near-field waves*. The existence of evanescent waves is reasonable for a beam since it has two boundary conditions at each end. Of course, some boundary conditions, such as a hinged end or sliding end, do not generate evanescent waves. This issue will be further discussed in Sect. 7.4.1.3.

(2) Traveling waves

The first and second terms in Eq. (7.4.5) are the right traveling wave and left traveling wave at phase velocity c_{BP}, or called *far-field waves*. The traveling waves of a uniform beam, different from those of a uniform rod, are dispersive waves. That is, the phase velocity defined in Eq. (7.4.6) depends on the wave frequency or wave number.

To check the relation between phase velocity c_{BP} and wave number κ of the traveling wave of the beam, solving Eq. (7.4.3) for ω leads to

$$\omega = \kappa^2 \sqrt{\frac{EI}{\rho A}} = \sqrt{\frac{E}{\rho}} \sqrt{\frac{I}{A}} \kappa^2 = c_0 r_g \kappa^2 \qquad\qquad (7.4.7)$$

where $c_0 \equiv \sqrt{E/\rho}$ is the longitudinal wave speed of a uniform beam, $r_g \equiv \sqrt{I/A}$ is the radius of gyration of the second moment of cross-sectional area of the beam, or named the *radius of gyration* for short. Substitution of Eq. (7.4.7) into Eq. (7.4.6) gives the phase velocity of the bending wave as follows

$$c_{BP} = \frac{\omega}{\kappa} = c_0 r_g \kappa \tag{7.4.8}$$

From Eq. (7.4.7), it is easy to derive the *group velocity* of the bending wave, namely,

$$c_{BG} = \frac{d\omega}{d\kappa} = 2c_0 r_g \kappa \tag{7.4.9}$$

As $c_{BP} < c_{BG}$ holds, the bending wave of an E-B beam exhibits abnormal dispersion.

Example 7.4.1 Study the phase velocity and group velocity of a hinged-hinged beam with a circular cross-section of radius $\bar{r}$.

The radius of gyration of the beam is

$$r_g = \sqrt{\frac{\pi(2\bar{r})^4/64}{\pi \bar{r}^2}} = \frac{\bar{r}}{2} \tag{a}$$

The wave number of the r-th natural vibration satisfies

$$\kappa_r = \frac{r\pi}{L}, \quad r = 1, 2, \dots \tag{b}$$

Substitution of Eq. (a) and Eq. (a) into Eqs. (7.4.8) and (7.4.9) leads to the phase velocity and group velocity of the r-th natural vibration as follows

$$c_{BP}(r) = c_0 \frac{r\pi\bar{r}}{2L}, \quad c_{BG} = c_0 \frac{r\pi\bar{r}}{L} \tag{c}$$

In practice, the E-B beam is a proper model for slender beams satisfying $\bar{r}/L \ll 1$. As will be shown in Sect. 7.4.3, the E-B beam is only applicable to describing the natural vibrations of lower order r, say, $r < L/(\pi\bar{r})$. Substituting $r < L/(\pi\bar{r})$ into Eq. (c) leads to $c_{BP}(r) < c_{BG}(r) < c_0$. That is, both phase velocity and group velocity of a natural bending vibration of lower order are lower than the longitudinal wave speed. This phenomenon is reasonable from the theory of three-dimensional waves in Appendix A.1.2. Yet, neither phase velocity nor group velocity in Eq. (c) makes sense, exhibiting $c_{BP} \to +\infty$ and $c_{BG} \to +\infty$, when $r \to +\infty$. This issue will be further discussed in Sect. 7.4.4.

7.4.1.2 Wave Solution of Real Functions

It seems feasible to take the real part of Eq. (7.4.5) and introduce new integration constants in order to derive the solution of real functions for the bending wave of an E-B beam, but the result will be very complicated. To simplify the result, denote the real expression of Eq. (7.4.2), with an initial phase angle θ introduced, as follows

$$v(x, t) = \tilde{v}(x) \cos(\omega t + \theta) \tag{7.4.10}$$

Thus, Eq. (7.4.3) becomes a differential equation of real unknown function $\tilde{v}(x)$

$$\frac{\mathrm{d}^4 \tilde{v}(x)}{\mathrm{d}x^4} - \kappa^4 \tilde{v}(x) = 0, \quad \kappa \equiv \sqrt[4]{\frac{\rho A \omega^2}{EI}} > 0 \tag{7.4.11}$$

The solution of Eq. (7.4.11), thus, takes the following form

$$\begin{aligned} \tilde{v}(x) &= a_1 \exp(-\mathrm{i}\kappa x) + a_2 \exp(\mathrm{i}\kappa x) + a_3 \exp(-\kappa x) + a_4 \exp(\kappa x) \\ &= b_1 \cos(\kappa x) + b_2 \sin(\kappa x) + b_3 \cosh(\kappa x) + b_4 \sinh(\kappa x) \end{aligned} \tag{7.4.12}$$

where a_k and b_k, $k = 1, 2, 3, 4$ are the complex constants and real constants respectively. Using Eulerian formula, we recast Eq. (7.4.12) as

$$\begin{aligned} \tilde{v}(x) &= \frac{b_1}{2}\big[\exp(\mathrm{i}\kappa x) + \exp(-\mathrm{i}\kappa x)\big] + \frac{b_2}{2\mathrm{i}}\big[\exp(\mathrm{i}\kappa x) - \exp(-\mathrm{i}\kappa x)\big] \\ &\quad + \frac{b_3}{2}\big[\exp(\kappa x) + \exp(-\kappa x)\big] + \frac{b_4}{2}\big[\exp(\kappa x) - \exp(-\kappa x)\big] \end{aligned} \tag{7.4.13}$$

A comparison between Eqs. (7.4.12) and (7.4.13) leads to the following relations for each complex constant and two real constants

$$a_1 = \frac{b_1 + \mathrm{i}b_2}{2}, \quad a_2 = \frac{b_1 - \mathrm{i}b_2}{2}, \quad a_3 = \frac{b_3 - b_4}{2}, \quad a_4 = \frac{b_3 + b_4}{2} \tag{7.4.14}$$

There follows

$$b_1 = a_1 + a_2, \quad b_2 = \mathrm{i}(a_2 - a_1), \quad b_3 = a_3 + a_4, \quad b_4 = a_4 - a_3 \tag{7.4.15}$$

Substitution of Eq. (7.4.15) into Eq. (7.4.12) gives the solution required.

7.4.1.3   Reflection of Bending Waves at a Beam Boundary

Intuitively, the reflection of bending waves of a beam at its boundary is much more complicated than that of longitudinal waves of a rod. Because of the decaying property of evanescent waves, it is sufficient to study the reflection of a traveling wave incident upon the boundary of an E-B beam. To be more specific, consider that the bending wave of an E-B beam travels in the right direction and undergoes a reflection at the right boundary of the beam as shown in Fig. 7.29, where the rectlinear spring and torsional spring have stiffness coefficients k_v and k_θ, respectively.

According to Eq. (7.4.5) and associated discussions, the superposition of the incident wave traveling in the right direction and two waves reflected at the right boundary can be written as

Fig. 7.29 A semi-infinitely long E-B beam with an elastic boundary

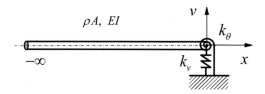

$$v_c(x, t) = a_1 \exp[i(\omega t - \kappa_1 x)] + a_2 \exp[i(\omega t + \kappa_1 x)]$$
$$+ a_4 \exp(i\omega t + \kappa_2 x), \quad x \in (-\infty, \ 0] \tag{7.4.16}$$

where $a_1 \exp[i(\omega t - \kappa_1 x)]$ is the incident wave, $a_2 \exp[i(\omega t + \kappa_1 x)]$ is the reflected traveling wave and $a_4 \exp(i\omega t + \kappa_2 x)$ is the reflected evanescent wave. In view of the bounded wave amplitude, $\kappa_2 > 0$ must hold for $x < 0$. It is worthy to notice that the wave numbers κ_2 and κ_1 for an evanescent wave and a traveling wave may not be identical in general.

As shown in Fig. 7.29, the bending moments of the beam and the shear forces of the beam satisfy the equilibrium equations at the right end as follows

$$EI\,v_{xx}(0, t) + k_\theta v_x(0, t) = 0, \quad EI\,v_{xxx}(0, t) - k_v v(0, t) = 0 \tag{7.4.17}$$

Substitution of Eq. (7.4.16) into Eq. (7.4.17) leads to

$$\begin{cases} (\kappa_1^2 + i\kappa_1\alpha_\theta)a_1 + (\kappa_1^2 - i\kappa_1\alpha_\theta)a_2 - (\kappa_2^2 + \alpha_\theta\kappa_2)a_4 = 0 \\ (i\kappa_1^3 - \alpha_v)a_1 - (i\kappa_1^3 + \alpha_v)a_2 + (\kappa_2^3 - \alpha_v)a_4 = 0 \end{cases} \tag{7.4.18}$$

where

$$\alpha_\theta \equiv \frac{k_\theta}{EI}, \quad \alpha_v \equiv \frac{k_v}{EI} \tag{7.4.19}$$

Solving Eq. (7.4.18) gives two ratios of the reflected wave amplitude to the incident wave amplitude. They are called the *reflection coefficient of traveling wave* and the *reflection coefficient of evanescent wave*, namely,

$$\begin{cases} \gamma_T \equiv \dfrac{a_2}{a_1} = \dfrac{(\kappa_2^3 - \alpha_v)(\kappa_1^2 + i\kappa_1\alpha_\theta) + (\kappa_2^2 + \alpha_\theta\kappa_2)(i\kappa_1^3 - \alpha_v)}{(\kappa_2^2 + \alpha_\theta\kappa_2)(i\kappa_1^3 + \alpha_v) + (\kappa_2^3 - \alpha_v)(i\kappa_1\alpha_\theta - \kappa_1^2)} \\[4mm] \gamma_E \equiv \dfrac{a_4}{a_1} = \dfrac{(i\kappa_1^3 + \alpha_v)(\kappa_1^2 + i\kappa_1\alpha_\theta) + (\kappa_1^2 - i\kappa_1\alpha_\theta)(i\kappa_1^3 - \alpha_v)}{(\kappa_2^2 + \alpha_\theta\kappa_2)(i\kappa_1^3 + \alpha_v) + (\kappa_2^3 - \alpha_v)(i\kappa_1\alpha_\theta - \kappa_1^2)} \end{cases} \tag{7.4.20}$$

To simplify the discussion about Eq. (7.4.20), we focus on four kinds of homogeneous boundary conditions and check the mechanisms of Eq. (7.4.20).

(1) Hinged boundary

For a hinged boundary, $\alpha_\theta = 0$ and $\alpha_v = +\infty$ hold true. Substituion of them into Eq. (7.4.20) leads to the following two reflection coefficients

$$\gamma_T = -1, \quad \gamma_E = 0 \tag{7.4.21}$$

As shown in Eq. (7.4.21), the hinged boundary reflects the harmonic traveling wave incident upon the beam to a harmonic traveling wave of the same amplitude but an opposite phase, and the hinged boundary does not generate any evanescent waves.

(2) Sliding boundary

Now, $\alpha_\theta = +\infty$ and $\alpha_v = 0$, the reflection coefficients in Eq. (7.4.20) become

$$\gamma_T = 1, \quad \gamma_E = 0 \tag{7.4.22}$$

That is, the harmonic traveling wave incident upon the beam undergoes a reflection at the sliding boundary and becomes a harmonic traveling wave of the same amplitude and the same phase, rather than any evanescent waves.

(3) Clamped boundary

In this case, $\alpha_\theta = +\infty$ and $\alpha_v = +\infty$ make the reflection coefficients be

$$\gamma_T = \frac{i\kappa_1 + \kappa_2}{i\kappa_1 - \kappa_2}, \quad \gamma_E = -\frac{2i\kappa_1}{i\kappa_1 - \kappa_2} \tag{7.4.23}$$

That is, the clamped boundary makes the harmonic traveling wave incident upon the beam be a harmonic traveling wave with different amplitude and a phase shift, and produces an evanescent wave reflected.

(4) Free boundary

Substituting $\alpha_\theta = 0$ and $\alpha_v = 0$ into Eq. (7.4.20) gives

$$\gamma_T = \frac{i\kappa_1 + \kappa_2}{i\kappa_1 - \kappa_2}, \quad \gamma_E = \left(\frac{\kappa_1}{\kappa_2}\right)^2 \frac{2i\kappa_1}{i\kappa_1 - \kappa_2} \tag{7.4.24}$$

That is, the free boundary also reflects a harmonic traveling wave incident upon the beam to a harmonic traveling wave with different amplitude and a phase shift, and also induces an evanescent reflected.

Remark 7.4.1 A comparison of the four kinds of boundary conditions shows that the free boundary and clamped boundary have the same reflection coefficient of traveling wave and their coefficients of evanescent wave are different by a factor $-(\kappa_1/\kappa_2)^2$, while both hinged boundary and sliding boundary do not produce evanescent waves and their reflection coefficients are different by a factor -1. Recalling Sect. 5.3, one can find the correspondence between the dual of boundaries and the reflection coefficients. For example, the traveling wave and evanescent wave of an E-B beam with a free boundary or a clamped boundary have the same wave number, namely, $\kappa_1 = \kappa_2$, so that the reflection coefficients of evanescent wave at the free and clamped boundaries are different by a factor -1.

Remark 7.4.2 Some previous studies[23,24] dealt with the reflection of a traveling wave at the right boundary of the beam, but their results seem not correct. The former book, for example, missed an imaginary number i in front of all κ_1 in the reflection coefficients of the free boundary, while the latter book missed a minus in front of κ_2 in all reflection coefficients.

7.4.2 Free Waves of a Timoshenko Beam

From 1911 to 1912, S. P. Timoshenko and P. Ehrenfest, an Ukrainian mechanicist and an Austria physicist in Russia, studied the natural vibrations of a relatively thick beam and improved the dynamic equation of an E-B beam by taking the shear deformation and rotary inertia of the beam into account.

As a matter of fact, either the shear deformation or the rotary inertia had been studied before[25]. For example, the research of the rotary inertia of a vibrating beam can be traced back to the study of Lord Rayleigh, a great British physicist, in 1877, or even back to an earlier research of J. A. C. Bresse, a French mechanicist, in 1859, while the earliest research on the shear deformation of a beam was due to the study of W. Y. M. Rankine in the UK in 1858. Yet, the beam model accounting for both shear deformation and rotary inertia has been referred to as *Timoshenko beam* in academia for almost one century. After recalling the above history, it is preferable to use the term *Timoshenko-Ehrenfest beam* or the term *T-E beam* for short in this text.

In the model of a T-E beam, the assumption of deformation of an E-B beam can be released. That is, the cross-section of a T-E beam remains plane during deformation, but does not need to be perpendicular to the neutral axis of beam.

7.4.2.1 Dynamic Equation

Consider a T-E beam with length L, cross-sectional area A, the second moment of cross-sectional area $I \equiv r_g^2 A$, material density ρ, Young's modulus of material E, and shear modulus of material G. Establish a frame of coordinates oxv in Fig. 7.30, where axis x coincides with the neutral axis of the undeformed beam. Let $v(x, t)$ be the dynamic deflection of the neutral axis of the beam at cross-section x and moment t, $\gamma(x, t)$ be the shear angle of the beam at cross-section x and moment t.

In Fig. 7.30, the solid quadrangle is an infinitesimal segment of the T-E beam. The rotation angle $\theta(x, t)$ of the normal line of the cross-section includes two parts. One is the conventional slope $\partial v(x, t)/\partial x$ of an E-B beam, i.e., the rotation angle of

[23] Doyle J F (1997). Wave Propagation in Structures. New York: Springer-Verlag, 84–86.

[24] Hagedorn P, Das Gupta A (2007). Vibrations and Waves in Continuous Mechanical Systems. Chichester: John Wiley & Sons Inc., 142–143.

[25] Elishakoff I (2020). Who developed the so-called Timoshenko beam theory. Mathematics and Mechanics of Solids, 25(1): 97–116.

Fig. 7.30 The deformation and internal forces of an infinitesimal segment of a T-E beam

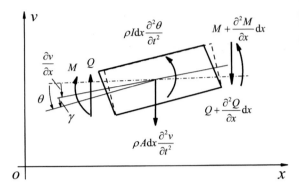

the normal line of the cross-section denoted by dashed quadrangle, while the other is the shear angle $\gamma(x, t)$ of the cross-section. There follows

$$\theta(x, t) = \frac{\partial v(x, t)}{\partial x} + \gamma(x, t) \tag{7.4.25}$$

From mechanics of material, the bending moment $M(x, t)$ and the slope $\partial v(x, t)/\partial x$ satisfy

$$M(x, t) = EI \frac{\partial^2 v(x, t)}{\partial x^2} \tag{7.4.26}$$

while the shear force $Q(x, t)$ and the shear angle $\gamma(x, t)$ yield

$$Q(x, t) = \beta GA\gamma(x, t) = \beta GA \left[\theta(x, t) - \frac{\partial v(x, t)}{\partial x} \right] \tag{7.4.27}$$

where β is the *coefficient of shear stress* associated with the cross-sectional shape. It is feasible to determine β from the equivalent strain energies between the exact solution of three-dimenal elasticity and the approximate solution of a beam based on the uniformly distribution of shear stress on a given shape of the cross-section. For instance, $\beta = 5/6$ holds for a rectangular cross-section and $\beta = 9/10$ holds for a circular cross-section.

The translation of the mass center of an infinitesimal segment with length dx and its rotation about the mass center satisfy the two coupled dynamic equations

$$\begin{cases} \rho A \dfrac{\partial^2 v(x, t)}{\partial t^2} dx = -\dfrac{\partial Q(x, t)}{\partial x} dx \\[2mm] \rho I \dfrac{\partial^2 \theta(x, t)}{\partial t^2} dx = \dfrac{\partial M(x, t)}{\partial x} dx - Q(x, t) dx - \dfrac{\partial Q(x, t)}{\partial x} dx^2 \end{cases} \tag{7.4.28}$$

Eliminating dx of the first order from Eq. (7.4.28) and letting the remaining $dx \to 0$, we have

$$\begin{cases} \rho A \dfrac{\partial^2 v(x,t)}{\partial t^2} + \dfrac{\partial Q(x,t)}{\partial x} = 0 \\[2mm] \rho I \dfrac{\partial^2 \theta(x,t)}{\partial t^2} - \dfrac{\partial M(x,t)}{\partial x} + Q(x,t) = 0 \end{cases} \tag{7.4.29}$$

Substitution of Eqs. (7.4.26) and (7.4.27) into Eq. (7.4.29) leads to the dynamic equations of a T-E beam as follows

$$\begin{cases} \rho A \dfrac{\partial^2 v(x,t)}{\partial t^2} + \beta G A \left[\dfrac{\partial \theta(x,t)}{\partial x} - \dfrac{\partial^2 v(x,t)}{\partial x^2} \right] = 0 \\[2mm] \rho I \dfrac{\partial^2 \theta(x,t)}{\partial t^2} - E I \dfrac{\partial^2 \theta(x,t)}{\partial x^2} + \beta G A \left[\theta(x,t) - \dfrac{\partial v(x,t)}{\partial x} \right] = 0 \end{cases} \tag{7.4.30}$$

Equation (7.4.30) is a set of partial differential equations, for deflection $v(x,t)$ and rotation angle $\theta(x,t)$, of the second order, which can be converted as a single partial differential equation, for deflection $v(x,t)$, of the fourth order. For this purpose, differentiating the first equation in Eq. (7.4.30) with respect to t twice, and differentiating the second equation with respect to x once, we have

$$\begin{cases} \rho A \dfrac{\partial^4 v(x,t)}{\partial t^4} + \beta G A \left[\dfrac{\partial^3 \theta(x,t)}{\partial x \partial t^2} - \dfrac{\partial^4 v(x,t)}{\partial x^2 \partial t^2} \right] = 0 \\[2mm] \rho I \dfrac{\partial^3 \theta(x,t)}{\partial x \partial t^2} - E I \dfrac{\partial^3 \theta(x,t)}{\partial x^3} + \beta G A \left[\dfrac{\partial \theta(x,t)}{\partial x} - \dfrac{\partial^2 v(x,t)}{\partial x^2} \right] = 0 \end{cases} \tag{7.4.31}$$

Eliminating the terms with mixed partial derivatives of $\theta(x,t)$ in Eq. (7.4.31) leads to

$$\frac{\rho}{\beta G} \frac{\partial^4 v(x,t)}{\partial t^4} - \frac{\partial^4 v(x,t)}{\partial x^2 \partial t^2} + \frac{E}{\rho} \frac{\partial^3 \theta(x,t)}{\partial x^3} + \frac{\beta G A}{\rho I} \left[\frac{\partial^2 v(x,t)}{\partial x^2} - \frac{\partial \theta(x,t)}{\partial x} \right] = 0 \tag{7.4.32}$$

Solving the first equation in Eq. (7.4.30) for $\partial\theta(x,t)/\partial x$ and then differentiating $\partial\theta(x,t)/\partial x$ twice with respect to x, we have the following two relations

$$\begin{cases} \dfrac{\partial \theta(x,t)}{\partial x} = \dfrac{\partial^2 v(x,t)}{\partial x^2} - \dfrac{\rho}{\beta G} \dfrac{\partial^2 v(x,t)}{\partial t^2} \\[2mm] \dfrac{\partial^3 \theta(x,t)}{\partial x^3} = \dfrac{\partial^4 v(x,t)}{\partial x^4} - \dfrac{\rho}{\beta G} \dfrac{\partial^4 v(x,t)}{\partial x^2 \partial t^2} \end{cases} \tag{7.4.33}$$

By substituting Eq. (7.4.33) into Eq. (7.4.32), we arrive at a partial differential equation of the fourth order in terms of deflection $v(x,t)$, namely,

$$\underbrace{\rho A \frac{\partial^2 v(x,t)}{\partial t^2} + EI \frac{\partial^4 v(x,t)}{\partial x^4}}_{A} \underbrace{- \rho I \frac{\partial^4 v(x,t)}{\partial x^2 \partial t^2}}_{B} \underbrace{- \frac{\rho I E}{\beta G} \frac{\partial^4 v(x,t)}{\partial x^2 \partial t^2}}_{C} + \underbrace{\frac{\rho^2 I}{\beta G} \frac{\partial^4 v(x,t)}{\partial t^4}}_{D} = 0$$

$$(7.4.34)$$

Remark 7.4.3 Equation (7.4.30) describes the dynamics of a T-E beam in terms of the deflection $v(x,t)$ and rotation angle $\theta(x,t)$ and includes both effects of shear deformation and rotary inertia of cross-section explicitly. Equation (7.4.34) lacks of the above intuitions since it involves the deflection $v(x,t)$ only. However, it is easy to see that term A in Eq. (7.4.34) is identical to the summation of both inertial term and elastic term for an E-B beam. Section 7.4.3 will demonstrate that, via the natural vibrations of a hinged-hinged T-E beam, term B comes from the rotary inertia, term C is due to the shear deformation, and term D is a coupled contribution of both rotary inertia and shear deformation.

Remark 7.4.4 The above formulation procedure can be found from many books, but some of them had typos or minor problems[26,27].

7.4.2.2   Analysis of Free Waves

Let the deflection $v(x,t)$ and rotation angle $\theta(x,t)$ be expressed as the real parts of two complex functions $v_c(x,t)$ and $\theta_c(x,t)$ as follows

$$\begin{cases} v(x,t) = \mathrm{Re}[v_c(x,t)], & v_c(x,t) = \tilde{v}_c(x)\exp(i\omega t) \\ \theta(x,t) = \mathrm{Re}[\theta_c(x,t)], & \theta_c(x,t) = \tilde{\theta}_c(x)\exp(i\omega t) \end{cases}$$

$$(7.4.35)$$

where $\tilde{v}_c(x)$ and $\tilde{\theta}_c(x)$ are the complex amplitudes to be determined. Substituting Eq. (7.4.35) into Eq. (7.4.30), we get a set of homogeneous ordinary differential equations, valid for an arbitrary time t, of complex functions $\tilde{v}_c(x)$ and $\tilde{\theta}_c(x)$, namely,

$$\begin{cases} -\omega^2 \rho A \tilde{v}_c(x) + \beta GA \left[\dfrac{\mathrm{d}\tilde{\theta}_c(x)}{\mathrm{d}x} - \dfrac{\mathrm{d}^2 \tilde{v}_c(x)}{\mathrm{d}x^2} \right] = 0 \\[3mm] -\omega^2 \rho I \tilde{\theta}_c(x) - EI \dfrac{\mathrm{d}^2 \tilde{\theta}_c(x)}{\mathrm{d}x^2} + \beta GA \left[\tilde{\theta}_c(x) - \dfrac{\mathrm{d}\tilde{v}_c(x)}{\mathrm{d}x} \right] = 0, \quad 0 \le x \le L \end{cases}$$

$$(7.4.36)$$

Let the solutions of Eq. (7.4.36) be expressed as

$$\tilde{v}_c(x) = \hat{v}\exp(sx), \quad \tilde{\theta}_c(x) = \hat{\theta}\exp(sx)$$

$$(7.4.37)$$

---

[26] Graff K F (1975). Wave Motion in Elastic Solids. Columbus: Ohio State University Press, 180–197.

[27] Dai H L (2014). Elasto-dynamics. Changsha: Press of Hunan University, 95–98 (in Chinese).

Substitution of Eq. (7.4.37) into Eq. (7.4.36) leads to an eigenvalue problem

$$\begin{bmatrix} -\beta GAs^2 - \rho A\omega^2 & \beta GAs \\ -\beta GAs & \beta GA - EIs^2 - \rho I\omega^2 \end{bmatrix} \begin{bmatrix} \hat{v} \\ \hat{\theta} \end{bmatrix} = 0 \qquad (7.4.38)$$

The non-zero solutions of Eq. (7.4.38) require a characteristic equation, namely,

$$\begin{aligned} &\det \begin{bmatrix} -\beta GAs^2 - \rho A\omega^2 & \beta GAs \\ -\beta GAs & \beta GA - EIs^2 - \rho I\omega^2 \end{bmatrix} \\ &= (\beta GAs^2 + \rho A\omega^2)(EIs^2 + \rho I\omega^2 - \beta GA) + (\beta GAs)^2 \\ &= \beta GEAIs^4 + \rho AI(E + \beta G)\omega^2 s^2 + \rho^2 AI\omega^4 - \rho\beta GA^2\omega^2 = 0 \end{aligned} \qquad (7.4.39)$$

Let Eq. (7.4.39) be substituted with $I = r_g^2 A$ and divided by $\beta GEr_g^2\omega^4$, we have

$$\left(\frac{s}{\omega}\right)^4 + \left(\frac{\rho}{\beta G} + \frac{\rho}{E}\right)\left(\frac{s}{\omega}\right)^2 + \frac{\rho^2}{\beta GE} - \frac{\rho}{Er_g^2\omega^2} = 0 \qquad (7.4.40)$$

Noting the following two relations

$$\frac{\rho}{E} = \frac{1}{c_0^2}, \quad \frac{G}{E} = \frac{1}{2(1+v)} \qquad (7.4.41)$$

we recast Eq. (7.4.40) as a simpler form as follows

$$\left(\frac{c_0 s}{\omega}\right)^4 + B\left(\frac{c_0 s}{\omega}\right)^2 + C = 0 \qquad (7.4.42)$$

where the dimensionless parameters B and C satisfy

$$B \equiv \frac{2(1+v)}{\beta} + 1, \quad C \equiv \frac{2(1+v)}{\beta} - \frac{c_0^2}{r_g^2\omega^2} \qquad (7.4.43)$$

Given Poisson's ratio v, coefficient of shear stress β, radius of gyration r_g and longitudinal wave speed c_0, the parameters in Eq. (7.4.43) satisfy $B > 0$ and $C < 0$. In this case, Eq. (7.4.42) has a pair of real roots with opposite signs if Eq. (7.4.42) is taken as a quadratic equation of $(c_0 s/\omega)^2$. That is, Eq. (7.4.42) has a pair of pure imaginary roots and a pair of real roots with opposite signs as follows

$$\begin{cases} s_{1,2} = \pm i\kappa, \quad s_{3,4} = \pm\lambda, \\ \kappa \equiv \frac{\omega}{c_0}\sqrt{\frac{B + \sqrt{B^2 - 4C}}{2}} > 0, \quad \lambda \equiv \frac{\omega}{c_0}\sqrt{\frac{\sqrt{B^2 - 4C} - B}{2}} > 0 \end{cases} \qquad (7.4.44)$$

Thus, the complex solutions of Eq. (7.4.30) take the following form

$$\begin{cases} \tilde{v}_c(x,t) = \left[a_1 \exp(-i\kappa x) + a_2 \exp(i\kappa x) + a_3 \exp(-\lambda x) + a_4 \exp(\lambda x)\right] \exp(i\omega t) \\ \tilde{\theta}_c(x,t) = \left[b_1 \exp(-i\kappa x) + b_2 \exp(i\kappa x) + b_3 \exp(-\lambda x) + b_4 \exp(\lambda x)\right] \exp(i\omega t) \end{cases}$$
$$(7.4.45)$$

Similar to the wave analysis of an E-B beam, Eq. (7.4.45) can be recast as the superposition of traveling waves and evanescent waves, namely,

$$\begin{cases} v_c(x,t) = a_1 \exp[i\kappa(c_{BP}t - x)] + a_2 \exp[i\kappa(c_{BP}t + x)] \\ \quad + a_3 \exp(-\kappa x)\exp(i\omega t) + a_4 \exp(\kappa L)\exp[-\kappa(L-x)]\exp(i\omega t) \\ \theta_c(x,t) = b_1 \exp[i\kappa(c_{BP}t - x)] + b_2 \exp[i\kappa(c_{BP}t + x)] \\ \quad + b_3 \exp(-\kappa x)\exp(i\omega t) + b_4 \exp(\kappa L)\exp[-\kappa(L-x)]\exp(i\omega t) \end{cases}$$
$$(7.4.46)$$

where the phase velocity of bending wave of the T-B beam is

$$c_{BP} \equiv \frac{\omega}{\kappa} = c_0 \sqrt{\frac{2}{B + \sqrt{B^2 - 4C}}} \tag{7.4.47}$$

In Eq. (7.4.46), the third and fourth terms of $v_c(x,t)$ and $\theta_c(x,t)$ are the evanescent waves, which get the maximum at the left end $x = 0$ and the right end $x = L$, and decrease with increasing of x and $L - x$, respectively. The first term and the second term are the right traveling wave and the left traveling wave with phase velocity c_{BP}, which will be discussed further.

Remark 7.4.5 From the theory of elasto-dynamics in Appendix A.1.2, the shear wave speed of three-dimensional continua is $c_s = \sqrt{G/\rho}$. Given a T-E beam with a rectangular cross-section or a circular cross-section, the coefficient of shear stress is $\beta = 5/6$ or $\beta = 9/10$ such that $\sqrt{\beta} \in (0.912, \ 0.949)$ holds. As such, it is reasonable to define the *approximate shear wave speed* of a T-E beam as follows

$$c_S \equiv \sqrt{\frac{\beta G}{\rho}} \approx \sqrt{\frac{G}{\rho}} = c_s \tag{7.4.48}$$

From the longitudinal wave speed c_0 of a uniform beam and the approximate shear wave speed c_S in Eq. (7.4.48), we recast Eq. (7.4.40) as

$$c_0^2 c_S^2 s^4 + (c_0^2 + c_S^2)\omega^2 s^2 + \omega^2 \left(\omega^2 - \frac{c_S^2}{r_g^2}\right) = 0 \tag{7.4.49}$$

Solving Eq. (7.4.49) leads to

$$s^2 = \frac{1}{2c_0^2 c_S^2}\left[-(c_0^2 + c_S^2)\omega^2 \pm \sqrt{(c_0^2 + c_S^2)^2\omega^4 - 4c_0^2 c_S^2\omega^2\left(\omega^2 - \frac{c_S^2}{r_g^2}\right)}\right]$$

$$
= -\frac{1}{2}\left(\frac{\omega^2}{c_S^2} + \frac{\omega^2}{c_0^2}\right) \pm \sqrt{\frac{\omega^4}{4c_0^4}\left(\frac{c_0^2}{c_S^2} - 1\right)^2 + \frac{\omega^2}{r_g^2 c_0^2}}
$$

$$
= -\frac{\eta^2}{2}(\sigma^2 + 1) \pm \sqrt{\frac{\eta^4}{4}(\sigma^2 - 1)^2 + \frac{\eta^2}{r_g^2}} \tag{7.4.50}
$$

where

$$
\sigma \equiv \frac{c_0}{c_S} = \sqrt{\frac{E}{\beta G}} = \sqrt{\frac{2(1+v)}{\beta}} > 1, \quad \eta \equiv \frac{\omega}{c_0} > 0 \tag{7.4.51}
$$

They are the ratio of the longitudinal wave speed to the approximate shear wave speed, and the ratio of the bending wave frequency to the longitudinal wave speed. It is easy, using Eq. (7.4.50), to recast Eq. (7.4.44) as a formulation in many references, namely,

$$
\begin{cases}
s_{1,2} = \pm i\kappa, \quad \kappa \equiv \eta\left\{\left[\left(\frac{\sigma^2-1}{2}\right)^2 + \frac{1}{r_g^2\eta^2}\right]^{1/2} + \frac{\sigma^2+1}{2}\right\}^{1/2} > 0 \\[4mm]
s_{3,4} = \pm\lambda, \quad \lambda \equiv \eta\left\{\left[\left(\frac{\sigma^2-1}{2}\right)^2 + \frac{1}{r_g^2\eta^2}\right]^{1/2} - \frac{\sigma^2+1}{2}\right\}^{1/2} > 0
\end{cases} \tag{7.4.52}
$$

7.4.2.3 Dispersion of Bending Traveling Waves

In the analysis of bending waves, although the phase velocity c_{BP} is defined, the coefficient C in Eq. (7.4.47) depends on the squared wave frequency ω^2 and gives difficulty of checking the dispersion relation of phase velocity c_{BP} and wave number κ. If we focus on the bending traveling waves only, it is reasonable to assume the solution of traveling waves so as to derive the relation of phase velocity c_{BP} and wave number κ[28].

For bending traveling waves, substitution of $s = \pm i\kappa$ into Eq. (7.4.40) leads to

$$
\beta EG r_g^2 \kappa^4 - \rho r_g^2(E + \beta G)\omega^2\kappa^2 + \rho^2 r_g^2\omega^4 - \rho\beta G\omega^2 = 0 \tag{7.4.53}
$$

Let Eq. (7.4.53) be divided by $\rho^2 r_g^2\kappa^4$ and then written as

$$
\left(\frac{\omega}{\kappa}\right)^4 - \left[\left(\frac{E}{\rho} + \frac{\beta G}{\rho}\right) + \frac{\beta G}{\rho r_g^2\kappa^2}\right]\left(\frac{\omega}{\kappa}\right)^2 + \frac{\beta EG}{\rho^2} = 0 \tag{7.4.54}
$$

---

[28] Graff K F (1975). Wave Motion in Elastic Solids. Columbus: Ohio State University Press, 185.

From the definitions of c_0 and c_S, we recast Eq. (7.4.54) as

$$\left(\frac{\omega}{\kappa}\right)^4 - \left[(c_0^2 + c_S^2) + \frac{c_S^2}{r_g^2\kappa^2}\right]\left(\frac{\omega}{\kappa}\right)^2 + c_0^2 c_S^2 = 0 \tag{7.4.55}$$

Substitution of $c_{BP} = \omega/\kappa$ in Eq. (7.4.47) into Eq. (7.4.55) leads to

$$\left(\frac{c_{BP}}{c_0}\right)^4 - \tilde{B}\left(\frac{c_{BP}}{c_0}\right)^2 + \tilde{C} = 0, \quad \tilde{B} \equiv 1 + \frac{c_S^2}{c_0^2}\left(1 + \frac{1}{r_g^2\kappa^2}\right) > 0, \quad \tilde{C} \equiv \frac{c_S^2}{c_0^2} > 0 \tag{7.4.56}$$

Solving Eq. (7.4.56) for c_{BP}/c_0, we have a pair of positive real roots of phase velocity as follows

$$\frac{c_{BPa}}{c_0} = \sqrt{\frac{\tilde{B} - \sqrt{\tilde{B}^2 - 4\tilde{C}}}{2}}, \quad \frac{c_{BPb}}{c_0} = \sqrt{\frac{\tilde{B} + \sqrt{\tilde{B}^2 - 4\tilde{C}}}{2}} \tag{7.4.57}$$

When wave number $\kappa \to +\infty$, Eq. (7.4.56) becomes

$$\left(\frac{c_{BP}}{c_0}\right)^4 - \left(1 + \frac{c_S^2}{c_0^2}\right)\left(\frac{c_{BP}}{c_0}\right)^2 + \frac{c_S^2}{c_0^2} = 0 \tag{7.4.58}$$

Solving Eq. (7.4.58) gives two asymptotic solutions of phase velocity as following

$$\frac{c_{BPa}}{c_0} = \frac{c_S}{c_0}, \quad \frac{c_{BPb}}{c_0} = 1 \tag{7.4.59}$$

That is, two phase velocities $c_{BPa} \to c_s$ and $c_{BPb} \to c_0$ when $\kappa \to +\infty$. This looks more reasonable than the phase velocity of an E-B beam since it goes to infinity when $\kappa \to +\infty$.

If both material and cross-section of a beam are given, Poisson's ratio ν and coefficient of shear stress β are fixed parameters, and so is the ratio c_S^2/c_0^2. From Eqs. (7.4.56) and (7.4.57), the dimensionless phase velocities c_{BPa}/c_0 and c_{BPb}/c_0 only depend on dimensionless wave number $r_g\kappa$. For the T-E beam with a rectangular cross-section, the coefficient of shear stress is $\beta = 5/6$. Figure 7.31 shows the variations of both dimensionless phase velocities c_{BPa}/c_0 and c_{BPb}/c_0 of the beam with increasing of $r_g\kappa$ for Poisson's ratio $\nu = 0.28$. As shown in Fig. 7.31, the lower frequency branch c_{BPa}/c_0 is monotonically increasing and getting close to $c_S/c_0 \approx 0.571$, while the higher frequency branch c_{BPb}/c_0 monotonically decreases and approaches $c_0/c_0 = 1$. As a reference, the dashed line gives the dimensionless phase velocity c_{BP}/c_0 of an E-B beam.

Fig. 7.31 The relations of phase velocities and wave number of the beam with a rectangular cross-section (thick solid line: c_{BPa}/c_0, thin solid line: c_{BPb}/c_0, dashed line c_{BP}/c_0, dot lines: the asymptotes)

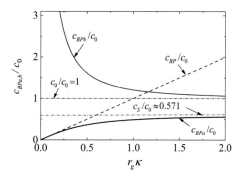

Remark 7.4.6 It is worthy to point out that the dimensionless phase velocities c_{BPa}/c_0 and c_{BPb}/c_0 in Fig. 7.31 are very close to those of a beam with a circular cross-section. More significantly, the lower frequency branch of phase velocity c_{BPa}/c_0 of a beam with a circular cross-section well agrees with the exact solution of three-dimensional elasto-dynamics. As such, the phase velocity c_{BPa} of lower frequency branch approaches the approximate shear wave speed c_S, but the phase velocity c_{BPb}/c_0 of higher frequency branch has aroused some arguments and not seen enough applications[29,30].

From Remark 7.4.6, the lower frequency branch of phase velocity of a T-E beam with a circular or rectangular cross-section can serve as an accurate solution to check the validity range of an E-B beam. Consider a beam with a rectangular cross-section, height h and radius of gyration $r_g = h/\sqrt{12}$. Let $\kappa = 2\pi/\lambda$ be the wave number and λ be the wave length. Then, we get the dimensionless wave number $r_g\kappa = (2\pi/\sqrt{12}) \cdot (h/\lambda) \approx 1.814h/\lambda$ in Fig. 7.31. In a range of $r_g\kappa \approx 1.814h/\lambda < 0.1$, the dimensionless phase velocity c_{BP}/c_0 of an E-B beam is close to the phase velocity c_{BPa}/c_0 of the lower frequency branch of a T-E beam, and the corresponding wave length is $\lambda > 18.14h$. In engineering, the validity range of an E-B beam assumed is that the cross-sectional height is short than five percent of the wave length, that is, $h < \lambda_r/20$. As the wave length λ_r of the r-th mode is doubled distance d_r of two nodes of the mode shape, the above requirement is equivalent to $d_r > 10h$.

Finally, we discuss the group velocity of a T-E beam, which has two branches, similar to the behavior of phase velocity, as follows

$$\begin{cases} c_{BGa} = \dfrac{d(\kappa c_{BPa})}{d\kappa} = c_{BPa} + \kappa \dfrac{dc_{BPa}}{d\kappa} \\ c_{BGb} = \dfrac{d(\kappa c_{BPb})}{d\kappa} = c_{BPb} + \kappa \dfrac{dc_{BPb}}{d\kappa} \end{cases} \tag{7.4.60}$$

[29] Elishakoff I, Kaplunov J, Nolde E (2015). Celebrating the centenary of Timoshenko's study of effects of shear deformation and rotary inertia. Applied Mechanics Reviews, 67(6): 060802.

[30] Jinsberg J H (2001). Mechanical and Structural Vibrations: Theory and Applications. Baffins Lane: John Wiley & Sons Inc., 487–490.

In Fig. 7.31, the lower frequency branch of phase velocity yields $dc_{BPa}/d\kappa > 0$ such that $c_{BGa} > c_{BPa}$ holds in Eq. (7.4.60) and this branch exhibits the abnormal dispersion, while the higher frequency branch of phase velocity satisfies $dc_{BPb}/d\kappa < 0$, which results in $c_{BGb} < c_{BPb}$ in Eq. (7.4.60) and exhibits the normal dispersion as well.

7.4.3 Natural Vibrations of a Timoshenko Beam

This subsection presents how to establish the simple form of an eigenvalue problem from the boundary conditions of a T-B beam so as to study the natural vibrations.

7.4.3.1   Basic Idea of Computation

According to Eq. (7.4.35), the natural vibration of a T-E beam with length L yields the following real expressions

$$v(x, t) = \tilde{v}(x) \cos(\omega t), \quad \theta(x, t) = \tilde{\theta}(x) \cos(\omega t), \quad x \in [0, \ L] \tag{7.4.61}$$

Following Eq. (7.4.45), we write amplitude $\tilde{v}(x)$ in Eq. (7.4.61) as

$$\tilde{v}(x) = a_1 \cos(\kappa x) + a_2 \sin(\kappa x) + a_3 \cosh(\lambda x) + a_4 \sinh(\lambda x) \tag{7.4.62}$$

where a_r, $r = 1, 2, 3, 4$ are the four parameters to be determined. If we express amplitude $\tilde{\theta}(x)$ in Eq. (7.4.61) similarly, the other four unknown parameters appear and greatly increase the order of the eigenvalue problem to be solved.

To avoid this burden, substituting Eq. (7.4.61) into the first equation in Eq. (7.4.33) and eliminating $\cos(\omega t)$, we get

$$\frac{d\tilde{\theta}(x)}{dx} = \frac{d^2 \tilde{v}(x)}{dx^2} + \frac{\rho \omega^2}{\beta G} \tilde{v}(x) \tag{7.4.63}$$

Let Eq. (7.4.63) be substituted with Eq. (7.4.62) and integrated with respect to x, we have

$$\begin{aligned}
\tilde{\theta}(x) &= -a_1 \kappa \sin(\kappa x) + a_2 \kappa \cos(\kappa x) + a_3 \lambda \sinh(\lambda x) + a_4 \lambda \cosh(\lambda x) \\
&\quad + \frac{\rho \omega^2}{\beta G} \left[\frac{a_1}{\kappa} \sin(\kappa x) - \frac{a_2}{\kappa} \cos(\kappa x) + \frac{a_3}{\lambda} \sinh(\lambda x) + \frac{a_4}{\lambda} \cosh(\lambda x) \right] \\
&= \frac{\omega^2 - \kappa^2 c_S^2}{\kappa c_S^2} [a_1 \sin(\kappa x) - a_2 \cos(\kappa x)]
\end{aligned}$$

$$+ \frac{\omega^2 + \lambda^2 c_S^2}{\lambda c_S^2}[a_3 \sinh(\lambda x) + a_4 \cosh(\lambda x)] \tag{7.4.64}$$

where the integration constants are neglected since the subsequent studies will not touch upon the rigid-body motions. Substituting Eq. (7.4.62) into Eq. (7.4.26), we get the bending moment at cross-section x as follows

$$\tilde{M}(x) = EI \frac{\partial^2 \tilde{v}(x)}{\partial x^2}$$
$$= EI\left[-a_1 \kappa^2 \cos(\kappa x) - a_2 \kappa^2 \sin(\kappa x) + a_3 \lambda^2 \cosh(\lambda x) + a_4 \lambda^2 \sinh(\lambda x)\right] \tag{7.4.65}$$

In view of Eqs. (7.4.27), (7.4.62) and (7.4.64), the shear force at cross-section x satisfies

$$\tilde{Q}(x) = \beta AG\left[\tilde{\theta}(x) - \frac{d\tilde{v}(x)}{dx}\right]$$
$$= \beta AG\left\{\left(\frac{\omega^2 - \kappa^2 c_S^2}{\kappa c_S^2} + 1\right)[a_1 \sin(\kappa x) - a_2 \cos(\kappa x)]\right.$$
$$\left. + \left(\frac{\omega^2 + \lambda^2 c_S^2}{\lambda c_S^2} - 1\right)[a_3 \sinh(\lambda x) + a_4 \cosh(\lambda x)]\right\} \tag{7.4.66}$$

Substituting Eqs. (7.4.62), (7.4.64), (7.4.65) and (7.4.66) into the four boundary conditions of two ends of the T-B beam and postulating the condition of a non-trivial solutions, we have the eigenvalue problem of wave numbers. After solving wave numbers κ_r, $r = 1, 2, \ldots$ and substituting them into the expression of $\tilde{B}$ in Eq. (7.4.56), we obtain two phase velocities c_{BPa} and c_{BPb} from Eq. (7.4.57) and get the corresponding natural frequencies $\omega_{ra} = c_{BPa}\kappa_r$, $\omega_{rb} = c_{BPb}\kappa_r$, $r = 1, 2, \ldots$
It is worthy to point out that the above procedure involves very complicated algebraic manipulations and requires some software, such as Maple, even for homogeneous boundaries. The exceptional case is the hinged-hinged beam to be discussed as follows.

7.4.3.2 Natural Vibrations of a Hinged-Hinged T-E Beam

Example 7.4.2 Study the natural vibrations of a hinged-hinged T-E beam of length L.

Substituting Eqs. (7.4.62) and (7.4.65) into the left boundary conditions of the beam leads to

$$\begin{cases} \tilde{v}(0) = a_1 + a_3 = 0 \\ \tilde{M}(0) = EI(a_3 \lambda^2 - a_1 \kappa^2) = 0 \end{cases} \tag{a}$$

Substituting Eqs. (7.4.62) and (7.4.65), through solution $a_1 = 0$ and $a_3 = 0$ of Eq. (a), into the right boundary conditions of the beam, we have

$$
\begin{cases}
\tilde{v}(L) = a_2 \sin(\kappa L) + a_4 \sinh(\lambda L) = 0 \\
\tilde{M}(L) = EI\left[a_4 \lambda^2 \sinh(\lambda L) - a_2 \kappa^2 \sin(\kappa L)\right] = 0
\end{cases}
\tag{b}
$$

The non-trivial solutions a_2 and a_4 for Eq. (b) require

$$
\det\begin{bmatrix}
\sin(\kappa L) & \sinh(\lambda L) \\
-\kappa^2 EI \sin(\kappa L) & \lambda^2 EI \sinh(\lambda L)
\end{bmatrix} = (\lambda^2 + \kappa^2) EI \sin(\kappa L) \sinh(\lambda L) = 0
\tag{c}
$$

which yields

$$
\sin(\kappa L) = 0
\tag{d}
$$

Solving Eq. (d) gives the following wave numbers

$$
\kappa_r = \frac{r \pi}{L}, \quad r = 1, 2, \ldots
\tag{e}
$$

By substituting Eq. (e), together with $a_2 \neq 0$ and $a_4 = 0$, into Eq. (7.4.62), we get the mode shapes of the T-E beam as follows

$$
\tilde{v}_r(x) = a_2 \sin(\kappa_r x), \quad r = 1, 2, \ldots
\tag{f}
$$

Although the mode shapes of a hinged-hinged T-E beam are the same as those of an E-B beam, their associated natural frequencies are different. Substituting Eq. (e) into the expression of $\tilde{B}$ in Eq. (7.4.56) and solving Eq. (7.4.57) for phase velocities, we get the natural frequencies as follows

$$
\begin{cases}
\omega_{ra} = \kappa_r c_0 \sqrt{\dfrac{\tilde{B} - \sqrt{\tilde{B}^2 - 4\tilde{C}}}{2}}, \quad \omega_{rb} = \kappa_r c_0 \sqrt{\dfrac{\tilde{B} + \sqrt{\tilde{B}^2 - 4\tilde{C}}}{2}} \\[2mm]
\tilde{B} = 1 + \dfrac{c_S^2}{c_0^2}\left(1 + \dfrac{1}{r_g^2 \kappa_r^2}\right) = 1 + \tilde{C}\left(1 + \dfrac{1}{r_g^2 \kappa_r^2}\right) > 0, \quad \tilde{C} = \dfrac{c_S^2}{c_0^2} = \dfrac{\beta}{2(1 + \nu)} > 0
\end{cases}
\tag{g}
$$

In addition, we can also recast Eq. (g) as a popular form, namely,

$$\omega_{ra,b} = \kappa_r c_0 \left\{ \frac{(1 + c_S^2/c_0^2)}{2} + \frac{c_S^2/c_0^2}{2r_g^2\kappa_r^2} \right.$$

$$\left. \mp \left[\frac{(1 - c_S^2/c_0^2)^2}{4} + \frac{c_S^2/c_0^2(1 + c_S^2/c_0^2)}{2r_g^2\kappa_r^2} + \left(\frac{c_S^2/c_0^2}{2r_g^2\kappa_r^2} \right)^2 \right]^{1/2} \right\}^{1/2}, \quad r = 1, 2, \dots. \tag{h}$$

It is worthy to notice that Eq. (h) provides two branches of natural frequency corresponding to the lower and higher frequency branches of phase velocity. We keep the higher frequency branch, albeit some arguments as stated in Remark 7.4.6, for subsequent comparisons.

In what follows, we study two hinged-hinged T-E beams of a rectangular cross-section with different heights to make a comparison of natural frequencies of them with an E-B beam. Figure 7.32a and Fig. 7.32b display the dimensionless natural frequencies of the first 10 orders for two T-E beams with dimensionless heights $h/L = 0.05$ and $h/L = 0.1$, respectively, with an increase of mode order. Here, the solid circle line presents the lower frequency branch $\omega_{ra}/(c_0L)$, the solid square line gives the higher frequency branch $\omega_{rb}/(c_0L)$, and the dashed circle line represents the dimensionless natural frequencies of an E-B beam defined as

$$\frac{\omega_{re}}{c_0L} = \frac{r_g\kappa_r^2}{L}, \quad r = 1, 2, \dots \tag{i}$$

As shown in Fig. 7.32a, the first six natural frequencies of the E-B beam with $h/L = 0.05$ and those of the corresponding T-E beam are very close, while the first three natural frequencies of the E-B beam and the T-E beam are close enough for $h/L = 0.1$ in Fig. 7.32b. It is easy to understand the above phenomenon since the abscissa in Fig. 7.31 represents the dimensionless wave number $r_g\kappa = \kappa h/\sqrt{12}$ for the beam with a rectangular cross-section. When the natural frequencies of an E-B beam and a T-E beam are getting close, their wave numbers must be close such that

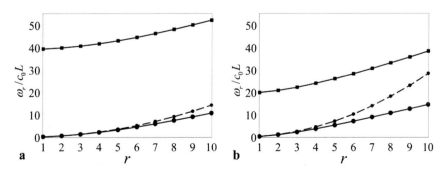

Fig. 7.32 The dimensionless natural frequencies of two hinged-hinged beams with a rectangular cross-section ($\nu = 0.28$, $\beta = 5/6$) with an increase of mode order (solid circle line: $\omega_{ra}/(\omega_0L)$, solid square line: $\omega_{rb}/(\omega_0L)$, dashed circle line: $\omega_{re}/(\omega_0L)$). **a** $h/L = 0.05$; **b** $h/L = 0.1$

the doubled cross-sectional height of the beam requires a half reduction of either the wave number or the mode order.

The other difference between a T-E beam and an E-B beam is the mode shapes of rotation angle. From Eq. (7.4.64), we have the mode shapes of rotation angle of a hinged-hinged T-E beam as follows

$$\tilde{\theta}_{ra,b}(x) = \frac{\kappa_r^2 c_S^2 - \omega_{ra,b}^2}{\kappa_r c_S^2} a_2 \cos(\kappa_r x) = a_2 \kappa_r \left(1 - \frac{\omega_{ra,b}^2}{\kappa_r^2 c_S^2}\right) \cos(\kappa_r x), \quad r = 1, 2, ...$$

(j)

while the mode shapes of rotation angle of a hinged-hinged E-B beam are simply $\theta_r(x) = a_2 \kappa_r \cos(\kappa_r x)$, $r = 1, 2, ...$ To demonstrate their difference, we define the difference ratio of amplitude of $\tilde{\theta}_{ra,b}(x)$ to that of $\theta_r(x)$, namely,

$$\eta_{ra,b} \equiv 1 - \frac{\omega_{ra,b}^2}{\kappa_r^2 c_S^2}, \quad r = 1, 2, ...$$

(k)

Using Eq. (g), we can recast Eq. (k) as the following form for easy computation

$$\eta_{ra,b} = 1 - \frac{\omega_{ra,b}^2/\kappa_r^2 c_0^2}{c_S^2/c_0^2} = \frac{\tilde{B} \mp \sqrt{\tilde{B}^2 - 4\tilde{C}}}{2\tilde{C}}, \quad r = 1, 2, ...$$

(l)

Figure 7.33 shows the difference ratio of the mode shapes of rotation angle of a T-E beam with respect to an E-B beam for $h/L = 0.05$ and $h/L = 0.1$ when $\nu = 0.28$ and $\beta = 5/6$. As shown in Fig. 7.33a, the difference ratio of the lower frequency branch increases with increasing of mode order, and the larger dimensionless height h/L further enhances the above increase. In Fig. 7.33b, the difference ratio of the higher frequency branches is much larger than that of lower frequency branches, but the influence of dimensionless height h/L becomes less significant when the mode order increases.

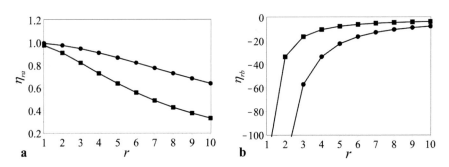

Fig. 7.33 The difference ratio of mode shape of rotation angle of a hinged-hinged T-E beam with a rectangular cross-section with an increase of mode order (solid circle line: $h/L = 0.05$, solid square line: $h/L = 0.1$). **a.** the lower frequency branch ω_{ra}; **b.** the higher frequency branch ω_{rb}

7.4.3.3 Influences of Rotary Inertia and Shear Deformation

This subsection is about the influence of either rotary inertia or shear deformation of cross-section on the lower natural frequency branch of a T-E beam, respectively.

(1)   Influence of rotary inertia

As recalled in the beginning of the section, Bresse in France and Rayleigh in the UK had studied the influence of rotary inertia on the natural vibrations of a beam with shear deformation neglected before the joint research of Timoshenko and Erhenfest. To study the effect of rotary inertia only, we substitute the natural frequency ω_r, with a simplified subscript for the lower branch, into Eq. (7.4.55) and recast it as

$$\frac{1}{c_S^2}\left(\frac{\omega_r}{\kappa_r}\right)^4 - \left[\left(1+\frac{c_0^2}{c_S^2}\right)+\frac{1}{r_g^2\kappa_r^2}\right]\left(\frac{\omega_r}{\kappa_r}\right)^2 + c_0^2 = 0 \qquad (7.4.67)$$

Letting the approximate shear wave speed $c_S \to +\infty$ in Eq. (7.4.67) removes the shear deformation accordingly and simplifies Eq. (7.4.67) as

$$c_0^2 - \left(1+\frac{1}{\kappa_r^2 r_g^2}\right)\left(\frac{\omega_r}{\kappa_r}\right)^2 = 0 \qquad (7.4.68)$$

Solving Eq. (7.4.68) for ω_r gives

$$\omega_r = c_0\kappa_r\left(1+\frac{1}{r_g^2\kappa_r^2}\right)^{-\frac{1}{2}} = c_0\kappa_r\left(\frac{r_g^2\kappa_r^2}{1+r_g^2\kappa_r^2}\right)^{\frac{1}{2}} = \frac{c_0 r_g\kappa_r^2}{\sqrt{1+r_g^2\kappa_r^2}} \qquad (7.4.69)$$

Noting that the numerator in Eq. (7.4.69) happens to be the natural frequency $\omega_{re} \equiv c_0 r_g\kappa_r^2$ of an E-B beam, we get

$$\omega_r = \frac{\omega_{re}}{\sqrt{1+r_g^2\kappa_r^2}} \qquad (7.4.70)$$

Equation (7.4.70) indicates the fact that the rotary inertia reduces the natural frequencies of a beam. This fact coincides with intuitions. Obviously, the effect of rotary inertia becomes significant for the natural frequencies of higher order.

In some previous studies, the condition $G \to +\infty$ was imposed on Eq. (7.4.34) such that Eq. (7.4.34) was simplified as

$$\rho A\frac{\partial^2 v(x,t)}{\partial t^2} + EI\frac{\partial^4 v(x,t)}{\partial x^4} - \rho I\frac{\partial^4 v(x,t)}{\partial x^2\partial t^2} = 0 \qquad (7.4.71)$$

In this case, let the natural vibration be expressed as

$$v_r(x, t) = a_{2r} \sin(\kappa_r x) \sin(\omega_r t), \quad r = 1, 2, \ldots \quad (7.4.72)$$

Substitution of Eq. (7.4.72) into Eq. (7.4.71) leads to

$$\omega_r = \sqrt{\frac{EI\kappa_r^4}{\rho A(1 + r_g^2 \kappa_r^2)}} = \frac{c_0 r_g \kappa_r^2}{\sqrt{1 + r_g^2 \kappa_r^2}} = \frac{\omega_{re}}{\sqrt{1 + r_g^2 \kappa_r^2}}, \quad r = 1, 2, \ldots \quad (7.4.73)$$

In a recent study[31], however, the above treatment was criticized to violate the constitutive law of elastic materials since $G \to +\infty$ implies $E \to +\infty$. As a matter of fact, the former treatment required that the longitudinal wave speed c_0 remains unchanged when $c_S \to +\infty$. This also implies independence between elastic moduli E and G.

(2)   Influence of shear deformation

As stated in the beginning of this subsection, Rankine studied the influence of shear deformation of a beam with rotary inertia neglected. Following his idea, we neglect the term associated with $\rho I \omega^2$ in Eq. (7.4.39) and simplify Eq. (7.4.40) as

$$\beta EG r_g^2 s^4 + \rho r_g^2 E \omega^2 s^2 - \rho \beta G \omega^2 = 0 \quad (7.4.74)$$

Substitution of $s_{1,2} = \pm i\kappa_r$ into Eq. (7.4.74) leads to the characteristic equation of the r-th natural frequency as follows

$$\beta EG r_g^2 \kappa_r^4 - \rho r_g^2 E \kappa_r^2 \omega^2 - \rho \beta G \omega^2 = 0, \quad r = 1, 2, \ldots \quad (7.4.75)$$

Solving Eq. (7.4.75) gives the natural frequencies

$$\omega_r = \sqrt{\frac{\beta EG r_g^2 \kappa_r^4}{\rho E r_g^2 \kappa_r^2 + \rho \beta G}} = \frac{c_0 c_S r_g \kappa_r^2}{\sqrt{c_0^2 r_g^2 \kappa_r^2 + c_S^2}} = \frac{\omega_{re}}{\sqrt{1 + (E/\beta G) r_g^2 \kappa_r^2}}, \quad r = 1, 2, \ldots \quad (7.4.76)$$

As shown in Eq. (7.4.76), the consideration of the shear deformation of a beam reduces all natural frequencies and agrees with intuitions. Accordingly, the influence becomes significant for the natural frequencies of higher order.

To check the effect of shear deformation directly, one can remove terms B and D in (7.4.34) simultaneously and then arrives at

$$\rho A \frac{\partial^2 v(x, t)}{\partial t^2} + EI \frac{\partial^4 v(x, t)}{\partial x^4} - \frac{\rho I E}{\beta G} \frac{\partial^4 v(x, t)}{\partial x^2 \partial t^2} = 0 \quad (7.4.77)$$

---

[31] Li L X, Pei Y L, Duan T C (2018). Study on physical essence of the Timoshenko beam theory and its relation with the Euler beam theory. Science China: Technological Science, 48(4): 360–368 (in Chinese).

Substitution of Eq. (7.4.72) into Eq. (7.4.77) leads to

$$EI\kappa_r^4 - \left(\frac{\rho I E}{\beta G}\kappa_r^2 + \rho A\right)\omega_r^2 = 0, \quad r = 1, 2, \ldots \tag{7.4.78}$$

There follows the same result as Eq. (7.4.76), namely,

$$\omega_r = \sqrt{\frac{EI\kappa_r^4}{\dfrac{\rho I E}{\beta G}\kappa_r^2 + \rho A}} = \frac{\omega_{re}}{\sqrt{1 + (E/\beta G)r_g^2\kappa_r^2}}, \quad r = 1, 2, \ldots \tag{7.4.79}$$

Equation (7.4.79) also implies that term D in Eq. (7.4.34) is due to the coupled influences of both shear deformation and rotary inertia of a T-E beam.

Remark 7.4.7 In some elementary textbooks, a simper treatment is to drop off the first term in Eq. (7.4.55) since it is much smaller than the second and third terms with wave speeds. As such, it is easier to get the approximate natural frequencies of the lower frequency branch of a T-E beam as follows[32]

$$\omega_{ra} = \frac{\omega_{re}}{\sqrt{1 + r_g^2\kappa_r^2 + r_g^2\kappa_r^2(c_0^2/c_S^2)}} \approx \omega_{re}\left[1 - \frac{r_g^2\kappa_r^2}{2}\left(1 + \frac{c_0^2}{c_S^2}\right)\right], \quad r = 1, 2, \ldots \tag{7.4.80}$$

A comparison of Eq. (7.4.80) with Eqs. (7.4.70) and (7.4.76) shows that Eq. (7.4.80) only accounts for the linear superposition of effects of shear deformation and rotary inertia, but does not take their coupled term in Eq. (h) in Example 7.4.2 into consideration.

7.4.4 Concluding Remarks

This section presented the detailed study on the free waves and vibrations of both E-B beams and T-E beams so as to address Problem 6A in part, and led to the major conclusions as following.

(1) The bending waves of a beam are dispersive waves, including two traveling waves in the left and the right directions, and two evanescent waves decaying away from the left and right boundaries. Different homogeneous boundaries of a beam exhibit different reflection properties for incident waves. For example, the incident traveling wave of an E-B beam is only superposed with a reflected traveling wave at a hinged boundary or a sliding boundary, but is mixed with a

---

[32] Timoshenko S, Young D H, Weaver Jr W (1974). Vibration Problems in Engineering. 4th Edition. New York: John Wiley and Sons Inc., 434–435.

reflected evanescent wave at a free clamped boundary or a free boundary. This property well agrees with those of the dual of boundaries in Sect. 5.3.

(2) The study of bending waves of an E-B beam is a simplified theory of one-dimensional waves based on both static and dynamic assumptions, and hence greatly deviates from three-dimensional elasto-dynamics when the wave number becomes large. For example, both phase velocity and group velocity of an E-B beam approach infinity with increasing of wave number, thus, the E-B beam can only be used for the dynamic problem of a slender beam with a few of wave numbers.

(3) The T-B beam with rotary inertia and shear deformation taken into account provides a significant improvement for E-B beams. The phase velocity of lower frequency branch of a T-B beam with a circular cross-section well agrees with the exact solution of three-dimensional elasto-dynamics and can predict the bending waves of a beam when the wave number is large. Yet, the phase velocity of higher frequency branch may not be reliable for the accurate computations of a waveguide.

(4) A comparison of wave dispersions between E-B beams and T-E beams shows that E-B beams well predict the wave dispersions when $r_g \kappa < 0.1$ holds. Here, r_g is the radius of gyration and κ is the wave number. As such, E-B beams are applicable to the dynamic problems with the shortest wave length larger than 20 times of the cross-sectional height.

7.5 Forced Waves and Vibrations of Beams

Based on the studies in Sect. 7.4, this section deals with the forced waves and forced vibrations, including an impacting response, of uniform E-B beams, or called beams for short. The section begins with the steady-state waves of an infinitely long uniform beam subject to a transverse harmonic force, and then turns to the steady-state vibration of a finitely long uniform beam to a transverse force. Finally, the section presents how to compute the impacting response of a clamped-free uniform beam by using the modal superposition method so as to address Problem 6A and 6B proposed in Chap. 1.

7.5.1 Waves of an Infinitely Long Beam to a Transverse Harmonic Force

In principle, the methods in Sects. 7.1.1 and 7.1.2 are available to study the harmonic waves and vibrations of an infinitely long uniform beam. This subsection prefers the complex function method and does not make any comparison of different methods.

7.5.1.1 Bending Waves to a Transverse Harmonic Force

As shown in Fig. 7.34, an infinitely long uniform beam is subject to a transverse harmonic force $F\cos(\omega_0 t)$ at the origin of axis x, which coincides with the neutral axis of the beam.

Let $v(x, t)$ be the dynamic deflection of the beam at cross-section x and moment t, which yields the dynamic equation as follows

$$\rho A \frac{\partial^2 v(x,t)}{\partial t^2} + EI \frac{\partial^4 v(x,t)}{\partial x^4} = \delta(x)F\cos(\omega_0 t), \quad x \in (-\infty, +\infty), \quad t \in [0, +\infty)$$

$$(7.5.1)$$

Recasting Eq. (7.5.1) as a partial differential equation of complex function $v_c(x, t)$ under a complex exponential excitation, we have

$$\rho A \frac{\partial^2 v_c(x,t)}{\partial t^2} + EI \frac{\partial^4 v_c(x,t)}{\partial x^4} = \delta(x)F\exp(i\omega_0 t), \quad x \in (-\infty, +\infty), \quad t \in [0, +\infty)$$

$$(7.5.2)$$

Accordingly, the real part of Eq. (7.5.2) is Eq. (7.5.1) and the real part of solution of Eq. (7.5.2) is the solution of Eq. (7.5.1).

Let the solution of Eq. (7.5.2) be expressed as

$$v_c(x, t) = \tilde{v}_c(x)\exp(i\omega_0 t), \quad x \in (-\infty, +\infty), \quad t \in [0, +\infty) \tag{7.5.3}$$

where $\tilde{v}_c(x)$ is the complex amplitude to be determined. Substitution of Eq. (7.5.3) into Eq. (7.5.2) leads to an ordinary differential equation in terms of complex amplitude $\tilde{v}_c(x)$, namely,

$$\frac{d^4 \tilde{v}_c(x)}{dx^4} - \kappa^4 \tilde{v}_c(x) = \frac{F}{EI}\delta(x), \quad x \in (-\infty, +\infty), \quad \kappa \equiv \sqrt[4]{\frac{\rho A \omega_0^2}{EI}} > 0 \quad (7.5.4)$$

Here, κ is the wave number corresponding to excitation frequency ω_0.

Fig. 7.34 An infinitely long uniform beam subject to a transverse harmonic force

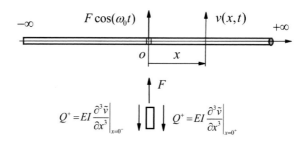

When $x \neq 0$, Eq. (7.5.4) is the same homogeneous differential equation in Eq. (7.4.3), which has the general solution given in Eq. (7.4.4). As discussed for Eq. (7.4.5), in the right part of beam, i.e., $x > 0$, Eq. (7.4.4) has only right traveling wave $a_1 \exp(-i\kappa x)$ and right evanescent wave $a_3 \exp(-\kappa x)$ such that $a_2 = 0$ and $a_4 = 0$ hold for Eq. (7.4.4), while in the left part of beam, i.e., $x < 0$, $a_1 = 0$ and $a_3 = 0$ hold in Eq. (7.4.4). That is, Eq. (7.4.4) has only left traveling wave $a_2 \exp(i\kappa x)$ and left evanescent wave $a_4 \exp(\kappa x)$, which decays with increasing of $|-x|$. As a consequence, the solution of Eq. (7.5.4) reads

$$\tilde{v}_c(x) = \begin{cases} a_1 \exp(-i\kappa x) + a_3 \exp(-\kappa x), & x \in (0, +\infty) \\ a_2 \exp(i\kappa x) + a_4 \exp(\kappa x), & x \in (-\infty, 0) \end{cases} \tag{7.5.5}$$

where constants a_r, $r = 1, 2, 3, 4$ are to be determined from the compatibility conditions at the excitation location in two steps.

The first step is to notice that the infinitely long beam is symmetric about the cross-section $x = 0$ such that the slope of the beam vanishes at $x = 0$. From Eq. (7.5.5), thus, we have

$$\left. \frac{d\tilde{v}_c(x)}{dx} \right|_{x=0} = \begin{cases} -i\kappa a_1 - \kappa a_3 = 0, & x \in (0, +\infty) \\ i\kappa a_2 + \kappa a_4 = 0, & x \in (-\infty, 0) \end{cases} \tag{7.5.6}$$

There follows

$$a_1 = ia_3, \quad a_2 = ia_4 \tag{7.5.7}$$

Next, we take an infinitesimal segment of the beam with length dx at $x = 0$ as shown in Fig. 7.34, where the inertial force of the infinitesimal segment yields $\rho A \omega^2 v_c(0) \exp(i\omega_0 t) dx \to 0$ when $dx \to 0$, while the shear forces $Q^+ \exp(i\omega_0 t)$ and $Q^- \exp(i\omega_0 t)$ applied to both sides of the infinitesimal segment are balanced with the excitation force $F \exp(i\omega_0 t)$ as follows

$$\begin{cases} \dfrac{F}{2} = Q^+ = EI \left. \dfrac{d^3 \tilde{v}_c(x)}{dx^3} \right|_{x=0^+} = EI(i\kappa^3 a_1 - \kappa^3 a_3), & x = 0^+ \\ \dfrac{F}{2} = Q^- = -EI \left. \dfrac{d^3 \tilde{v}_c(x)}{dx^3} \right|_{x=0^-} = EI(i\kappa^3 a_2 - \kappa^3 a_4), & x = 0^- \end{cases} \tag{7.5.8}$$

Substituting Eq. (7.5.7) into Eq. (7.5.8) leads to

$$\begin{cases} a_1 = -\dfrac{iF}{4EI\kappa^3}, & a_2 = -\dfrac{iF}{4EI\kappa^3} \\ a_3 = -\dfrac{F}{4EI\kappa^3}, & a_4 = -\dfrac{F}{4EI\kappa^3} \end{cases} \tag{7.5.9}$$

In view of Eq. (7.5.5) and Eq. (7.5.9), we have the complex solution of Eq. (7.5.4) as follows

$$\tilde{v}_c(x) = -\frac{F}{4EI\kappa^3} \begin{cases} i\exp(-i\kappa x) + \exp(-\kappa x), & x \in (0, +\infty) \\ i\exp(i\kappa x) + \exp(\kappa x), & x \in (-\infty, 0) \end{cases}$$

$$= -\frac{F}{4EI\kappa^3}\left[i\exp(-i\kappa|x|) + \exp(-\kappa|x|)\right] \tag{7.5.10}$$

Substitution of Eq. (7.5.10) into Eq. (7.5.3) leads to the complex solution of Eq. (7.5.2), namely,

$$v_c(x,t) = -\frac{F}{4EI\lambda^3}\{i\exp[i(\omega_0 t - \kappa|x|)] + \exp(-\kappa|x|)\exp(i\omega_0 t)\} \tag{7.5.11}$$

The real part of Eq. (7.5.11) is the solution of Eq. (7.5.1), that is, the harmonic bending wave of the beam as follows

$$v(x,t) = \frac{F}{4EI\kappa^3}\left[\sin(\omega_0 t - \kappa|x|) - \exp(-\kappa|x|)\cos(\omega_0 t)\right] \tag{7.5.12}$$

Given $F = 4EI\kappa^3$, $\omega_0 = \pi$ and $x_0 = 0$, Fig. 7.35 depicts the temporal-spatial structure of the harmonic bending wave in Eq. (7.5.12), exhibiting the left and right traveling waves along two characteristic lines.

Accordingly, if the transverse harmonic force is applied to the cross-section x_0, Eqs. (7.5.10) and (7.5.12) can be recast as

$$\tilde{v}_c(x) = -\frac{F}{4EI\kappa^3}\left[i\exp(-i\kappa|x - x_0|) + \exp(-\kappa|x - x_0|)\right] \tag{7.5.13}$$

$$v(x,t) = \frac{F}{4EI\kappa^3}\left[\sin(\omega_0 t - \kappa|x - x_0|) - \exp(-\kappa|x - x_0|)\cos(\omega_0 t)\right] \tag{7.5.14}$$

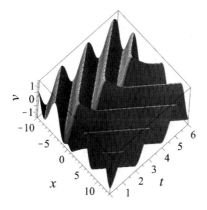

Fig. 7.35 The temporal-spatial structure of the harmonic bending waves of the beam

7.5.1.2 Bending Vibration at Excitation Location

As studied in Sects. 7.1.1 and 7.1.2, the steady-state vibration of an infinitely long rod at the position of an axial harmonic force has a phase delay $\pi/2$ compared with the harmonic force. The steady-state vibration of an infinitely long beam at the position of a transverse harmonic force looks more complicated.

When $x = 0$, Eq. (7.5.12) leads to the steady-state vibration of an infinitely long beam at the position of a transverse harmonic force, or called the direct steady-state vibration for short, namely,

$$v(0, t) = \frac{F}{4EI\kappa^3}[\sin(\omega_0 t) - \cos(\omega_0 t)] = \frac{F}{2\sqrt{2}EI\kappa^3}\cos\left(\omega_0 t - \frac{3\pi}{4}\right) \quad (7.5.15)$$

That is, the direct steady-state vibration has a delayed phase of $3\pi/4$ with respect to the transverse harmonic force.

Remark 7.5.1 It is worthy to notice that Eq. (7.5.3) describes the steady-state vibration, rather than any transient response, of an infinitely long beam. According to the formulation procedure from Eqs. (7.5.12) to (7.5.15), the direct steady-state vibration has a phase shift of $\pi/2$ with respect to the harmonic force if the vibration includes traveling waves only. That is, the phase shift $3\pi/4$ includes the contribution of evanescent waves.

As Eq. (7.5.15) describes the direct steady-state vibration, one may wonder if Eq. (7.5.15) covers the static case when $\omega_0 \to 0^+$. Noting, from Eq. (7.4.7), that the harmonic wave number of the beam is $\kappa = O(\sqrt{\omega_0})$, we derive the direct steady-state vibration of the beam from Eq. (7.5.15), when $\omega_0 \to 0^+$, as

$$\lim_{\omega_0 \to 0} v(0, t) = \frac{F}{4EI} \lim_{\omega_0 \to 0^+} \frac{1}{\kappa^3}[\sin(\omega_0 t) - \cos(\omega_0 t)]$$

$$= \frac{F}{4EI(\rho A/EI)^{3/4}} \lim_{\omega_0 \to 0^+} \frac{\sin(\omega_0 t) - \cos(\omega_0 t)}{\omega_0^{3/2}} \quad (7.5.16)$$

Given a finite time, the limit of Eq. (7.5.16) is minus infinity and contradicts any intuition of mechanics.

As a matter of fact, the settle-down time of the steady-state vibration yields $t \to +\infty$ when $\omega_0 \to 0^+$ so that the direct steady-state vibration does not approach minus infinity. Given $F = 4EI(\rho A/EI)^{3/4}$, for example, Fig. 7.36 shows the three-dimensional contours of Eq. (7.5.16) with a variation of (ω_0, t) and indicates that there exists a moment t so that $v(0, t) > 0$ holds when $\omega_0 \to 0^+$. The direct steady-state vibration $v(0, t)$ oscillates around the equilibrium position when $t \to +\infty$.

Fig. 7.36 The displacement history of an infinitely long uniform beam at the excitation position with a variation of excitation frequency

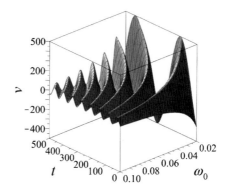

7.5.2 Vibrations of a Finitely Long Beam to a Transverse Harmonic Force

As studied in Sects. 7.4.1 and 7.5.1, Eq. (7.4.4) describes the amplitude of free waves of a beam with arbitrary boundaries, while Eq. (7.5.13) is the amplitude of the forced wave of an infinitely long beam under a transverse harmonic force at cross-section x_0. As such, their superposition describes the wave amplitude of a beam with arbitrary boundary conditions and subject to a transverse harmonic excitation at cross-section x_0 as follows

$$\tilde{v}_c(x) = a_1 \exp(-i\kappa x) + a_2 \exp(i\kappa x) + a_3 \exp(-\kappa x) + a_4 \exp(\kappa x)$$
$$-\frac{F}{4EI\kappa^3}\left[i\exp(-i\kappa|x - x_0|) + \exp(-\kappa|x - x_0|)\right] \quad (7.5.17)$$

where $a_k, k = 1, 2, 3, 4$ are the complex constants to be determined from four boundary conditions at two ends of the beam.

Given the excitation position x_0 and the response position x on the beam, we can select the excitation frequency ω_0 and derive a set of linear algebraic equations in view of $\kappa = \sqrt[4]{\rho A \omega_0^2/(EI)}$ and four boundary conditions, and then solve the four complex constants $a_k(\omega_0), k = 1, 2, 3, 4$. Substituting them into Eq. (7.5.17) with $F = 1$ leads to the frequency response function $H_{xx_0}(\omega_0)$ of the beam. The following three examples will demonstrate the above procedure.

7.5.2.1 Two Case Studies of a Hinged-Hinged Beam

Example 7.5.1 Consider a hinged-hinged steel beam with a rectangular cross-section and subject to a transverse harmonic force at the mid-span of the beam. Let b and h be the width and height of the cross-section, and the parameters of the beam be as following

$$L = 1 \text{ m}, \quad b = 0.01 \text{ m}, \quad h = 0.005 \text{ m}, \quad E = 210 \text{ GPa}, \quad \rho = 7800 \text{ kg/m}^3 \tag{a}$$

Given the excitation amplitude $F = 1$ N, compute the direct frequency response function.

As the problem is mirror-symmetric about the mid-span cross-section of the beam, setting the origin of coordinate at the mid-span and letting $x_0 = 0$, $a_1 = a_2$ and $a_3 = a_4$, we simplify Eq. (7.5.17) as

$$\tilde{v}_c(x) = 2a_1 \cos(\kappa x) + 2a_3 \cosh(\kappa x) - \frac{F}{4EI\kappa^3}\left[i\exp(-i\kappa|x|) + \exp(-\kappa|x|)\right] \tag{b}$$

From the mirro-symmetry of Eq. (b), we only need to consider the boundary conditions at one of ends of the beam, e.g.,

$$\tilde{v}_c\left(\frac{L}{2}\right) = 0, \quad \left.\frac{d^2\tilde{v}_c(x)}{dx^2}\right|_{x=L/2} = 0 \tag{c}$$

Substitution of Eq. (b) into Eq. (c) leads to a set of linear algebraic equations of complex constants a_1 and a_3 as follows

$$\begin{bmatrix} 2\cos(\kappa L/2) & 2\cosh(\kappa L/2) \\ -2\cos(\kappa L/2) & 2\cosh(\kappa L/2) \end{bmatrix}\begin{bmatrix} a_1 \\ a_3 \end{bmatrix}$$
$$= \frac{F}{4EI\kappa^3}\begin{bmatrix} \exp(-\kappa L/2) - i\exp(-i\kappa L/2) \\ \kappa^2 \exp(-\kappa L/2) - i\kappa^2 \exp(-i\kappa L/2) \end{bmatrix} \tag{d}$$

The condition of singular coefficient matrix in Eq. (d) is

$$\Delta = 4\kappa^2 \cos\left(\frac{\kappa L}{2}\right)\cosh\left(\frac{\kappa L}{2}\right) = 0 \tag{e}$$

It is easy to get, from Eq. (e), the wave number and natural frequencies of the hinged-hinged beam as following

$$\kappa_r = \frac{(2r-1)\pi}{L}, \quad \omega_r = \frac{(2r-1)^2\pi^2 c_0 r_g}{L^2}, \quad r = 1, 2, \ldots \tag{f}$$

Except for the above wave numbers, Eq. (d) has a set of unique solutions. Solving Eq. (d) for a_1 and a_3, and substituting them into Eq. (b), we get the frequency response function $\tilde{v}_c(x)$.

Now, setting $x = x_0 = 0$ and replacing wave number κ with excitation frequency f, which yields

Fig. 7.37 The amplitude of the direct frequency response function at the mid-span of a hinged-hinged beam

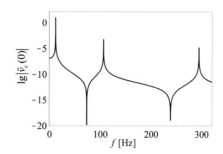

$$\kappa = \frac{\sqrt{\omega_0}}{\sqrt[4]{EI/\rho A}} = \frac{2\pi f}{\sqrt[4]{EI/\rho A}}, \quad f \in [0\,\text{Hz}, \ 320\,\text{Hz}] \tag{g}$$

we have the amplitude of direct frequency response function as shown in Fig. 7.37, where only three resonant peaks of odd orders appear because of the mirror-symmetry of the problem.

Example 7.5.2 If the transverse harmonic force in Example 7.5.1 is moved to a quarter-span measured from the left end of the beam, compute the direct frequency response function and make a comparison with the solution of modal superposition.

Let the left end of the beam be the origin of coordinate, then the excitation position and response position become $x = x_0 = L/4$. As this problem does not have mirror-symmetry, it is necessary to derive a set of linear algebraic equations of four unknowns $a_k(f)$, $k = 1, 2, 3, 4$ from Eq. (7.5.17) under four boundary conditions. Using software Maple to compute $a_k(f)$, $k = 1, 2, 3, 4$ for excitation frequency $f \in [0\,\text{Hz}, \ 320\,\text{Hz}]$, we get the amplitude of the direct frequency response function as shown in Fig. 7.38, where the fourth resonant peak disappears since $x_0 = L/4$ is a node of the fourth mode shape of the hinged-hinged beam.

Accordingly, the above result is the exact solution of the direct frequency response function with the accuracy governed by computers, while the solution of modal superposition has truncation errors. In Fig. 7.38, the thin solid line and dashed line represent

Fig. 7.38 The amplitufde of the direct frequency response function of a hinged-hinged beam at a quarter-span (thick solid line: the exact solution, thin solid line: a modal solution of 5 modes, dashed line: a modal solution of 7 modes)

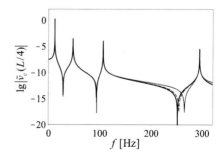

the solutions of modal superposition with five and seven modes, respectively. The solution with seven modes exhibits a more accurate anti-resonance around 250 Hz.

This example well supports the assertion that the modal superposition method gives the reliable dynamic response of a beam at a lower frequency band if the mode shapes involved are accurate enough. Of course, it is necessary to check the validity of the E-B beam model or use a T-E beam model directly if the dynamic response of a beam involves any mode shapes of higher order.

7.5.2.2   A Case Study of a Clamped-Free Beam

Example 7.5.3 Given a clamped-free beam subject to a transverse harmonic force at the free end, compute the cross frequency response function when the beam parameters are the same as those in Example 7.5.1.

Let the left end of the beam be the origin of coordinate such that the right end of the beam corresponds to $x = L$. As we have not yet studied the dynamics of a beam under an excitation at the free end, we take $x_0 \to L^-$ in Eq. (7.5.17) as the free end and compute the dynamic response of the beam.

Figure 7.39 presents the amplitude of the cross frequency response function measured at $x = L/4$ from the left end. As analyzed in Sect. 4.3, the cross frequency response function here does not exhibit any anti-resonance between two resonant peaks in a lower frequency band $f \in [0\,\text{Hz}, \quad 200\,\text{Hz}]$.

7.5.3   Impact Response of a Clamped-Free Beam

Compared with the impact problem of a uniform rod, it is a tough work to compute the impact response of a finitely long uniform beam because the bending wave of a simple E-B beam exhibits dispersion. The available choice of impact computations for beams is either the modal superposition method or the direct numerical integration. Among the studies based on the modal superposition method, some used several hundreds

Fig. 7.39 The amplitude of the cross frequency response function of a clamped-free beam $(x_0 \to L^-, \; x = L/4)$

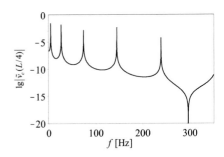

Fig. 7.40 The schematic of a clamped-free beam impacting a rigid wall

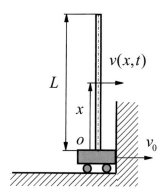

or even one thousand of mode shapes of a beam to compute the impact response[33], while some others used only a few mode shapes of the beam[34]. One may question the proper selection of modal truncation. In addition, the mode shapes of higher order of an E-B beam do not carry proper information about the bending waves as studied in Sect. 7.4 and make some troubles to modal superposition. This subsection demonstrates how to deal with the above issues through an impact problem of a clamped-free beam.

As shown in Fig. 7.40, a clamped-free E-B beam horizontally impacts a rigid wall with a velocity v_0 and then keeps contact with the wall. We take the beginning of the contact as the initial moment of the problem and establish the dynamic equation of the beam as following

$$\rho A \frac{\partial^2 v(x, t)}{\partial t^2} + EI \frac{\partial^4 v(x, t)}{\partial x^4} = 0, \quad x \in [0, L], \quad t \in [0, +\infty) \qquad (7.5.18)$$

Equation (7.5.18) yields both boundary conditions and initial conditions as follows

$$\begin{cases} v(0, t) = 0, \quad v_x(0, t) = 0, \quad v_{xx}(L, t) = 0, \quad v_{xxx}(L, t) = 0 \\ v(x, 0) = 0, \quad v_t(x, 0) = v_0 \end{cases} \qquad (7.5.19)$$

The computational experience indicates that the following three issues should be well treated if the modal superposition method is used to get the approximate solution of an impacting problem. The first issue is the modal expansion, the second is the approximate expressions for the mode shapes of higher order of an E-B beam, and the third is the right modal truncation.

---

[33] Su Y C, Ma C C (2012). Transient wave analysis of a cantilever Timoshenko beam subjected to impact loading by Laplace transform and normal mode methods. International Journal of Solids and Structures, 49(9): 1158–1176.

[34] Jin D P, Hu H Y (2005). Vibro-impacts and Controls. Beijing: Science Press, 67–74, 158–170 (in Chinese).

7.5.3.1 Modal Series Solution

Let the impact response of the E-B beam be expressed as the following modal series

$$v(x, t) = \sum_{r=1}^{+\infty} \overline{\varphi}_r(x) q_r(t), \quad x \in [0, L], \quad t \in [0, +\infty) \qquad (7.5.20)$$

where $\overline{\varphi}_r(x)$, $r = 1, 2, \ldots$ are the normalized mode shapes of a clamped-free beam with respect to the displacement of the free end. Their expressions are to be discussed in Sect. 7.5.3.2.

Substituting Eq. (7.5.20) into Eq. (7.5.18) leads to

$$\rho A \sum_{r=1}^{+\infty} \overline{\varphi}_r(x) \frac{\mathrm{d}^2 q_r(t)}{\mathrm{d}t^2} + EI \sum_{r=1}^{+\infty} \frac{\mathrm{d}^4 \overline{\varphi}_r(x)}{\mathrm{d}x^4} q_r(t) = 0 \qquad (7.5.21)$$

Let Eq. (7.5.21) be multiplied by $\rho A \overline{\varphi}_s(x)$ and integrated over the beam length, we derive, from the orthogonality of mode shapes, a set of decoupled ordinary differential equations

$$M_r \frac{\mathrm{d}^2 q_r(t)}{\mathrm{d}t^2} + K_r q_r(t) = 0, \quad r = 1, 2, \ldots \qquad (7.5.22)$$

where M_r and K_r are the r-th modal mass and stiffness coefficients as follows

$$\begin{cases} M_r \equiv \displaystyle\int_0^L \rho A \, \overline{\varphi}_r^2(x) \mathrm{d}x, \\[3mm] K_r \equiv \displaystyle\int_0^L EI \frac{\mathrm{d}^4 \overline{\varphi}_r(x)}{\mathrm{d}x^4} \varphi_r(x) \mathrm{d}x = \int_0^L EI \left[\frac{\mathrm{d}^2 \overline{\varphi}_r(x)}{\mathrm{d}x^2} \right]^2 \mathrm{d}x, \quad r = 1, 2, \ldots \end{cases}$$
$$(7.5.23)$$

The general solution of Eq. (7.5.22) yields

$$q_r(t) = a_r \cos(\omega_r t) + b_r \sin(\omega_r t), \quad r = 1, 2, \ldots \qquad (7.5.24)$$

where ω_r is the r-th natural frequency to be discussed in Sect. 7.5.3.2. Substituting Eq. (7.5.24), through Eq. (7.5.20), into the initial conditions in Eq. (7.5.19), we have

$$\sum_{r=1}^{+\infty} \overline{\varphi}_r(x) a_r = 0, \quad \sum_{r=1}^{+\infty} \overline{\varphi}_r(x) b_r \omega_r = v_0, \quad r = 1, 2, \ldots \qquad (7.5.25)$$

Let Eq. (7.5.25) be multiplied by $\rho A \overline{\varphi}_s(x)$ and then integrated over the beam length, we derive, using the orthogonality of mode shapes again, the coefficients in Eq. (7.5.25), namely,

$$a_r = 0, \quad b_r = \frac{v_0 \rho A}{M_r \omega_r} \int_0^L \overline{\varphi}_r(x) \mathrm{d}x, \quad r = 1, 2, \ldots \tag{7.5.26}$$

To this end, substituting Eq. (7.5.26), through Eq. (7.5.24), into Eq. (7.5.20), we derive the modal series solution of displacement as follows

$$v(x, t) = v_0 \rho A \sum_{r=1}^{+\infty} \frac{1}{M_r \omega_r} \overline{\varphi}_r(x) \sin(\omega_r t) \tag{7.5.27}$$

The corresponding modal series solutions of velocity and bending moment, which will be used for computing the dynamic stress, are as follows

$$v_t(x, t) = v_0 \rho A \sum_{r=1}^{+\infty} \frac{1}{M_r} \overline{\varphi}_r(x) \cos(\omega_r t) \tag{7.5.28}$$

$$M(x, t) = EI v_{xx}(x, t) = v_0 \rho A EI \sum_{r=1}^{+\infty} \frac{1}{M_r \omega_r} \overline{\varphi}_{rxx}(x) \sin(\omega_r t) \tag{7.5.29}$$

Accordingly, they are formal solutions to be justified.

7.5.3.2 Asymptotic Expression of Mode Shapes of Higher Order

From Table 2.1, it is easy to write the natural frequencies and normalized mode shapes of the clamped-free beam with respect to the free end displacement as follows

$$\begin{cases} \omega_r \equiv \kappa_r^2 \sqrt{\dfrac{EI}{\rho A}}, \quad \overline{\varphi}_r(x) \equiv \dfrac{\varphi_r(x)}{\varphi_r(L)}, \quad r = 1, 2, \ldots \\[4mm] \varphi_r(x) = \cosh(\kappa_r x) - \cos(\kappa_r x) + \dfrac{\cos(\kappa_r L) + \cosh(\kappa_r L)}{\sin(\kappa_r L) + \sinh(\kappa_r L)} [\sin(\kappa_r x) - \sinh(\kappa_r x)] \end{cases}$$
$$\tag{7.5.30}$$

where the wave number κ_r yields the following characteristic equation

$$\cos(\kappa_r L) \cosh(\kappa_r L) + 1 = 0 \tag{7.5.31}$$

The eigenvalues of Eq. (7.5.31) are

$$\begin{cases} \kappa_1 L = 1.8751, \quad \kappa_2 L = 4.6941, \quad \kappa_3 L = 7.8548 \\[2mm] \kappa_r L = \dfrac{(2r-1)\pi}{2}, \quad r = 4, 5, \ldots \end{cases} \tag{7.5.32}$$

The computation of modal series solution has to deal with the mode shapes of higher order, but the absolute values of hyperbolic triangle functions are much larger than those of triangle functions so that the subtraction of two hyperbolic triangle functions losses accuracy of computation and may lead to wrong numerical results. For example, software Maple fails to provide right results from Eq. (7.5.30) when the mode order reaches $r > 8$.

To make a rough estimation, let the mode shapes in Eq. (7.5.30) be approximated as

$$\lim_{\kappa_r \to +\infty} \varphi_r(x) = \lim_{\kappa_r \to +\infty} [\cosh(\kappa_r x) - \cos(\kappa_r x) + \sin(\kappa_r x) - \sinh(\kappa_r x)]$$

$$= \lim_{\kappa_r \to +\infty} \left[\sin(\kappa_r x) - \cos(\kappa_r x) + \exp(-\kappa_r x) \right] \qquad (7.5.33)$$

Here, Eq. (7.5.33) dose not have term $\exp(\kappa_r x)$ and can not satisfy the free boundary condition of an E-B beam. As such, it is necessary to derive a more accurate estimation, or called an asymptotic expression, for the mode shapes with large wave numbers.

To study the proper asymptotic expressions of the mode shapes when the wave number $\kappa_r \to +\infty$, we take the terms associated with the hyperbolic triangle functions of x from Eq. (7.5.30) and denote it as

$$\delta \equiv \cosh(\kappa_r x) - \frac{\cos(\kappa_r L) + \cosh(\kappa_r L)}{\sin(\kappa_r L) + \sinh(\kappa_r L)} \sinh(\kappa_r x) \qquad (7.5.34)$$

From Eq. (7.5.31), it is easy to see that $\cos(\kappa_r L) = -1/\cosh(\kappa_r L) \to 0$ holds when $\kappa_r \to +\infty$. There follows $\sin(\kappa_r L) \to (-1)^{r-1}$ when $\kappa_r \to +\infty$. Keeping these results in mind and using the asymptotic relations $\exp(-\kappa_r L) \to 0$ and $\exp(-\kappa_r x - \kappa_r L) \to 0$ when $\kappa_r \to +\infty$, we can formulate

$$\delta = \frac{\cosh(\kappa_r x)[\sin(\kappa_r L) + \sinh(\kappa_r L)] - [\cos(\kappa_r L) + \cosh(\kappa_r L)]\sinh(\kappa_r x)}{\sin(\kappa_r L) + \sinh(\kappa_r L)}$$

$$\approx \frac{[\exp(\kappa_r x) + \exp(-\kappa_r x)][2\sin(\kappa_r L) + \exp(\kappa_r L)] - \exp(\kappa_r L)[\exp(\kappa_r x) - \exp(-\kappa_r}{2\exp(\kappa_r L)}$$

$$= \frac{\sin(\kappa_r L)[\exp(\kappa_r x) + \exp(-\kappa_r x)] + \exp(\kappa_r L)\exp(-\kappa_r x)}{\exp(\kappa_r L)}$$

$$\approx (-1)^r \exp[\kappa_r(x - L)] + \exp(-\kappa_r x) \qquad (7.5.35)$$

Hence, the mode shape of higher order has an *asymptotic expression* as follows

$$\varphi_r(x) = \sin(\kappa_r x) - \cos(\kappa_r x) + \exp(-\kappa_r x) + (-1)^r \exp[\kappa_r(x - L)] \qquad (7.5.36)$$

Remark 7.5.2 Except for hinged-hinged E-B beams and sliding-sliding E-B beams, the mode shapes of other E-B beams include hyperbolic triangle functions. Hence, it is universally important to derive the asymptotic expressions for the mode shapes of higher order before computing the impacting response of E-B beams via the modal superposition method.

7.5.3.3 Modal Truncation

Compared with the studies of impacting rods, the computation of modal series solutions of an impacting E-B beam should deal with the following two issues. First, the mode shapes of most E-B beams include hyperbolic triangle functions and the coefficients of modal series solution are so complicated that the theory of Fourier series is not applicable to checking the convergence of modal series solution. Second, it is necessary to avoid using the mode shapes of higher order since the mode shapes of an E-B beam are only valid for the natural vibrations of small wave number. In what follows, these two issues are discussed, respectively.

We first follow the study in Sect. 7.4.2 to determine the maximal truncation order. In Fig. 7.31, the dimensionless phase velocity c_B/c_0 of an E-B beam is close to the dimensionless phase velocity c_{BPa}/c_0 of lower frequency branch of a T-E beam when $r_g\kappa < 0.1$, where κ is the wave number and r_g is the radius of gyration of cross-section of the beam. We use this inequality as a criterion and let the maximal wave number be $\kappa_{max} < 0.1/r_g$. From Eq. (7.5.32), then, the maximal truncation order n_{max} yields

$$\frac{(2n_{max} - 1)\pi}{2L} = \kappa_{max} < \frac{0.1}{r_g} \qquad (7.5.37)$$

There follows the maximal truncation order

$$n_{max} < \frac{0.1L}{\pi r_g} + \frac{1}{2} \qquad (7.5.38)$$

Of course, the maximal truncation order here is not a rigid bound, and a slightly excessive order does not bring large errors and essential mistakes in applications.

Next, we discuss the convergence of the modal series solution. In the discussion about the modal series solution of impacting rods in Sect. 7.3.5, the convergence rate was expressed as $O(1/n^\alpha)$ and the case when $\alpha > 3/2$ was said to be the fast uniform convergence. Now, we do not have a simple expression for M_r in Eq. (7.5.27), but can make a comparison between the coefficient $1/(M_r\omega_r)$ in Eq. (7.5.27) and $1/(M_1\omega_1r^\alpha)$ to study the convergence rate.

To make a graphic comparison, we define the *decreasing rate function* of the displacement series solution as

$$p_d(r) \equiv \lg\left(\frac{1}{M_r\omega_r}\right) = -\lg(M_r\omega_r) \qquad (7.5.39)$$

and a reference function with power α, namely,

$$p_\alpha(r) \equiv \lg\left(\frac{1}{M_1\omega_1 r^\alpha}\right) = -\alpha \lg(r) - \lg(M_1\omega_1) \qquad (7.5.40)$$

7.5.3.4 A Case Study

Example 7.5.4 Study the dynamic response of a clamped-free uniform E-B beam impacting the rigid wall with velocity $v_0 = 1$ m/s as shown in Fig. 7.40, where the cross-section of the beam is a rectangular with width b and height h. The parameters of the beam are

$$L = 1 \text{ m}, \quad b = 0.01\text{m} \quad h = 0.005 \text{ m}, \quad E = 210 \text{ GPa}, \quad \rho = 7800 \text{ kg/m}^3 \quad (a)$$

At first, we substitute the radius of gyration $r_g = h/\sqrt{12}$ and Eq. (a) into Eq. (7.5.38), and get the maximal modal truncation order as follows

$$n_{\max} < \frac{0.1L}{\pi r_g} + \frac{1}{2} = \frac{0.1\sqrt{12}L}{\pi h} + \frac{1}{2} = 22.55 \qquad (b)$$

Thus, we take the modal truncation order as $n_c = 20$.

Next, we check the convergence rate of the displacement seriese solution in Eq. (7.5.27). Figure 7.41 presents a comparison among decreasing rate function $p_d(r)$ and referene function $p_\alpha(r)$ for $\alpha = 3/2$ and $\alpha = 7/2$, respectively. It is easy to see that the displacement series solution has a convergence rate about $\alpha = 7/2$, exhibiting the fast uniform convergence.

Afterwards, we determine the wave velocities and the moment when the bending wave first arrives at the top end of the beam. From Eqs. (7.4.8) and (7.4.9), we get both phase velocities and group velocities of bending waves, corresponding to the natural vibrations of the first order and the 20-th order, respectively, as follows

Fig. 7.41 A comparion among decreasing rate function and reference functions for the displacement series solution of an impacting beam (thick solid line: $p_{3/2}(r)$, thin solid line: $p_{7/2}(r)$, solid circles: $p_d(r)$)

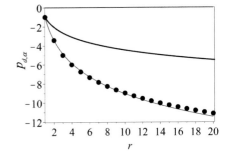

$$
\begin{cases}
c_{BPL} = c_0 r_g \kappa_1 = \sqrt{\dfrac{E}{\rho}} \cdot \dfrac{h}{\sqrt{12}} \cdot \dfrac{\pi}{2L} \approx 11.76 \text{ m/s}, \\[2mm]
c_{BGL} = 2 c_{BPL} \approx 23.53 \text{ m/s} \\[2mm]
c_{BPH} = c_0 r_g \kappa_{nc} = \sqrt{\dfrac{E}{\rho}} \cdot \dfrac{h}{\sqrt{12}} \cdot \dfrac{(2 n_c - 1)\pi}{2L} = 458.8 \text{ m/s}, \\[2mm]
c_{BGH} = 2 c_{BPH} \approx 917.6 \text{ m/s}
\end{cases}
\tag{c}
$$

Using Eq. (c), we derive the moments when the bending waves of the first order and the 20-th order arrive at the top end of the beam with group velocity and phase velocity as follows

$$
\begin{cases}
t_{GH} = \dfrac{L}{c_{BGH}} \approx 0.001 \text{ s}, \quad t_{BH} = 2 t_{GH} \approx 0.002 \text{ s} \\[3mm]
t_{GL} = \dfrac{L}{c_{BGL}} \approx 0.043 \text{ s}, \quad t_{BL} = 2 t_{GL} \approx 0.086 \text{ s}
\end{cases}
\tag{d}
$$

Now, we discuss the dynamic displacement $v(x, t)$ of the clamped-free E-B beam during the contact with the rigid wall in Eq. (7.5.27).

We first take the moment of $t_{GH} \approx 0.001$ s when the bending wave of the natural vibration of the 20-th order arrives at the top end of the beam with group velocity and check the temporal-spatial surface of wave $v(x, t)$ for $t \in [0, t_{GH}]$ as shown in Fig. 7.42a, where the intersection of surface $v(x, t)$ and plane $t = t_{GH}$, i.e., curve $v(x, t_{GH})$, depicts the dynamic configuration of the beam at moment $t = t_{GH}$. In Fig. 7.42a, the peak of the natural vibration of the first order arrives at a vicinity of $x = 0.1L$, while that of the 20-th order reaches the top end of the beam at $x = L$. However, the contribution of natural vibrations of higher order is negligible such that the displacement history of the beam at the top end, i.e., $v(L, t)$, looks like a straight line, which implies the rigid-body motion of the top end due to the beam deflection near the bottom end of beam.

Next, we take the moment of $t_{BH} \approx 0.002$ s when the natural vibration of the 20-th order first arrives at the top end of the beam with phase velocity and check the temporal-spatial surface of wave $v(x, t)$ for $t \in [0, t_{BH}]$. At this moment, the waves of natural vibrations with an order higher than four have been reflected at the top end of the beam such that the dynamic configuration $v(x, t_{BH})$ presents stronger oscillations and the displacement history $v(L, t)$ is not a straight line as shown in Fig. 7.42b, exhibiting an elastic deformation.

As time proceeds, the bending waves including more natural vibrations undergo a reflection at the top end of the beam such that the temporal-spatial surface looks more and more complicated. Figure 7.42c displays surface $v(x, t)$ for $t \in [0, t_{GL}]$ and the bending wave of the first order arrives at the top end of the beam with group velocity, while Fig. 7.42d shows surface $v(x, t)$ for $t \in [0, t_{BL}]$ and the bending wave of the first mode reaches the top end of the beam with phase velocity. As shown in 7.42d,

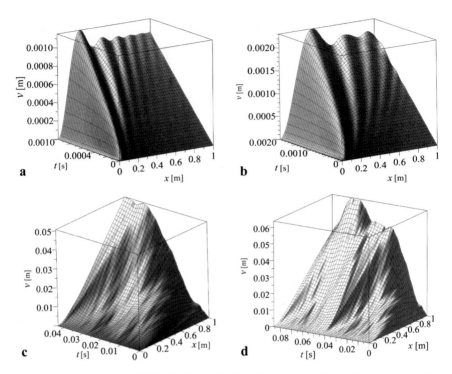

Fig. 7.42 The temporal-spatial distributions of the bending wave of a clamped-free beam impacting a rigid wall during the initial stage. **a** $t = t_{GH} \approx 0.001$ s; **b** $t = t_{BH} \approx 0.002$ s; **c** $t = t_{GL} \approx 0.043$ s; **d** $t = t_{BL} \approx 0.086$ s

the maximum of the tip displacement is about 0.006 m and happens at a moment about $t \approx 0.07$ s between t_{GL} and t_{BL}.

After the bending waves of natural vibrations of lower order undergo a reflection, it is hard to discuss the temporal-spatial distribution of bending wave. In this stage, we check the displacement of the top end of the beam and the bending moment of the bottom end of the beam as shown in Fig. 7.43.

As shown in Fig. 7.43, the two time histories get their maximal peaks simultaneously. It is worthy to point out that the bending moment computed via 20 modes exhibits many fluctuations of higher frequencies, which may be damped out by structural damping. To clearly show the result, Fig. 7.43b gives the bending moment computed via the first three modes. From this point of view, using more mode shapes, i.e., $n_c > n_{max}$, in Eq. (7.5.27) will smooth the approximate displacement solution in flat regions, but may not be a proper choice for computing the bending moment.

The studies and results in this subsection provide a reference for the dynamic analysis of slender space structures introduced in Sect. 1.2.6. For instance, the assembly of space structures in orbit will deal with various impact problems of slender structures, such as the transportation braking of a slender structure. If an aluminum truss with length $L = 50$ m is equivalent to an E-B beam with a square cross-section of

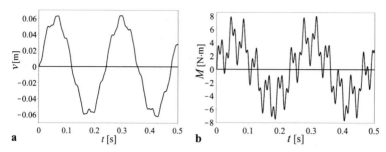

Fig. 7.43 The displacement of the top end and the bending moment of the bottom end of a clamped-free beam impacting a rigid wall during the initial stage. **a** the displacement of the top end $v(L, t)$; **b** the bending moment of the bottom end $M(0, t)$

edge length $b = h = 0.2$ m, it is possible to estimate the braking dynamics of the truss moving with velocity $v_0 = 0.1$ m/s. Following Example, 7.5.4, we get the moments when the wave of the first mode arrives at the free end of the truss with group velocity and phase velocity are about $2.7 \sim 5.4$ s, while the vibration period and vibration amplitude at the free end of the truss are about 15 s and 0.4 m. As a consequence, it is necessary to unitize advanced materials and control measures to reduce the vibration amplitude and vibration period.

7.5.4   Concluding Remarks

This section presented how to study the forced waves and vibrations of various E-B beams, with an emphasis on the impact waves, to address Problems 6A and 6B associated with beams. The major conclusions are as follows.

(1)   The complex function method enables one to derive the analytic expression for the steady-state wave of an infinitely long E-B beam to a transverse harmonic force. The traveling wave and evanescent wave shift the phase of the direct steady-state vibration by $3\pi/4$ with respect to the harmonic force. The direct steady-state vibration undergoes oscillations for long duration when the excitation frequency approaches zero.

(2)   Using the complex function method, one can easily compute the exact steady-state vibration of a finitely long E-B beam to a transverse harmonic force. The computation only solves a set of linear complex algebraic equations for four unknowns at a given excitation frequency, and does not deal with any transcendental eigenvalue problems for modal analysis. Thus, software Maple enables one to analyze the frequency response of an E-B beam with a few parameters, such as beam parameters, the excitation position and measurement position to be designed, and to verify the accuracy of modal series solutions.

(3)   The section demonstrated the impact computation of an E-B beam via the modal superposition method and addressed three important computational

Fig. 7.44 The schematic of
a single walled carbon
nanotube

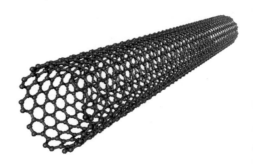

issues. That is, how to get the asymptotic expressions for the mode shapes
of higher order, how to determine the truncated mode order from the maximal
wave number in view of wave dispersion, and how to evaluate the convergence
rate of a modal series solution. The numerical examples validated the studies
on the above issues, which are helpful for computing other impact problmes.
The numerical results of the E-B beam impacting a rigid wall clearly show
the temporal-spatial distribution of a bending wave of the beam in terms of
different natural vibrations.

7.6   Further Reading and Thinking

(1)   Read chapters of one book among the references[35,36,37] and sort out the computational methods of structural impacts with wave effects neglected. Discuss how
to simplify the dynamic analysis of the rod horizontally impacting a rigid wall
in Sect. 7.3.

(2)   To study the bending waves of a single walled carbon nanotube, or called a
SWCN for short, shown in Fig. 7.44, it is possible to simplify the SWCN as
a T-E beam with periodic micro-structures, the equivalent material of which
yields a non-locally elastic constitutive law as follows

$$\sigma_x = E\left(\varepsilon_x + r^2 \frac{\partial^2 \varepsilon_x}{\partial x^2}\right) \qquad (7.6.1)$$

where σ_x and ε_x represent the normal stress and normal strain of a cross-
section of the beam, E is Young's modulus, r is a constant to describe the
micro-structure of an equivalent material for CWCN[38]. Establish the dynamic

---

[35] Stronge W J (200). Impact Mechanics. Cambridge: Cambridge University Press, 1–115.

[36] Hu H Y (1998). Mechanical Vibrations and Shocks. Beijing: Press of Aviation Industry, 242–262
(in Chinese).

[37] Jin D P, Hu H Y (2005). Vibro-impacts and Control. Beijing: Science Press, 1–74 (in Chinese).

[38] Wang L F, Hu H Y (2005). Flexural wave propagation in single-walled carbon nanotubes. Physical
Review B, 71(19): 195412.

equation of the T-E beam for a SWCN, study the group velocity, wave dispersion and cut-off frequency of the SWCN[39] and discuss the above studies from the viewpoint of beauty of science.

(3) Consider a clamped-clamped E-B beam and a clamped-clamped T-E beam, both of which have length L and a uniform rectangular cross-section with dimensionless height $h/L = 0.05$. Compute the first ten natural vibrations of two beams and derive the asymptotic expressions for their mode shapes of higher order. Assume that both beams are subject to a transverse harmonic force at position $x = L/5$ measured from the left end, compute the steady-state vibrations of the beams at a frequency band between the first six and ten natural frequencies and make a comparison of direct frequency response functions for two beams.

(4) Read the references[40,41,42] and discuss the coupling problem of bending wave and axial wave of an infinitely long E-B beam with an initial small curvature.

(5) Given a uniform rod vertically impacting a rigid ground at low velocity under gravity, study the influence of gravity on the rebounding moment of the rod and analyze the motion of the rod after rebounding.

[39] Wang L F, Guo W L, Hu H Y (2008). Group velocity of wave propagation in carbon nanotubes. Proceedings of the Royal Society A, 464(2094): 1423–1438.

[40] Rao S S (2007). Vibration of Continuous Systems. Hoboken: John Wiley & Sons Inc., 393–419.

[41] Doyle J F (1997). Wave Propagation in Structures. New York: Springer-Verlag, 43–104.

[42] Graff K F (1975). Wave Motion in Elastic Solids. Columbus: Ohio State University Press, 75–212.

Appendix
Three-Dimensional Waves in an Elastic Medium

Numerous references have been available so far for the three-dimensional waves in various continua. This appendix, hence, serves as an introduction to the three-dimensional waves in a linear homogeneous elastic medium, or called a medium for short, in order to assist readers to gain an insight into the waves and vibrations of one-dimensional structures in Chap. 7. Different from the appendices in most textbooks, the appendix includes the detailed deduction procedure for the above purpose.

A.1 Description of Three-Dimensional Waves

This section deals with the dynamic equations of three-dimensional waves in a medium and presents how to derive their general solutions via separation of variables in order to lay a theoretical foundation for subsequent sections.

A.1.1 Elasto-Dynamic Equations of a Three-Dimensional Medium

A.1.1.1 Basic Description

Given a medium and a fixed frame of Cartesian coordinates $ox_1x_2x_3$, the study begins with an infinitesimal cube centered at an arbitrary material point at (x_1, x_2, x_3) in the frame. Let $u_i(x_1, x_2, x_3, t)$, $i = 1, 2, 3$ be the displacement components of the material point at moment t in the frame of coordinates such that the linear strain components of the infinitesimal cube satisfy the geometric equations as follows

$$\varepsilon_{ij}(x_1, x_2, x_3, t) \equiv \frac{1}{2}\left(\frac{\partial u_i}{\partial x_j} + \frac{\partial u_j}{\partial x_i}\right), \quad i, j = 1, 2, 3 \tag{A.1.1}$$

© Science Press 2022
H. Hu, *Vibration Mechanics*,
https://doi.org/10.1007/978-981-16-5457-2

Let $\sigma_{ij}(x_1, x_2, x_3, t)$ be the stress components, $\rho f_i(x_1, x_2, x_3, t)$ be the body-force components with material density ρ, and $-\rho \partial^2 u_i(x_1, x_2, x_3, t)/\partial t^2$ be the inertial force components of the cube at moment t, respectively. According to d'Alembert's principle, the above force components applied to the cube satisfy the following equilibrium equations

$$\sum_{j=1}^{3} \frac{\partial \sigma_{ij}}{\partial x_j} + \rho f_i - \rho \frac{\partial^2 u_i}{\partial t^2} = 0, \quad i = 1, 2, 3 \tag{A.1.2}$$

In addition, the linear constitutive law of the medium of concern reads

$$\sigma_{ij} = \lambda \delta_{ij} e + 2\mu \varepsilon_{ij}, \quad e \equiv \sum_{k=1}^{3} \varepsilon_{kk}, \quad i, j = 1, 2, 3 \tag{A.1.3}$$

where λ and μ are *Lamé's constants*, e is the *bulk strain*, δ_{ij} is the Kronecker symbol satisfying $\delta_{ii} = 1$, $\delta_{ij} = 0$, $i \neq j$ as defined in Eq. (4.1.24). Sometimes, it may be preferable to use more intuitive material parameters, such as Young's modulus E, shear modulus G and Poisson's ratio ν, instead of Lamé's constants, which can be expressed as

$$\lambda = \frac{\nu E}{(1+\nu)(1-2\nu)}, \quad \mu = G = \frac{E}{2(1+\nu)} \tag{A.1.4}$$

From Eq. (A.1.3), the averaged stress σ_m yields

$$\sigma_m \equiv \frac{1}{3} \sum_{k=1}^{3} \sigma_{kk} = \frac{3\lambda + 2\mu}{3} \sum_{k=1}^{3} \varepsilon_{kk} = \frac{3\lambda + 2\mu}{3} e \tag{A.1.5}$$

whereby, the term of *bulk modulus* is defined as a ratio K of σ_m to e, namely,

$$K \equiv \frac{\sigma_m}{e} = \frac{3\lambda + 2\mu}{3} = \frac{E}{3(1-2\nu)} \tag{A.1.6}$$

Hence, Eqs. (A.1.4) and (A.1.6) result in the following two relations for later use

$$\lambda + 2\mu = K + \frac{4G}{3}, \quad \lambda + 2\mu = \frac{E(1-\nu)}{(1+\nu)(1-2\nu)} \tag{A.1.7}$$

A.1.1.2 Navier's Equations

According to the geometric equations, equilibrium equations and constitutive law, it is easy to derive the dynamic equations of a three-dimensional medium in terms of displacement components. For this purpose, substituting Eq. (A.1.1), through Eq. (A.1.3), into Eq. (A.1.2), we have

$$\lambda \delta_{ij} \frac{\partial}{\partial x_j} \sum_{k=1}^{3} \frac{\partial u_k}{\partial x_k} + \mu \sum_{j=1}^{3} \frac{\partial}{\partial x_j} \left(\frac{\partial u_i}{\partial x_j} + \frac{\partial u_j}{\partial x_i} \right) + \rho f_i - \rho \frac{\partial^2 u_i}{\partial t^2} = 0, \quad i = 1, 2, 3$$

(A.1.8)

Using the property of Kronecker symbol to simplify the first term in Eq. (A.1.8) and replacing subscript j in the second term in Eq. (A.1.8) with k, we derive

$$\lambda \frac{\partial}{\partial x_i} \sum_{k=1}^{3} \frac{\partial u_k}{\partial x_k} + \mu \sum_{k=1}^{3} \left(\frac{\partial^2 u_i}{\partial x_k^2} + \frac{\partial^2 u_k}{\partial x_i \partial x_k} \right) + \rho f_i - \rho \frac{\partial^2 u_i}{\partial t^2} = 0, \quad i = 1, 2, 3$$

(A.1.9)

By recasting Eq. (A.1.9), we have the dynamic equations of a three-dimensional medium in terms of displacement components as follows

$$(\lambda + \mu) \frac{\partial}{\partial x_i} \sum_{k=1}^{3} \frac{\partial u_k}{\partial x_k} + \mu \sum_{k=1}^{3} \frac{\partial^2 u_i}{\partial x_k^2} + \rho f_i = \rho \frac{\partial^2 u_i}{\partial t^2}, \quad i = 1, 2, 3 \qquad (A.1.10)$$

This is a set of partial differential equations first derived by C. L. Navier, a French scientist, in 1821 and has been called *Navier's equations* after his name.

It is straightforward to recast Navier's equations as a vector form, namely,

$$(\lambda + \mu) \nabla [\nabla \cdot \boldsymbol{u}(\boldsymbol{x}, t)] + \mu \nabla^2 \boldsymbol{u}(\boldsymbol{x}, t) + \rho \boldsymbol{f}(\boldsymbol{x}, t) = \rho \boldsymbol{u}_{tt}(\boldsymbol{x}, t) \qquad (A.1.11)$$

where $\boldsymbol{x} \equiv [x_1 \ x_2 \ x_3]^{\mathrm{T}}$ represents the position vector, the components in displacement vector $\boldsymbol{u}(\boldsymbol{x}, t)$ and body-force vector $\rho \boldsymbol{f}(\boldsymbol{x}, t)$ have been defined in Sect. A.1.1.1, while ∇ and ∇^2 are respectively *Nabla operator* and *Laplace operator* defined as

$$\nabla \equiv \sum_{j=1}^{3} \frac{\partial}{\partial x_j} \boldsymbol{i}_j, \quad \nabla^2 \equiv \nabla \cdot \nabla = \left(\sum_{k=1}^{3} \frac{\partial}{\partial x_j} \boldsymbol{i}_j \right) \cdot \left(\sum_{k=1}^{3} \frac{\partial}{\partial x_k} \boldsymbol{i}_k \right) = \sum_{j=1}^{3} \frac{\partial^2}{\partial x_j^2}$$

(A.1.12)

where $\boldsymbol{i}_j, \ j = 1, 2, 3$ are the unit base vectors of the frame of Cartesian coordinates.

A.1.2 Helmholtz Decomposition of a Displacement Field

We consider the displacement vector $u(x, t)$ governed by Navier's equation as a temporal-spatially varying vector field of describing the motions of a mass point x, including the dilatations, compressions, rotations and shear deformations of an infinitesimal cube centered at x. The following analysis shows that different motions exhibit their wave properties, such as their wave speeds, in particular.

For simplicity, we neglect body-force vector $\rho f(x, t)$ in Eq. (A.1.11) and get

$$(\lambda + \mu)\nabla[\nabla \cdot u(x, t)] + \mu\nabla^2 u(x, t) = \rho u_{tt}(x, t) \qquad (A.1.13)$$

The following analysis begins with two special cases, and then deals with the general case.

A.1.2.1 Two Special Cases

(1) Irrotational case

In this case, the displacement field has no rotations so that $\text{rot}(u) \equiv \nabla \times u = 0$ holds. The property of cross product of three vectors leads to

$$0 = \nabla \times (\nabla \times u) = \nabla(\nabla \cdot u) - (\nabla \cdot \nabla)u \Rightarrow \nabla(\nabla \cdot u) = (\nabla \cdot \nabla)u = \nabla^2 u$$
$$(A.1.14)$$

Thus, the two terms in the left-hand side of Eq. (A.1.13) can merge so that Eq. (A.1.13) becomes a three-dimensional wave equation as follows

$$c_p^2 \nabla^2 u(x, t) = u_{tt}(x, t), \quad c_p \equiv \sqrt{\frac{\lambda + 2\mu}{\rho}} = \sqrt{\frac{3K + 4G}{3\rho}} \qquad (A.1.15)$$

As the wave described by Eq. (A.1.15) does not undergo rotations, it is called an *irrotational wave*. Figure A.1a presents the schematic of the wave, which undergoes dilatations and compressions along axis x_1. Thus, the wave is also called a *pressure wave* or a *P-wave* for short. In addition, as an arbitrary mass point in the wave vibrates in the direction of wave propagation, the wave is also named a *longitudinal wave*.

(2) Rotational case

In this case, the displacement vector has no divergence such that $\text{div}(u) \equiv \nabla \cdot u = 0$ holds. Now, the medium does not undergo any volume changes due to either dilatations or compressions, but rotations and shear deformations. Thus, the first term in the left-hand side of Eq. (A.1.13) disappears and Eq. (A.1.13) becomes the following three-dimensional wave equation

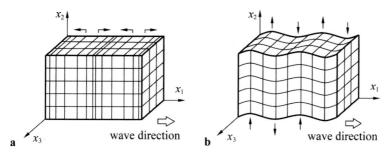

Fig. A.1 The schematic of two types of waves in a medium. **a** an irrotational wave, i.e., a P-wave; **b** an equivoluminal wave, i.e., an S-wave

$$c_s^2 \nabla^2 \boldsymbol{u}(\boldsymbol{x}, t) = \boldsymbol{u}_{tt}(\boldsymbol{x}, t), \quad c_s \equiv \sqrt{\frac{\mu}{\rho}} = \sqrt{\frac{G}{\rho}} \tag{A.1.16}$$

The wave governed by Eq. (A.1.16) does not have any volume changes and is called an *equivoluminal wave*. Figure A.1b shows the schematic of the wave undergoing a shear deformation in plane (x_2, x_3). Hence, the wave is also called a *shear wave* or an *S-wave* for short. Furthermore, the vibration of an arbitrary mass point in the wave is perpendicular to the direction of wave propagation. Thus, the wave is also named a *transverse wave*.

To check the relation between wave speeds c_p and c_s defined above, substituting Eq. (A.1.7) into Eq. (A.1.15) and Eq. (A.1.16), we have

$$\begin{cases} c_p \equiv \sqrt{\dfrac{\lambda + 2\mu}{\rho}} = \sqrt{\dfrac{E(1 - v)}{\rho(1 + v)(1 - 2v)}} = c_0 \sqrt{\dfrac{1 - v}{(1 + v)(1 - 2v)}} \geq c_0 \\[2ex] c_s \equiv \sqrt{\dfrac{\mu}{\rho}} = \sqrt{\dfrac{E}{2\rho(1 + v)}} = c_0 \sqrt{\dfrac{1}{2(1 + v)}} \end{cases} \tag{A.1.17}$$

where $c_0 \equiv \sqrt{E/\rho}$ is the longitudinal wave speed of a uniform rod in the elementary theory of one-dimensional waves.

Now, define the *ratio of wave speeds* as follows

$$D \equiv \frac{c_p}{c_s} = \sqrt{\frac{\lambda + 2\mu}{\mu}} = \sqrt{\frac{2(1 - v)}{1 - 2v}} > 1 \tag{A.1.18}$$

Equation (A.1.18) shows that any P-wave has a higher propagation speed than S-waves in the same medium. That is, the P wave due to an initial disturbance always arrives at a given position in the medium first, and then comes the S-wave. Hence, the P-wave and the S-wave also have the names of the *primary wave* and the *secondary wave*, respectively. This property will play an important role in subsequent studies.

Fig. A.2 The relation of
dimensionless wave speeds
and Poisson's ratio (thick
solid line: P-wave, thin solid
line: S-wave)

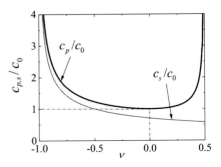

Figure A.2 presents the wave speeds in Eq. (A.1.17) scaled by c_0 with a variation of Poisson's ratio $\nu \in (-1, \quad 0.5)$. For a conventional elastic medium, Poisson's ratio yields $\nu \in [0, \quad 0.5)$ such that the relation $c_s/c_0 < 1 \leq c_p/c_0$ holds. That is, the longitudinal wave speed c_0 of a rod in the elementary theory is between the speeds of an S-wave and a P-wave. In particular, $\nu = 0$ leads to $c_p = c_0$ and $c_s = 0.707c_0$, while $\nu = 0.28$ gives $c_p = 1.13\,c_0$ and $c_s = 0.625c_0$.

Remark A.1.1: In three-dimensional elasto-dynamics, the P-wave includes dynamic dilatation, compression, and shear deformation since wave speed c_p depends on both bulk modulus K and shear modulus G. In the elementary theory of one-dimensional waves, however, the simplified waves in a rod exhibit only tension and compression such that the longitudinal wave speed c_0 depends on Young's modulus E only. It is worthy to notice that the P-wave always includes shear deformations even when Poisson's ratio vanishes and $c_p = c_0$ holds true. In fact, $\nu = 0$ leads to $\lambda = 0$ in Eq. (A.1.4) so that the first term in the left-hand side of Eq. (A.1.8) disappears, but the second term still represents shear deformations.

Remark A.1.2: In three-dimensional waves, the S-wave speed c_s depends on shear modulus $\mu = G$ and indicates the contribution of shear deformation to S-waves. For example, the torsional wave of an infinitely long shaft of circular cross-section exhibits pure shear deformations such that the torsional wave speed is c_s. As studied in Sect. 7.4.4, the bending wave speed of the lower frequency branch of a T-E beam monotonically increases and approaches the approximate shear wave speed $\sqrt{\beta G/\rho}$, indicating that the bending wave of high wave number is mainly a shear wave. Of course, the approximate shear wave speed is $0.571c_0$, lower than the shear wave speed $c_s \approx 0.625c_0$ when $\nu = 0.28$. As for an E-B beam, the bending wave looks like a transverse wave, but the bending wave speed greatly deviates from the shear wave speed in a three-dimensional medium since the shear deformation of cross-section of an E-B beam has not been taken into account.

A.1.2.2 Decomposition in a General Case

In 1829, S. D. Poisson, a French scientist, found that the displacement field $u(x, t)$ governed by Eq. (A.1.13) includes two parts, that is, an irrotational part and a rotational part. Based on *Helmholtz's theorem of decomposition* for a vector field in mathematics, G. Lamé, a French mathematician, introduced a scalar potential function $\varphi(x, t)$ and a vectorial potential function $\psi(x, t)$ in 1852 and decomposed the vector field $u(x, t)$ as

$$u(x, t) = \nabla\varphi(x, t) + \nabla \times \psi(x, t) \tag{A.1.19}$$

He proved that if scalar potential function $\varphi(x, t)$ and vectorial potential function $\psi(x, t)$ satisfy the following two wave equations

$$c_p^2 \nabla^2 \varphi(x, t) = \varphi_{tt}(x, t), \quad c_s^2 \nabla^2 \psi(x, t) = \psi_{tt}(x, t) \tag{A.1.20}$$

as well as a normalization condition

$$\nabla \cdot \psi(x, t) = 0 \tag{A.1.21}$$

Eq. (A.1.19) is the solution of Eq. (A.1.13).

To verify the above assertion, substitute Eq. (A.1.19) into Eq. (A.1.13) and get

$$(\lambda + \mu)\nabla\{\nabla \cdot [\nabla\varphi(x, t) + \nabla \times \psi(x, t)]\} + \mu\nabla^2[\nabla\varphi(x, t) + \nabla \times \psi(x, t)]$$
$$= \rho[\nabla\varphi_{tt}(x, t) + \nabla \times \psi_{tt}(x, t)] \tag{A.1.22}$$

In view of the property of a mixed product of three vectors, the term $\nabla \cdot [\nabla \times \psi(x, t)]$ in the first pair of braces in Eq. (A.1.22) disappears. Exchanging the differentiation sequence, we recast Eq. (A.1.22) as follows

$$\nabla[(\lambda + 2\mu)\nabla^2\varphi(x, t) - \rho\varphi_{tt}(x, t)] + \nabla \times [\mu\nabla^2\psi(x, t) - \rho\psi_{tt}(x, t)] = 0 \tag{A.1.23}$$

If Eq. (A.1.20) serves as preconditions, Eq. (A.1.23) becomes an identity so that Eq. (A.1.19) is the solution of Eq. (A.1.13).

It is worthy to notice that a vectorial potential function is equivalent to three scalar potential functions and the total number of potential functions is more than the number of displacement components in $u(x, t)$. Hence, it is necessary to use the normalization condition in Eq. (A.1.21) to eliminate a redundant potential function.

In practice, one usually uses the component forms of Eq. (A.1.19), Eq. (A.1.20) and Eq. (A.1.21), including the displacement components described by potential functions, the normalization condition, and four wave equations of three-dimensions as follows

$$\begin{cases} u_i = \dfrac{\partial \varphi}{\partial x_i} + \displaystyle\sum_{j=1}^{3}\sum_{k=1}^{3} e_{ijk}\dfrac{\partial \psi_k}{\partial x_j}, \quad \displaystyle\sum_{j=1}^{3}\dfrac{\partial \psi_j}{\partial x_j} = 0 \\[3mm] c_p^2 \displaystyle\sum_{j=1}^{3}\dfrac{\partial^2 \varphi}{\partial x_j^2} = \dfrac{\partial^2 \varphi}{\partial t^2}, \quad c_s^2 \displaystyle\sum_{j=1}^{3}\dfrac{\partial^2 \psi_i}{\partial x_j^2} = \dfrac{\partial^2 \psi_i}{\partial t^2}, \quad i = 1, 2, 3 \end{cases} \tag{A.1.24}$$

where e_{ijk} is the *Ricci symbol* defined as

$$e_{ijk} \equiv \begin{cases} 1, & i, j, k \text{ are even permutations of } 1, 2, 3 \\ -1, & i, j, k \text{ are odd permutations of } 1, 2, 3 \\ 0, & \text{two subscripts among } i, j, k \text{ are identical} \end{cases} \tag{A.1.25}$$

When solving Eq. (A.1.24), one only needs to solve three wave equations and to get, say, φ, ψ_1 and ψ_2, then to determine redundant potential function ψ_3 from the normalization condition in Eq. (A.1.24). Finally, substitution of all potential functions into the first equation in Eq. (A.1.24) leads to the displacement components. As all potential functions satisfy wave equations, the three-dimensional displacement waves are non-dispersive.

Remark A.1.3: The wave problem involving body-force vector $\rho f(x, t)$ is more complicated. One treatment is to take $\rho f(x, t)$ as a vector field and use Lamé's potential functions to describe $\rho f(x, t)$. The treatment begins with Eq. (A.1.11) and extends Eq. (A.1.15) and Eq. (A.1.16) to inhomogeneous wave equations, and then checks the effect of body-force vector $\rho f(x, t)$[1]. In the case of gravitational force, i.e., $\rho f(x, t) = \rho g$, which does not depend on position vector x, it is feasible to introduce an auxiliary vector $\tilde{u}(x, t) = u(x, t) - \rho g$ and to convert the inhomogeneous wave equations to homogeneous wave equations. This is the simplest case with a body-force vector.

A.1.3   Solutions of Three-Dimensional Wave Equations

As shown in Sect. A.1.2.2, the key issue of studying three-dimensional waves is to solve the wave equations governing both scalar and vectorial potential functions in a unified form as follows

$$c^2 \nabla^2 \phi(x, t) - \phi_{tt}(x, t) = 0, \quad c > 0 \tag{A.1.26}$$

where $\phi(x, t)$ is either scalar function $\phi(x, t)$, or the components ψ_1, ψ_2, ψ_3 of vectorial potential function $\psi(x, t)$, while c is the wave speeds c_p or c_s.

---

[1] Achenbach J D (1973). Wave Propagation in Elastic Solids. Amsterdam: North-Holland Publishing Company, 88–93.

Let the solution of Eq. (A.1.26) be in a form of separated variables

$$\phi(x, t) = \tilde{\phi}(x)q(t) \tag{A.1.27}$$

Substitution of Eq. (A.1.27) into Eq. (A.1.26) leads to

$$\frac{c^2}{\tilde{\phi}(x)}\nabla^2\tilde{\phi}(x) = \frac{1}{q(t)}\frac{d^2q(t)}{dt^2} \tag{A.1.28}$$

Following the idea of separation of variables for a one-dimensional wave equation in Sect. 7.1, function $q(t)$ describes a harmonic vibration if and only if both sides of Eq. (A.1.28) are equal to a constant $-\omega^2 < 0$. Under this condition, Eq. (A.1.28) becomes two differential equations as follows

$$\begin{cases} \dfrac{d^2q(t)}{dt^2} + \omega^2 q(t) = 0 \\ \nabla^2\tilde{\phi}(x) + \kappa^2\tilde{\phi}(x) = 0, \quad \kappa \equiv \dfrac{\omega}{c} > 0 \end{cases} \tag{A.1.29a, b}$$

The general solution of Eq. (A.1.29a) reads

$$q(t) = b_1\exp(-i\omega t) + b_2\exp(i\omega t) \tag{A.1.30}$$

which describes the time-varying property of a harmonic wave of frequency ω. Eq. (A.1.29b) is named the *Helmholtz equation*, which describes the spatial-varying property of a harmonic wave with a wave number κ.

The method of separation of variables enables one to solve Helmholtz equation in several steps. The first step is to let the solution of Eq. (A.1.29b) be in a form of separated variables as follows

$$\tilde{\phi}(x) = \phi_1(x_1)\phi_2(x_2)\phi_3(x_3) \tag{A.1.31}$$

Substituting Eq. (A.1.31) into Eq. (A.1.29b) leads to

$$\begin{aligned} &\phi_2(x_2)\phi_3(x_3)\frac{d^2\phi_1(x_1)}{dx_1^2} + \phi_1(x_1)\phi_3(x_3)\frac{d^2\phi_2(x_2)}{dx_2^2} \\ &+\phi_1(x_1)\phi_2(x_2)\frac{d^2\phi_3(x_3)}{dx_3^2} + \kappa^2\phi_1(x_1)\phi_2(x_2)\phi_3(x_3) = 0 \end{aligned} \tag{A.1.32}$$

or an equivalent form

$$\frac{1}{\phi_1(x_1)}\frac{d^2\phi_1(x_1)}{dx_1^2} + \frac{1}{\phi_2(x_2)}\frac{d^2\phi_2(x_2)}{dx_2^2} + \kappa^2 = -\frac{1}{\phi_3(x_3)}\frac{d^2\phi_3(x_3)}{dx_3^2} \tag{A.1.33}$$

To study traveling waves in axis x_3, one can assume that both sides of Eq. (A.1.33) are equal to a constant $k_3^2 > 0$ so that Eq. (A.1.33) becomes

$$\begin{cases} \dfrac{\mathrm{d}^2\phi_3(x_3)}{\mathrm{d}x_3^2} + \kappa_3^2\phi_3(x_3) = 0 \\[3mm] \dfrac{1}{\phi_1(x_1)}\dfrac{\mathrm{d}^2\phi_1(x_1)}{\mathrm{d}x_1^2} + \kappa^2 - \kappa_3^2 = -\dfrac{1}{\phi_2(x_2)}\dfrac{\mathrm{d}^2\phi_2(x_2)}{\mathrm{d}x_2^2} \end{cases} \qquad \text{(A.1.34a, b)}$$

As the second step, it is reasonable to let both sides in Eq. (A.1.34b) be equal to a constant $k_2^2 > 0$ such that Eq. (A.1.34b) becomes two ordinary differential equations as follows

$$\begin{cases} \dfrac{\mathrm{d}^2\phi_2(x_2)}{\mathrm{d}x_2^2} + \kappa_2^2\phi_2(x_2) = 0 \\[3mm] \dfrac{\mathrm{d}^2\phi_1(x_1)}{\mathrm{d}x_1^2} + \kappa_1^2\phi_1(x_1) = 0 \end{cases} \qquad \text{(A.1.35a, b)}$$

where

$$\sum_{j=1}^{3}\kappa_j^2 = \kappa^2 \qquad \text{(A.1.36)}$$

In the third step, as Eqs. (A.1.34a), (A.1.35a) and (A.1.35b) take the same form of a one-dimensional wave equation, they have the general solutions in the same form, namely,

$$\phi_j(x_j) = c_j\exp(\mathrm{i}\kappa_j x_j) + d_j\exp(-\mathrm{i}\kappa_j x_j), \quad \kappa_j > 0, \quad j = 1, 2, 3 \qquad \text{(A.1.37)}$$

If the conditions of $\kappa_j > 0$ are released, Eq. (A.1.37) can be recast as a more general form as follows

$$\phi_j(x_j) = c_j(\kappa_j)\exp(\mathrm{i}\kappa_j x_j), \quad j = 1, 2, 3 \qquad \text{(A.1.38)}$$

Substitution of Eq. (A.1.37) into Eq. (A.1.31) leads to

$$\begin{cases} \tilde{\phi}(\boldsymbol{x}) = \prod_{j=1}^{3} c_j(\kappa_j)\exp(\mathrm{i}\kappa_j x_j) = c\exp\left(\mathrm{i}\sum_{j=1}^{3}\kappa_j x_j\right) = c\exp(\mathrm{i}\boldsymbol{\kappa}^{\mathrm{T}}\boldsymbol{x}) \\[3mm] c \equiv c_1(\kappa_1)c_2(\kappa_2)c_3(\kappa_3), \quad \boldsymbol{\kappa} \equiv \begin{bmatrix} \kappa_1 & \kappa_2 & \kappa_3 \end{bmatrix}^{\mathrm{T}} \end{cases}$$

$$\text{(A.1.39)}$$

In Eq. (A.1.39), vector $\boldsymbol{\kappa}$ is called a *wave vector* defined as

$$\boldsymbol{\kappa} = \kappa \hat{\boldsymbol{\kappa}}, \quad \hat{\boldsymbol{\kappa}} \equiv \frac{\boldsymbol{\kappa}}{\kappa}, \quad \kappa \equiv \sqrt{\kappa_1^2 + \kappa_2^2 + \kappa_3^2} \tag{A.1.40}$$

where the unit vector $\hat{\boldsymbol{\kappa}}$ points the direction of wave propagation, κ is the wave number along vector $\hat{\boldsymbol{\kappa}}$, that is, the wave density in direction $\hat{\boldsymbol{\kappa}}$. Now, the integration constant c depends on the components of wave vector and can be denoted as $c(\boldsymbol{\kappa})$.

Finally, substituting Eqs. (A.1.30) and (A.1.39) into Eq. (A.1.27) and using integration constants $a_j(\boldsymbol{\kappa}) = b_j c(\boldsymbol{\kappa}), \quad j = 1, 2$, we derive

$$\phi_{\boldsymbol{\kappa}}(\boldsymbol{x}, t) = a_1(\boldsymbol{\kappa}) \exp\left[i(\boldsymbol{\kappa}^T \boldsymbol{x} - \omega t)\right] + a_2(\boldsymbol{\kappa}) \exp\left[i(\boldsymbol{\kappa}^T \boldsymbol{x} + \omega t)\right] \tag{A.1.41}$$

Equation (A.1.41) gives a solution of Eq. (A.1.26) along a given wave vector $\boldsymbol{\kappa}$. For arbitrary components of a wave vector, the superposition principle of linear differential equations enables one to integrate Eq. (A.1.41) with respect to wave vector $\boldsymbol{\kappa}$ and to derive the general solution of Eq. (A.1.26) as follows

$$\phi(\boldsymbol{x}, t) = \int_{-\infty}^{+\infty} \left\{ a_1(\boldsymbol{\kappa}) \exp\left[i(\boldsymbol{\kappa}^T \boldsymbol{x} - \omega t)\right] + a_2(\boldsymbol{\kappa}) \exp\left[i(\boldsymbol{\kappa}^T \boldsymbol{x} + \omega t)\right] \right\} d\boldsymbol{\kappa}$$
$$\tag{A.1.42}$$

Remark A.1.4: Compared with d'Alembert's solution of a one-dimensional wave equation in Sect. 7.1.1, Eq. (A.1.41) also includes two traveling waves in positive and negative directions along wave vector $\boldsymbol{\kappa}$. One may notice that the right traveling wave in Sect. 7.1.1 is $u_R(c_0 t - x)$ or $u_R(\omega t - \kappa x)$, while the corresponding traveling wave here is $a_1(\boldsymbol{\kappa}) \exp[i(\boldsymbol{\kappa}^T \boldsymbol{x} - \omega t)]$. The sign difference between their arguments is not essential.

Remark A.1.5: In the above separation of variables, an assumption was made for $k_j^2 > 0, \quad j = 1, 2, 3$ to derive the wave solutions described by complex exponential functions of constant amplitude. If no assumption is made for $k_j^2 > 0, \quad j = 1, 2, 3$, the wave amplitudes in Eq. (A.1.37) are no longer constants, exhibiting the evanescent waves in Chap. 7 or the inhomogeneous planar waves to be discussed in Sect. A.2.1.

A.2 Two Kinds of Simple Waves

Among three-dimensional waves, planar waves and spherical waves are two kinds of simple, but essential waves in studying three-dimensional elasto-dynamics. To well understand the assumptions of planar waves for both rods and beams in Chap. 7, this section focuses on the theory of planar waves.

A.2.1 Planar Waves

Planar wave refers to the simplified wave having a planar wavefront at any moment. In practice, it is reasonable to approximate the wavefront of a wave far from an initial disturbance as a plane in a small range.

Consider a planar wave in three-dimensional space $\mathfrak{R}^3$, where an arbitrary fixed point o is taken as the origin of the frame of coordinates. Let $\boldsymbol{n} \in \mathfrak{R}^3$ be the normal vector of the wavefront of a planar wave, $\boldsymbol{x} \in \mathfrak{R}^3$ be the position vector of an arbitrary point on the planar wavefront. At an initial moment, the wavefront satisfies the plane equation in $\mathfrak{R}^3$ as follows

$$\boldsymbol{n}^\mathrm{T}\boldsymbol{x} = d, \quad t = 0 \tag{A.2.1}$$

where $d > 0$ is the distance from the planar wavefront to origin o. When $t > 0$, the wavefront propagates along vector $\boldsymbol{n}$ with a constant speed $c > 0$ such that the wavefront yields a kinematic relation as follows

$$\boldsymbol{n}^\mathrm{T}\boldsymbol{x} - ct = d, \quad t > 0 \tag{A.2.2}$$

If the following function is introduced

$$\phi(\boldsymbol{x}, t) = \phi(\boldsymbol{n}^\mathrm{T}\boldsymbol{x} - ct) \tag{A.2.3}$$

it is easy to see that function $\phi(\boldsymbol{n}^\mathrm{T}\boldsymbol{x} - ct)$ takes the constant $\phi(d)$ in the wavefront and satisfies the following wave equation

$$c^2 \nabla^2 \phi(\boldsymbol{x}, t) - \phi_{tt}(\boldsymbol{x}, t) = 0 \tag{A.2.4}$$

Hence, function $\phi(\boldsymbol{x}, t)$ is the general form of a potential function of planar waves in the three-dimensional space.

According to Eq. (A.2.4), the scalar potential function, the vectorial potential function and the normalization condition of a three-dimensional planar wave satisfy

$$\varphi(\boldsymbol{x}, t) = \varphi(\boldsymbol{n}^\mathrm{T}\boldsymbol{x} - c_p t), \quad \boldsymbol{\psi}(\boldsymbol{x}, t) = \boldsymbol{\psi}(\boldsymbol{n}^\mathrm{T}\boldsymbol{x} - c_s t), \quad \nabla \cdot \boldsymbol{\psi}(\boldsymbol{n}^\mathrm{T}\boldsymbol{x} - c_s t) = 0 \tag{A.2.5}$$

The following two subsections will deal with the planar waves in a given direction and an arbitrary direction, respectively.

A.2.1.1 Planar Waves Traveling in a Given Direction

Let the positive direction of axis x_1 points the propagation direction of the planar wave such that $\boldsymbol{n} = [1\ 0\ 0]^\mathrm{T}$ holds. Then, the above potential functions and the normalization condition become

Fig. A.3 The schematic of three-dimensional planar waves traveling in direction x_1

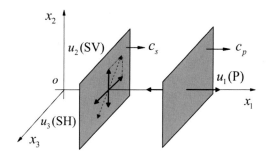

$$\varphi(\boldsymbol{x}, t) = \varphi(x_1 - c_p t), \quad \boldsymbol{\psi}(\boldsymbol{x}, t) = \boldsymbol{\psi}(x_1 - c_s t), \quad \psi_{1\xi}(x_1 - c_s t) = 0 \quad \text{(A.2.6)}$$

where subscript ξ stands for the derivative with respect to argument $\xi = x_1 - c_s t$. Substitution of Eq. (A.2.6) into Eq. (A.1.24) leads to

$$
\begin{cases}
u_1 = \dfrac{\partial \varphi}{\partial x_1} + \dfrac{\partial \psi_3}{\partial x_2} - \dfrac{\partial \psi_2}{\partial x_3} = \dfrac{\partial \varphi}{\partial x_1} = \varphi_\xi(x_1 - c_p t) \\[2mm]
u_2 = \dfrac{\partial \varphi}{\partial x_2} + \dfrac{\partial \psi_1}{\partial x_3} - \dfrac{\partial \psi_3}{\partial x_1} = -\psi_{3\xi}(x_1 - c_s t) \\[2mm]
u_3 = \dfrac{\partial \varphi}{\partial x_3} + \dfrac{\partial \psi_2}{\partial x_1} - \dfrac{\partial \psi_1}{\partial x_2} = \psi_{2\xi}(x_1 - c_s t)
\end{cases}
\quad \text{(A.2.7)}
$$

Here, subscript ξ represents the derivative with respect to either argument $\xi = x_1 - c_s t$ or argument $\xi = x_1 - c_p t$.

As shown in Fig. A.3, $u_1(x_1, t)$ represents the longitudinal planar wave traveling in the right direction with speed c_p and the vibration direction of an arbitrary mass point denoted by a short arrow in the wavefront is parallel to the direction of wave propagation, $u_2(x_1, t)$ and $u_3(x_1, t)$ represent two transverse planar waves traveling in the right direction with speed c_s and the two short arrows in the wavefront give two vibration directions of an arbitrary mass point perpendicular to the direction of wave propagation. As discussed in Sect. A.1.2, $c_p > c_s$. In seismology, the above three waves are named the *pressure wave*, *horizontal shear wave* and *vertical shear wave*, or the *P-wave*, *SH-wave* and *SV-wave* for short. The SH-wave and the SV-wave in Fig. A.3 can be composed as an S-wave with the vibration direction denoted by the dashed arrow.

A.2.1.2 Planar Waves Traveling in an Arbitrary Direction in a Plane

This subsection deals with the propagation of a planar wave along an arbitrary direction in plane (x_1, x_2) in order to make a preparation for studying the wave reflections at the boundary of a semi-infinite medium in Sect. A.3.

In this case, the potential functions of the planar wave do not depend on x_3 and take the forms $\varphi(x_1, x_2, t)$ and $\psi_j(x_1, x_2, t)$, $j = 1, 2, 3$. The displacement components and normalization condition of the potential functions are

Fig. A.4 The vectorial descriptions of a P-wave, an SV-wave and an SH-wave

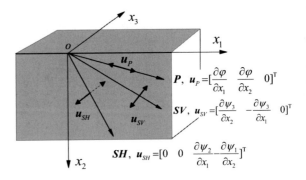

$$u_1 = \frac{\partial \varphi}{\partial x_1} + \frac{\partial \psi_3}{\partial x_2}, \quad u_2 = \frac{\partial \varphi}{\partial x_2} - \frac{\partial \psi_3}{\partial x_1}, \quad u_3 = \frac{\partial \psi_2}{\partial x_1} - \frac{\partial \psi_1}{\partial x_2}, \quad \frac{\partial \psi_1}{\partial x_1} + \frac{\partial \psi_2}{\partial x_2} = 0$$

$$(A.2.8)$$

Substitution of Eq. (A.2.8) into Eq. (A.1.1) gives the strain components of the planar wave as follows

$$\begin{cases} \varepsilon_{11} = \dfrac{\partial^2 \varphi}{\partial x_1^2} + \dfrac{\partial^2 \psi_3}{\partial x_1 \partial x_2}, & \varepsilon_{12} = \dfrac{1}{2}\left(2\dfrac{\partial^2 \varphi}{\partial x_1 \partial x_2} + \dfrac{\partial^2 \psi_3}{\partial x_2^2} - \dfrac{\partial^2 \psi_3}{\partial x_1^2} \right) \\[3mm] \varepsilon_{13} = \dfrac{1}{2}\left(\dfrac{\partial^2 \psi_2}{\partial x_1^2} - \dfrac{\partial^2 \psi_1}{\partial x_1 \partial x_2} \right), & \varepsilon_{22} = \dfrac{\partial^2 \varphi}{\partial x_2^2} - \dfrac{\partial^2 \psi_3}{\partial x_1 \partial x_2} \\[3mm] \varepsilon_{23} = \dfrac{1}{2}\left(\dfrac{\partial^2 \psi_2}{\partial x_1 \partial x_2} - \dfrac{\partial^2 \psi_1}{\partial x_2^2} \right), & \varepsilon_{33} = 0 \end{cases} \qquad (A.2.9)$$

According to Eq. (A.2.9) and the constitutive law in Eq. (A.1.3), we derive the stress components as follows

$$\begin{cases} \sigma_{11} = (\lambda + 2\mu)\left(\dfrac{\partial^2 \varphi}{\partial x_1^2} + \dfrac{\partial^2 \varphi}{\partial x_2^2} \right) - 2\mu\left(\dfrac{\partial^2 \varphi}{\partial x_2^2} - \dfrac{\partial^2 \psi_3}{\partial x_1 \partial x_2} \right), \\[3mm] \sigma_{12} = \mu\left(2\dfrac{\partial^2 \varphi}{\partial x_1 \partial x_2} + \dfrac{\partial^2 \psi_3}{\partial x_2^2} - \dfrac{\partial^2 \psi_3}{\partial x_1^2} \right), \quad \sigma_{13} = \mu\left(\dfrac{\partial^2 \psi_2}{\partial x_1^2} - \dfrac{\partial^2 \psi_1}{\partial x_1 x_2} \right) \\[3mm] \sigma_{22} = (\lambda + 2\mu)\left(\dfrac{\partial^2 \varphi}{\partial x_1^2} + \dfrac{\partial^2 \varphi}{\partial x_2^2} \right) - 2\mu\left(\dfrac{\partial^2 \varphi}{\partial x_1^2} + \dfrac{\partial^2 \psi_3}{\partial x_1 \partial x_2} \right) \\[3mm] \sigma_{23} = \mu\left(\dfrac{\partial^2 \psi_2}{\partial x_1 \partial x_2} - \dfrac{\partial^2 \psi_1}{\partial x_2^2} \right) \quad \sigma_{33} = \lambda\left(\dfrac{\partial^2 \varphi}{\partial x_1^2} + \dfrac{\partial^2 \varphi}{\partial x_2^2} \right) \end{cases} \qquad (A.2.10)$$

From Eqs. (A.2.8), (A.2.9) and (A.2.10), the three types of planar waves can be divided into two groups as follows.

In the first group, the potential functions of waves are $\varphi(x_1, x_2, t)$ and $\psi_3(x_1, x_2, t)$, while the displacement components are $u_1(x_1, x_2, t)$ and $u_2(x_1, x_2, t)$. Here, potential function $\varphi(x_1, x_2, t)$ describes a P-wave traveling in direction $\boldsymbol{P}$ in Fig. A.4

and the displacement vector u_P denoted by a short arrow is parallel to vector P, while potential function $\psi_3(x_1, x_2, t)$ describes an SV-wave traveling in direction SV in Fig. A.4 and the displacement vector u_{SV} normal to vector SV falls into plane (x_1, x_2). The above P-wave and SV-wave produce the strain components $\varepsilon_{11}(x_1, x_2, t)$, $\varepsilon_{22}(x_1, x_2, t)$ and $\varepsilon_{12}(x_1, x_2, t)$, and the corresponding stress components $\sigma_{11}(x_1, x_2, t)$, $\sigma_{22}(x_1, x_2, t)$, $\sigma_{12}(x_1, x_2, t)$ and $\sigma_{33}(x_1, x_2, t)$. Section A.3 will demonstrate that a P-wave and an SV-wave can convert to each other through a reflection at the boundary of a semi-infinite medium.

In the second group, the potential functions $\psi_1(x_1, x_2, t)$ and $\psi_2(x_1, x_2, t)$ lead to the single displacement component $u_3(x_1, x_2, t)$ of an SH-wave traveling in direction SH in Fig. A.4, and the displacement vector u_{SH} denoted by a short arrow is parallel to axis x_3 and normal to vector SH. The SH-wave produces the shear strain components $\varepsilon_{13}(x_1, x_2, t)$ and $\varepsilon_{23}(x_1, x_2, t)$, as well as the shear stress components $\sigma_{13}(x_1, x_2, t)$ and $\sigma_{23}(x_1, x_2, t)$. Thus, the SH-wave is a pure shear wave.

Now we discuss the above three types of planar waves independently and derive their simplified descriptions and associated wave equations, respectively.

(1) P-waves

In this case, the only potential function $\varphi(x_1, x_2, t)$ in Eq. (A.2.8) gives the three displacement components as follows

$$u_1 = \frac{\partial \varphi}{\partial x_1}, \quad u_2 = \frac{\partial \varphi}{\partial x_2}, \quad u_3 \equiv 0 \qquad (A.2.11)$$

They correspond to vector u_P in Fig. A.4 and describe a longitudinal wave in plane (x_1, x_2). The potential function $\varphi(x_1, x_2, t)$ yields the following wave equation

$$c_p^2 \left(\frac{\partial^2 \varphi}{\partial x_1^2} + \frac{\partial^2 \varphi}{\partial x_2^2} \right) = \frac{\partial^2 \varphi}{\partial t^2} \qquad (A.2.12)$$

In view of Eq. (A.2.9), the strain components of the P-wave are

$$\varepsilon_{11} = \frac{\partial^2 \varphi}{\partial x_1^2}, \quad \varepsilon_{12} = \frac{\partial^2 \varphi}{\partial x_1 \partial x_2}, \quad \varepsilon_{22} = \frac{\partial^2 \varphi}{\partial x_2^2}, \quad \varepsilon_{13} = 0, \quad \varepsilon_{23} = 0, \quad \varepsilon_{33} = 0$$

$$(A.2.13)$$

From Eq. (A.2.10), we have the stress components of the P-wave as follows

$$\begin{cases} \sigma_{11} = (\lambda + 2\mu)\nabla^2 \varphi - 2\mu \frac{\partial^2 \varphi}{\partial x_2^2}, \quad \sigma_{12} = 2\mu \frac{\partial^2 \varphi}{\partial x_1 \partial x_2} \\[2mm] \sigma_{22} = (\lambda + 2\mu)\nabla^2 \varphi - 2\mu \frac{\partial^2 \varphi}{\partial x_1^2}, \quad \sigma_{13} = 0, \quad \sigma_{23} = 0, \quad \sigma_{33} = \lambda \nabla^2 \varphi \end{cases} \qquad (A.2.14)$$

As shown in Eqs. (A.2.13) and (A.2.14), although all strain components in direction x_3 vanish, the stress component σ_{33} in direction x_3 may not be zero such that the P-wave is independent of axis x_3 and in a planar stress status.

(2) SV-waves

In this case, the only potential function $\psi_3(x_1, x_2, t)$ in Eq. (A.2.8) leads to the displacement components of the SV-wave as follows

$$u_1 = \frac{\partial \psi_3}{\partial x_2}, \quad u_2 = -\frac{\partial \psi_3}{\partial x_1}, \quad u_3 \equiv 0 \tag{A.2.15}$$

They correspond to vector $\mathbf{u}_{SV}$ in Fig. A.4 and describe a transverse wave in plane (x_1, x_2). The potential function yields the following wave equation

$$c_s^2 \left(\frac{\partial^2 \psi_3}{\partial x_1^2} + \frac{\partial^2 \psi_3}{\partial x_2^2} \right) = \frac{\partial^2 \psi_3}{\partial t^2} \tag{A.2.16}$$

From Eq. (A.2.9), the strain components of the SV-wave satisfy

$$\begin{cases} \varepsilon_{11} = \dfrac{\partial^2 \psi_3}{\partial x_1 \partial x_2}, \quad \varepsilon_{12} = \dfrac{1}{2} \left(\dfrac{\partial^2 \psi_3}{\partial x_2^2} - \dfrac{\partial^2 \psi_3}{\partial x_1^2} \right), \quad \varepsilon_{22} = -\dfrac{\partial^2 \psi_3}{\partial x_1 \partial x_2} \\ \varepsilon_{13} = 0, \quad \varepsilon_{23} = 0, \quad \varepsilon_{33} = 0 \end{cases} \tag{A.2.17}$$

There follow the stress components of SV-wave from Eq. (A.2.10), namely,

$$\begin{cases} \sigma_{11} = 2\mu \dfrac{\partial^2 \psi_3}{\partial x_1 \partial x_2}, \quad \sigma_{12} = \mu \left(\dfrac{\partial^2 \psi_3}{\partial x_2^2} - \dfrac{\partial^2 \psi_3}{\partial x_1^2} \right), \quad \sigma_{22} = -2\mu \dfrac{\partial^2 \psi_3}{\partial x_1 \partial x_2} \\ \sigma_{13} = 0, \quad \sigma_{23} = 0, \quad \sigma_{33} = 0 \end{cases}$$

$$\tag{A.2.18}$$

Equations (A.2.17) and (A.2.18) show that the SV-wave is independent of axis x_3, and in both planar strain status and planar stress status.

(3) SH-waves

In this case, the wave has two potential functions $\psi_1(x_1, x_2, t)$ and $\psi_2(x_1, x_2, t)$ such that the displacement components and the normalization condition of two potential functions are as follows

$$u_1 = 0, \quad u_2 = 0, \quad u_3 = -\frac{\partial \psi_1}{\partial x_2} + \frac{\partial \psi_2}{\partial x_1}, \quad \frac{\partial \psi_1}{\partial x_1} + \frac{\partial \psi_2}{\partial x_2} = 0 \tag{A.2.19}$$

The above displacement components correspond to vector $\mathbf{u}_{SH}$ parallel to axis x_3 in Fig. A.4, where $\mathbf{u}_{SH}$ is perpendicular to both vectors $\mathbf{u}_P$ and $\mathbf{u}_{SV}$. As the two potential functions are not independent and describe a single displacement component $u_3(x_1, x_2, t)$, it is feasible to use $u_3(x_1, x_2, t)$ to describe the SH-wave. From Eq. (A.1.16), the displacement of SH-wave yields

$$c_s^2 \left(\frac{\partial^2 u_3}{\partial x_1^2} + \frac{\partial^2 u_3}{\partial x_2^2} \right) = \frac{\partial^2 u_3}{\partial t^2} \tag{A.2.20}$$

In view of Eq. (A.1.1), the strain components of the SH-wave satisfy

$$
\begin{cases}
\varepsilon_{11} = \dfrac{\partial u_1}{\partial x_1} = 0, \quad \varepsilon_{12} = 0, \quad \varepsilon_{22} = \dfrac{\partial u_2}{\partial x_2} = 0 \\[2mm]
\varepsilon_{13} = \dfrac{1}{2}\dfrac{\partial u_3}{\partial x_1}, \quad \varepsilon_{23} = \dfrac{1}{2}\dfrac{\partial u_3}{\partial x_2}, \quad \varepsilon_{33} = \dfrac{\partial u_3}{\partial x_3} = 0
\end{cases}
\tag{A.2.21}
$$

The stress components of the SH-wave, from Eq. (A.1.3), are as follows

$$
\begin{cases}
\sigma_{11} = 0, \quad \sigma_{12} = 0, \quad \sigma_{22} = 0 \\[2mm]
\sigma_{13} = \mu \dfrac{\partial u_3}{\partial x_1}, \quad \sigma_{23} = \mu \dfrac{\partial u_3}{\partial x_2}, \quad \sigma_{33} = 0
\end{cases}
\tag{A.2.22}
$$

As shown in Eqs. (A.2.21) and (A.2.22), both non-zero shear strain and non-zero shear stress are in direction x_3.

A.2.1.3   Inhomogeneous Planar Waves

In an infinite medium, the potential functions of a planar wave in Eq. (A.2.5) satisfy Eq. (A.1.41) such that the wave has the constant amplitude and is called a *homogeneous planar wave*. In a semi-finite or a finite medium, however, when a homogeneous planar wave undergoes a reflection at the boundary of the medium, it may become an *inhomogeneous planar wave*.

When the amplitudes of inhomogeneous planar P-wave and SV-wave vary along axis x_2, their potential functions can be assumed, respectively, as

$$
\begin{cases}
\varphi(x_1, x_2, t) = \tilde{\varphi}(x_2) \exp[i\kappa(x_1 - ct)] \\[2mm]
\psi_3(x_1, x_2, t) = \tilde{\psi}_3(x_2) \exp[i\kappa(x_1 - ct)]
\end{cases}
\tag{A.2.23}
$$

where wave speed c and wave number κ are constants to be determined. Substituting Eq. (A.2.23) into Eq. (A.1.24) and eliminating exponential functions, we derive two ordinary differential equations of $\tilde{\varphi}(x_2)$ and $\tilde{\psi}_3(x_2)$, namely,

$$
\begin{cases}
\dfrac{d^2 \tilde{\varphi}(x_2)}{dx_2^2} + \kappa^2 r_p^2 \tilde{\varphi}(x_2) = 0, \quad r_p \equiv \sqrt{\dfrac{c^2}{c_p^2} - 1} \\[4mm]
\dfrac{d^2 \tilde{\psi}_3(x_2)}{dx_2^2} + \kappa^2 r_s^2 \tilde{\psi}_3(x_2) = 0, \quad r_s \equiv \sqrt{\dfrac{c^2}{c_s^2} - 1}
\end{cases}
\tag{A.2.24}
$$

From the relation among wave speed c, P-wave speed c_p and SV-wave speed c_s, detailed discussions cover the following five cases.

Case (1): When $c > c_p, 0 < r_p < r_s$ holds such that the solutions of Eq. (A.2.24) take the following form

$$\begin{cases} \tilde{\varphi}(x_2) = a_1 \exp(-i\kappa r_p x_2) + a_2 \exp(i\kappa r_p x_2) \\ \tilde{\psi}_3(x_2) = b_1 \exp(-i\kappa r_s x_2) + b_2 \exp(i\kappa r_s x_2) \end{cases} \qquad \text{(A.2.25)}$$

According to Eq. (A.1.37), both P-wave and SV-wave are homogeneous planar waves discussed before.

Case (2): When $c = c_p, r_p = 0$ and $r_s = \sqrt{c_p^2/c_s^2 - 1} > 0$ hold so that the solutions of Eq. (A.2.24) have the following form

$$\begin{cases} \tilde{\varphi}(x_2) = a_1 x_2 + a_2 \\ \tilde{\psi}_3(x_2) = b_1 \exp\left[-i\kappa (c_p^2/c_s^2 - 1)^{1/2} x_2\right] + b_2 \exp\left[i\kappa (c_p^2/c_s^2 - 1)^{1/2} x_2\right] \end{cases} \qquad \text{(A.2.26)}$$

where the first term in the first equation approaches infinity with an increase of x_2 and needs to be removed to make sense in physics. As such, the P-wave does not depend on x_2 and the SV-wave is a homogeneous planar wave. As will be discussed in Sect. A.3.1.4, the P-wave without $a_1 x_2$ is a grazing wave reflected by an incident SV-wave at the free boundary $x_2 = 0$.

Case (3): When $c_s < c < c_p, r_s > 0$ holds and r_p is an imaginary number, which can be expressed as $r_p \equiv i v_p$ with $v_p \equiv \sqrt{1 - c^2/c_p^2}$. In this case, the solutions of Eq. (A.2.24) look like

$$\begin{cases} \tilde{\varphi}(x_2) = a_1 \exp(-\kappa v_p x_2) + a_2 \exp(\kappa v_p x_2) \\ \tilde{\psi}_3(x_2) = b_1 \exp(-i\kappa r_s x_2) + b_2 \exp(i\kappa r_s x_2) \end{cases} \qquad \text{(A.2.27)}$$

In Eq. (A.2.27), the second term of the first equation approaches infinity when x_2 keeps increasing and needs to be removed to make sense in physics. In this case, the SV-wave is a homogeneous planar wave, while the P-wave is an inhomogeneous planar wave with an exponentially decreasing amplitude when x_2 goes on. As will be discussed in Sect. A.3.1.4, the inhomogeneous P-wave comes from the reflection of an incident SV wave at the free boundary $x_2 = 0$ and only exists near the free surface of a medium. As such, the inhomogeneous P-wave is named a *surface wave*, while the homogeneous SV-wave traveling in a medium is called a *body wave*.

Case (4): When $c = c_s, r_s = 0$ holds and r_p is an imaginary number expressed as $r_p \equiv i v_{ps}$ with $v_{ps} \equiv \sqrt{1 - c_s^2/c_p^2}$ such that the solutions of Eq. (A.2.24) take the following form

$$\begin{cases} \tilde{\varphi}(x_2) = a_1 \exp(-\kappa v_{ps} x_2) + a_2 \exp(\kappa v_{ps} x_2) \\ \tilde{\psi}_3(x_2) = b_1 x_2 + b_2 \end{cases} \qquad \text{(A.2.28)}$$

In Eq. (A.2.28), the second term of the first equation and the first term of the second equation approach infinity when x_2 increases so that they ought to be removed. Now, the P-wave is an inhomogeneous wave, while the SV-wave has nothing to do with x_2. As discussed in the reference[2], this phenomenon does not occur in the conventional elastic medium with Poisson's ratio $v \in (0.0, 0.5)$.

Case (5): When $c < c_s$, both r_p and r_s are imaginary numbers and can be expressed as $r_p \equiv iv_p$ and $r_s \equiv iv_s$, where $v_p \equiv \sqrt{1 - c^2/c_p^2}$ and $v_s \equiv \sqrt{1 - c^2/c_s^2}$. Thus, the solutions of Eq. (A.2.24) take the following form

$$\begin{cases} \tilde{\varphi}(x_2) = a_1 \exp(-\kappa v_p x_2) + a_2 \exp(\kappa v_p x_2) \\ \tilde{\psi}_3(x_2) = b_1 \exp(-\kappa v_s x_2) + b_2 \exp(\kappa v_s x_2) \end{cases} \tag{A.2.29}$$

where the second terms in both expressions approach infinity with an increase of x_2, and have to be removed to make sense. In this case, as will be studied in Sect. A.3.3, Eq. (A.2.29) describes an inhomogeneous P-wave and an inhomogeneous SV-wave, and their superposition may be a surface wave near the free boundary $x_2 = 0$ and propagates freely in direction x_1.

A.2.2 Spherical Waves

In an infinite medium, the wave due to a point disturbance propagates outward in all directions and the wavefront is a spherical surface. This wave, hence, refers to a *spherical wave*.

Now, we take the center of the initial disturbance as the origin to establish a frame of Cartesian coordinates (x_1, x_2, x_3) and a frame of spherical coordinates (r, θ_1, θ_2). The transform between the two frames yields

$$\begin{cases} x_1 = r \sin(\theta_1) \cos(\theta_2), \quad x_2 = r \sin(\theta_1) \sin(\theta_2), \quad x_3 = r \cos(\theta_1) \\ r \in [0, +\infty), \quad \theta_1 \in [0, \pi], \quad \theta_2 \in [0, 2\pi) \end{cases} \tag{A.2.30}$$

From the spherical symmetry of the wave, the only non-zero displacement component is the radial displacement $u_r(r, t)$ in the direction of wave propagation so that the spherical wave is a P-wave. From the derivative relation between the two frames and Laplace operator in the frame of spherical coordinates[3], the radial displacement and the potential function satisfy

$$u_r(r, t) = \frac{\partial \varphi(r, t)}{\partial r}, \quad \nabla^2 \varphi(r, t) = \frac{1}{r^2} \frac{\partial}{\partial r} \left[r^2 \frac{\partial \varphi(r, t)}{\partial r} \right] = \frac{1}{c_p^2} \varphi_{tt}(r, t) \tag{A.2.31}$$

---

[2] Eringen A C, Suhubi E S (1975). Elastodynamics. Volume II. New York: Academic Press, 524.
[3] Courant R, Hilbert D (1989). Methods of Mathematical Physics. Volume I. New York: John Wiley & Sons Inc., 535.

Multiplying both sides of the wave equation in Eq. (A.2.31) by r and using the following relation

$$\frac{1}{r}\frac{\partial}{\partial r}\left[r^2\frac{\partial\varphi(r,t)}{\partial r}\right] = r\frac{\partial^2\varphi(r,t)}{\partial r^2} + 2\frac{\partial\varphi(r,t)}{\partial r} = \frac{\partial^2[r\varphi(r,t)]}{\partial r^2} \qquad (A.2.32)$$

we derive the one-dimensional wave equation of function $r\varphi(r,t)$ as follows

$$\frac{\partial^2[r\varphi(r,t)]}{\partial r^2} = \frac{1}{c_p^2}\frac{\partial^2[r\varphi(r,t)]}{\partial t^2} \qquad (A.2.33)$$

In an infinite medium, the spherical wave undergoes dilatation in radial direction and does not have reflection. Following d'Alembert's solution of a one-dimensional wave equation in Sect. 7.1.1, we get the potential function of the spherical wave as following

$$\varphi(r,t) = \frac{1}{r}\varphi_R(r - c_p t) \qquad (A.2.34)$$

Substituting Eq. (A.2.34) into the first equation in Eq. (A.2.31) and denoting the derivative of $\varphi_R(\xi)$ with respect to argument $\xi = r - c_p t$ as $\varphi_{R\xi}$, we derive the radial displacement of the spherical wave as follows

$$u_r(r,t) = \frac{\partial\varphi(r,t)}{\partial r} = \frac{1}{r^2}\left[r\varphi_{R\xi}(r - c_p t) - \varphi_R(r - c_p t)\right] \qquad (A.2.35)$$

As shown in Eq. (A.2.35), the amplitude of the spherical wave decreases with an increase of radial distance r measured from the center of initial disturbance.

A.3 Wave Reflections at Boundary of a Semi-infinite Medium

This section presents how to study the reflection of a planar harmonic wave at the boundary of a semi-infinite medium shown in Fig. A.5, where the medium occupies a half space $x_2 \geq 0$, and the other half space $x_2 < 0$ is vacuum. The incident planar harmonic wave travels in the medium and undergoes a reflection at boundary $x_2 = 0$.

As analyzed in Sect. A.2.1, a planar wave can be decomposed into a P-wave, an SV-wave and an SH-wave. In Fig. A.5, let axis x_3 be parallel to the vibration direction of the SH-wave, then both vibration directions of the P-wave and the SV-wave are in plane (x_1, x_2). Therefore, we first analyze the P-wave and the SV-wave together since they affect each other, and then analyze the SH-wave separately.

Fig. A.5 The reflections of a
P-wave, an SV-wave and an
SH-wave at the boundary of
a semi-infinite medium

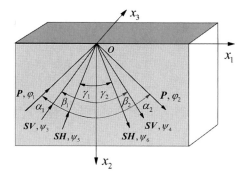

A.3.1 Reflections of a P-wave and an SV-Wave

In Fig. A.5, the medium is infinite in direction x_3 such that the potential functions
of a P-wave and an SV-wave do not depend on x_3. Let an incident P-wave and an
incident SV-wave be described by potential functions $\varphi_1(x_1, x_2, t)$ and $\psi_3(x_1, x_2, t)$,
and their incidence angles be $\alpha_1 \in [0, \pi/2]$ and $\beta_1 \in [0, \pi/2]$, respectively. The two
waves undergo the reflection at boundary $x_2 = 0$, and produce a reflected P-wave
and a reflected SV-wave with potential functions $\varphi_2(x_1, x_2, t)$ and $\psi_4(x_1, x_2, t)$, as
well as reflecting angles α_2 and β_2 to be determined, respectively.

Assume that the above P-waves and SV-waves are homogeneous harmonic waves
and their potential functions are as follows

$$\begin{cases} \varphi_1(x_1, x_2, t) = a_1 \exp\left[i\kappa_1(l_1x_1 - m_1x_2 - c_pt)\right] \\ \varphi_2(x_1, x_2, t) = a_2 \exp\left[i\kappa_2(l_2x_1 + m_2x_2 - c_pt)\right] \\ \psi_3(x_1, x_2, t) = a_3 \exp[i\kappa_3(l_3x_1 - m_3x_2 - c_st)] \\ \psi_4(x_1, x_2, t) = a_4 \exp[i\kappa_4(l_4x_1 + m_4x_2 - c_st)] \end{cases} \quad \text{(A.3.1)}$$

where κ_j, $j = 1, 2, 3, 4$ are the wave numbers, l_j, m_j, $j = 1, 2, 3, 4$ are the cosine
functions and each of their arguments is an angle between the direction of wave
propagation and an axis of the frame of coordinates. There follow the relations

$$\begin{cases} l_j = \sin\alpha_j, \quad m_j = \cos\alpha_j, \quad l_{j+2} = \sin\beta_j, \quad m_{j+2} = \cos\beta_j, \quad j = 1, 2 \\ l_j^2 + m_j^2 = 1, \quad j = 1, 2, 3, 4 \end{cases}$$

$$\text{(A.3.2)}$$

As shown in Fig. A.5, the vertical component of each incident wave is in the negative
direction of axis x_2. Thus, the coefficients $-m_1$ and $-m_3$ appear in Eq. (A.3.1).

According to Eq. (A.2.8), the resultant displacement components due to both
incident waves and reflected waves can be written as

$$\begin{cases} u_1 = \dfrac{\partial \varphi_1}{\partial x_1} + \dfrac{\partial \varphi_2}{\partial x_1} + \dfrac{\partial \psi_3}{\partial x_2} + \dfrac{\partial \psi_4}{\partial x_2} = i(\kappa_1 l_1 \varphi_1 + \kappa_2 l_2 \varphi_2 - \kappa_3 m_3 \psi_3 + \kappa_4 m_4 \psi_4) \\[2mm] u_2 = \dfrac{\partial \varphi_1}{\partial x_2} + \dfrac{\partial \varphi_2}{\partial x_2} - \dfrac{\partial \psi_3}{\partial x_1} - \dfrac{\partial \psi_4}{\partial x_1} = i(-\kappa_1 m_1 \varphi_1 + \kappa_2 m_2 \varphi_2 - \kappa_3 l_3 \psi_3 - \kappa_4 l_4 \psi_4) \end{cases}$$

$$\text{(A.3.3)}$$

From Eq. (A.2.9), the resultant strain components superposed by incident waves and reflected waves satisfy

$$\begin{cases} \varepsilon_{11} = -(\kappa_1^2 l_1^2 \varphi_1 + \kappa_2^2 l_2^2 \varphi_2 - \kappa_3^2 l_3 m_3 \psi_3 + \kappa_4^2 l_4 m_4 \psi_4) \\[2mm] \varepsilon_{22} = -(\kappa_1^2 m_1^2 \varphi_1 + \kappa_2^2 m_2^2 \varphi_2 + \kappa_3^2 l_3 m_3 \psi_3 - \kappa_4^2 l_4 m_4 \psi_4) \\[2mm] \varepsilon_{21} = \kappa_1^2 l_1 m_1 \varphi_1 - \kappa_2^2 l_2 m_2 \varphi_2 - \dfrac{1}{2}\kappa_3^2 (m_3^2 - l_3^2)\psi_3 - \dfrac{1}{2}\kappa_4^2 (m_4^2 - l_4^2)\psi_4 \end{cases}$$

$$\text{(A.3.4)}$$

In view of Eq. (A.2.10) and the last row in Eq. (A.3.2), we get the resultant stress components associated with the subsequent analysis, namely,

$$\begin{cases} \sigma_{22} = -(\lambda + 2\mu m_1^2)\kappa_1^2 \varphi_1 - (\lambda + 2\mu m_2^2)\kappa_2^2 \varphi_2 - 2\mu\kappa_3^2 l_3 m_3 \psi_3 + 2\mu\kappa_4^2 l_4 m_4 \psi_4 \\[2mm] \sigma_{21} = \mu\left[2\kappa_1^2 l_1 m_1 \varphi_1 - 2\kappa_2^2 l_2 m_2 \varphi_2 + \kappa_3^2 (l_3^2 - m_3^2)\psi_3 + \kappa_4^2 (l_4^2 - m_4^2)\psi_4\right] \end{cases}$$

$$\text{(A.3.5)}$$

Given an incident wave, it is feasible to determine the reflected waves with consideration of the compatibility conditions at boundary $x_2 = 0$. The boundary conditions include the resultant displacement conditions of $u_1(x_1, 0, t) = 0$ and $u_2(x_1, 0, t) = 0$, and the resultant stress conditions of $\sigma_{22}(x_1, 0, t) = 0$ and $\sigma_{21}(x_1, 0, t) = 0$. To determine a reflection wave, we require only two conditions. Thus, there are four kinds of possible boundaries as follows.

(1) A clamped boundary: $u_1(x_1, 0, t) = 0$ and $u_2(x_1, 0, t) = 0$.
(2) A free boundary: $\sigma_{22}(x_1, 0, t) = 0$ and $\sigma_{21}(x_1, 0, t) = 0$.
(3) A smooth rigid boundary I: $u_2(x_1, 0, t) = 0$ and $\sigma_{21}(x_1, 0, t) = 0$.
(4) A smooth rigid boundary II: $u_1(x_1, 0, t) = 0$ and $\sigma_{22}(x_1, 0, t) = 0$.

From the practical viewpoint, we study the first two boundaries hereinafter and present how they reflect an incident P-wave and an incident SV-wave, respectively.

A.3.1.1   An Incident P-wave Reflected at a Clamped Boundary

At the clamped boundary, the resultant displacement components due to an incident P-wave and the reflected waves satisfy

$$u_1(x_1, 0, t) = 0, \quad u_2(x_1, 0, t) = 0 \qquad \text{(A.3.6)}$$

As no incident SV-wave shows up, then $\psi_3 = 0$ holds in Eq. (A.3.3). Substitution of Eq. (A.3.3) and $\psi_3 = 0$ into Eq. (A.3.6) leads to

$$\begin{cases} i\{\kappa_1 l_1 a_1 \exp[i\kappa_1(l_1 x_1 - c_p t)] + \kappa_2 l_2 a_2 \exp[i\kappa_2(l_2 x_1 - c_p t)] \\ \quad + \kappa_4 m_4 a_4 \exp[i\kappa_4(l_4 x_1 - c_s t)]\} = 0 \\ i\{-\kappa_1 m_1 a_1 \exp[i\kappa_1(l_1 x_1 - c_p t)] + \kappa_2 m_2 a_2 \exp[i\kappa_2(l_2 x_1 - c_p t)] \\ \quad - \kappa_4 l_4 a_4 \exp[i\kappa_4(l_4 x_1 - c_s t)]\} = 0 \end{cases} \tag{A.3.7}$$

Equation (A.3.7) holds for arbitrary coordinate x_1 and arbitrary time t such that the coefficients of x_1 in Eq. (A.3.7) are identical, and so are those of t. Thus, we have

$$\begin{cases} \kappa_1 c_p = \kappa_2 c_p = \kappa_4 c_s \\ \kappa_1 l_1 = \kappa_2 l_2 = \kappa_4 l_4 \end{cases} \Rightarrow \begin{cases} \kappa_1 = \kappa_2, \quad l_1 = l_2 \\ \kappa_1 l_1 = \kappa_4 l_4 \end{cases} \tag{A.3.8}$$

Using the following parameters with clear meanings, we recast Eq. (A.3.8) as

$$\kappa_1 = \kappa_2 \equiv \kappa_p, \quad \kappa_4 \equiv \kappa_s, \quad \alpha_1 = \alpha_2 \equiv \alpha, \quad \beta_2 \equiv \beta, \quad \kappa_p \sin\alpha = \kappa_s \sin\beta \tag{A.3.9}$$

The first four equations in Eq. (A.3.9) show that the wave number of the P-wave remains unchanged before and after reflection, and the incidence angle and reflecting angle are identical. In addition, the last equation in Eq. (A.3.9) gives well-known *Snell's law* as follows

$$D = \frac{c_p}{c_s} = \frac{\kappa_s}{\kappa_p} = \frac{\sin\alpha}{\sin\beta} > 1, \quad D \equiv \sqrt{\frac{2(1-\nu)}{1-2\nu}} \tag{A.3.10}$$

where D was defined as the ratio of wave speeds in Eq. (A.1.18). Equation (A.3.10) is an important relation between the incidence angle α of a P-wave and the reflecting angle β of an SV-wave in the same medium.

Now, substituting Eqs. (A.3.9) and (A.3.10) into Eq. (A.3.7) and eliminating exponential functions, we have

$$\begin{cases} (\sin\alpha)a_2 + D(\cos\beta)a_4 = -(\sin\alpha)a_1 \\ (\cos\alpha)a_2 - D(\sin\beta)a_4 = (\cos\alpha)a_1 \end{cases} \tag{A.3.11}$$

Solving Eq. (A.3.11) for a_2/a_1 and a_4/a_1 leads to two reflection coefficients, that is, the ratios of reflected wave amplitude to incident wave amplitude, namely,

$$\frac{a_2}{a_1} = \frac{\cos(\alpha+\beta)}{\cos(\alpha-\beta)}, \quad \frac{a_4}{a_1} = -\frac{\sin(2\alpha)}{D\cos(\alpha-\beta)} \tag{A.3.12}$$

Remark A.3.1: Some books dealt with this reflection problem and presented the reflection coefficients in terms of displacement amplitudes. In this case, the reflection coefficient of P-wave yields $\gamma_{21} \equiv a_2 \kappa_p / (a_1 \kappa_p) = a_2 / a_1$ while the reflection coefficient of SV-wave reads $\gamma_{41} \equiv a_4 \kappa_s / (a_1 \kappa_p) = -\sin(2\alpha)/\cos(\alpha - \beta)$, which looks different from the result in the reference[4] by a minus. The reason is that the direction of coordinate x_2 there was upward so that all signs in front of m_j, $j = 1, 2, 3, 4$ in Eq. (A.3.2) changed, whereas $\sin(2\alpha)$ became $-\sin(2\alpha)$.

The discussion about Eq. (A.3.12) begins with a given Poisson's ratio ν for the medium under the condition of $D > 1$ in Eq. (A.3.10). For an incidence angle $\alpha \in [0, \pi/2]$ of the P-wave, there must be a reflecting angle of the SV-wave satisfying $\beta = \arcsin[\sin(\alpha)/D] < \alpha$ so that one has two reflection coefficients from Eq. (A.3.12). Figure A.6 presents the variations of reflection coefficients a_2/a_1 and a_4/a_1, as well as reflecting angle β, with an increase of incidence angle α for four typical values of Poisson's ratio. In Fig. A.6, both units of angles α and β are scaled in degrees for easy understanding. Here are the discussions about some typical phenomena in Eq. (A.3.12) and Fig. A.6.

(1) Normal incidence

Given $\alpha = 0$, Snell's law gives $\beta = 0$ such that Eq. (A.3.12) leads to $a_2 = a_1$ and $a_4 = 0$. In this case, the incident P-wave and reflected P-wave are identical in amplitude, wave number and phase, and the reflected wave does not include any SV-wave. Substitution of $a_2 = a_1$ and $a_4 = 0$ into Eq. (A.3.3) leads to the fact that both horizontal and vertical displacements are zeros and satisfy the boundary conditions in Eq. (A.3.6).

In addition, substituting $a_2 = a_1$ and $a_4 = 0$ into Eq. (A.3.5), we get the resultant stress components due to an incident P-wave and the reflected P-wave in the medium as follows

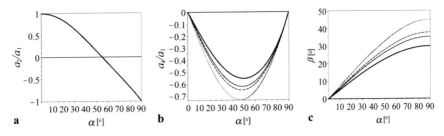

Fig. A.6 The reflection relations of an incident P-wave at a clamped boundary vs. incidence angle α (thick solid line: $\nu = 1/3$, thin solid line: $\nu = 1/4$, dashed line: $\nu = 1/5$, dot line: $\nu = 0$). **a** a_2/a_1 for the reflected P-wave; **b** a_4/a_1 for the reflected S-wave; **c** Reflecting angle β

---

[4] Achenbach J D (1973). Wave Propagation in Elastic Solids. Amsterdam: North-Holland Publishing Company, 177.

$$\begin{cases} \sigma_{22}(x_1, x_2, t) = -(\lambda + 2\mu)\kappa_p^2 a_1 \{\exp[i\kappa_p(-x_2 - c_p t)] + \exp[i\kappa_p(x_2 - c_p t)]\} \\ \sigma_{21}(x_1, x_2, t) = 0 \end{cases}$$

$$(A.3.13)$$

When $x_2 = 0$, the normal stress of an incident P-wave and the normal stress of the reflected P-wave are identical. Thus, the resultant normal stress is doubled at the clamped boundary compared with the normal stress of the incident P-wave, i.e.,

$$\sigma_{22}(x_1, 0, t) = -2(\lambda + 2\mu)\kappa_p^2 a_1 \exp(-i\kappa_p c_p t) \qquad (A.3.14)$$

(2) Mode conversion

In view of Eq. (A.3.12), $\alpha + \beta = \pi/2$ leads to $a_2 = 0$ such that an incident P-wave fully becomes a reflected SV-wave. This phenomenon is called the *mode conversion* for two types of waves. From Snell's law, the corresponding incidence angle yields

$$\alpha + \arcsin\left(\frac{\sin\alpha}{D}\right) = \frac{\pi}{2} \qquad (A.3.15)$$

In Fig. A.6a, there is a unique incidence angle α satisfying $a_2/a_1 = 0$ such that the mode conversion occurs. In this case, the corresponding reflection coefficient a_4/a_1 slightly differs from its extreme value as shown in Fig. A.6b.

A.3.1.2 An Incident P-wave Reflected at a Free Boundary

In a free boundary, the resultant normal stress and shear stress due to both incident and reflected waves satisfy the following boundary conditions

$$\sigma_{22}(x_1, 0, t) = 0, \quad \sigma_{21}(x_1, 0, t) = 0 \qquad (A.3.16)$$

By substituting Eq. (A.3.5) and $\psi_3 = 0$ into Eq. (A.3.16) and using the last row in Eq. (A.3.2) to simplify the result, we get

$$\begin{cases} (\lambda + 2\mu m_1^2)\kappa_1^2 \varphi_1 + (\lambda + 2\mu m_2^2)\kappa_1^2 \varphi_2 - 2\mu\kappa_4^2 l_4 m_4 \psi_4 = 0 \\ 2\kappa_1^2 l_1 m_1 \varphi_1 - 2\kappa_2^2 l_2 m_2 \varphi_2 - \kappa_4^2 (m_4^2 - l_4^2)\psi_4 = 0 \end{cases} \qquad (A.3.17)$$

Following a similar procedure from Eqs. (A.3.7) to (A.3.8), we get Eqs. (A.3.8) and (A.3.9) again. Substitution of them into Eq. (A.3.17) leads to

$$\begin{cases} (\lambda + 2\mu\cos^2\alpha)(a_1 + a_2) - \mu D^2 \sin(2\beta)a_4 = 0 \\ \sin(2\alpha)(a_1 - a_2) - D^2 \cos(2\beta)a_4 = 0 \end{cases} \qquad (A.3.18)$$

To solve Eq. (A.3.18), we use Eq. (A.3.10) to simplify the coefficient of $(a_1 + a_2)$ in Eq. (A.3.18) as following

$$\lambda + 2\mu \cos^2 \alpha = \lambda + 2\mu - 2\mu \sin^2 \alpha = \mu D^2 \left(1 - \frac{2 \sin^2 \alpha \cos^2 \beta}{\sin^2 \alpha} \right)$$

$$= \mu D^2 \cos(2\beta) \tag{A.3.19}$$

Substituting Eq. (A.3.19) into Eq. (A.3.18) gives

$$\begin{cases} \cos(2\beta) \left(1 + \dfrac{a_2}{a_1} \right) - \sin(2\beta) \left(\dfrac{a_4}{a_1} \right) = 0 \\[3mm] \sin(2\alpha) \left(1 - \dfrac{a_2}{a_1} \right) - D^2 \cos(2\beta) \left(\dfrac{a_4}{a_1} \right) = 0 \end{cases} \tag{A.3.20}$$

Solving Eq. (A.3.20) for a_2/a_1 and a_4/a_1, we have two reflection coefficients

$$\begin{cases} \dfrac{a_2}{a_1} = \dfrac{\sin(2\alpha) \sin(2\beta) - D^2 \cos^2(2\beta)}{\sin(2\alpha) \sin(2\beta) + D^2 \cos^2(2\beta)} \\[3mm] \dfrac{a_4}{a_1} = \dfrac{2 \sin(2\alpha) \cos(2\beta)}{\sin(2\alpha) \sin(2\beta) + D^2 \cos^2(2\beta)} \end{cases} \tag{A.3.21}$$

Remark A.3.2: Many books presented the solutions for this reflection problem, but had some typos. As discussed in Remark A.3.1, the upward direction of axis x_2 in the references[5,6] should lead to $\gamma_{41} = -Da_4/a_1$, instead of $\gamma_{41} = Da_4/a_1$. The deduction procedure from Eqs. (A.3.17) to (A.3.21) and related results in the references[7,8] are also subject to corrections.

Figure A.7 presents the variations of reflection coefficients a_2/a_1 and a_4/a_1, as well as reflecting angle β, with increasing incidence angle α for four values of Poisson's ratio. Here again, both units of angles α and β are in degrees. As discussed below, Eq. (A.3.21) and Fig. A.7 exhibit some typical phenomena

(1) Normal incidence

Given $\alpha = 0$, Snell's law leads to $\beta = 0$ such that Eq. (A.3.21) becomes $a_2/a_1 = -1$ and $a_4/a_1 = 0$. That is, the reflected wave includes a P-wave only, rather than any SV-waves. The reflected P-wave and the incident P-wave are identical in both amplitude and wave number, but opposite in phase.

Substituting $a_2/a_1 = -1$ and $a_4/a_1 = 0$ into Eq. (A.3.3), we get the resultant displacements due to both incident P-wave and reflected P-wave as follows

---

[5] Achenbach J D (1973). Wave Propagation in Elastic Solids. Amsterdam: North-Holland Publishing Company, 177.

[6] Hagedorn P, Das Gupta A (2007). Vibrations and Waves in Continuous Mechanical Systems. Chichester: John Wiley & Sons Inc., 2007: 349–351.

[7] Du Q H, Yang X A, Xu X Q, et al. (1994). Handbook of Engineering Mechanics. Beijing: Press of Higher Education, 2331 (in Chinese).

[8] Dai H L (2014). Elasto-dynamics. Changsha: Press of Hunan University, 68–69 (in Chinese).

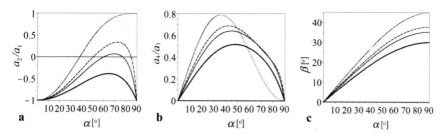

Fig. A.7 The reflection relations of an incident P-wave at a free boundary vs. incidence angle α(thick solid line: $\nu = 1/3$, thin solid line: $\nu = 1/4$, dashed line: $\nu = 1/5$, dot line: $\nu = 0$). **a** a_2/a_1 for the reflected P-wave; **b** a_4/a_1 for the reflected S-wave; **c** reflecting angle β

$$\begin{cases} u_1(x_1, x_2, t) = 0 \\ u_2(x_1, x_2, t) = -i\kappa_p a_1 \left\{ \exp\left[i\kappa_p(-x_2 - c_p t)\right] + \exp\left[i\kappa_p(x_2 - c_p t)\right]\right\} \end{cases} \tag{A.3.22}$$

At the free boundary $x_2 = 0$, both horizontal displacements due to an incident P-wave and the reflected P-wave are always zeros, while the vertical displacements due to them are identical such that the resultant vertical displacement is doubled as follows

$$u_2(x_1, 0, t) = -2i\kappa_p a_1 \exp(-i\kappa_p c_p t) \tag{A.3.23}$$

In this case, both incident and reflected P-waves do not produce shear stress. The vertically incident tension stress wave is reflected to the compression stress wave, and vice versa. As such, the normal stress of an incident wave and that of the reflected wave counteract each other to satisfy the free boundary condition.

(2) Mode conversion

In Fig. A.7a, the reflection coefficient a_2/a_1 intersects the abscissa twice for $\nu = 1/4$ or $\nu = 1/5$ such that $a_2/a_1 = 0$ holds. In these intersections, an incident P-wave becomes an SV-wave after a reflection and the phenomenon of mode conversion happens. Consider, as an example, $\nu = 1/4$, the two intersections correspond to incidence angles $\alpha = \pi/3$ and $\alpha = 0.429\pi$, respectively. The distance between two intersections increases with decreasing of ν, while the two intersections merge or disappear when ν is large enough.

(3) Grazing incidence

Grazing incidence means that the incidence angle yields $\alpha \to \pi/2$ such that the incidence direction is parallel to the free boundary $x_2 = 0$. Substitution of $\alpha = \pi/2$ into Eq. (A.3.21) leads to

$$a_2 = -a_1, \quad a_4 = 0 \tag{A.3.24}$$

When $x_2 > 0$, the rigorous analysis of the reflected P-wave leads to checking the limit process and indicates that the amplitude of the reflected P-wave decreases with an increase of the distance measured from the boundary surface[9].

(4) Experimental measurement

Compared with a clamped boundary, it is easy to observe and measure the reflection of P-wave at a free boundary. As $\alpha > 0$ and $\beta > 0$ hold, we define an *apparent wave number*, from Snell's law, as follows

$$\kappa \equiv \kappa_p \sin \alpha = \kappa_s \sin \beta \tag{A.3.25}$$

and then define an *apparent velocity* as

$$c \equiv \frac{\omega}{\kappa} = \frac{\kappa_p c_p}{\kappa_p \sin \alpha} = \frac{c_p}{\sin \alpha} = \frac{c_s}{\sin \beta} \tag{A.3.26}$$

It is easy to measure the wave frequency, apparent wave number and apparent velocity at the free boundary and then derive other information, such as the incidence angle and reflecting angle, according to Eq. (A.3.25), Eq. (A.3.26) and the wave speeds.

Remark A.3.3: As analyzed above, when an incident P-wave vertically arrives at a clamped boundary or a free boundary, the reflected wave is a P-wave with the same amplitude and wave number. The reflected P-wave at the clamped boundary keeps the same phase, but that at the free boundary has an opposite phase. The resultant wave of an incident P-wave and the reflected P-wave is a standing wave. The special example is the reflection of a longitudinal wave of a uniform rod at a fixed end or a free end. If the cross-section of the rod at the free end is not perpendicular to the rod axis, the reflected wave from an incident longitudinal wave may involve a transverse wave such that the free end of the rod exhibits a bending vibration.

A.3.1.3   An Incident SV-Wave Reflected at a Clamped Boundary

At a clamped boundary, the resultant displacements due to an incident SV-wave and the reflected wave satisfy the following boundary conditions

$$u_1(x_1, 0, t) = 0, \quad u_2(x_1, 0, t) = 0 \tag{A.3.27}$$

Substitution of Eq. (A.3.3) and $\varphi_1 = 0$ into Eq. (A.3.27) leads to

$$\begin{cases} i(\kappa_2 l_2 \varphi_2 - \kappa_3 m_3 \psi_3 + \kappa_4 m_4 \psi_4) = 0 \\ i(\kappa_2 m_2 \varphi_2 - \kappa_3 l_3 \psi_3 - \kappa_4 l_4 \psi_4) = 0 \end{cases} \tag{A.3.28}$$

---

[9] Achenbach J D (1973). Wave Propagation in Elastic Solids. Amsterdam: North-Holland Publishing Company, 176.

Following the procedure of studying a P-wave incident upon a clamped boundary, we have

$$\begin{cases} (\sin\alpha)a_2 + D(\cos\beta)a_4 = D(\cos\beta)a_3 \\ (\cos\alpha)a_2 - D(\sin\beta)a_4 = D(\sin\alpha)a_3 \end{cases} \qquad \text{(A.3.29)}$$

whereby we derive two reflection coefficients as follows

$$\frac{a_2}{a_3} = \frac{D\sin(2\beta)}{\cos(\alpha - \beta)}, \quad \frac{a_4}{a_3} = \frac{\cos(\alpha + \beta)}{\cos(\alpha - \beta)} \qquad \text{(A.3.30)}$$

Remark A.3.4: To discuss the reflection of an SV-wave above, we should determine an allowable range of incidence angle β first. From Eq. (A.3.10), $\sin(\beta) = \sin(\alpha)/D < \sin(\alpha)$ holds such that the reflecting angle α of the P-wave increases with increasing of incidence angle β of the SV-wave. As the upper bound of reflecting angle of the P-wave is $\alpha_{cr} = \pi/2$, the incidence angle of the SV-wave has an upper bound $\beta_{cr} = \arcsin(1/D)$. In the range of $\beta \in (\beta_{cr}, \pi/2)$, Eq. (A.3.30) fails to work and the reflected P-wave becomes an inhomogeneous wave, the amplitude of which needs to be determined from Eq. (A.2.27). The next subsection will analyze a similar case when an incident SV-wave undergoes a reflection at a free boundary.

Given Poisson's ratio v, Eq. (A.3.10) gives the ratio of wave speeds D. From the range of incidence angle $0 \le \beta \le \beta_{cr} = \arcsin(1/D)$ for the SV-wave, we derive the reflecting angle $\alpha = \arcsin(D\sin\beta) > \beta$ and then get the reflection coefficients from Eq. (A.3.30). Figure A.8 presents the variations of reflection coefficients a_2/a_3 and a_4/a_3, as well as reflecting angle α, with an increase of incidence angle β for four typical values of Poisson's ratio. Both units of angles α and β in Fig. A.8 are in degrees.

As shown in Fig. A.8, the incidence angle of the SV-wave has a bound β_{cr}, and all curves in Fig. A.8 exhibit asymptotic behaviors when $\beta \to \beta_{cr}$. For instance, the reflection coefficient a_4/a_3 approaches -1 and the reflecting angle α goes to 90°. Some other properties of Eq. (A.3.30) and Fig. A.8 are discussed as follows.

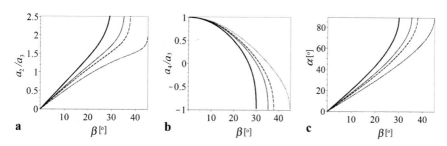

Fig. A.8 The reflection relations of an incident SV-wave at a clamped boundary vs. incidence angle β (thick solid line: $v = 1/3$, thin solid line: $v = 1/4$, dashed line: $v = 1/5$, dot line: $v = 0$). **a** a_2/a_3 for the reflected P-wave; **b** a_4/a_3 for the reflected S-wave; **c** reflecting angle α

(1) Normal incidence

Given $\beta = 0$, Snell's law leads to $\alpha = 0$ so that the reflected wave amplitudes are $a_2 = 0$ and $a_4 = a_3$. Substituting them into Eqs. (A.3.3) and (A.3.5) respectively, we verify that both horizontal displacement and vertical displacement satisfy the boundary conditions, while the resultant stress components due to both incident SV-wave and reflected SV-wave satisfy

$$\begin{cases} \sigma_{22}(x_1, x_2, t) = 0 \\ \sigma_{21}(x_1, x_2, t) = -\mu\kappa_s^2 a_3\{\exp[i\kappa_s(-x_2 - c_s t)] + \exp[i\kappa_s(x_2 - c_s t)]\} \end{cases} \quad \text{(A.3.31)}$$

At the clamped boundary $x_2 = 0$, the shear stress of the reflected SV-wave is as same as that of an incident wave SV-wave such that the resultant shear stress is doubled compared with the shear stress of the incident SV-wave as follows

$$\sigma_{21}(x_1, 0, t) = -2\mu\kappa_s^2 a_3 \exp(-i\kappa_s c_s t) \quad \text{(A.3.32)}$$

(2) Mode conversion

In view of Eq. (A.3.30), $\alpha + \beta = \pi/2$ leads to $a_4 = 0$. That is, the mode conversion occurs. From Snell's law, the incidence angle in this case yields

$$\beta + \arcsin(D \sin \beta) = \frac{\pi}{2} \quad \text{(A.3.33)}$$

As shown in Fig. A.8b, there is a unique incidence angle β satisfying $a_4/a_3 = 0$ so as to realize the mode conversion.

A.3.1.4 An Incident SV-Wave Reflected at a Free Boundary

At the free boundary, the resultant stress components due to an incident SV-wave and the reflected wave superposed satisfy the following boundary conditions

$$\sigma_{22}(x_1, 0, t) = 0, \quad \sigma_{21}(x_1, 0, t) = 0 \quad \text{(A.3.34)}$$

Substitution of Eq. (A.3.5) and $\varphi_1 = 0$ into Eq. (A.3.34) leads to

$$\begin{cases} (\lambda + 2\mu m_2^2)\kappa_2^2 \varphi_2 + 2\mu\kappa_3^2 l_3 m_3 \psi_3 - 2\mu\kappa_4^2 l_4 m_4 \psi_4 = 0 \\ 2\kappa_2^2 l_2 m_2 \varphi_2 + \kappa_3^2(m_3^2 - l_3^2)\psi_3 + \kappa_4^2(m_4^2 - l_4^2)\psi_4 = 0 \end{cases} \quad \text{(A.3.35)}$$

Following the similar procedure of a P-wave incident on a free boundary, we have

$$\begin{cases} (\lambda + 2\mu \cos^2 \alpha)a_2 + \mu D^2 \sin(2\beta)(a_3 - a_4) = 0 \\ \sin(2\alpha)a_2 + D^2 \cos(2\beta)(a_3 + a_4) = 0 \end{cases} \quad \text{(A.3.36)}$$

Using Eq. (A.3.19), we can recast Eq. (A.3.36) as

$$
\begin{cases}
\cos(2\beta)\left(\dfrac{a_2}{a_3}\right) + \sin(2\beta)\left(1 - \dfrac{a_4}{a_3}\right) = 0 \\[3mm]
\sin(2\alpha)\left(\dfrac{a_2}{a_3}\right) + D^2 \cos(2\beta)\left(1 + \dfrac{a_4}{a_3}\right) = 0
\end{cases}
\tag{A.3.37}
$$

Solving Eq. (A.3.37) for the two reflection coefficients gives

$$
\begin{cases}
\dfrac{a_2}{a_3} = -\dfrac{D^2 \sin(4\beta)}{\sin(2\alpha)\sin(2\beta) + D^2 \cos^2(2\beta)} \\[4mm]
\dfrac{a_4}{a_3} = \dfrac{\sin(2\alpha)\sin(2\beta) - D^2 \cos^2(2\beta)}{\sin(2\alpha)\sin(2\beta) + D^2 \cos^2(2\beta)}
\end{cases}
\tag{A.3.38}
$$

Remark A.3.5: Many books dealt with this reflection problem, but had some typos. As stated in Remark A.3.1 and Remark A.3.2, the upward direction of axis x_2 in the references[10,11] should lead to $\gamma_{23} = -Da_2/a_3$, rather than $\gamma_{23} = Da_2/a_3$. The results corresponding to Eq. (A.3.38), as well as Eq. (A.3.42) later, in the references[12,13] are also subject to corrections.

In view of Remark A.3.4, an incidence angle β needs to be determined for the discussion about Eq. (A.3.38) first since the incidence angle of an SV-wave has a bound $\beta_{cr} = \arcsin(1/D)$. Thus, given Poisson's ratio v, we get the ratio of wave speeds D from Eq. (A.3.10) and the range of incidence angle $0 \le \beta \le \beta_{cr} = \arcsin(1/D)$ for an incident SV-wave such that the reflecting angle of the P-wave yields $\alpha = \arcsin(D \sin \beta) > \beta$. Finally, we get two reflection coefficients from Eq. (A.3.38).

Figure A.9 shows the variations of reflection coefficients a_2/a_3 and a_4/a_3, as well as reflecting angle α, with increasing incidence angle β for four values of Poisson's ratio. Here, the units of angles α and β are in degrees. As shown in Fig. A.9, the curves except for the case of $v = 0$ exhibit asymptotic behaviors when $\beta \to \beta_{cr}$. For example, the reflecting angle of the P-wave yields $\alpha \to 90°$ and the reflection coefficient of the SV-wave becomes $a_4/a_3 \to -1$ for $v > 0$. Here are the discussions about some other properties of Eq. (A.3.38) and Fig. A.9.

---

[10] Achenbach J D (1973). Wave Propagation in Elastic Solids. Amsterdam: North-Holland Publishing Company, 177–179.

[11] Hagedorn P, Das Gupta A (2007). Vibrations and Waves in Continuous Mechanical Systems. Chichester: John Wiley & Sons Inc., 351–352.

[12] Dai H L (2014). Elasto-dynamics. Changsha: Press of Hunan University, 70–71 (in Chinese).

[13] Du Q H, Yang X A, Xu X Q, et al. (1997). Handbook of Engineering Mechanics. Beijing: Press of Higher Education, 2331 (in Chinese).

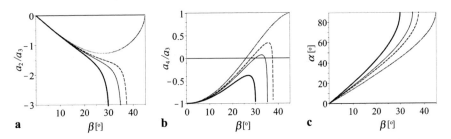

Fig. A.9 The reflection relations of an incident SV-wave at a free boundary vs. incidence angle β (thick solid line: $\nu = 1/3$, thin solid line: $\nu = 1/4$, dashed line: $\nu = 1/5$, dot line: $\nu = 0$). **a** a_2/a_3 for the reflected P-wave; **b** a_4/a_3 for the reflected SV-wave; **c** reflecting angle α

(1) Normal incidence

When $\beta = 0$, Snell's law leads to $\alpha = 0$ such that $a_2/a_3 = 0$ and $a_4/a_3 = -1$ hold for Eq. (A.3.38). That is, an incident SV-wave is fully reflected as the SV-wave with the same amplitude and wave number, but an opposite phase. Substituting $a_2/a_3 = 0$ and $a_4/a_3 = -1$ into Eq. (A.3.5), we find that, in the free boundary, the normal stress due to either an incident SV-wave or the reflected SV-wave is zero, while the resultant shear stress due to an incident SV-wave and the reflected SV-wave is zero.

Substitution of $a_2/a_3 = 0$ and $a_4/a_3 = -1$ into Eq. (A.3.3) leads to the resultant displacement components due to both incident SV-wave and reflected SV-wave as follows

$$\begin{cases} u_1(x_1, x_2, t) = -i\kappa_s a_3\{\exp[i\kappa_s(x_2 - c_st)] + \exp[i\kappa_s(-x_2 - c_st)]\} \\ u_2(x_1, x_2, t) = 0 \end{cases} \quad \text{(A.3.39)}$$

At the free boundary $x_2 = 0$, the vertical displacement keeps zero and the horizontal displacements due to an incident SV-wave or the reflected SV-wave are identical so that the resultant horizontal displacement is doubled, namely,

$$u_1(x_1, 0, t) = -2i\kappa_s a_3 \exp(-i\kappa_s c_st) \quad \text{(A.3.40)}$$

(2) Wave type unchanged

As shown in Eq. (A.3.38), Figs. A.9a and A.9b, $a_2/a_3 = 0$ and $a_4/a_3 = 1$ hold when $\beta = \beta_{cr} = \pi/4$ under the condition of $\nu = 0$. In this case, the reflected wave is still an SV-wave with both wave number and phase unchanged.

(3) Mode conversion

In Eq. (A.3.38), $a_4/a_3 = 0$ leads to a mode conversion. That is, an incident SV-wave fully becomes a P-wave after the reflection. The numerator of the second equation in Eq. (A.3.38) gives the following condition of the mode conversion

$$\sin(2\alpha)\sin(2\beta) = D^2\cos^2(2\beta) \quad \text{(A.3.41)}$$

For a given D, Eq. (A.3.41) corresponds to the two intersections of a curve with the abscissa in Fig. A.9b. That is, two incidence angles satisfy Eq. (A.3.41) for proper Poisson's ratio v. Substitution of Eq. (A.3.41) into Eq. (A.3.38) leads to the reflection coefficient of P-wave as follows

$$\frac{a_2}{a_3} = -\frac{D^2 \sin(4\beta)}{2D^2 \cos^2(2\beta)} = -\tan(2\beta) \qquad \text{(A.3.42)}$$

(4) Grazing incidence

Let the incidence angle $\beta \to \pi/2$, that is, the incidence direction is parallel to the free boundary $x_2 = 0$. Substituting $\beta = \pi/2$ into Eq. (A.3.38) gives

$$a_2 = 0, \quad a_4 = -a_3 \qquad \text{(A.3.43)}$$

As shown in Eq. (A.3.43), the reflected wave is an SV-wave with an opposite phase along the free boundary $x_2 = 0$.

(5) Critical reflection

In view of Remark A.3.4, when the reflecting angle of a P-wave is $\alpha = \pi/2$, the P-wave is in a critical case named a *critical reflection*, the corresponding incidence angle β of the SV-wave is named the *critical incidence angle* and denoted as β_{cr}. As $\beta_{cr} = \arcsin(1/D)$ holds, the SV-wave with the critical incidence angle β_{cr} leads to $\sin\alpha = 1$ such that the apparent velocity is $c = c_p$. According to Eqs. (A.2.23), (A.2.26) and associated discussions, the reflected P-wave is a grazing wave and the reflected SV-wave is a homogeneous S-wave. Their potential functions satisfy

$$\begin{cases} \varphi_2(x_1, x_2, t) = a_2 \exp\left[i\kappa_p(x_1 - c_p t)\right], \\ \psi_4(x_1, x_2, t) = a_4 \exp\{i\kappa_s[(\sin\beta_{cr})x_1 + (\cos\beta_{cr})x_2 - c_s t]\} \end{cases} \qquad \text{(A.3.44)}$$

where the expression of $\psi_4(x_1, x_2, t)$ comes from the second equation in Eq. (A.2.26) with $\kappa = \kappa_p = \kappa_s \sin\beta_{cr}$ and $\kappa r_s = \kappa_p r_s = \kappa_s \cos\beta_{cr}$ substituted. From Eq. (A.3.38), we have the reflection coefficients

$$\begin{cases} \dfrac{a_2}{a_3} = -\dfrac{\sin(4\beta_{cr})}{\cos^2(2\beta_{cr})} = -\dfrac{4(\sin\beta_{cr})\sqrt{1 - \sin^2\beta_{cr}}}{1 - 2\sin^2\beta_{cr}} = -\dfrac{4\sqrt{D^2 - 1}}{D^2 - 2} \\ \dfrac{a_4}{a_3} = -1 \end{cases} \qquad \text{(A.3.45)}$$

(6) Inhomogeneous planar wave

If the incidence angle of an SV-wave yields $\beta \in (\beta_{cr}, \pi/2)$, Eq. (A.3.38) based on the precondition of homogeneous planar waves can not predict any reflected P-waves, and thus, we ought to study inhomogeneous planar waves. As discussed in

Sect. A.2.1, we use Eq. (A.2.23) and the meaningful terms in Eq. (A.2.27) to describe the reflected P-wave and SV-wave governed by two potential functions as follows

$$
\begin{cases}
\varphi_2(x_1, x_2, t) = a_2 \exp(-\kappa v_p x_2) \exp[i\kappa(x_1 - ct)] \\
\psi_4(x_1, x_2, t) = a_4 \exp(\ i\kappa r_s x_2) \exp[i\kappa(x_1 - ct)]
\end{cases}
\tag{A.3.46}
$$

where $\kappa = \kappa_s \sin \beta$ and $c = c_s / \sin \beta$ are the apparent wave number in Eq. (A.3.25) and the apparent velocity in Eq. (A.3.26).

Substitution of Eq. (A.3.46) into Eq. (A.2.8) leads to the resultant displacement components due to an incident SV-wave and the reflected wave as follows

$$
\begin{cases}
u_1(x_1, x_2, t) = \dfrac{\partial \varphi_2}{\partial x_1} + \dfrac{\partial \psi_3}{\partial x_2} + \dfrac{\partial \psi_4}{\partial x_2} = i(\kappa \varphi_2 - \kappa_3 m_3 \psi_3 + \kappa r_s \psi_4) \\
u_2(x_1, x_2, t) = \dfrac{\partial \varphi_2}{\partial x_2} - \dfrac{\partial \psi_3}{\partial x_1} - \dfrac{\partial \psi_4}{\partial x_1} = -\kappa v_p \varphi - i(\kappa_3 l_3 \psi_3 + \kappa \psi_4)
\end{cases}
\tag{A.3.47}
$$

The non-zero strain components, from Eq. (A.2.9), are

$$
\begin{cases}
\varepsilon_{11} = -\kappa^2 \varphi_2 + \kappa_s^2 l_3 m_3 \psi_3 - \kappa^2 r_s \psi_4 \\
\varepsilon_{22} = \kappa^2 v_p^2 \varphi_2 - \kappa_3^2 l_3 m_3 \psi_3 + \kappa^2 r_s \psi_4 \\
\varepsilon_{21} = -2i\kappa^2 v_p \varphi_2 + \kappa_s^2(l_3^2 - m_3^2)\psi_3/2 + \kappa^2(1 - r_s^2)\psi_4/2
\end{cases}
\tag{A.3.48}
$$

Substituting Eq. (A.3.38) into Eq. (A.1.3) gives the stress conditions at the free boundary $x_2 = 0$ as follows

$$
\begin{cases}
\sigma_{22} = [(\lambda + 2\mu)\kappa^2 v_p^2 - \lambda\kappa^2]\varphi_2 - 2\mu\kappa_s^2 l_3 m_3 \psi_3 + 2\mu\kappa^2 r_s \psi_4 = 0 \\
\sigma_{12} = \mu[-2i\kappa^2 v_p \varphi_2 + \kappa_s^2(l_3^2 - m_3^2)\psi_3 + \kappa^2(1 - r_s^2)\psi_4] = 0
\end{cases}
\tag{A.3.49}
$$

From the following relations derived already

$$
\begin{cases}
l_3 = \sin \beta, \quad m_3 = \cos \beta, \quad \kappa = \kappa_s \sin \beta, \quad \lambda = \mu(D^2 - 2) \\
D^2 = \dfrac{c_p^2}{c_s^2}, \quad v_p^2 = 1 - \dfrac{c^2}{c_p^2}, \quad r_s^2 = \dfrac{c^2}{c_s^2} - 1 = \dfrac{1}{\sin^2 \beta} - 1 = \cot^2 \beta
\end{cases}
\tag{A.3.50}
$$

simplifying Eq. (A.3.49) and eliminating exponential functions, we have

$$
\begin{cases}
(1 - r_s^2)a_2 + 2r_s a_4 = 2r_s a_3 \\
2iv_p a_2 + (r_s^2 - 1)a_4 = (1 - r_s^2)a_3
\end{cases}
\tag{A.3.51}
$$

By solving Eq. (A.3.51), we get two complex reflection coefficients as follows

$$\frac{a_2}{a_3} = \frac{4r_s(1 - r_s^2)}{(1 - r_s^2)^2 + 4iv_pr_s}, \quad \frac{a_4}{a_3} = -\frac{(1 - r_s^2)^2 - 4iv_pr_s}{(1 - r_s^2)^2 + 4iv_pr_s} \qquad \text{(A.3.52)}$$

As shown in the above analysis, the reflected P-wave travels along axis x_1 with an apparent velocity $c = c_s/\sin\beta$ and an exponentially decreasing amplitude along axis x_2. This is an inhomogeneous planar wave. The reflected SV-wave propagates at shear wave speed c_s and has reflecting angle β. The numerator and denominator of a_4/a_3 is a pair of conjugate complex numbers such that $|a_4/a_3| = 1$ holds true. Hence, the reflected SV-wave has the same amplitude as that of the incident SV-wave, but a delayed phase shift $2\text{arccot}[(1 - r_s^2)^2/(4v_pr_s)]$. This phenomenon is called a *total reflection*.

Remark A.3.6: When an incident SV-wave normally arrives at a clamped or free boundary, the reflected wave is an SV-wave with the same amplitude and wave number, but with the same or an opposite phase. The resultant wave of the incident SV-wave and the reflected SV-waves is a standing SV-wave. The reflection of a bending wave of an E-B beam at a clamped end or a free end is a simplified case of the above results. If the cross-section of the beam end is not perpendicular to the beam axis, the reflected wave of an incident bending wave of the beam may include a longitudinal wave.

A.3.2 Reflection of an SH-Wave

As shown in Fig. A.5, let $u_3(x_1, x_2, t)$ be the displacement component of an SH-wave. From Eq. (A.2.8), $u_3(x_1, x_2, t)$ depends on two potential functions $\psi_1(x_1, x_2, t)$ and $\psi_2(x_1, x_2, t)$, and satisfies

$$u_3(x_1, x_2, t) = \frac{\partial\psi_1}{\partial x_2} - \frac{\partial\psi_2}{\partial x_1} \qquad \text{(A.3.53)}$$

while the two potential functions yield the normalization condition as follows

$$\frac{\partial\psi_1}{\partial x_1} + \frac{\partial\psi_2}{\partial x_2} = 0 \qquad \text{(A.3.54)}$$

Thus, it is possible to use a single potential function to describe the displacement $u_3(x_1, x_2, t)$.

For this purpose, define a new potential function $\psi(x_1, x_2, t)$ satisfying the following wave equation

$$c_s^2\nabla^2\psi(x_1, x_2, t) - \frac{\partial^2\psi(x_1, x_2, t)}{\partial t^2} = 0 \qquad \text{(A.3.55)}$$

Using $\psi(x_1, x_2, t)$ to express the two potential functions $\psi_1(x_1, x_2, t)$ and $\psi_2(x_1, x_2, t)$ gives the normalization condition

$$\psi_1 = \frac{\partial \psi}{\partial x_2}, \quad \psi_2 = -\frac{\partial \psi}{\partial x_1} \quad \Rightarrow \quad \frac{\partial \psi_1}{\partial x_1} + \frac{\partial \psi_2}{\partial x_2} = \frac{\partial^2 \psi}{\partial x_1 \partial x_2} - \frac{\partial^2 \psi}{\partial x_2 \partial x_1} = 0$$

$$(A.3.56)$$

It is easy to verify that $\psi_1(x_1, x_2, t)$ and $\psi_2(x_1, x_2, t)$ satisfy the following two wave equations

$$\begin{cases} c_s^2 \nabla^2 \psi_1(x_1, x_2, t) - \dfrac{\partial^2 \psi_1(x_1, x_2, t)}{\partial t^2} \\ \quad = \dfrac{\partial}{\partial x_2}\left[c_s^2 \nabla^2 \psi(x_1, x_2, t) - \dfrac{\partial^2 \psi(x_1, x_2, t)}{\partial t^2}\right] = 0 \\ c_s^2 \nabla^2 \psi_2(x_1, x_2, t) - \dfrac{\partial^2 \psi_2(x_1, x_2, t)}{\partial t^2} \\ \quad = -\dfrac{\partial}{\partial x_1}\left[c_s^2 \nabla^2 \psi(x_1, x_2, t) - \dfrac{\partial^2 \psi(x_1, x_2, t)}{\partial t^2}\right] = 0 \end{cases}$$

$$(A.3.57)$$

Hence, Eq. (A.3.53) becomes

$$u_3(x_1, x_2, t) = \frac{\partial \psi_1}{\partial x_2} - \frac{\partial \psi_2}{\partial x_1} = \frac{\partial^2 \psi}{\partial x_2^2} + \frac{\partial^2 \psi}{\partial x_1^2} = \nabla^2 \psi(x_1, x_2, t) \qquad (A.3.58)$$

As shown in Fig. A.5, let γ_1 and γ_2 be an incidence angle and a reflecting angle, respectively. Assume that the new potential functions of the incident SH-wave and the reflected SH-wave are

$$\begin{cases} \psi_5(x_1, x_2, t) = a_5 \exp[i\kappa_5(l_5 x_1 - m_5 x_2 - c_s t)] \\ \psi_6(x_1, x_2, t) = a_6 \exp[i\kappa_6(l_6 x_1 + m_6 x_2 - c_s t)] \end{cases} \qquad (A.3.59)$$

where κ_j, $j = 5, 6$ are the wave numbers, l_j, m_j, $j = 5, 6$ are the cosine functions, and each of their arguments is an angle between the direction of wave propagation and an axis of the frame of coordinates. Therefore, they have a relation with angles γ_1 and γ_2 as follows

$$l_{j+4} = \sin \gamma_j, \quad m_{j+4} = \cos \gamma_j, \quad l_{j+4}^2 + m_{j+4}^2 = 1, \quad j = 1, 2 \qquad (A.3.60)$$

In Fig. A.5, the vertical component of an incident SH-wave is in the negative direction of axis x_2 so that coefficient $-m_5$ appears in Eq. (A.3.59).

According to Eqs. (A.3.58), (A.3.59) and (A.3.60), the resultant displacement components due to both incident SH-wave and reflected SH-wave is

$$u_3(x_1, x_2, t) = \nabla^2[\psi_5(x_1, x_2, t) + \psi_6(x_1, x_2, t)]$$
$$= -\{\kappa_5^2 a_5 \exp[i\kappa_5(l_5 x_1 - m_5 x_2 - c_s t)]$$
$$+ \kappa_6^2 a_6 \exp[i\kappa_6(l_6 x_1 + m_6 x_2 - c_s t)]\} \tag{A.3.61}$$

The corresponding non-zero resultant strain components are as follows

$$
\begin{cases}
\varepsilon_{13} = -\dfrac{i}{2}\{\kappa_5^3 l_5 a_5 \exp[i\kappa_5(l_5 x_1 - m_5 x_2 - c_s t)] \\
\quad + \kappa_6^3 l_6 a_6 \exp[i\kappa_6(l_6 x_1 + m_6 x_2 - c_s t)]\} \\
\varepsilon_{23} = -\dfrac{i}{2}\{-\kappa_5^3 m_5 a_5 \exp[i\kappa_5(l_5 x_1 - m_5 x_2 - c_s t)] \\
\quad + \kappa_6^3 m_6 a_6 \exp[i\kappa_6(l_6 x_1 + m_6 x_2 - c_s t)]\}
\end{cases} \tag{A.3.62}
$$

There follow the non-zero resultant stress components

$$
\begin{cases}
\sigma_{13} = -i\mu\{\kappa_5^3 l_5 a_5 \exp[i\kappa_5(l_5 x_1 - m_5 x_2 - c_s t)] \\
\quad + \kappa_6^3 l_6 a_6 \exp[i\kappa_6(l_6 x_1 + m_6 x_2 - c_s t)]\} \\
\sigma_{23} = -i\mu\{-\kappa_5^3 m_5 a_5 \exp[i\kappa_5(l_5 x_1 - m_5 x_2 - c_s t)] \\
\quad + \kappa_6^3 m_6 a_6 \exp[i\kappa_6(l_6 x_1 + m_6 x_2 - c_s t)]\}
\end{cases} \tag{A.3.63}
$$

To determine the reflection of SH-wave in a semi-infinite medium, one needs only one boundary condition, that is, the compatibility condition along axis x_3.

A.3.2.1 An Incident SH-Wave Reflected at a Clamped Boundary

At the clamped boundary, the resultant horizontal displacement due to both incident SH-wave and reflected SH-wave is zero, namely,

$$u_3(x_1, 0, t) = 0 \tag{A.3.64}$$

Substitution of Eq. (A.3.61) into Eq. (A.3.64) leads to

$$\kappa_5^2 a_5 \exp[i\kappa_5(l_5 x_1 - c_s t)] + \kappa_6^2 a_6 \exp[i\kappa_6(l_6 x_1 - c_s t)] = 0 \tag{A.3.65}$$

The above relation holds for an arbitrary time t and an arbitrary position coordinate x_1 such that all corresponding coefficients must be identical, i.e.,

$$\kappa_5 c_s = \kappa_6 c_s, \quad \kappa_5 l_5 = \kappa_6 l_6 \quad \Rightarrow \quad \kappa_5 = \kappa_6, \quad l_5 = l_6 \tag{A.3.66}$$

Therefore, the incident and reflected SH-waves have the same wave number, and the incidence angle and the reflecting angle are identical.

Take the following parameters with physical meanings

$$\kappa_5 = \kappa_6 \equiv \kappa_s, \quad \gamma_1 = \gamma_2 \equiv \gamma \tag{A.3.67}$$

Substituting Eq. (A.3.67) into Eq. (A.3.65) and eliminating exponential functions, we have

$$\kappa_s^2(a_5 + a_6) = 0 \quad \Rightarrow \quad a_6 = -a_5 \tag{A.3.68}$$

That is, the incident SH wave and the reflected SH wave have the same amplitude, but an opposite phase.

To check the normal incidence of an SH-wave, let $\gamma = 0$, that is, $l_5 = l_6 = 0$ and $m_5 = m_6 = 1$. Substituting them, together with Eqs. (A.3.67) and (A.3.68), into Eqs. (A.3.61) and (A.3.63), we immediately verify that the resultant horizontal displacements due to both incident SH-wave and reflected SH-wave is zero, while the resultant shear stresses due to both incident and reflected SH-waves satisfy

$$\begin{cases} \sigma_{13}(x_1, x_2, t) = 0 \\ \sigma_{23}(x_1, x_2, t) = i\mu\kappa_s^3 a_5\{\exp[i\kappa_s(-x_2 - c_s t)] + \exp[i\kappa_s(x_2 - c_s t)]\} \end{cases} \tag{A.3.69}$$

At the clamped boundary $x_2 = 0$, the resultant shear stress is doubled compared with the shear stress due to the incident SH-wave as follows

$$\sigma_{23}(x_1, 0, t) = 2i\mu\kappa_s^3 a_5 \exp(-i\kappa_s c_s t) \tag{A.3.70}$$

A.3.2.2 An Incident SH-Wave Reflected at a Free Boundary

At the free boundary $x_2 = 0$, the resultant shear stresses due to an incident SH-wave and the reflected SH-wave should be zero, namely,

$$\sigma_{23}(x_1, 0, t) = 0 \tag{A.3.71}$$

Substitution of the second equation in Eq. (A.3.63) into Eq. (A.3.71) leads to

$$-\kappa_5^3 m_5 a_5 \exp[i\kappa_5(l_5 x_1 - c_s t)] + \kappa_6^3 m_6 a_6 \exp[i\kappa_6(l_6 x_1 - c_s t)] = 0 \tag{A.3.72}$$

Similarly, we get Eqs. (A.3.66) and (A.3.67) again. Substituting them into Eq. (A.3.72) and eliminating exponential functions, we have

$$\kappa_s^3(\cos \gamma)(a_6 - a_5) = 0 \tag{A.3.73}$$

Regardless of grazing incidence, the incidence angle yields $0 \le \gamma < \pi/2$ so that $\cos \gamma \ne 0$ holds. Equation (A.3.73) leads to $a_6 = a_5$, that is, the reflected SH-wave is identical to the incident SH-wave. Substituting $a_6 = a_5$ into Eq. (A.3.61), we get the resultant displacement due to both incident and reflected SH-waves as follows

$$u_3(x_1, x_2, t) = -\kappa_s^2 a_5 \{ \exp[i\kappa_s(x_1 \sin \gamma - x_2 \cos \gamma - c_s t)]$$
$$+ \exp[i\kappa_s(x_1 \sin \gamma + x_2 \cos \gamma - c_s t)] \} \qquad (A.3.74)$$

For a normal incidence, substitution of $\gamma = 0$ into Eq. (A.3.74) leads to

$$u_3(x_1, x_2, t) = -\kappa_s^2 a_5 \{ \exp[i\kappa_s(-x_2 - c_s t)] + \exp[i\kappa_s(x_2 - c_s t)] \} \qquad (A.3.75)$$

At the free boundary $x_2 = 0$, the resultant wave of both incident and reflected SH-waves makes the total horizontal displacement be doubled as

$$u_3(x_1, 0, t) = -2\kappa_s^2 a_5 \exp(-i\kappa_s c_s t) \qquad (A.3.76)$$

To summarize the above results, the major reflection rules of three types of planar harmonic waves are given in Table A.1, where the doubled displacement or stress components occur at the boundary and the superscript (I) represents incident wave.

To this end, this section has completed the reflection analysis of three types of planar harmonic waves incident upon the boundary of a semi-infinite medium. The analysis here can be extended to dealing with the problem of both reflection and refraction at an interface of two different media. At such an interface of media, an incident wave produces not only reflected waves but also refracted waves, and leads to more complicated mathematical manipulations in analysis. For example, it is necessary to account for four compatibility conditions of both reflected and refracted P-waves, and both reflected and refracted SV-waves at the interface simultaneously, and then solve a set of linear algebraic equations of four unknowns to determine two reflection coefficients and two refraction coefficients. Yet, this case still enjoys some similarities as the analysis in this section. For instance, both reflected and refracted waves satisfy Snell's laws, that is, the ratio of wave speeds of each medium determines the relation between a reflecting angle and a refracting angle. In addition, the reflection and refraction of an SH-wave are independent of both P-waves and SV-waves.

Table A.1 Reflection rules of three types of incident planar waves at the boundary of a semi-infinite medium

Wave type	Normal incident upon a clamped boundary	Normal incident upon a free boundary	Arbitrary incident upon a free boundary
P	P, $a_2 = a_1$, $\sigma_{22} = 2\sigma_{22}^{(I)}$	P, $a_2 = -a_1$, $u_2 = 2u_2^{(I)}$	P and SV, or mode conversion
SV	SV, $a_4 = a_3$, $\sigma_{21} = 2\sigma_{21}^{(I)}$	SV, $a_4 = -a_3$, $u_1 = 2u_1^{(I)}$	P and SV, or mode conversion
SH	SH, $a_6 = -a_5$, $\sigma_{23} = 2\sigma_{23}^{(I)}$	SH, $a_6 = a_5$, $u_3 = 2u_3^{(I)}$	SH

A.3.3 Rayleigh Surface Waves

As discussed for inhomogeneous waves in Sect. A.2.1, there exist both inhomogeneous P-wave and SV-wave in a medium when $c < c_s$. These two waves seem not important at first glance since their amplitudes exponentially decrease. In 1887, nevertheless, Lord Rayleigh, a great British physicist, found that the above two waves can form a surface wave traveling freely. This finding is named a *Rayleigh surface wave* or a *Rayleigh wave* for short.

Now, we discuss Rayleigh surface wave in the frame of coordinates in Fig. A.5, where $x_2 = 0$ represents the interface between a semi-infinite medium and a half space of vacuum. In view of Eqs. (A.2.23) and (A.2.29), the inhomogeneous P-wave and SV-wave traveling in the medium ($x_2 \geq 0$) yield the potential functions as follows

$$\begin{cases} \varphi(x_1, x_2, t) = a \exp\left[-\kappa\sqrt{1 - (c/c_p)^2}\,x_2\right] \exp[i\kappa(x_1 - ct)] \\ \psi_3(x_1, x_2, t) = b \exp\left[-\kappa\sqrt{1 - (c/c_s)^2}\,x_2\right] \exp[i\kappa(x_1 - ct)] \end{cases} \tag{A.3.77}$$

where c and κ are the apparent velocity and wave number to be determined. The existence of Rayleigh surface wave depends on whether a real number c exists as an apparent velocity such that $c < c_s$ holds true.

To study the existence of such a real number, define the following two constants

$$\begin{cases} r \equiv \kappa\sqrt{1 - (c/c_p)^2} \\ s \equiv \kappa\sqrt{1 - (c/c_s)^2} \end{cases} \tag{A.3.78}$$

and recast Eq. (A.3.77) as

$$\begin{cases} \varphi(x_1, x_2, t) = a \exp[-rx_2 + i\kappa(x_1 - ct)] \\ \psi_3(x_1, x_2, t) = b \exp[-sx_2 + i\kappa(x_1 - ct)] \end{cases} \tag{A.3.79}$$

From Eq. (A.3.79), we have the displacement components of the above waves as follows

$$\begin{cases} u_1(x_1, x_2, t) = \dfrac{\partial\varphi}{\partial x_1} + \dfrac{\partial\psi_3}{\partial x_2} = i\kappa\varphi - s\psi_3 \\ u_2(x_1, x_2, t) = \dfrac{\partial\varphi}{\partial x_2} - \dfrac{\partial\psi_3}{\partial x_1} = -r\varphi - i\kappa\psi_3 \end{cases} \tag{A.3.80}$$

Substitution of Eq. (A.3.80) into Eq. (A.1.1) leads to the corresponding strain components

$$\begin{cases} \varepsilon_{11}(x_1, x_2, t) = -\kappa^2 \varphi - i\kappa s \psi_3 \\ \varepsilon_{22}(x_1, x_2, t) = r^2 \varphi + i\kappa s \psi_3 \\ \varepsilon_{21}(x_1, x_2, t) = -ir\kappa \varphi + (s^2 + \kappa^2)\psi_3/2 \end{cases} \tag{A.3.81}$$

By substituting Eq. (A.3.81) into Eq. (A.1.3), we get the following stress components

$$\begin{cases} \sigma_{22} = -(\lambda + 2\mu)(r^2\varphi + i\kappa s\psi_3) + \lambda(\kappa^2\varphi + i\kappa s\psi_3) \\ \sigma_{21} = -2i\mu r\kappa\varphi + \mu(s^2 + \kappa^2)\psi_3 \end{cases} \tag{A.3.82}$$

Substituting Eqs. (A.3.82) and (A.3.79) into the stress conditions of the free boundary $x_2 = 0$, namely,

$$\sigma_{22}(x_1, 0, t) = 0, \quad \sigma_{21}(x_1, 0, t) = 0 \tag{A.3.83}$$

and eliminating exponential functions, we have

$$\begin{cases} [(\lambda + 2\mu)r^2 - \lambda\kappa^2]a + 2i\kappa\mu s b = 0 \\ -2ir\kappa a + (s^2 + \kappa^2)b = 0 \end{cases} \tag{A.3.84}$$

The sufficient and necessary condition of any non-trivial solution (a, b) of Eq. (A.3.84) yields

$$[(\lambda + 2\mu)r^2 - \lambda\kappa^2](s^2 + \kappa^2) - 4\kappa^2\mu rs = 0 \tag{A.3.85}$$

For convenience of discussion, let Eq. (A.3.85) be divided by $\mu\kappa^4$ and then be substituted with Eq. (A.3.10), we have

$$\left[D^2 \left(\frac{r}{\kappa}\right)^2 - \frac{\lambda}{\mu} \right] \left[1 + \left(\frac{s}{\kappa}\right)^2 \right] - 4\left(\frac{r}{\kappa}\right)\left(\frac{s}{\kappa}\right) = 0 \tag{A.3.86}$$

Substituting Eq. (A.3.78) into Eq. (A.3.86) and utilizing Eq. (A.3.10) again, we derive

$$\left\{ D^2 \left[1 - \left(\frac{c}{c_p}\right)^2 \right] - \frac{\lambda}{\mu} \right\} \left[1 + 1 - \left(\frac{c}{c_s}\right)^2 \right] - 4\left[1 - \left(\frac{c}{c_p}\right)^2 \right]^{1/2} \left[1 - \left(\frac{c}{c_s}\right)^2 \right]^{1/2}$$

$$= \left[2 - \left(\frac{c}{c_s}\right)^2 \right]^2 - 4\left[1 - \frac{1}{D^2}\left(\frac{c}{c_s}\right)^2 \right]^{1/2} \left[1 - \left(\frac{c}{c_s}\right)^2 \right]^{1/2} = 0 \tag{A.3.87}$$

The above equation can be recast, through a new parameter $\xi^2 \equiv (c/c_s)^2$, as follows

$$(2 - \xi^2)^2 = 4\left(1 - \frac{1}{D^2}\xi^2 \right)^{1/2} (1 - \xi^2)^{1/2} \tag{A.3.88}$$

Let both sides of Eq. (A.3.88) be squared, we have a quartic equation of ξ^2, namely,

$$\xi^2 f(\xi^2) = 0, \quad f(\xi^2) \equiv \left[\xi^6 - 8\xi^4 + \left(24 - \frac{16}{D^2} \right) \xi^2 + 16 \left(\frac{1}{D^2} - 1 \right) \right] = 0$$
(A.3.89)

From $f(0) = 16(1/D^2 - 1) < 0$ and $f(1) = 1 > 0$, the cubic equation $f(\xi^2) = 0$ in Eq. (A.3.89) has a real root $\xi_R^2 \in (0, 1)$ such that

$$\begin{cases} \xi_R = \left[\dfrac{8}{3} - \dfrac{\delta}{3(1 - \nu)} + \dfrac{3(5\nu - 2)}{8\delta} \right]^{1/2} \\[3mm] \delta \equiv \left[(1 - \nu)^2 \left(224\nu - 44 + 12\sqrt{\dfrac{96\nu^3 - 48\nu^2 + 63\nu - 15)}{1 - \nu}} \right) \right]^{1/3} \end{cases} \quad \text{(A.3.90)}$$

Thus, a real number $\xi_R c_s \in (0, \ c_s)$ exists and can serve as the wave speed so that both Eqs. (A.3.78) and (A.3.77) hold true. For Poisson's ratio $\nu \in [0, \ 0.5]$ of a conventional elastic medium, ξ_R is a unique root of cubic equation $f(\xi) = 0$ for $\xi \in (0, 1)^{14}$.

To this end, we define the *Rayleigh wave speed* $c_R \equiv \xi_R c_s$, which yields

$$c_R < c_s < c_p \tag{A.3.91}$$

In Eq. (A.3.90), ξ_R only depends on Poisson's ratio and has nothing to with wave frequency. Hence, Rayleigh waves are dispersive. Because of the complicated expressions in Eq. (A.3.90), we turn to the discussion about numerical results in Fig. A.10.

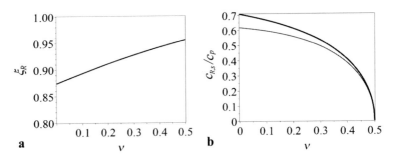

Fig. A.10 The variation of Rayleigh wave speed versus Pisson's ratio ν. **a** the dimensionless wave speed $\xi_R = c_R/c_s$; **b** the dimensionless wave speeds (thick solid line: c_s/c_p, thin solid line: c_R/c_p)

---

[14] Achenbach J D (1973). Wave Propagation in Elastic Solids. Amsterdam: North-Holland Publishing Company, 189–191.

As shown in Fig. A.10a, ξ_R monotonically increases with an increase of Poisson's ratio ν. Take $\nu = 0.25$ as an example, we have $c_R = 0.9194c_s$. Figure A.10b presents the comparison of dimensionless SV-wave speed and the Rayleigh wave speed with increasing Poisson's ratio ν. Among the waves due to an earthquake, for example, the P-wave is the fastest and the Rayleigh wave is the slowest, while the SV-wave is between them. Their speed differences vanish when Poisson's ratio $\nu \rightarrow 0.5$.

Finally, we prove that the trajectory of an arbitrary mass point in Rayleigh waves is a counter clockwise ellipse with an increase of time. Solving the second equation in Eq. (A.3.84) for wave amplitude b leads to

$$b = \frac{2ir\kappa a}{\kappa^2 + s^2} \tag{A.3.92}$$

Substitution of Eq. (A.3.92) into Eq. (A.3.80) gives

$$\begin{cases} u_1(x_1, x_2, t) = i\kappa a \left[\exp(-rx_2) - \frac{2rs}{\kappa^2 + s^2} \exp(-sx_2) \right] \exp[i\kappa(x_1 - c_R t)] \\ u_2(x_1, x_2, t) = \kappa a \left[-\frac{r}{\kappa} \exp(-rx_2) + \frac{2r\kappa}{\kappa^2 + s^2} \exp(-sx_2) \right] \exp[i\kappa(x_1 - c_R t)] \end{cases} \tag{A.3.93}$$

Given the position (x_1, x_2) of an arbitrary material point in the medium, the real part of Eq. (A.3.93) corresponds to the parametric equation of an ellipse as follows

$$\begin{cases} u_1(x_1, x_2, t) = -\kappa a \left[\exp(-rx_2) - \frac{2rs}{\kappa^2 + s^2} \exp(-sx_2) \right] \sin[\kappa(x_1 - c_R t)] \\ u_2(x_1, x_2, t) = \kappa a \left[-\frac{r}{\kappa} \exp(-rx_2) + \frac{2r\kappa}{\kappa^2 + s^2} \exp(-sx_2) \right] \cos[\kappa(x_1 - c_R t)] \end{cases} \tag{A.3.94}$$

To check the motion of the mass point at the free boundary, we substitute $x_2 = 0$ into Eq. (A.3.94) and derive the parametric equation of an elliptic motion described by two displacement components, namely,

$$\begin{cases} u_1(x_1, 0, t) = \kappa a \left(\frac{2rs}{\kappa^2 + s^2} - 1 \right) \sin[\kappa(x_1 - c_R t)] \\ u_2(x_1, 0, t) = \kappa a \left(\frac{2r\kappa}{\kappa^2 + s^2} - \frac{r}{\kappa} \right) \cos[\kappa(x_1 - c_R t)] \end{cases} \tag{A.3.95}$$

Let $\nu = 0.25$, as an example, we get $c_p = 1.732c_s$ and $c_R = 0.9194c_s$ such that Eq. (A.3.95) becomes

$$\begin{cases} u_1(x_1, 0, t) = -0.4227\kappa a \sin[\kappa(x_1 - 0.9194c_s t)] \\ u_2(x_1, 0, t) = 0.6204\kappa a \cos[\kappa(x_1 - 0.9194c_s t)] \end{cases} \tag{A.3.96}$$

For an arbitrary Poisson's ratio ν of the medium, the S-wave speed is $c_s = \sqrt{G/\rho}$, and Eq. (A.3.10) gives the P-wave speed c_p. From Eqs. (A.3.90) and (A.3.91), we get Rayleigh wave speed c_R, and then compute r/κ and s/κ from Eq. (A.3.78), and finally obtain, from Eq. (A.3.95), the dimensionless amplitudes of horizontal and vertical displacements as follows

$$\left| \frac{u_{1\,\max}}{\kappa a} \right| = \left| \frac{2(r/\kappa)(s/\kappa)}{1+(s/\kappa)^2} - 1 \right|, \quad \left| \frac{u_{2\,\max}}{\kappa a} \right| = \left| \frac{2(r/\kappa)}{1+(s/\kappa)^2} - \frac{r}{\kappa} \right| \qquad (A.3.97)$$

Figure A.11 presents the variation of the above dimensionless displacement amplitudes of the Rayleigh wave with an increase of Poisson's ratio. As shown in Fig. A.11, the vertical displacement amplitude is much larger than the horizontal displacement amplitude. In an earthquake, for instance, the vertical displacement of Rayleigh wave exhibits large amplitude and is the main source of disasters. As the Rayleigh wave travels more slowly than both P-wave and SV-wave in the same medium, there has been an attempt to detect the arrival of a P-wave due to the earthquake and send out emergence messages before the arrival of any Rayleigh waves

Remark A.3.7: In 1924, R. Stoneley, a British seismologist, found a new surface wave traveling near an interface between two different media with close wave speeds. This surface wave also consists of two inhomogeneous planar P-wave and SV-wave, and is called a *Stoneley wave*. The speed of Stoneley wave is higher than the Rayleigh wave speed, but lower than the shear wave speeds of the two media. Like Rayleigh waves, Stoneley waves are also a type of non-dispersive waves.

Remark A.3.8: In 1911, A. E. H. Love, a British scientist, found that if a semi-infinite medium has a layer with identical thickness attached, the wave refracted into the layer may undergo repeated reflections back and forth between the two surfaces of the layer and becomes a new surface wave, called a *Love surface wave* or a *Love wave* for short. Let c_s be the shear wave speed of the medium, c_s' be the shear wave speed of the layer. If $c_s' < c_s$ holds, then an SH-wave in the medium can enter into the layer and becomes a Love wave traveling within the layer at a wave number dependent speed $c(\kappa)$, which falls into the interval (c_s', c_s). Hence, different from Rayleigh waves and Stoneley waves, Love waves are dispersive.

Fig. A.11 The variation of dimensionless amplitudes of the Rayleigh wave with an increase of Poisson's ratio (thick solid line: the horizontal displacement, thin solid line: the vertical displacement)

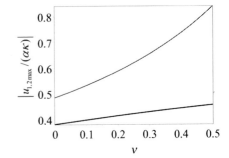

Bibliography

1. Abrate S (1995). Vibration of non-uniform rods and beams. Journal of Sound and Vibration, 185(4): 703–716.
2. Achenbach J D (1973). Wave Propagation in Elastic Solids. Amsterdam: North-Holland Publishing Company.
3. Arnold V I (1978). Mathematical Methods of Classical Mechanics. New York: Springer-Verlag.
4. Ardema M D (2005). Analytical Dynamics: Theory and Applications. New York: Kluwer Academic Publisher.
5. Balachandran B, Magrab E B (2019). Vibrations. 3rd Edition. Cambridge: Cambridge University Press.
6. Braun S, Ewins D, Rao S S (2002). Encyclopedia of Vibration. Volume I. San Diego: Academic Press.
7. Chen B (2012). Analytical Dynamics. 2nd Edition. Beijing: Peking University Press, (in Chinese).
8. Craig Jr. R R (1981). Structural Dynamics. New York: John Wiley and Sons Inc.
9. Courant R, Hilbert D (1989). Methods of Mathematical Physics. Volume I. New York: John Wiley & Sons Inc.
10. Courant R, Hilbert D (1989). Methods of Mathematical Physics. Volume II. New York: John Wiley & Sons Inc.
11. Den Hartog J P (1984). Mechanical Vibrations. 4th Edition. New York: Dover Publications.
12. Doyle J F (1997). Wave Propagation in Structures. New York: Springer-Verlag.
13. Du M L, Wang Z H, Hu H Y (2013). Measuring memory with the order of fractional derivative. Scientific Reports, 3: 3431.
14. Elishakoff I (2020). Who developed the so-called Timoshenko beam theory. Mathematics and Mechanics of Solids, 25(1): 97–116.
15. Elishakoff I, Kaplunov J, Nolde E (2015). Celebrating the centenary of Timoshenko's study of effects of shear deformation and rotary inertia. Applied Mechanics Reviews, 67(6): 060802.
16. Eringen A C, Suhubi E S (1975). Elastodynamics, Volume II. New York: Academic Press.
17. Evensen D A (1976). Vibration analysis of multi-symmetric structures. AIAA Journal, 14(4): 446–453.
18. Gantmacher F R, Krein M G (2002). Oscillation Matrices and Kernels and Small Vibrations of Mechanical Systems. 2nd Edition. Rhode Island: AMS Chelsea Publishing.
19. Graff K F (1975). Wave Motion in Elastic Solids. Columbus: Ohio State University Press.
20. Greenwood D T (2003). Advanced Dynamics. Cambridge: Cambridge University Press.
21. Guo S Q, Yang S P (2012). Wave motions in non-uniform one-dimensional waveguides. Journal of Vibration and Control, 18(1): 92–100.
22. Hagedorn P, Das Gupta A (2007). Vibrations and Waves in Continuous Mechanical Systems. Chichester: John Wiley & Sons Inc.

© Science Press 2022
H. Hu, *Vibration Mechanics*,
https://doi.org/10.1007/978-981-16-5457-2

23. Hu H Y, Cheng D L (1986). Generalized mode synthesis of cyclosymmetrical structures. Vibration and Shock, 5(4): 1–7 (in Chinese).

24. Hu H Y, Cheng D L, Wang L (1986). Vibration analysis of a cyclosymmetric structure with central axis. Shanghai: The 2nd National Conference on Computational Mechanics, No.2-236 (in Chinese).

25. Hu H Y, Cheng D L (1988). Investigation on modal characteristics of cyclosymmetric structures. Chinese Journal of Applied Mechanics, 5(3): 1–8 (in Chinese).

26. Hu H Y (1993). Structural damping model and system response in time domain. Chinese Journal of Applied Mechanics, 10(1): 34–43 (in Chinese).

27. Hu H Y (1998). Mechanical Vibrations and Shock. Beijing: Press of Aviation Industry, (in Chinese).

28. Hu H Y (2000). Aesthetical considerations of vibration theory and its development. Journal of Vibration Engineering, 13(2): 161–169 (in Chinese).

29. Hu H Y (2018). On the degree of freedom of a mechanical system. Chinese Journal of Theoretical and Applied Mechanics, 50(5): 1135–1144 (in Chinese).

30. Hu H Y (2018). On the node number of a natural mode shape. Journal of Dynamics and Control, 16(3): 193–200 (in Chinese).

31. Hu H Y (2018). On the anti-resonance problem of a linear system. Journal of Dynamics and Control, 16(5): 385–390 (in Chinese).

32. Hu H Y (2020). Duality relations of rods in natural vibrations. Journal of Dynamics and Control, 18(2): 1–8 (in Chinese).

33. Hu H Y (2020). Duality relations of beams in natural vibrations. Chinese Journal of Theoretical and Applied Mechanics, 52(1): 139–149 (in Chinese).

34. Jin D P, Hu H Y (2005). Vibro-impacts and Controls. Beijing: Science Press, (in Chinese).

35. Jinsberg J H (2001). Mechanical and Structural Vibrations: Theory and Applications. Baffins Lane: John Wiley & Sons Inc.

36. Joshi A W (1977). Elements of Group Theory for Physicists. New Delhi: Wiley Eastern Inc.

37. Lee T, Leok M, McClamroch N H (2018). Global Formulations of Lagrangian and Hamiltonian Dynamics on Manifolds. New York: Springer-Verlag.

38. Li Y F, Hu H Y, Wu Y J (1988). Vibration attenuation design of complicated structures based on their original responses. Journal of Nanjing Aeronautical Institute, English Edition, 5(1): 133–142.

39. Marsden J E, Ratiu T S (1999). Introduction to Mechanics and Symmetry. 2nd Edition. New York: Springer-Verlag.

40. Meirovitch L (2001). Fundamentals of Vibrations. International Edition. Singapore: McGraw Hill Book Company.

41. Olson B J, Shaw S W, Shi C Z, et al (2014). Circulant matrices and their application to vibration analysis. Applied Mechanics Review, 66(4): 040803.

42. Radman Z (2004). Towards aesthetics of science. Journal of the Faculty Letters, University of Tokyo, Aesthetics, 29: 1–16.

43. Ram Y M, Elhay S (1995). Dualities in vibrating rods and beams: continuous and discrete models. Journal of Sound and Vibration, 184(5): 648–655.

44. Rao S S (2007). Vibration of Continuous Systems. Hoboken: John Wiley & Sons Inc.

45. Scanlan R H (1970). Linear damping models and causality in vibrations. Journal of Sound and Vibration, 13(4): 499–509.

46. Stronge W J (2000). Impact Mechanics. Cambridge: Cambridge University Press.

47. Stewart I, Golubitsky M (1992). Fearful Symmetry. Oxford: Blackwell Publisher.

48. Thomas D L (1974). Standing waves in rotationally periodic structures. Journal of Sound and Vibration, 37(2): 288–290.

49. Thomas D L (1979). Dynamics of rotationally periodic structures. International Journal of Numerical Methods in Engineering, 14(1): 81–102.

50. Timoshenko S, Young D H, Weaver Jr W (1974). Vibration Problems in Engineering. 4th Edition. New York: John Wiley and Sons Inc.

51. Tveito A, Winther R (2005). Introduction to Partial Differential Equations. Berlin: Springer-Verlag.
52. Wang D J, Wang Q S, He B C (2019). Qualitative Theory in Structural Mechanics. Singapore: Springer Nature.
53. Wang L F, Hu H Y (2005). Flexural wave propagation in single-walled carbon nanotubes. Physical Review B, 71(19): 195412.
54. Wang L F, Guo W L, Hu H Y (2008). Group velocity of wave propagation in carbon nanotubes. Proceedings of the Royal Society A, 464(2094): 1423–1438.
55. Xu B S, Yin Q Z (2008). Aesthetic Methods in Mathematics. Dalian: Press of Dalian University of Science and Technology, (in Chinese).
56. Zheng G T (2016). Sequel of Structural Dynamics: Applications to Flight Vehicles. Beijing: Science Press, (in Chinese).

Index

© Science Press 2022
H. Hu, *Vibration Mechanics*,
https://doi.org/10.1007/978-981-16-5457-2

495

Printed in the United States
by Baker & Taylor Publisher Services